AF324110

MISSILE AERODYNAMICS

(Jack N. Nielsen)

MISSILE AERODYNAMICS

JACK N. NIELSEN

CHIEF SCIENTIST, NASA AMES RESEARCH CENTER
MOFFETT FIELD, CALIFORNIA

Formerly President, Nielsen Engineering & Research, Inc.
Mountain View, California

NIELSEN ENGINEERING & RESEARCH, INC.
Mountain View, California
1988

This Book Has Been Reprinted on
The Occasion of the 70th Birthday
of Jack N. Nielsen

MISSILE AERODYNAMICS

© 1960 by Nielsen Engineering & Research, Inc.
510 Clyde Avenue
Mountain View, California

Originally published 1960 by McGraw-Hill Book Company, Inc.
Reprinted 1988 by NEAR, Inc.

Printed in the United States of America

Library of Congress Cataloging-in-Publication Data

Nielsen, Jack Norman.
    Missile aerodynamics.

    Reprint. Originally published: New York :
McGraw-Hill, 1960. (McGraw-Hill series in missile
and space technology)
    Bibliography: p.
    Includes index.
    1. Guided missiles. 2. Aerodynamics. I. Title.
II. Series: McGraw-Hill series in missile and space
technology.
UG1310.N48   1988   623.4'519   88-15223

ISBN 0-9620629-0-1

TO GISELA AND DAGMAR

PREFACE TO THE NEAR EDITION

The author of *Missile Aerodynamics,* Dr. Jack N. Nielsen, has been the foremost expert in the field for more than 35 years, and it is fitting that this new edition of his book should coincide with his 70th birthday. It is now 28 years since the initial publication of *Missile Aerodynamics* by McGraw-Hill, and in this period considerable advances have been made in aerodynamic prediction methods.

Most of these advances have been made in parallel with the growth in the use and efficiency of computers. Today, *in principle,* the Navier-Stokes equations can be solved to simulate the complete air flow around a missile; however, in practice, deficiencies in algorithms and turbulence modeling have prevented the realization of this goal. One of the unfortunate, but predictable, outcomes of this reliance on "brute force" computations of aerodynamic phenomena is that present, and possibly future, generations of aerodynamicists may not develop a sound background and understanding of the basic theory of aerodynamics. This can result in blind acceptance of computationally generated numbers. It is hoped by the republication of *Missile Aerodynamics* that this basic aerodynamic theory, especially as applied to missiles, will be more readily available and thus can provide the background and insight essential to critical examination of computational results.

Jack Nielsen has been the intellectual force behind most of the recent advances in missile aerodynamics, and those of us who still work in the company which he founded remember him with a mixture of awe and affection.

David Nixon, President
Nielsen Engineering &
Research, Inc.

PREFACE

In recent years the great many persons who have become actively connected with missile science and engineering have had to rely principally on technical journals and papers for aerodynamic information. The literature in missile aerodynamics is extensive and in many respects complete, but an over-all view of the field is reserved to those few specialists familiar with the hundreds of excellent technical papers available. However, a large group of persons who would find such an over-all view useful in the performance of their duties cannot, for one reason or another, review the numerous technical papers. It is principally for this group that the present volume has been written. The book attempts to present a rational and unified account of the principal results of missile aerodynamics.

A *missile* is described by Webster as a weapon or object capable of being thrown, hurled, or projected so as to strike a distant object. One distinction between a missile and an airplane is that, unlike an airplane, a missile is usually expendable in the accomplishment of its mission. From a configurational point of view, the distinction is frequently made that a missile is more slender than an airplane and tends to possess smaller wings in proportion to its body. These distinctions are, however, subject to many exceptions. In fact, the configurational distinctions between missiles and airplanes seem to narrow as the operational speeds increase. Therefore much of the missile aerodynamics contained herein will be directly applicable to airplanes.

Since this book draws on a large number of technical papers for much of its content, it is important that the policy with regard to credit for technical material be clear. The author would like to quote original sources in all cases. Such a course of action is, however, impractical because original sources are often impossible to ascertain, or not readily available. Thus the references to technical papers herein are those most convenient from the standpoints of availability or pedagogical usefulness, or simply those most familiar to the author.

The book attempts to present a rational account of the principal subjects of missile aerodynamics. It further attempts to present adequate mathematical treatment of the subjects for use in design. The alternative approach, of compiling a handbook of missile design data, was not attempted for several reasons in addition to the author's natural disinclinations. First, the wide range of missile configurations and their continuous evolution render it difficult to specify design data of general utility. Second, design data are often classified.

The author has been influenced in his choice of subject matter by consideration of his special competencies. However, in the interests of completeness, he has included many subjects in which he has no particular competence. Many subjects are treated extensively from a mathematical point of view, but many other subjects of equal importance are either not amenable to mathematical treatment or are imperfectly understood. Nevertheless the author has chosen to treat such subjects qualitatively, even though such treatment may not enhance the elegance of the book. The emphasis in the main is on supersonic speeds, although much material applicable to subsonic speeds is included. Such emphasis is consistent with the facts that missiles fly mostly at supersonic speeds and that many excellent books on subsonic aerodynamics are already available. Though certain subjects have been included in the interests of completeness, no claim for completeness is made. The sin of omission is considered preferable to inadequate treatment of more material.

Readers frequently wonder what motivates the author in his arrangement of material. The first chapter is purely introductory in character, and the second chapter collects together for convenient use many of the results used repeatedly in subsequent chapters. The third chapter treats the subject of slender-body theory which the author considers the backbone of missile aerodynamics. Slender-body theory has the great advantage that it is mathematically tractable for a very wide range of missile configurations. In Chaps. 4 to 8, inclusive, an attempt is made to present missile aerodynamics in an orderly building up of a missile from its component parts, the body alone, the wing panels, the tail, and the control surfaces. Since the aerodynamics of a tail behind the wings of a missile depends on the flow field of the wing-body combination, such flow fields are discussed in Chap. 6 before the discussion of wing-body-tail combinations in Chap. 7. The final two chapters of the book treat the important subjects of drag and stability derivatives. The nature of aerodynamic drag makes desirable a separate chapter devoted to drag. The chapter on stability derivatives attempts to treat all forces and moments on a missile (other than drag) from a general and unified point of view.

The author would like to acknowledge the many contributions made by others to the book. Professors Holt Ashley, J. C. Hunsaker, and Arthur Bryson reviewed parts of the manuscript and made a number of helpful suggestions. I should like to thank those members of the staff of the Ames Laboratory of the National Advisory Committee for Aeronautics, Dean Chapman, Max Heaslet, Robert T. Jones, Morris Rubesin, Murray

Tobak, and Milton Van Dyke, who willingly reviewed those parts of the book of particular interest to them.   Also, the author would like to pay tribute to those members of the staff of the 1- by 3-foot supersonic wind tunnel branch with whom he has worked in the field of missile aerodynamics for many years, and particularly to Wallace Davis, branch chief.   These co-workers of the author include Wallace Davis, Elliott Katzen, Richard Spahr, William Pitts, Leland Jorgensen, George Kaattari, Frederick Goodwin, and others.   The exacting job of preparing the final manuscript was faithfully undertaken by Virginia Stalder.   H. Guyford Stever has been very kind in seeking out the book for his series and in lending general encouragement and advice to the author.   In conclusion, the author would like to acknowledge his debt to the National Advisory Committee for Aeronautics, in whose laboratory much of the knowledge in this book was originated, and without whose cooperation this book would not have been possible.

Jack N. Nielsen

ERRATA

(Indicated by * in page margin of text)

		ERROR	**CORRECTION**
Page 5	Fig. 1-3	Angle α_t measures to V_0 vector	Angle α_t should measure to projection of V_0 vector in the x-z plane
Page 5	Line 23	Table 1-1	Table 1-2
Page 24	Eqn. (2-52)	$\dfrac{1 - 2/3BA}{1 - 1/2BA}$	$\dfrac{1 - (2/3BA)}{1 - (1/2BA)}$
Page 28	Eqn. (2-62) 3rd line	$\dfrac{-\partial\phi}{\partial y}$	$\dfrac{-\partial\psi}{\partial y}$
Page 37	Eqn. (3-12) 2nd line	$\displaystyle\int_0^x S''(\xi)$	$\displaystyle\int_0^x S'(\xi)$
Page 50	Eqn. (3-65)	$F'(\xi)\,d\xi$	$F(\xi)\,d\xi$
Page 91	Fig. (4-19) coordinate next to Γ inside circle at left	$\dfrac{a^2}{\delta_0}$	$\dfrac{-a^2}{\delta_0}$
Page 95	Eqn. (4-91) in e exponents	$i\theta$	$i\phi$
Page 112	Line 6 from bottom	...bank of the...	...bank on the...
Page 127	Line 4 from bottom	pressure of distribution	distribution of pressure
Page 129	Line 5	left panel	right panel
Page 221	Eqn. (8-30)	$(s^2 + a^2/s)$	$(s + a^2/s)$

xiii

CONTENTS

Preface to NEAR Edition vii
Preface to Original Edition ix
List of Errata xiii

Chapter 1. INTRODUCTION 1

1-1. Missile Aerodynamics versus Airplane Aerodynamics 1
1-2. Classification of Missiles 1
1-3. Axes; Angle of Bank and Included Angle 3
1-4. Angles of Attack and Sideslip 4
1-5. Glossary of Special Terms 6

Chapter 2. SOME FORMULAS COMMONLY USED IN MISSILE AERODYNAMICS 8

2-1. Nonlinear Potential Equation 8
2-2. Linearization of Potential Equation 10
2-3. Bernoulli's Equation; Pressure Coefficient as a Power Series in Velocity Components 12
2-4. Classification of Various Theories Used in Succeeding Chapters . 14
2-5. Line Pressure Source 17
2-6. Aerodynamic Characteristics of Rectangular and Triangular Lifting Surfaces on the Basis of Supersonic Wing Theory 18
2-7. Simple Sweep Theory 24
2-8. Conformal Mapping; Notation; Listings of Mappings and Flows . 25
2-9. Elliptic Integrals 31

Chapter 3. SLENDER-BODY THEORY AT SUPERSONIC AND SUBSONIC SPEEDS 34

Slender Bodies of Revolution

3-1. Slender Bodies of Revolution at Zero Angle of Attack at Supersonic Speeds; Sources 34
3-2. Slender Bodies of Revolution at Angle of Attack at Supersonic Speeds; Doublets 37
3-3. Slender-body Theory for Angle of Attack 39

Slender Bodies of General Cross Section at Supersonic Speeds

3-4. Solution of Potential Equation by the Method of Ward 40
3-5. Boundary Conditions; Accuracy of Velocity Components . . . 45
3-6. Determination of $a_0(\bar{x})$ and $b_0(\bar{x})$ 47
3-7. Pressure Coefficient 48
3-8. Lift, Sideforce, Pitching Moment, and Yawing Moment . . . 48
3-9. Drag Force 51
3-10. Drag Due to Lift 52
3-11. Formula Explicitly Exhibiting Dependence of Drag on Mach Number 54

Slender Bodies of General Cross Section at Subsonic Speeds

3-12. Solution of the Potential Equation 55
3-13. Determination of $a_0(\bar{x})$ and $b_0(\bar{x})$ 58
3-14. Drag Formula for Subsonic Speeds; d'Alembert's Paradox 59

Chapter 4.   AERODYNAMICS OF BODIES; VORTICES 66

Inviscid Flow

4-1. Lift and Moment of Slender Bodies of Revolution 66
4-2. Pressure Distribution and Loading of Slender Bodies of Revolution; Circular Cones 70
4-3. Slender Bodies of Elliptical Cross Section; Elliptical Cones 74
4-4. Quasi-cylindrical Bodies 80

Vortices

4-5. Positions and Strengths of Body Vortices 85
4-6. Forces and Moments Due to Body Vortices; Allen's Crossflow Theory . 89
4-7. Motion of Symmetrical Pair of Crossflow Vortices in Presence of Circular Cylinder 91
4-8. Motion of Vortices in Presence of a Noncircular Slender Configuration . 94
4-9. Lift and Sideforce on Slender Configuration Due to Free Vortices . . 96
4-10. Rolling Moment of Slender Configuration Due to Free Vortices . . . 101

Chapter 5.   WING-BODY INTERFERENCE 112

5-1. Definitions; Notation 113
5-2. Planar Wing and Body Interference 114
5-3. Division of Lift between Wing and Body; Panel Center of Pressure . . 118
5-4. Cruciform Wing and Body Interference 121
5-5. Effect of Angle of Bank on Triangular Panel Characteristics; Panel-Panel Interference 125
5-6. Summary of Results; Afterbody Effects 129
5-7. Application to Nonslender Configurations; Calculative Example . . . 134
5-8. Simplified Vortex Model of Wing-Body Combination 138

Chapter 6.   DOWNWASH, SIDEWASH, AND THE WAKE 144

6-1. Vortex Model Representing Slender Wing with Trailing Edge Normal to Flow 145
6-2. Rolling Up of Vortex Sheet behind a Slender Wing 148
6-3. Calculation of Induced Velocities of Trailing-vortex System . . . 153
6-4. Vortex Model of Planar Wing and Body Combination 156
6-5. Factors Influencing Vortex Paths and Wake Shape behind Panels of Planar Wing and Body Combination 166
6-6. Factors Influencing Downwash Field behind Panels of Planar Wing and Body Combination 169
6-7. Cruciform Arrangements 171

Chapter 7.   WING-TAIL INTERFERENCE 181

7-1. Wing-Tail Interference; Flat Vortex Sheet 182
7-2. Pressure Loading on Tail Section Due to Discrete Vortices in Plane of Tail 184

7-3. Lift on Tail Section and Tail Efficiency for Discrete Vortices in Plane of Tail 189
7-4. Tail Interference Factor 192
7-5. Calculation of Tail Lift Due to Wing Vortices 194
7-6. Use of Reverse-flow Method for Calculating Aerodynamic Forces on Tail Section in Nonuniform Flow 198
7-7. Shock-expansion Interference 201

Chapter 8. AERODYNAMIC CONTROLS 208

8-1. Types of Controls; Conventions 209
8-2. All-movable Controls for Planar Configurations 213
8-3. All-movable Controls for Cruciform Configurations 225
8-4. Coupling Effects in All-movable Controls 228
8-5. Trailing-edge Controls 234
8-6. Some Nonlinear Effects in Aerodynamic Control 242
8-7. Notes on Estimating Hinge Moments 247
8-8. Change in Missile Attitude Due to Impulsive Pitch Control; Altitude Effects 250

Chapter 9. DRAG 261

9-1. General Nature of Drag Forces; Components of Drag 262
9-2. Analytical Properties of Drag Curves 265

Pressure Foredrag

9-3. Pressure Foredrag of Slender Bodies of Given Shape; Drag Due to Lift 269
9-4. Pressure Foredrag of Nonslender Missile Noses at Zero Angle of Attack 275
9-5. Shapes of Bodies of Revolution for Least Pressure Foredrag at Zero Angle of Attack 280
9-6. Pressure Drag of Wings Alone 287
9-7. Pressure Foredrag of Wing-Body Combinations of Given Shape at Zero Angle of Attack 294
9-8. Wings and Wing-Body Combinations of Least Pressure Foredrag at Zero Angle of Attack 296
9-9. Minimizing Pressure Drag of Wings and Wing-Body Combinations beyond That Due to Thickness 302

Base Drag

9-10. Physical Features of Flow at a Blunt Base; Types of Flow 311
9-11. Basis for Correlation of Base-pressure Measurements 313
9-12. Correlation of Base-pressure Measurements for Blunt-trailing-edge Airfoils and Blunt-base Bodies of Revolution 317
9-13. Other Variables Influencing Base Pressure 321

Skin Friction

9-14. General Considerations of Skin Friction at Supersonic Speeds . . . 323
9-15. Laminar Skin Friction; Mean-enthalpy Method 330
9-16. Turbulent Skin Friction 334
9-17. Other Variables Influencing Skin Friction 336

Chapter 10. STABILITY DERIVATIVES 349

10-1. Reference Axes; Notation 350
10-2. General Nature of Aerodynamic Forces; Stability Derivatives . . 353

10-3. Properties of Stability Derivatives Resulting from Missile Symmetries; Maple-Synge Analysis for Cruciform Missiles 358

10-4. Maple-Synge Analysis for Triform Missiles and Other Missiles . . . 362

10-5. General Expressions for Stability Derivatives in Terms of Inertia Coefficients; Method of Bryson 363

10-6. Stability Derivatives of Slender Flat Triangular Wing 374

10-7. General Method of Evaluating Inertia Coefficients and Apparent Masses. 378

10-8. Table of Apparent Masses with Application to the Stability Derivatives of Cruciform Triangular Wings 386

10-9. Further Examples of the Use of Apparent-mass Table 391

10-10. Effects of Aspect Ratio on Stability Derivatives of Triangular Wings . 394

10-11. Contribution of the Empennage to Certain Stability Derivatives; Empennage Interference Effects 402

Name Index 433

Subject Index. 437

CHAPTER 1

INTRODUCTION

One purpose of this chapter is to point out some of the differences between airplanes and missiles by virtue of which missile aerodynamics embraces subjects not formerly of great interest in airplane aerodynamics. Another purpose is to collect in one place for ready reference many of the symbols, definitions, and conventions used throughout the book.

1-1. Missile Aerodynamics versus Airplane Aerodynamics

One of the principal differences between missiles and airplanes is that the former are usually expendable, and consequently are usually uninhabited. For this reason increased ranges of speed, altitude, and maneuvering accelerations have been opened up to missile designers, and these increased ranges have brought with them new aerodynamic problems. For instance, the higher allowable altitudes and maneuvering accelerations permit operation in the nonlinear range of high angles of attack. A missile may be ground-launched or air-launched and in consequence can undergo large longitudinal accelerations, can utilize very high wing loadings, and can dispense with landing gear. In the absence of a pilot the missile can sometimes be permitted to roll and thereby to introduce new dynamic stability phenomena. The problem of guiding the missile without a pilot introduces considerable complexity into the missile guidance system. The combination of an automatic guidance system and the air frame acting together introduces problems in stability and control not previously encountered. Many missiles tend to be slender, and many utilize more than the usual two wing panels. These trends have brought about the importance of slender-body theory and cruciform aerodynamics for missiles.

1-2. Classification of Missiles

Missiles can be classified on the basis of points of launching and impact, type of guidance system, trajectory, propulsive system, trim and control device, etc. An important classification on the basis of points of launching and impact is given in Table 1-1.

Another source of distinction among missiles is the guidance system. In a *command system* the missile and the target are continuously tracked

1

from one or more vantage points, and the necessary path for the missile to intercept the target is computed and relayed to the missile by some means such as radio. A *beam-riding missile* contains a guidance system to constrain it to a beam. The beam is usually a radar illuminating the target so that, if the missile stays in the beam, it will move toward the target. A *homing missile* has a seeker, which sees the target and gives the necessary directions to the missile to intercept the target. The homing missile can be subdivided into classes having active, semiactive, and passive guidance systems. In the *active* class the missile illuminates the target and receives the reflected signals. In the *semiactive* class the missile receives reflected signals from a target illuminated by means external to the missile. The *passive* type of guidance system depends on a receiver in the missile sensitive to the radiation of the target itself.

TABLE 1-1. CLASSIFICATION OF MISSILES

AAM	Air-to-air missile
ASM	Air-to-surface missile
AUM	Air-to-underwater missile
SAM	Surface-to-air missile
SSM	Surface-to-surface missile
UUM	Underwater-to-underwater missile

Another method of classifying missiles is with regard to the type of trajectory taken by the missile. A *ballistic missile* follows the usual ballistic trajectory of a hurled object. A *glide missile* is launched at a steep angle to an altitude depending on the range, and then glides down on the target. A *skip missile* is launched to an altitude where the atmosphere is very rare, and then skips along on the atmospheric shell.

On the basis of propulsive systems missiles fall into the categories of *turbojet, ram-jet, rocket*, etc. If the missile receives a short burst of power that rapidly accelerates it to top speed and then glides to its target, it is a *boost-glide missile*. Sometimes a missile is termed *single-stage, double-stage*, etc., depending on the number of stages of its propulsive system.

Further differentiation among missiles can be made on the basis of trim and control devices. A *canard missile* has a small forward lifting surface that can be used for either trim or control similar to a tail-first airplane. A missile controlled by deflecting the wing surfaces is termed a *wing-control missile*, and one controlled by deflecting the tail surfaces is termed a *tail-control missile*. It is to be noted that these definitions depend on which set of lifting surfaces is taken as the wing and which is taken as the tail. For missiles with two sets of lifting surfaces, we will specify the wing to be the main lifting surfaces and the tail to be the balancing surfaces, a distinction maintained throughout the book. In a *cruciform missile*, sets of controls at right angles permit the missile to turn immediately in any plane without the necessity of its banking. On the other hand a *bank-to-turn* missile, like an airplane, banks into the turn to bring the normal acceleration vector as close to the vertical plane of symmetry as possible.

1-3. Axes; Angle of Bank and Included Angle

Of the two general systems of axes used in the present book, the second system does not appear until the final chapter. The first system, shortly to be described, is one well adapted for use with the theory of complex variables and, as such, is useful in slender-body theory. The second axis system is the NACA standard used in such fields as stability derivatives and dynamic stability. It is described in detail in the final chapter. It would simplify matters if one set of axes were used in place of the two sets. Consideration was given to defining such a compromise set of axes, but the idea was discarded because the net effect would probably be to add another system, where too many systems already exist. Also, a single system of axes represents too great a departure from usage in the technical literature.

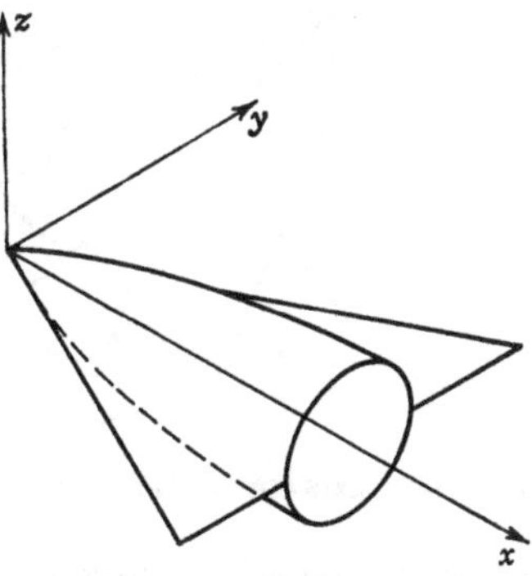

Fig. 1-1. Body axes.

The basic set of axes used in the first nine chapters is a set of body axes x, y, and z fixed in the missile with minor notational differences for various missile positions. The x axis is positive rearward and coincides with the longitudinal missile axis. The y axis is positive to the right, facing forward, and lies in the horizontal plane of symmetry when one exists. The z axis is positive vertically upward and lies in the vertical plane of symmetry if one exists. The x, y, and z axes shown in Fig. 1-1 form a right-handed system.

The body axes x, y, z take on all the possible orientations a missile can assume in a uniform air stream. The angles which conveniently specify the orientation of a missile with respect to its flight direction depend on the use to which such angles are to be put. For the purposes of this book a set of angles α_c and φ are convenient. Consider a missile mounted in a wind tunnel on a sting coincident with the prolongation of its longitudinal axis. Let the missile be aligned parallel to the wind velocity with the wing panels in the zero bank attitude. Denote the body axes in this initial position by $\bar{x}$, $\bar{y}$, and $\bar{z}$. Now rotate (pitch) the missile about the $\bar{y}$ axis by an angle α_c as shown in Fig. 1-2, so that $\bar{x}$ and $\bar{z}$ occupy the positions x' and z'. The angle α_c will be termed the *included angle* and is the angle included between the missile's longitudinal axis and the free-stream velocity. Now let the missile be rotated in a clockwise direction facing forward about the x' axis so that y' and z' go into y and z. The axes are related by the following equations:

$$x' = \bar{x} \cos \alpha_c - \bar{z} \sin \alpha_c$$
$$y' = \bar{y} \tag{1-1}$$
$$z' = \bar{x} \sin \alpha_c + \bar{z} \cos \alpha_c$$

$$x = \bar{x} \cos \alpha_c - \bar{z} \sin \alpha_c$$
$$y = \bar{y} \cos \varphi - \bar{x} \sin \alpha_c \sin \varphi - \bar{z} \cos \alpha_c \sin \varphi \tag{1-2}$$
$$z = \bar{y} \sin \varphi + \bar{x} \sin \alpha_c \cos \varphi + \bar{z} \cos \alpha_c \cos \varphi$$

From Eq. (1-2) the direction cosines between the $\bar{x}$, $\bar{y}$, $\bar{z}$ and the x, y, z axes can be readily found (Table 1-2). It is important to note that the

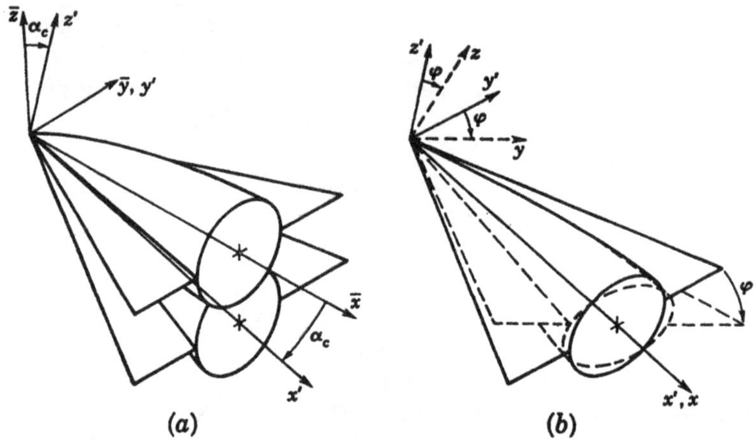

Fig. 1-2. Axis conventions for pitch and bank. (a) Pitch about $\bar{y}$; (b) bank about x'.

angle α_c must be applied to the missile before φ is applied for the above direction cosines to be valid. Thus, the pitch and bank operations are not commutative. In particular, if the missile is first banked about $\bar{x}$ and

TABLE 1-2. DIRECTION COSINES OF BODY AXES FOR
COMBINED α_c AND φ DISPLACEMENTS

	$\bar{x}$	$\bar{y}$	$\bar{z}$
x	$\cos \alpha_c$	0	$-\sin \alpha_c$
y	$-\sin \alpha_c \sin \varphi$	$\cos \varphi$	$-\cos \alpha_c \sin \varphi$
z	$\sin \alpha_c \cos \varphi$	$\sin \varphi$	$\cos \alpha_c \cos \varphi$

then pitched about $\bar{y}$, the $\bar{y}$ axis will remain perpendicular to the air stream. In other words, the missile will remain in a position of zero sideslip.

1-4. Angles of Attack and Sideslip

The angles of attack and sideslip are defined here as purely kinematic quantities depending only on velocity ratios. As such, they measure velocity components along the body axes of the missile. Let the air-stream velocity relative to the missile center of gravity be V_0 with components u, v, and w along x, y, and z, respectively. As defined, u, v, and w are flow velocities, and $-u$, $-v$, and $-w$ are velocities of the center of gravity with respect to the air stream. The angles of attack and sideslip have been defined in at least three ways.

The *small angle* definitions are

$$\alpha = \frac{w}{V_0} \qquad \beta = \frac{-v}{V_0} \qquad (1\text{-}3)$$

The *sine* definitions are

$$\sin \alpha_s = \frac{w}{V_0} \qquad \sin \beta_s = \frac{-v}{V_0} \qquad (1\text{-}4)$$

The *tangent* definitions are

$$\tan \alpha_t = \frac{w}{u} \qquad \tan \beta_t = \frac{-v}{u} \qquad (1\text{-}5)$$

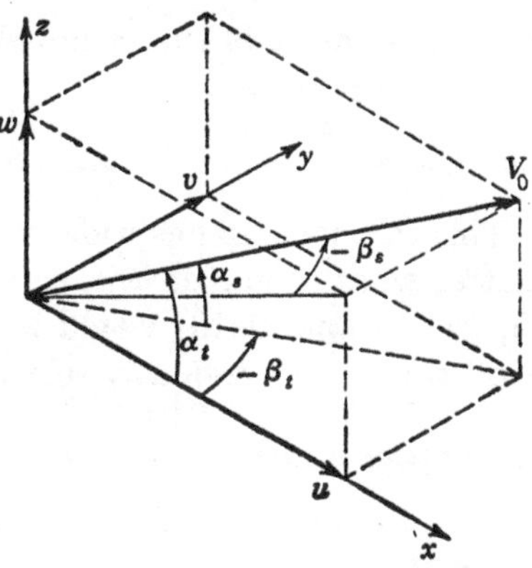

Fig. 1-3. Angles of attack and sideslip.

The subscripts s and t are used to differentiate between the sine and the tangent definitions. A graphic interpretation of the angles α_s, β_s, α_t, and β_t is shown in Fig. 1-3. Note that a positive sideslip angle occurs when the air stream approaches from the right facing forward. For small angles, the angles of attack and sideslip do not depend on which definition is used. For large angles, it is necessary to know which definitions have been adopted. Frequently, the sine definition is used for one quantity and the tangent definition for another.

It is a simple matter to relate the angles of attack and sideslip to the included angle and angle of bank. With the aid of Table 1-1, we have

$$\begin{aligned}
u &= V_0 \cos (\bar{x},x) = V_0 \cos \alpha_c \\
v &= V_0 \cos (\bar{x},y) = -V_0 \sin \alpha_c \sin \varphi \qquad (1\text{-}6) \\
w &= V_0 \cos (\bar{x},z) = V_0 \sin \alpha_c \cos \varphi
\end{aligned}$$

For given values of α_c and φ, the values of α_s and β_s are expressed by

$$\begin{aligned}
\sin \alpha_s &= \sin \alpha_c \cos \varphi \\
\sin \beta_s &= \sin \alpha_c \sin \varphi
\end{aligned} \qquad (1\text{-}7)$$

Conversely, the values of α_c and φ necessary to yield α_s and β_s are given by

$$\begin{aligned}
\sin^2 \alpha_c &= \sin^2 \alpha_s + \sin^2 \beta_s \\
\tan \varphi &= \frac{\sin \beta_s}{\sin \alpha_s}
\end{aligned} \qquad (1\text{-}8)$$

For the tangent definitions, a set of relationships exist similar to Eqs. (1-7) and (1-8):

$$\tan \alpha_t = \tan \alpha_c \cos \varphi$$
$$\tan \beta_t = \tan \alpha_c \sin \varphi \tag{1-9}$$

$$\tan^2 \alpha_c = \tan^2 \alpha_t + \tan^2 \beta_t$$
$$\tan \varphi = \frac{\tan \beta_t}{\tan \alpha_t} \tag{1-10}$$

For small values of the included angle α_c, both Eqs. (1-7) and (1-9) yield

$$\alpha = \alpha_c \cos \varphi$$
$$\beta = \alpha_c \sin \varphi \tag{1-11}$$

This relationship has wide use in cruciform aerodynamics. It does not matter what the angle φ is, so long as α_c is small. It is noteworthy that Eqs. (1-8) and (1-10) would be used to set a sting-mounted model in a wind tunnel to previously selected values of α_s, β_s, or of α_t, β_t.

Illustrative Example

Find the value of α_s, β_s, α_t, and β_t for an included angle of 30° and a bank angle of 25°.

From Eq. (1-7)

$$\sin \alpha_s = \sin \alpha_c \cos \varphi = 0.500(0.906) = 0.453$$
$$\alpha_s = 27°$$
$$\sin \beta_s = \sin \alpha_c \sin \varphi = (0.500)(0.423) = 0.212$$
$$\beta_s = 12.3°$$

From Eq. (1-9)

$$\tan \alpha_t = \tan \alpha_c \cos \varphi = (0.5774)(0.906) = 0.523$$
$$\alpha_t = 27.6°$$
$$\tan \beta_t = \tan \alpha_c \sin \varphi = (0.5774)(0.423) = 0.244$$
$$\beta_t = 13.7°$$

1-5. Glossary of Special Terms

Many special terms occur repeatedly in missile aerodynamics. Some of these terms are now listed for ready reference.

Body axes: a set of cartesian axes fixed in the missile and parallel to the axes of symmetry of the missile if such symmetry axes exist

Crossflow plane: a plane normal to the free-stream velocity

Cruciform wing: four similar wing panels mounted together at a common chord and displaced one from the next by $\pi/2$ radians of arc

Fineness ratio: ratio of body length to body diameter (calibers)

Horizontal plane of symmetry: the horizontal plane in which the lower half of the missile is the mirror image of the upper half

Included angle: angle between free-stream velocity and missile longitudinal axis

Interdigitation angle: angle between the plane of a lifting surface and the plane of another tandem lifting surface

Normal plane: a plane normal to the missile longitudinal axis

Subsonic leading edge: a leading edge such that the component of the free-stream Mach number normal to the edge is less than one

Supersonic leading edge: a leading edge such that the component of the free-stream Mach number normal to the edge is greater than one

Symmetrical wing: a wing possessing a horizontal plane of symmetry

Tangent ogive: a missile nose having constant radius of curvature in all planes through the longitudinal axis from the apex to the circular cylinder to which it is tangent

Trefftz plane: a fictitious crossflow plane infinitely far behind a missile or lifting surface to which the trailing vortex system extends without viscous dissipation

Vertical plane of symmetry: the vertical plane in which the left half of the missile is the mirror image of the right half

Wing panels: those parts of the main missile lifting surfaces exterior to the body

SYMBOLS

V_0	free-stream velocity
x, y, z	missile body axes; $\alpha_c \neq 0$, $\varphi \neq 0$
$\bar{x}, \bar{y}, \bar{z}$	missile body axes; $\alpha_c = 0$, $\varphi = 0$
x', y', z'	missile body axes; $\alpha_c \neq 0$, $\varphi = 0$
α	angle of attack
α_c	included angle
β	angle of sideslip
α_s, β_s	sine definitions of angles of attack and sideslip
α_t, β_t	tangent definitions of angles of attack and sideslip
φ	angle of bank

REFERENCE

1. Warren, C. H. E.: The Definitions of the Angles of Incidence and of Sideslip, *RAE Tech. Note Aero.* 2178, August, 1952.

CHAPTER 2

SOME FORMULAS COMMONLY USED
IN MISSILE AERODYNAMICS

The primary purpose of this chapter is to collect together for ready reference certain formulas of theoretical aerodynamics and mathematics commonly used in missile aerodynamics. These formulas are derived in detail and discussed in other works, and their rederivations here will not be attempted. Since repeated use is made of the formulas throughout the book, they are collected together in a single chapter for convenience, and to obviate repeated explanation of the formulas and notation. The formulas include the potential equation and Bernoulli's equation in their nonlinear and linearized forms. A listing and classification of the principal theories used in the book is provided. Some common aerodynamic formulas are included for line pressure sources, rectangular and triangular wings, and simple sweep theory. With regard to mathematical formulas, a list is given of conformal mappings used in the book, together with a list of the complex potentials of the flows to be used. The terminology and notation of elliptic integrals is also included.

2-1. Nonlinear Potential Equation

The common partial differential equation underlying the velocity fields of nearly all flows considered in this book is the potential equation. The *potential equation* is the partial differential equation for the velocity potential ϕ. The velocity potential is a scalar function of position and time, from which the flow velocities can be obtained by differentiation. For a discussion of the velocity potential, the reader is referred to Liepmann and Puckett.[1]* A number of conditions determine the actual form of the potential equation used in any particular case. Some of these conditions are (1) whether the fluid is compressible or incompressible, (2) the coordinate system used, (3) the velocity of the coordinate system with respect to the fluid far away, (4) whether the equation is linearized or retained in its nonlinear form, and (5) the basic flow about which the equation is linearized.

For the first case consider a compressible fluid stationary at infinity. Let the cartesian axes ξ, η, ζ (Fig. 2-1) be a set of axes fixed in the fluid. The pressure and density for the compressible fluid are related through

---

* Superior numbers refer to items in the bibliographies at the ends of chapters.

the isentropic law

$$\frac{p}{p_\infty} = \left(\frac{\rho}{\rho_\infty}\right)^\gamma \tag{2-1}$$

γ being the ratio of the specific heats.

Let Φ be the potential function. The full nonlinear equation[2] for Φ is

$$\left[c_\infty^2 - (\gamma - 1)\left(\Phi_\tau + \frac{\Phi_\xi^2 + \Phi_\eta^2 + \Phi_\zeta^2}{2}\right)\right](\Phi_{\xi\xi} + \Phi_{\eta\eta} + \Phi_{\zeta\zeta})$$
$$= \Phi_{\tau\tau} + (\Phi_\xi^2\Phi_{\xi\xi} + \Phi_\eta^2\Phi_{\eta\eta} + \Phi_\zeta^2\Phi_{\zeta\zeta}) + 2(\Phi_\xi\Phi_\eta\Phi_{\xi\eta} + \Phi_\xi\Phi_\zeta\Phi_{\xi\zeta}$$
$$+ \Phi_\eta\Phi_\zeta\Phi_{\eta\zeta}) + 2(\Phi_\xi\Phi_{\xi\tau} + \Phi_\eta\Phi_{\eta\tau} + \Phi_\zeta\Phi_{\zeta\tau}) \tag{2-2}$$

The symbol τ represents time and c_∞ is the speed of sound in the undisturbed air at ∞. Equation (2-2) can be considered as the nonlinear equation governing the pattern of the flow about a missile flying through still air as it would appear to an observer fixed on the ground. In many

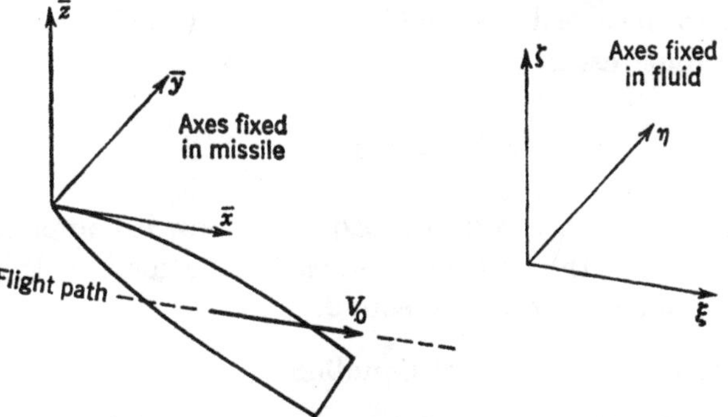

Fig. 2-1. Axes fixed in fluid and axes fixed in missile for uniform translation.

cases of interest in the theory of missile aerodynamics, the fluid velocity at infinity can be considered parallel and uniform, and the missile can be considered stationary with respect to the observer. It is now shown that the form of Eq. (2-2) is unchanged in this new frame of reference.

With reference to Fig. 2-1, let $\bar{x}$, $\bar{y}$, and $\bar{z}$ be axes fixed in the missile at time t with $\bar{x}$ parallel to the uniform velocity V_0 of the fluid at infinity as seen from the missile center of gravity. Also, choose the ξ axis of the ξ,η,ζ,τ system parallel to $\bar{x}$. To obtain the potential equation for the $\bar{x}$, $\bar{y}$, $\bar{z}$ axes with the fluid in uniform motion at infinity, we first convert the flow as seen by a ground observer from the ξ,η,ζ,τ system to the $\bar{x},\bar{y},\bar{z},t$ system with the fluid still stationary at infinity. Then we superimpose a velocity V_0 along the positive $\bar{x}$ axis to obtain the flow we seek. The transformation equations are

$$\begin{aligned}
\bar{x} &= \xi + V_0\tau \\
\bar{y} &= \eta \\
\bar{z} &= \zeta \\
t &= \tau
\end{aligned} \tag{2-3}$$

A new equation for Φ is then obtained in terms $\bar{x}$, $\bar{y}$, $\bar{z}$, t. Superimposing the velocity V_0 along the positive $\bar{x}$ axis changes the flow pattern and changes the potential Φ into the new potential ϕ in accordance with

$$\phi = V_0 \bar{x} + \Phi \tag{2-4}$$

The operations when carried out yield for ϕ

$$\left[c_\infty{}^2 + \frac{\gamma - 1}{2} V_0{}^2 - (\gamma - 1) \left(\phi_t + \frac{\phi_{\bar{x}}{}^2 + \phi_{\bar{y}}{}^2 + \phi_{\bar{z}}{}^2}{2} \right) \right]$$
$$(\phi_{\bar{x}\bar{x}} + \phi_{\bar{y}\bar{y}} + \phi_{\bar{z}\bar{z}}) = \phi_{tt} + (\phi_{\bar{x}}{}^2 \phi_{\bar{x}\bar{x}} + \phi_{\bar{y}}{}^2 \phi_{\bar{y}\bar{y}} + \phi_{\bar{z}}{}^2 \phi_{\bar{z}\bar{z}})$$
$$+ 2(\phi_{\bar{x}} \phi_{\bar{y}} \phi_{\bar{x}\bar{y}} + \phi_{\bar{x}} \phi_{\bar{z}} \phi_{\bar{x}\bar{z}} + \phi_{\bar{y}} \phi_{\bar{z}} \phi_{\bar{y}\bar{z}}) + 2(\phi_{\bar{x}} \phi_{\bar{x}t} + \phi_{\bar{y}} \phi_{\bar{y}t} + \phi_{\bar{z}} \phi_{\bar{z}t}) \tag{2-5}$$

A comparison of Eqs. (2-2) and (2-5) reveals the additional term $(\gamma - 1)V_0{}^2/2$ in the latter equation. A simple physical interpretation of this difference can be given. In Eq. (2-2), c_∞ is the speed of sound of the fluid at rest with respect to the ξ,η,ζ system. In the $\bar{x},\bar{y},\bar{z}$ system, c_∞ now corresponds to the speed of sound in fluid with velocity V_0. The speed of sound in fluid at rest in the $\bar{x},\bar{y},\bar{z}$ system, denoted by c_S, is given by

$$c_S{}^2 = c_\infty{}^2 + \frac{\gamma - 1}{2} V_0{}^2 \tag{2-6}$$

With this physical interpretation, Eq. (2-5) now is completely similar to Eq. (2-2). In fact, the first factor in each equation is nothing more than the square of the local speed of sound.

2-2. Linearization of Potential Equation

To linearize the potential equation, Eq. (2-5), we must reduce all terms greater than first order in the product of the potential and its derivatives to terms of first order or less. At the onset it should be stated that there are a number of ways of carrying out the linearization, and the correct way, if any, depends on the problem at hand. In any particular case the solution to a linearized problem should be examined to see if it fulfills the assumptions of the linearization. A particular way of linearizing the potential equation, which has proved particularly useful to the aerodynamicist, is to consider small changes in the velocity field from a uniform flow parallel to the $\bar{x}$ axis of speed V_0. The velocity components are then

$$\phi_{\bar{x}} = V_0 + \bar{u} \qquad \phi_{\bar{y}} = \bar{v} \qquad \phi_{\bar{z}} = \bar{w} \tag{2-7}$$

where $\bar{u}$, $\bar{v}$, and $\bar{w}$ are small perturbation velocities. With the possible exception of limited regions such as stagnation regions, our first assumption is that the perturbation velocities are small compared to V_0 so that

$$\frac{\bar{u}}{V_0}, \frac{\bar{v}}{V_0}, \frac{\bar{w}}{V_0} = O(\epsilon) \ll 1 \tag{2-8}$$

In this equation the symbol $O(\epsilon)$ stands for "of the order of magnitude of ϵ." In contrast to its more precise mathematical meaning, the symbol has the approximate physical meaning in the present connection that the velocity ratios have numerical values of a magnitude ϵ much less than unity. In the neighborhoods where the perturbation velocities are large, the solutions of the linear equation for small perturbation velocities cannot be accurate, but, if these regions are limited in extent and number, it can be hoped that the solutions will be representative of the flow in the large.

In connection with Eq. (2-8) we have also assumed that the perturbation velocities are of comparable magnitudes. If, as is frequently the case, the lateral extent of the region of influence of the body on the potential field is approximately the same length as the longitudinal extent, then, on the average, the gradients of the potential will be the same in all directions, and $\bar{u}$, $\bar{v}$, and $\bar{w}$ will be of comparable magnitude. The validity of this assumption must be adjudged for the particular problem at hand. The velocity components are then of orders

$$\phi_{\bar{x}} = V_0[1 + O(\epsilon)] \qquad \phi_{\bar{y}} = O(\epsilon V_0) \qquad \phi_{\bar{z}} = O(\epsilon V_0) \qquad (2\text{-}9)$$

If the lateral and longitudinal extents of the region of influence of the body on the potential field are characterized by length L, the second spatial derivatives of the potential are of order

$$\phi_{\bar{x}\bar{x}},\ \phi_{\bar{x}\bar{y}},\ \phi_{\bar{x}\bar{z}},\ \phi_{\bar{y}\bar{y}},\ \phi_{\bar{y}\bar{z}},\ \phi_{\bar{z}\bar{z}} = O\left(\frac{\epsilon V_0}{L}\right) \qquad (2\text{-}10)$$

We now need some measure of the orders of magnitude of the time derivatives of the potential. First, since

$$\phi - V_0\bar{x} = \int_{-\infty}^{\bar{x}} \bar{u}\, d\bar{x} \qquad (2\text{-}11)$$

we obtain

$$\phi - V_0\bar{x} = O(\epsilon V_0 L) \qquad (2\text{-}12)$$

Now consider the body to be undergoing some unsteady motion such as periodic oscillations characterized by frequency n per unit time. The perturbation potential will change $2n$ times per unit time so that

$$\phi_t = O(\epsilon V_0 L n)$$

Let us introduce the number of cycles per body length of travel

$$\chi = \frac{nL}{V_0} \qquad (2\text{-}13)$$

Then, in terms of this frequency parameter, the time derivatives are of orders of magnitude

$$\phi_t = O(\epsilon \chi V_0^2)$$

$$\phi_{tt} = O\left(\frac{\epsilon \chi^2 V_0^3}{L}\right)$$

$$\phi_{xt}, \phi_{yt}, \phi_{zt} = O\left(\frac{\epsilon \chi V_0^2}{L}\right) \tag{2-14}$$

The foregoing results permit us readily to determine the order of magnitude of all terms in Eq. (2-5) and to discard those of second and third order in ϵ. The resulting equation will be valid for all values of χ not greater than order of magnitude unity or, specifically, if

$$\chi \epsilon \ll 1 \tag{2-15}$$

The linearized equation is

$$\left(c_S^2 - \frac{\gamma - 1}{2} V_0^2\right) \nabla^2 \phi = \phi_{tt} + V_0^2 \phi_{xx} + 2 V_0 \phi_{xt} \tag{2-16}$$

In terms of the free-stream speed of sound

$$c_0^2 = c_S^2 - \frac{\gamma - 1}{2} V_0^2$$

and the free-stream Mach number

$$M_0 = \frac{V_0}{c_0} \tag{2-17}$$

Equation (2-16) becomes

$$\phi_{xx}(1 - M_0^2) + \phi_{yy} + \phi_{zz} = \frac{1}{c_0^2} \phi_{tt} + 2 \frac{M_0}{c_0} \phi_{xt} \tag{2-18}$$

This equation is the essential equation of linear aerodynamics.

2-3. Bernoulli's Equation; Pressure Coefficient as a Power Series in Velocity Components

Bernoulli's equation for the compressible unsteady potential flow of a fluid whose density is a function only of the pressure is in the ξ,η,ζ,τ system:

$$\int \frac{dp}{\rho} + \Phi_\tau + \frac{q^2}{2} = C(\tau) \tag{2-19}$$

where

$$q^2 = \Phi_\xi{}^2 + \Phi_\eta{}^2 + \Phi_\zeta{}^2$$

Some interpretation of the above form of Bernoulli's equation is interesting. In that form, it holds for each point in the unsteady flow for all times. The function C has the same value at all points in the flow at any particular time, but its value can change with time. However, if the flow at any point (such as at infinity) does not change with time, then C is constant with time also. Bernoulli's equation can be thought of as a relationship between the pressure field of the flow and the velocity field.

Thus, if Φ is known for a given flow, the pressure can be calculated from Eq. (2-19). For a steady flow with the pressure and density related by Eq. (2-1), Bernoulli's equation becomes

$$\frac{\gamma}{\gamma-1}\frac{p}{\rho} + \frac{q^2}{2} = \frac{\gamma}{\gamma-1}\frac{p_R}{\rho_R} + \frac{q_R^2}{2} \tag{2-20}$$

The quantities with no subscripts are for any general point, while those with subscripts R refer to quantities at some reference condition.

In linearized theory, Bernoulli's equation is generally used to obtain an expression for the pressure coefficient in terms of the velocity components $\bar{u}$, $\bar{v}$, $\bar{w}$ along the $\bar{x}$, $\bar{y}$, $\bar{z}$ axes. For this purpose we define the pressure coefficient P in terms of certain reference quantities

$$P = \frac{p - p_R}{\tfrac{1}{2}\rho_R V_R^2} \tag{2-21}$$

where p_R, ρ_R, and V_R are usually taken as the pressure, density, and velocity of the free stream (p_0,ρ_0,V_0) in the $\bar{x},\bar{y},\bar{z},t$ coordinate system or p_∞, ρ_∞, V_0 in the ξ,η,ζ,τ system for complete analogy between the two systems. To obtain the power series for P in velocity components let us perform the expansion in the ξ,η,ζ,τ system and then transform to the x,y,z,t system. With the subscript ∞ referring to the condition at ∞ in the ξ,η,ζ,τ system, integration of Eq. (2-19) yields

$$\frac{\gamma}{\gamma-1}\frac{p}{\rho} + \frac{q^2}{2} + \Phi_\tau = \frac{\gamma}{\gamma-1}\frac{p_\infty}{\rho_\infty} \tag{2-22}$$

where Φ and q are taken as zero at infinity. With the help of Eq. (2-1) and the Mach number relationship

$$M_\infty = \frac{V_0}{c_\infty} \tag{2-23}$$

we can put Eq. (2-22) into the following form after some manipulation:

$$P\frac{\gamma M_\infty^2}{2} = 1 - \frac{\gamma-1}{c_\infty^2}\left(\frac{q^2}{2} + \Phi_\tau\right)^{\gamma/(\gamma-1)} - 1 \tag{2-24}$$

Expansion of Eq. (2-24) yields the power series

$$P = \frac{p - p_\infty}{\tfrac{1}{2}\rho_\infty V_0^2} = -2\frac{q^2/2 + \Phi_\tau}{V_0^2} + M_\infty^2\left(\frac{q^2/2 + \Phi_\tau}{V_0^2}\right)^2 + M_\infty^4 O\left(\frac{q^2/2 + \Phi_\tau}{V_0^2}\right)^3 \tag{2-25}$$

where O designates order of magnitude. This series gives the pressure coefficient in powers of the derivatives of Φ in the ξ,η,ζ,τ system.

To convert Eq. (2-25) to the $\bar{x},\bar{y},\bar{z},t$ system, we note that the pressure coefficient has the same value and same physical significance in both systems since it is based on conditions at infinity. The only difference is the notational one that the subscript ∞ for p_∞, ρ_∞, and c_∞ must now be changed to 0. In particular M_∞ is now M_0. With these notational changes, we now introduce the potential ϕ in the x,y,z,t system which in accordance with Eqs. (2-3) and (2-4) is

$$\Phi(\xi,\eta,\zeta,\tau) = \phi(\bar{x} - V_0 t, \bar{y}, \bar{z}, t) - V_0 \bar{x} \tag{2-26}$$

For the derivatives of Φ we thus obtain

$$\begin{aligned}
\Phi_\xi &= \phi_{\bar{x}} - V_0 \\
\Phi_\eta &= \phi_{\bar{y}} \qquad\qquad \Phi_\tau = (\phi_{\bar{x}} - V_0)V_0 + \phi_t \\
\Phi_\zeta &= \phi_{\bar{z}}
\end{aligned} \tag{2-27}$$

If we further let $\bar{u}$, $\bar{v}$, and $\bar{w}$ be the *perturbation* velocities parallel to the $\bar{x}$, $\bar{y}$, and $\bar{z}$ axes, we have

$$\begin{aligned}
q^2 &= \Phi_\xi{}^2 + \Phi_\eta{}^2 + \Phi_\zeta{}^2 = \bar{u}^2 + \bar{v}^2 + \bar{w}^2 \\
\Phi_\tau &= \bar{u}V_0 + \phi_t
\end{aligned} \tag{2-28}$$

We thus interpret Eq. (2-25) as

$$P = \frac{p - p_0}{\frac{1}{2}\rho_0 V_0{}^2} = \frac{-2}{V_0{}^2}\left(\frac{\bar{u}^2 + \bar{v}^2 + \bar{w}^2}{2} + \bar{u}V_0 + \phi_t\right)$$
$$+ \frac{M_0{}^2}{V_0{}^4}\left(\frac{\bar{u}^2 + \bar{v}^2 + \bar{w}^2}{2} + \bar{u}V_0 + \phi_t\right)^2 + \cdots$$

so that as a final result we have

$$P = -\frac{2\bar{u}}{V_0} - \frac{2\phi_t}{V_0} + \frac{\bar{u}^2(M_0{}^2 - 1) - \bar{v}^2 - \bar{w}^2}{V_0{}^2}$$
$$+ \frac{M_0{}^2}{V_0{}^4}(\phi_t + 2\bar{u}V_0)\phi_t + \text{terms of third order in } \epsilon \tag{2-29}$$

Even though the potential equation is linearized, the square terms in Eq. (2-29) can be significant as, for instance, in slender-body theory.

2-4. Classification of Various Theories Used in Succeeding Chapters

Results from a number of aerodynamic theories are utilized in succeeding chapters. The theories to be used differ in a number of respects as follows:

(1) Potential or nonpotential
(2) Mach-number range of applicability
(3) Dimensionality of flow; i.e., two-dimensional, axially symmetric
(4) Shape of physical boundaries considered

All the theories we will consider are potential theories with the exception of the Newtonian theory [Eq. (9-50)] and the viscous crossflow theory

(Sec. 4-6).   With regard to the Mach-number range of applicability we will be concerned principally with theories valid in the supersonic speed range, although various of these theories are valid at subsonic speeds also. We will be interested in theories that apply to two-dimensional flows, axially symmetric flows, and three-dimensional flows.   As for the shape of the physical boundaries, such shapes as planar surfaces, bodies of revolution, airfoils, etc., are encountered in classifying the various theories.   Only steady flows are considered.

TABLE 2-1. CLASSIFICATION OF AERODYNAMIC THEORIES USED IN TEXT

Theory	Potential	Flow dimensionality	Typical shapes	Speed range	Class
Ackeret	Yes	Two-dimensional	Airfoils	$M > 1$	A
Busemann	Yes	Two-dimensional	Airfoils	$M > 1$	A
Shock-expansion	Yes	Two-dimensional	Airfoils	$M > 1$	A
Method of characteristics	Yes	Two-dimensional, axially symmetric	Airfoils and bodies of revolution	$M > 1$	A
Strip	Usually	Two-dimensional	Three-dimensional	Any M	B
Simple sweep	Usually	Two-dimensional	Swept wing and swept cylinders	Any M	B
Supersonic wing	Yes	Three-dimensional	Wings	$M > 1$	C
Conical flow	Yes	Three-dimensional	Wings, cones	$M > 1$ usually	C
Supersonic lifting line	Yes	Three-dimensional	Wings built of horseshoe vortices	$M > 1$	C
Quasi-cylinder	Yes	Three-dimensional	Quasi-cylinders	$M > 1$ usually	C
Slender body	Yes	Three-dimensional	"Slender" bodies	Any M	C
Newtonian impact	No	Three-dimensional	Any shape	Any M	D
Viscous crossflow	No	Three-dimensional	Slender bodies	Any M	D

A listing of the theories to be considered is given in Table 2-1.   The theories are classified in classes A, B, C, and D.   The first three classes are essentially potential theories but D is not.   Class A is a class of essentially two-dimensional theories; class B is the class of two-dimensional theories applied to three-dimensional shapes, and class C is a class of essentially three-dimensional theories.

The theories of class A are arranged in order of increasing exactitude. The first three theories have been treated in a form suitable for engineering calculations.[3]   The *Ackeret theory* embraces solutions of Eq. (2-18) specialized to two dimensions

$$(M_0{}^2 - 1)\phi_{xx} - \phi_{zz} = 0 \qquad (2\text{-}30)$$

and gives pressure coefficients linear in the flow-deflection angle. The *Busemann theory* is an application of the equations of oblique shock waves and Prandtl-Meyer flow expanded in a power series in the flow deflection to terms of the second degree. The actual formulas for the pressure coefficient are given in Sec. 8-6. The use of the equations of oblique shock waves and of Prandtl-Meyer flow in their full accuracy is termed *shock-expansion* theory. Calculation by the method of shock-expansion theory can conveniently be made by means of tables and charts.[3,4] As described by Sauer,[5] the *method of characteristics* is basically a two-dimensional graphical method for solving two-dimensional or axially symmetric potential flows. Though its use in three dimensions is not precluded on theoretical grounds, the graphical procedures are not convenient to carry out. In many instances the graphical procedures can be adapted to automatic computing techniques. In such cases the method

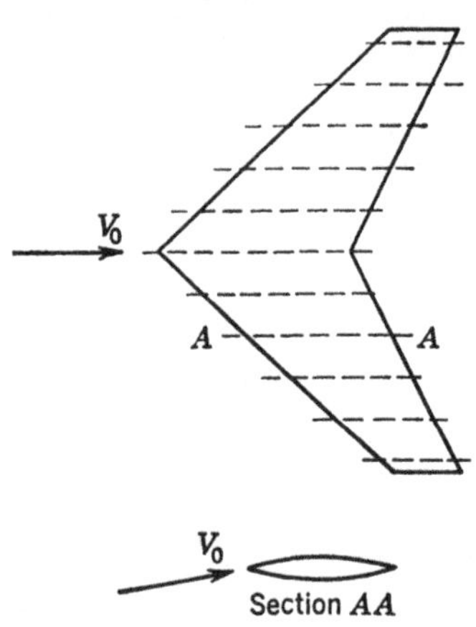

FIG. 2-2. Strip theory.

is not too time-consuming for common engineering use. Not the least of its many uses is as a standard of comparison for less accurate but more rapid methods, i.e., Sec. 9-4.

The theories of class *B* are two-dimensional methods applied directly to three-dimensional shapes. In *strip theory* any three-dimensional shape is sliced into strips by a series of parallel planes usually in the streamwise direction (Fig. 2-2). The primary assumption is that the flow in each strip is two-dimensional with no interaction between strips. To each strip is then applied any two-dimensional theory or even two-dimensional data. Simple sweep theory is a special form of strip theory applied normal to the leading edge of swept wings or cylinders. It is considered in Sec. 2-7.

With the exception of slender-body theory, the theories of class *C* all involve three independent space coordinates in their partial differential equations. (Conical flow theory can be put into a form with only two independent variables.) For slender-body theory, the third coordinate, streamwise distance, is manifest in the boundary conditions rather than in the partial differential equation. *Supersonic wing theory* is based on the linearized steady potential flow equation Eq. (2-18).

$$(M_0{}^2 - 1)\phi_{xx} - \phi_{yy} - \phi_{zz} = 0 \qquad (2\text{-}31)$$

and is discussed by Jones and Cohen.[7] In supersonic wing theory, the boundary conditions are applied in the $z = 0$ plane, the plane of the wing. Solutions are known for many different wing planforms for both lifting and nonlifting wings. Supersonic wing theory for lifting surfaces (no thickness) is termed *supersonic lifting-surface theory*, and some results for triangular and rectangular supersonic lifting surfaces are included in Sec. 2-6. *Conical flow theory* is a special form of linearized theory applicable to problems in which the flow quantities are constant along any line emanating from an apex. The supersonic flow over a cone or a triangular lifting surface are well-known examples of conical flow. The Jones's line pressure source described in Sec. 2-5 is another example. Lagerstrom[8] has listed a large number of conical flows. The utility of conical flow theory lies in the large number of wing flow fields that can be constructed by superimposing conical flow fields with different apex positions.

The counterpart at supersonic speeds of the Prandtl lifting-line theory will be termed *supersonic lifting-line theory*. The essential difference is that supersonic horseshoe vortices are used (Sec. 6-3) instead of subsonic horseshoe vortices. In this method, the lifting surface is replaced by one or more horseshoe vortices. In the process, the details of the flow in the vicinity of the wing are lost, but simplicity is gained in trying to calculate the flow field at distances remote from the wing. The calculation of downwash and sidewash velocities at distances remote from the wing is tractable only in a few cases with the full accuracy of supersonic wing theory. Again in the calculation of the flow field associated with wing-body combinations, the use of lifting-line theory is tractable where the full linearized theory is not. *Quasi-cylindrical theory* at supersonic speeds is analogous to *supersonic wing theory* in that both utilize the same partial differential equations, but in the former the boundary conditions are applied on a cylindrical surface, rather than the $z = 0$ plane as in the latter. In this connection the cylinder is any closed surface generated by a line moving parallel to a given line. Many lifting surfaces can be so generated. Herein we confine our applications of quasi-cylindrical theory to cylinders that are essentially circular.[9]

The remaining theory of class C (slender-body theory, about which we will have much to say) is particularly adapted to slender bodies such as many missiles. This theory, described in detail in Chap. 3, is based on solutions to Laplace's equations in two dimensions with the streamwise coordinate being manifest through the boundary conditions. The occurrence of Laplace's equation renders slender-body theory particularly amenable to mathematical treatment and makes its application to three-dimensional bodies tractable in many cases of interest. The theories of class D are not potential theories and are discussed in Secs. 4-6 and 9-5.

2-5. Line Pressure Source

As an example of a conical flow solution, we have the line pressure source of R. T. Jones,[10] which is useful in problems of controls, drag, etc. The general features of the flow are readily shown. Consider the infinite triangular cone shown in Fig. 2-3. Such a cone is the boundary formed by placing a line pressure source along the leading edge. The pressure coefficient for a subsonic leading edge is

$$P = \mathbf{RP}\, \frac{2\delta \cosh^{-1}\mu}{\pi(\tan^2 \Lambda - B^2)^{1/2}}$$

$$\mu = \frac{\tan \Lambda/B - B \tan \nu}{[(\tan \Lambda \tan \nu - 1)^2 + (z/x)^2(\tan^2 \Lambda - B^2)]^{1/2}} \qquad (2\text{-}32)$$

and for a supersonic leading edge is

$$P = \mathbf{RP}\, \frac{2\delta \cos^{-1}\mu}{\pi(B^2 - \tan^2 \Lambda)^{1/2}} \qquad (2\text{-}33)$$

Here the designation **RP** denotes the real part of the inverse cosine or inverse hyperbolic cosine. The equations show that the pressure coefficients depend only on tan ν, y/x and z/x quantities, which are constant along rays from the origin. The pressure field is therefore conical. The wedge and pressure field are symmetrical above and below the $z = 0$ plane.

The pressure field shown in Fig. 2-4 is that for a wedge with a subsonic leading edge. The pressure coefficient is zero along the left Mach line, increasing as we move from left to right. At the leading edge, the pressure coefficient is theoretically infinite. To the right of the leading edge, the pressure again falls from infinity to zero at the Mach line. The infinity can be viewed as high positive pressure corresponding to stagnation pressure. A wedge with a supersonic leading edge has a conical flow field of the type shown in Fig. 2-5. The distinctive feature is the region of constant pressure between the leading edge and the Mach line. By superimposing line pressure sources and sinks, a number of symmetrical wings of widely varying planform can be built up.

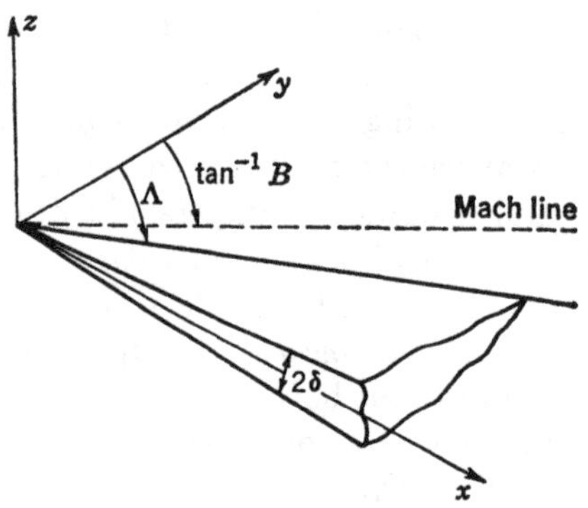

FIG. 2-3. Infinite wedge or infinite triangular cone.

2-6. Aerodynamic Characteristics of Rectangular and Triangular Lifting Surfaces on the Basis of Supersonic Wing Theory

In contrast to the symmetrical pressure fields of symmetrical wings at zero angle of attack, the pressure fields of lifting surfaces are asymmetrical; that is, the pressure changes sign between the upper and lower surfaces. Since we will deal extensively with lifting pressure fields, it is desirable to set up notation and terminology for loading coefficient, span

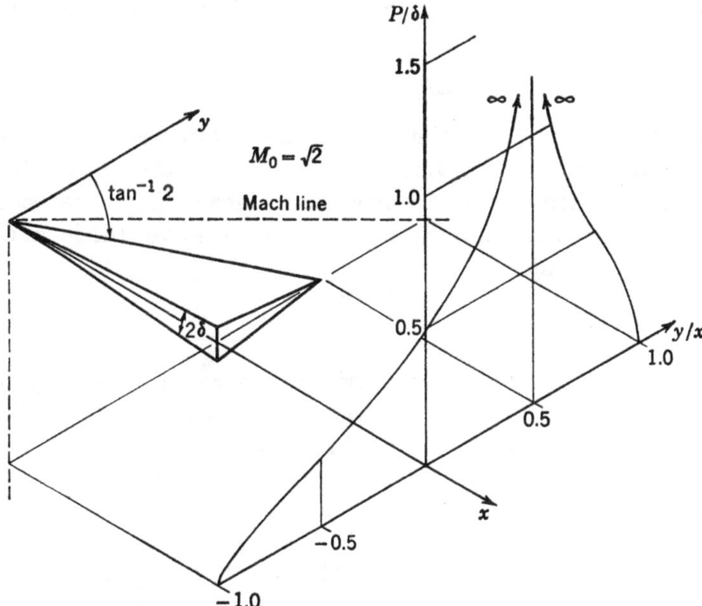

FIG. 2-4. Thickness pressure distribution on infinite wedge with subsonic leading edge.

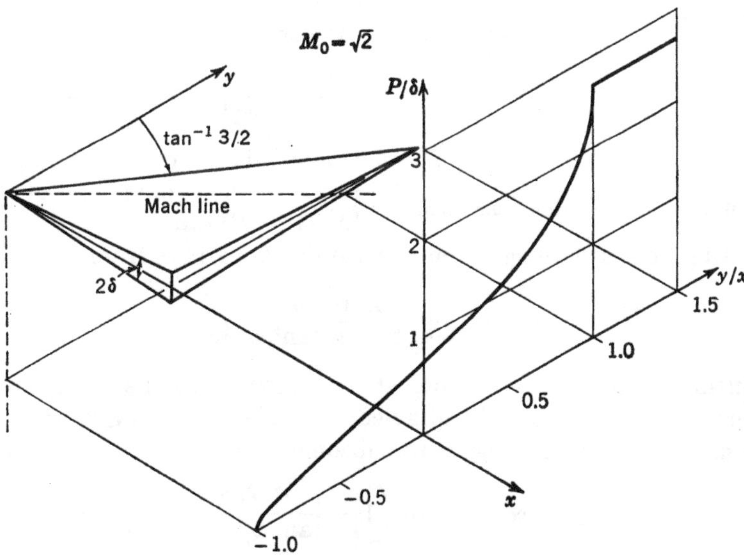

FIG. 2-5. Thickness pressure distribution on infinite wedge with supersonic leading edge.

loading, etc. By the *loading coefficient* of a wing or body, we mean the difference between the pressure coefficient at corresponding points on the upper and lower surfaces.

$$\Delta P = P^+ - P^- \tag{2-34}$$

The superscript plus $(+)$ refers to the impact pressures of the lower surface, while minus $(-)$ refers to the suction pressures of the upper surface. The distribution of ΔP over the surface is called the *loading distribution*. The *section lift coefficient* is the average over the local chord of the loading coefficient

$$c_l = \frac{1}{c} \int_{le}^{te} \Delta P \, dx \tag{2-35}$$

The *span-load distribution* is the distribution across the wing span of the product of the local chord and the section lift coefficient cc_l. The *center of pressure* is the position at which all the lift of a wing panel can be concentrated for the purpose of calculating moments.

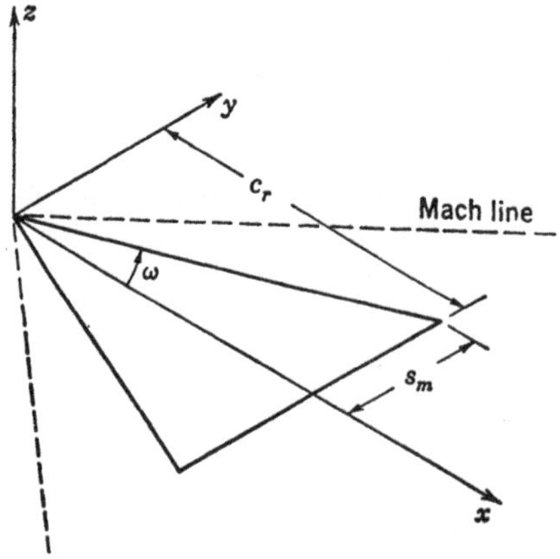

Let us now summarize some of the results of supersonic wing theory for triangular wings. For *triangular lifting surfaces with subsonic leading edges* (Fig. 2-6) the lift-curve slope has been determined by Stewart,[11] to be

$$\frac{dC_L}{d\alpha} = \frac{2\pi \tan \omega}{E(1 - B^2 \tan^2 \omega)^{1/2}} \tag{2-36}$$

where E is the complete elliptic integral of the second kind of modulus $(1 - B^2 \tan^2 \omega)^{1/2}$ (see Sec. 2-9).

Fig. 2-6. Notation for triangular wings.

The lifting pressure distribution is constant along rays from the apex

$$\Delta P = \frac{4\alpha \tan \omega}{(1 - \tan^2 \nu / \tan^2 \omega)^{1/2} E} \tag{2-37}$$

The lifting pressure field is conical with respect to the apex, and the pressures are infinite at the leading edges. The span-load distribution is *elliptical* for triangular wings with subsonic leading edges.

$$cc_l = (cc_l)_0 \left(1 - \frac{\tan^2 \nu}{\tan^2 \omega}\right)^{1/2} \tag{2-38}$$

The span loading at the root chord $(cc_l)_0$ is

$$(cc_l)_0 = \frac{4\alpha c_r \tan \omega}{E} \tag{2-39}$$

Because the flow is conical, each triangular element from the apex has its center of pressure at two-thirds the distance from the apex to the base. All triangular elements have their center of pressure at the two-thirds root-chord axial distance and so, therefore, does the wing. The lateral position of the center of pressure for an elliptical span loading is at the $4/3\pi$ semispan position.

The *triangular lifting surface with supersonic leading edges* also has simple aerodynamic properties. First, its lift-curve slope is the same as that of an infinite two-dimensional airfoil

$$\frac{dC_L}{d\alpha} = \frac{4}{B} \tag{2-40}$$

The loading distribution is conical and can be calculated directly from the results for line pressure sources in the preceding section since the upper and lower surfaces are independent. The slope τ is simply replaced by α in Eq. (2-33). On this basis with a line source along each leading edge we have for the wing loading

$$\Delta P = \frac{4\alpha}{\pi(B^2 - \text{ctn}^2\,\omega)^{1/2}}\left(\cos^{-1}\frac{\text{ctn}\,\omega/B - B\tan\nu}{|1 - \tan\nu/\tan\omega|}\right.$$
$$\left. + \cos^{-1}\frac{\text{ctn}\,\omega/B + B\tan\nu}{1 + \tan\nu/\tan\omega}\right) \tag{2-41}$$

Equation (2-41) yields a constant loading in the region between the Mach lines and the leading edges

$$\Delta P = \frac{4\alpha}{(B^2 - \text{ctn}^2\,\omega)^{1/2}} \tag{2-42}$$

For the region between the Mach lines, manipulation of Eq. (2-41) yields

$$\Delta P = \frac{4\alpha}{(B^2 - \text{ctn}^2\,\omega)^{1/2}}\left[1 - \frac{2}{\pi}\sin^{-1}\left(\frac{\text{ctn}^2\,\omega/B^2 - \text{ctn}^2\,\omega\tan^2\nu}{1 - \tan^2\nu\,\text{ctn}^2\,\omega}\right)^{1/2}\right] \tag{2-43}$$

The span loading in this case is not elliptical as for the lifting surface with subsonic leading edges but has a linear variation over the outboard section and a different variation between the Mach lines. For the linear part, we have with reference to Fig. 2-7

$$cc_l = \frac{4\alpha(s_m - y)\,\text{ctn}\,\omega}{(B^2 - \text{ctn}^2\,\omega)^{1/2}} \qquad \frac{s_m}{B\tan\omega} \leq |y| \leq s_m \tag{2-44}$$

and, over the inboard section,[12]

$$cc_l = \frac{4\alpha}{(B^2\tan^2\omega - 1)^{1/2}}\left[s_m + \frac{1}{\pi}(s_m - y)\sin^{-1}\frac{yB^2\tan^2\omega - s_m}{(s_m - y)B\tan\omega}\right.$$
$$\left. - \frac{s_m + y}{\pi}\sin^{-1}\frac{yB^2\tan^2\omega + s_m}{(s_m + y)B\tan\omega}\right] \qquad 0 \leq y \leq \frac{s_m}{B\tan\omega} \tag{2-45}$$

The center of pressure is still at the two-thirds root-chord axial location since the lifting pressure field is conical.

Turning now to the aerodynamic characteristics of *rectangular lifting surfaces* at supersonic speeds, we must differentiate several different cases, depending on the *effective aspect ratio BA*. For $BA > 2$ the tip Mach waves do not intersect, for $1 < BA < 2$ the tip Mach waves intersect

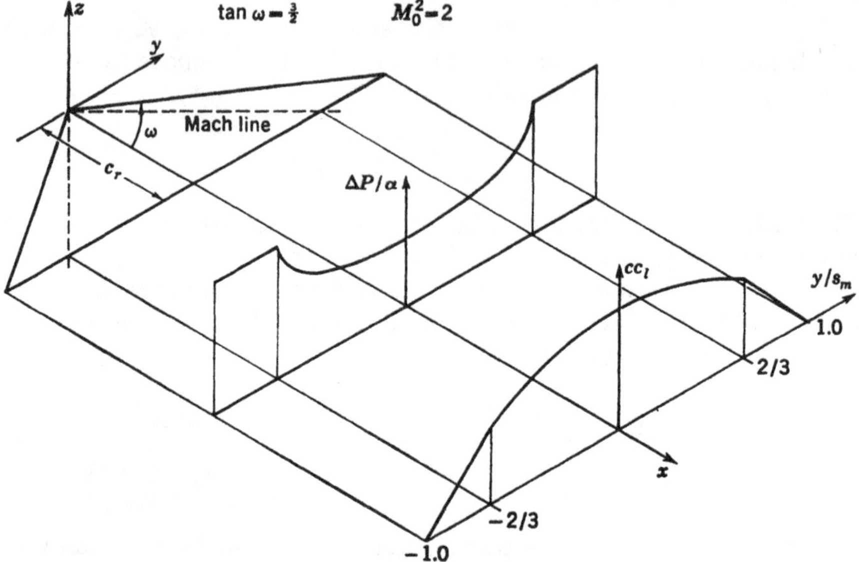

Fig. 2-7. Loading distribution along trailing edge, and span-load distribution for triangular wing with supersonic leading edges.

each other but do not intersect the wing tips, and for $\frac{1}{2} < BA < 1$ the tip Mach waves intersect the wing tips only once. The lift-curve slope for cases 1 and 2 (Fig. 2-8) has the same analytical form.

$$\frac{dC_L}{d\alpha} = \frac{4}{B}\left(1 - \frac{1}{2BA}\right) \qquad BA \geq 1 \qquad (2\text{-}46)$$

For case 3 the lift-curve slope is

$$\frac{dC_L}{d\alpha} = \frac{4}{\pi B}\left[\left(2 - \frac{1}{BA}\right)\sin^{-1} BA + (BA - 2)\cosh^{-1}\frac{1}{BA} \right.$$
$$\left. + \left(1 + \frac{1}{BA}\right)(1 - B^2A^2)^{\frac{1}{2}}\right] \qquad \frac{1}{2} \leq BA \leq 1$$

For cases 1, 2, and 3, characteristic regions I, II, and III are specified. The analytic form of the loading is different in each of the three regions. In region I there is no influence of the wing tips, and the loading coefficient has the two-dimensional value

$$\Delta P_{\mathrm{I}} = \frac{4\alpha}{B} \qquad (2\text{-}47)$$

The loading distribution within region II is conical from the extremities of the leading edges as calculated by Busemann.[13]   For the right region, we have with reference to Fig. 2-8

$$\Delta P_{\text{II}} = \frac{4\alpha}{B} \left\{ 1 - \left[1 - \frac{\cos^{-1}(1 - 2B\tan\theta_1)}{\pi} \right] \right\} \tag{2-48}$$

The loading is written here as the two-dimensional loading minus a decrement due to the wing tip.   The decrement due to the wing tips is shown

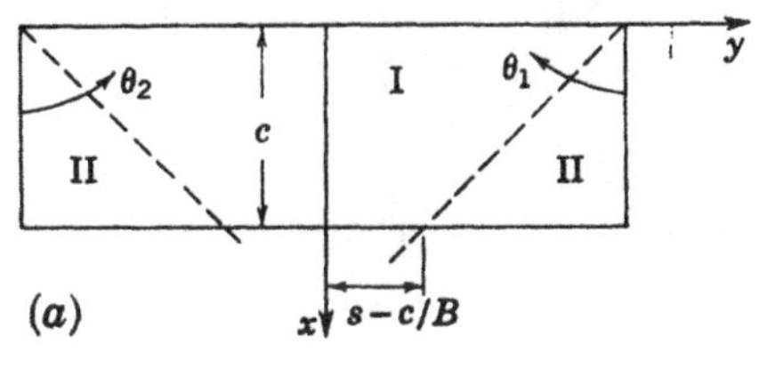

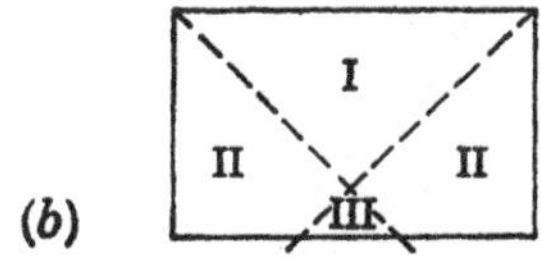

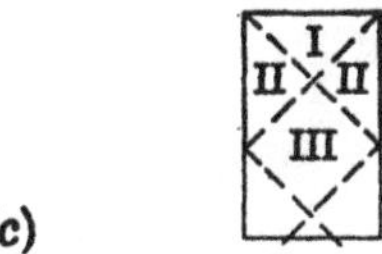

FIG. 2-8. Cases for rectangular wings.   (a) Case 1, $BA \geq 2$; (b) case 2, $1 \leq BA \leq 2$; (c) case 3, $\frac{1}{2} \leq BA \leq 1$.

in the loading diagram of Fig. 2-9.   In region III, the influence of both wing tips is felt so that both decrements apply

$$\Delta P_{\text{III}} = \frac{4\alpha}{B} \left\{ 1 - \left[1 - \frac{\cos^{-1}(1 - 2B\tan\theta_1)}{\pi} \right] - \left[1 - \frac{\cos^{-1}(1 - 2B\tan\theta_2)}{\pi} \right] \right\} \tag{2-49}$$

Span-loading results are now given for the case of $BA \geq 2$ so that only the influence of one tip is felt over the length of any chord:

$$cc_l = \frac{4\alpha c_r}{B} \qquad 0 \leq y \leq s - \frac{c_r}{B} \tag{2-50}$$

$$(cc_l) = \frac{4\alpha c_r}{B} \left\{ \frac{1}{\pi} \cos^{-1} \left[1 - \frac{2B(s-y)}{c} \right] + \frac{2}{\pi} \left[\frac{B(s-y)}{c} - B^2 \left(\frac{s-y}{c} \right)^2 \right]^{\frac{1}{2}} \right\} \tag{2-51}$$

The second equation gives the span loading in the tip regions, and the slope of the over-all span loading is shown in Fig. 2-9. The distance behind the wing leading edges of the center of pressure is

$$\frac{x_{\text{cp}}}{c} = \frac{1}{2}\frac{1 - 2/3BA}{1 - 1/2BA} \qquad 1 \leq BA \qquad (2\text{-}52)$$

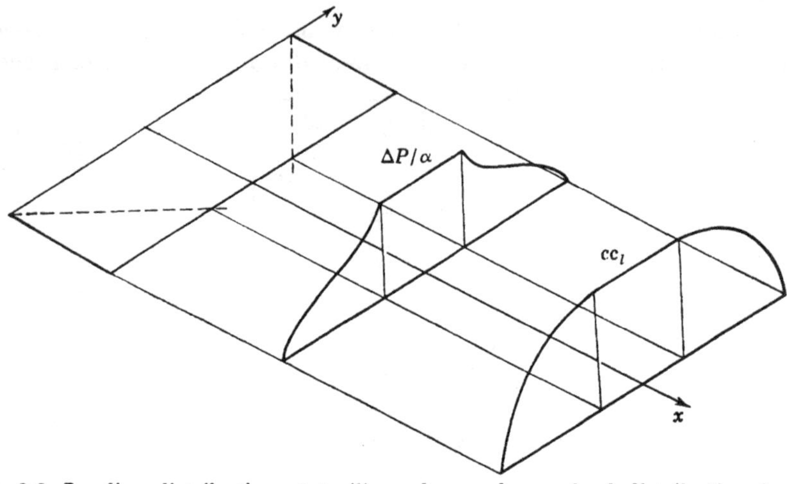

FIG. 2-9. Loading distribution at trailing edge, and span-load distribution for rectangular wing.

2-7. Simple Sweep Theory

For swept cylinders or swept wings, simple sweep theory offers an easy method of obtaining the flow field in many instances. Consider a swept wing of infinite span as shown in Fig. 2-10. Let the free-stream Mach number M_0 be resolved into a component M_p parallel to the leading edge and a component M_n perpendicular to it. The first thing to note is that the velocity component parallel to the leading edge does not influence the flow as viewed in planes perpendicular to the leading edge. The only role of M_p is to move the row of particles in one plane into the next as shown in Fig. 2-10. The flow in the normal planes thus depends only on the angle of attack and Mach number in the plane. In particular, if M_n is subsonic, the pressure distribution is typically subsonic in the normal planes even though M_0 may be supersonic. For further details the reader is referred to Jones.[6]

Illustrative Example

As an example in the use of simple sweep theory, consider the determination of the pressure distribution in the neighborhood of the leading edge of a triangular wing with supersonic leading edges. The use of simple sweep theory in this connection can be simply seen from Fig. 2-10, where the upstream Mach cone from point Q of the triangular wing intersects precisely the same planform area as if the triangular wing were part of an infinite wing.   The answer on the basis of simple sweep theory will therefore be exact to the order of linear theory.

For the wing we have

$$M_n = M_0 \cos \Lambda$$
$$V_n = V_0 \cos \Lambda \qquad (2\text{-}53)$$

The angle of attack by definition is

$$\alpha = \frac{-w}{V_0}$$

where $-w$ is the uniform downwash over the planform.   The corresponding definition for the angle of attack in the normal direction is

$$\alpha_n = \frac{-w}{V_n} = \frac{\alpha}{\cos \Lambda} \qquad (2\text{-}54)$$

since the downwash $-w$ is unchanged. By a direct application of Ackeret's two-dimensional theory the pressure coefficient is

$$\frac{p - p_0}{q_n} = \frac{2\alpha_n}{B_n}$$
$$= \frac{2\alpha}{\cos \Lambda (M_0{}^2 \cos^2 \Lambda - 1)^{\frac{1}{2}}} \qquad (2\text{-}55)$$

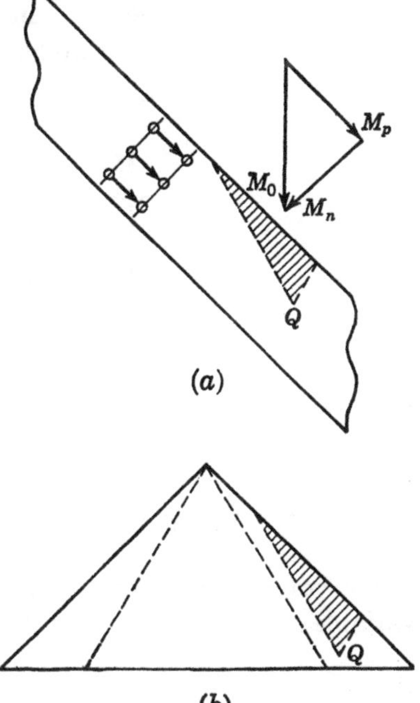

Fig. 2-10. Simple sweep theory with application to triangular wing with supersonic leading edges.   (a) Infinite swept wing; (b) triangular wing.

where B_n is $(M_n{}^2 - 1)^{\frac{1}{2}}$.   Equation (2-55) is valid between the wing leading edges and the Mach lines.   Referring the pressure coefficient to q_0 rather than q_n yields

$$P = \frac{p - p_0}{q_0} = \frac{2\alpha}{(B^2 - \tan^2 \Lambda)^{\frac{1}{2}}} \qquad (2\text{-}56)$$

The result of Eq. (2-56) for the pressure coefficient is in accord with Eq. (2-42) for the loading coefficient since $\omega = \pi/2 - \Lambda$.   The lift-curve slope of an infinite swept wing varies with sweepback angle in the same way as the pressure coefficient in Eq. (2-56).

2-8. Conformal Mapping; Notation; Listings of Mapping and Flows

We will have occasion to use conformal mapping to a considerable extent, so that it becomes desirable to gather together for ready reference the notations and formulas to be used.   This section is not intended to be an introduction to the subject such as, for instance, Milne-Thompson's discussion.[14] Conformal mapping is useful for finding incompressible potential flow about various missile cross sections from the known flow about other sections. The plane in which the flow is to be found is called the *physical plane* of the complex variable $\mathfrak{z} = y + iz$. The plane in which the flow is known will be termed the *transformed plane* with complex variable $\sigma = \xi + i\eta$. An example of the two planes is shown in Fig. 2-11 for a missile at angle of attack α_c and zero bank angle.

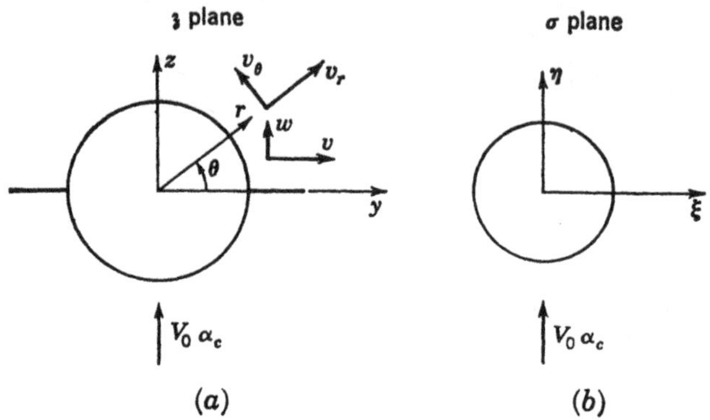

FIG. 2-11. Notation for (a) physical and (b) transformed planes.

The transformation equation is the relationship connecting the complex variables $\mathfrak{z}$ and σ. The transformation written in the following form

$$\sigma = \sigma(\mathfrak{z}) \tag{2-57}$$

can be interpreted to mean that any point in the $\mathfrak{z}$ plane can be transformed into a corresponding point in the σ plane. Likewise the inverse relationship

$$\mathfrak{z} = \mathfrak{z}(\sigma) \tag{2-58}$$

can be interpreted to mean that any point in the σ plane can be transformed into a corresponding point in the $\mathfrak{z}$ plane. The transformations we use will be ones causing no distortion of the planes at infinity. In such cases the transformation can be written

$$\sigma = \mathfrak{z} + \sum_{n=1}^{\infty} \frac{c_n}{\mathfrak{z}^n} \tag{2-59}$$

$$\mathfrak{z} = \sigma + \sum_{n=1}^{\infty} \frac{k_n}{\sigma^n} \tag{2-60}$$

The constants c_n or k_n may be complex. Several transformations which we will use are listed in Table 2-2.

Two-dimensional incompressible flows are described analytically by the two functions of a real variable, the potential function and the stream function, or by a single function of a complex variable, the complex potential.   The *complex potential* $W(\mathfrak{z})$ is

$$W(\mathfrak{z}) = \phi + i\psi \tag{2-61}$$

In accordance with Fig. 2-11 the velocity components parallel to the y and z axes are denoted v and w, respectively, while the radial and tangential velocity components are v_r and v_θ.   The v and w velocity com-

TABLE 2-2. CONFORMAL TRANSFORMATIONS WITH FIELD AT INFINITY
UNDISTORTED IN TRANSFORMATION

A. Circle into an ellipse:

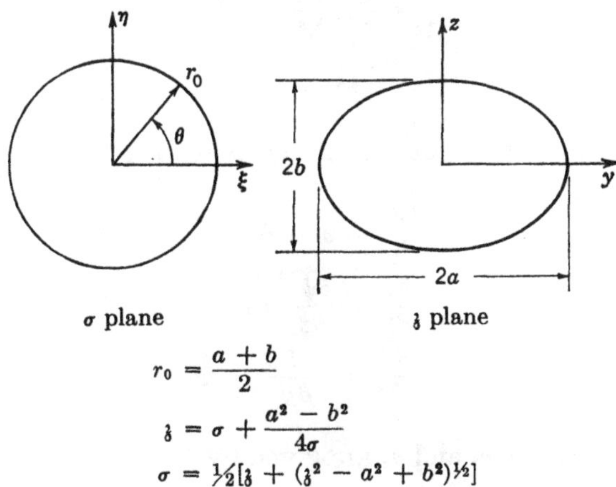

$$r_0 = \frac{a + b}{2}$$

$$\mathfrak{z} = \sigma + \frac{a^2 - b^2}{4\sigma}$$

$$\sigma = \tfrac{1}{2}[\mathfrak{z} + (\mathfrak{z}^2 - a^2 + b^2)^{\frac{1}{2}}]$$

B. Circle into planar midwing and body combination:

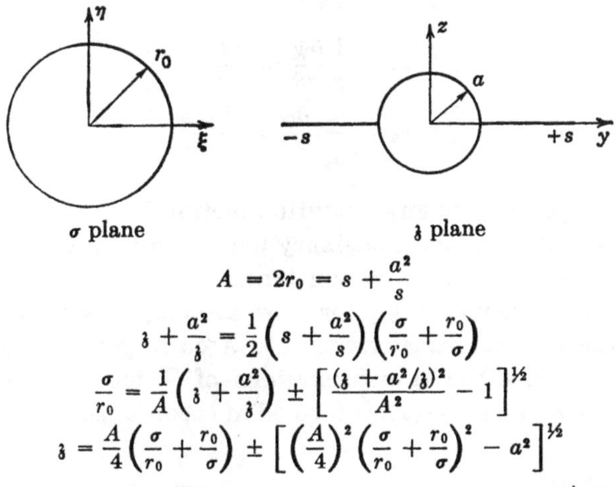

$$A = 2r_0 = s + \frac{a^2}{s}$$

$$\mathfrak{z} + \frac{a^2}{\mathfrak{z}} = \frac{1}{2}\left(s + \frac{a^2}{s}\right)\left(\frac{\sigma}{r_0} + \frac{r_0}{\sigma}\right)$$

$$\frac{\sigma}{r_0} = \frac{1}{A}\left(\mathfrak{z} + \frac{a^2}{\mathfrak{z}}\right) \pm \left[\frac{(\mathfrak{z} + a^2/\mathfrak{z})^2}{A^2} - 1\right]^{\frac{1}{2}}$$

$$\mathfrak{z} = \frac{A}{4}\left(\frac{\sigma}{r_0} + \frac{r_0}{\sigma}\right) \pm \left[\left(\frac{A}{4}\right)^2\left(\frac{\sigma}{r_0} + \frac{r_0}{\sigma}\right)^2 - a^2\right]^{\frac{1}{2}}$$

+ upper half space
− lower half space

TABLE 2-2. CONFORMAL TRANSFORMATIONS WITH FIELD AT INFINITY
UNDISTORTED IN TRANSFORMATION (*Continued*)

C. Circle into planar wing:

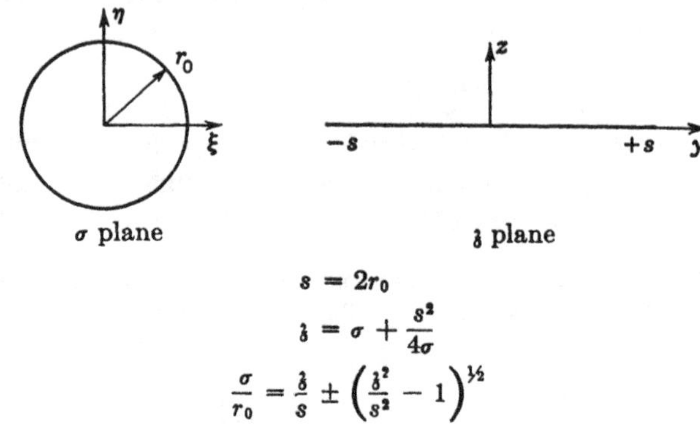

$$s = 2r_0$$

$$ȝ = \sigma + \frac{s^2}{4\sigma}$$

$$\frac{\sigma}{r_0} = \frac{ȝ}{s} \pm \left(\frac{ȝ^2}{s^2} - 1\right)^{\frac{1}{2}}$$

+ upper half space
− lower half space

ponents are related to various functions as follows:

$$\frac{dW}{dȝ} = v - iw$$

$$\frac{\partial \phi}{\partial y} = \frac{\partial \psi}{\partial z} = v \qquad\qquad (2\text{-}62)$$

$$\frac{\partial \phi}{\partial z} = \frac{-\partial \phi}{\partial y} = w$$

while the components v_r and v_θ are given by

$$\frac{dW}{dȝ} = (v_r - iv_\theta)e^{-i\theta}$$

$$v_r = \frac{1}{r}\frac{\partial \psi}{\partial \theta} = \frac{\partial \phi}{\partial r} \qquad\qquad (2\text{-}63)$$

$$v_\theta = -\frac{\partial \psi}{\partial r} = \frac{1}{r}\frac{\partial \phi}{\partial \theta}$$

The flow corresponding to any analytic function $W(ȝ)$ can be constructed by splitting W into real and imaginary parts and investigating the shape of the streamlines given by $\psi =$ constant.

The complex potential $W(ȝ)$ for flow associated with a given shape in the $ȝ$ plane can be transformed into a corresponding flow in the σ plane by employing the transformations of Table 2-2. The complex potential in the σ plane, $W_1(\sigma)$, is formed in accordance with the following relationship

$$W_1(\sigma) = W_1(\sigma(ȝ)) = W(ȝ) \qquad\qquad (2\text{-}64)$$

If the velocities in the $\mathfrak{z}$ plane are v and w and if those in the σ plane are v_1 and w_1, then

$$v_1 - iw_1 = \frac{dW_1}{d\sigma} = \frac{dW}{d\mathfrak{z}}\frac{d\mathfrak{z}}{d\sigma}$$

$$v_1 - iw_1 = (v - iw)\frac{d\mathfrak{z}}{d\sigma} = (v - iw)\left|\frac{d\mathfrak{z}}{d\sigma}\right|\exp\left(i\arg\frac{d\mathfrak{z}}{d\sigma}\right) \quad (2\text{-}65)$$

The conjugate complex velocity, $v - iw$, is thus magnified in the transformation by the factor $|d\mathfrak{z}/d\sigma|$ and rotated by the angle $\arg(d\mathfrak{z}/d\sigma)$. By making the transformation equations of the same form as Eq. (2-60), the value of $d\mathfrak{z}/d\sigma$ is unity for $\mathfrak{z}\to\infty$ and $\sigma\to\infty$, and the $\arg(d\mathfrak{z}/d\sigma)$ is zero under the same conditions. The flow field at infinity is thus undistorted. If the flow past a body B_1 in a parallel stream is known, the flow past a body B_2 in a parallel stream is obtained by use of Eq. (2-64) through the transformation of the type given by Eq. (2-59), which converts B_1 into B_2. In the present case the flow velocities are considered tangent to fixed surface boundaries, and Eq. (2-65) insures that this tangency condition is maintained during the transformation. Another case arises for bodies whose shapes are functions of time. Some of the complex potentials we will use are listed in Table 2-3.

TABLE 2-3. COMPLEX POTENTIALS FOR VARIOUS FLOWS

A. *Circular cylinder in uniform flow:*

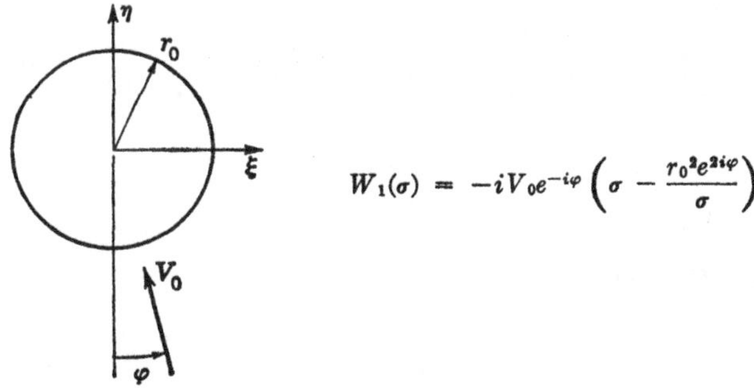

$$W_1(\sigma) = -iV_0 e^{-i\varphi}\left(\sigma - \frac{r_0^2 e^{2i\varphi}}{\sigma}\right)$$

B. *Uniformly expanding circle:*

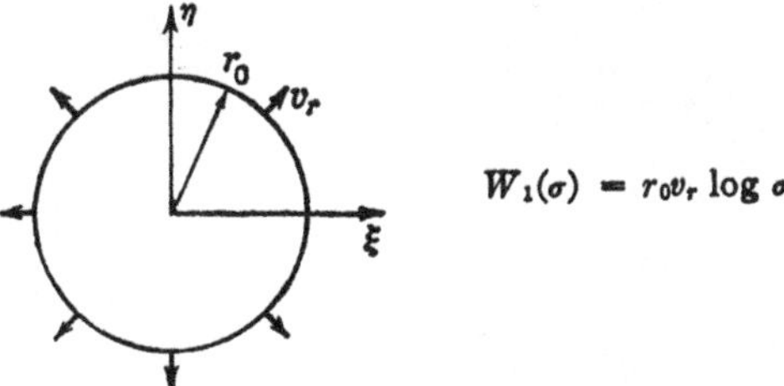

$$W_1(\sigma) = r_0 v_r \log \sigma$$

TABLE 2-3. COMPLEX POTENTIALS FOR VARIOUS FLOWS (*Continued*)

C. Expanding ellipse of constant a/b ratio:

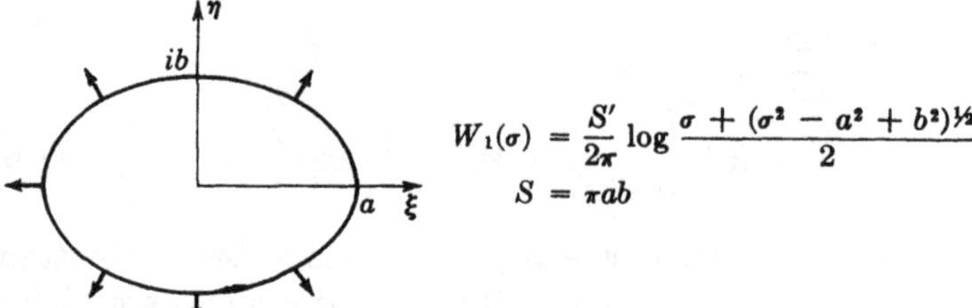

$$W_1(\sigma) = \frac{S'}{2\pi} \log \frac{\sigma + (\sigma^2 - a^2 + b^2)^{1/2}}{2}$$

$$S = \pi a b$$

D. Ellipse in uniform flow:

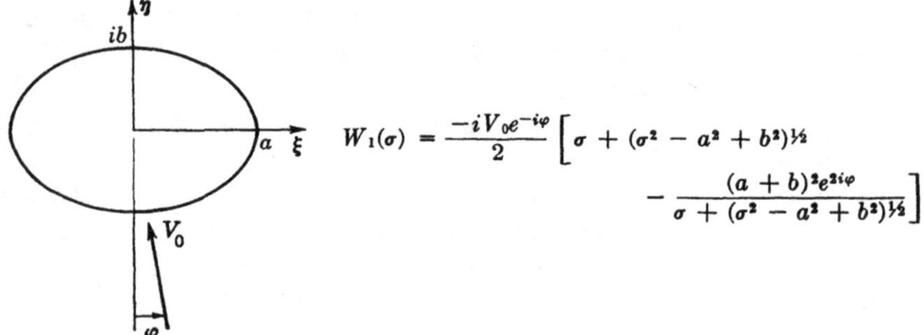

$$W_1(\sigma) = \frac{-iV_0 e^{-i\varphi}}{2}\left[\sigma + (\sigma^2 - a^2 + b^2)^{1/2} - \frac{(a+b)^2 e^{2i\varphi}}{\sigma + (\sigma^2 - a^2 + b^2)^{1/2}} \right]$$

E. Planar midwing and body combination:

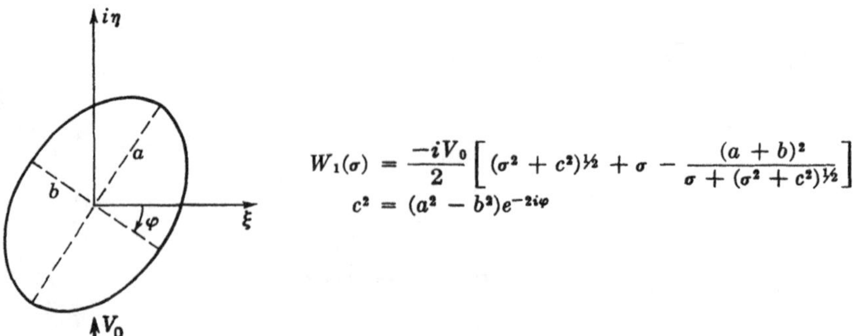

$$W_1(\sigma) = -iV_0 \left[\left(\sigma + \frac{a^2}{\sigma}\right)^2 - \left(s + \frac{a^2}{s}\right)^2 \right]^{1/2}$$

F. Ellipse banked with respect to lateral axis:

$$W_1(\sigma) = \frac{-iV_0}{2}\left[(\sigma^2 + c^2)^{1/2} + \sigma - \frac{(a+b)^2}{\sigma + (\sigma^2 + c^2)^{1/2}} \right]$$

$$c^2 = (a^2 - b^2)e^{-2i\varphi}$$

2-9. Elliptic Integrals

We shall have occasion to use elliptic integrals a number of times, so it is desirable to define notation and usage. The elliptic integrals of the first kind $F(k,\phi)$ and of the second kind $E(k,\phi)$ are defined as definite integrals

$$
\begin{aligned}
F(k,\phi) &= \int_0^\phi \frac{dz}{(1 - k^2 \sin^2 z)^{1/2}} = \int_0^{\sin^{-1}\phi} \frac{dx}{(1 - x^2)^{1/2}(1 - k^2 x^2)^{1/2}} \\
E(k,\phi) &= \int_0^\phi (1 - k^2 \sin^2 z)^{1/2}\, dz = \int_0^{\sin^{-1}\phi} \left(\frac{1 - k^2 x^2}{1 - x^2}\right)^{1/2} dx
\end{aligned}
\tag{2-66}
$$

The angle ϕ, which will usually lie between 0 and $\pi/2$, is termed the *amplitude*, and the parameter k is termed the *modulus*. The elliptic integrals are functions of amplitude and modulus only. If the amplitude is $\pi/2$, we call the elliptic integrals *complete*, and use the notation

$$
K(k) = F\left(k, \frac{\pi}{2}\right)
$$

$$
E(k) = E\left(k, \frac{\pi}{2}\right)
$$

Thus if the amplitude is not specified, it is assumed to be $\pi/2$, and the elliptic integral is complete. Tables of the elliptic integrals can be found in Byrd and Friedman.[15]

SYMBOLS

A	aspect ratio
B	$(M_0^2 - 1)^{1/2}$
c	local wing chord
c_0	velocity of sound in free stream in $\bar{x}$, $\bar{y}$, $\bar{z}$ system
c_l	section lift coefficient
c_n	complex constant
c_r	root chord
c_∞	velocity of sound at infinity in ξ,η,ζ system
c_S	velocity of sound at stagnation point in $\bar{x},\bar{y},\bar{z}$ system
$(cc_l)_0$	span loading at root chord
C	function depending only on τ
C_L	lift coefficient based on wing planform area
E	elliptic integral of second kind
F	elliptic integral of first kind
k	modulus of elliptic integral
k_n	complex constant
K	complete elliptic integral of first kind
le	leading edge
L	characteristic length

M_n	Mach number normal to leading edge
M_0	free-stream Mach number in $\bar{x},\bar{y},\bar{z}$ system
M_∞	V_0/c_∞
n	cycles per second
p	static pressure
p_0	free-stream static pressure in $\bar{x},\bar{y},\bar{z}$ system
p_∞	static pressure at infinity in ξ,η,ζ system
P	pressure coefficient
$P+$	pressure coefficient on impact surface
$P-$	pressure coefficient on suction surface
ΔP	loading coefficient
q	magnitude of velocity
q_0	free-stream dynamic pressure, $\frac{1}{2}\rho_0 V_0^2$
q_n	component of q_0 normal to leading edge
r,θ	polar coordinates; $y = r\cos\theta$, $z = r\sin\theta$
R	subscript, at reference condition
$\mathbf{RP}$	real part of
s_m	maximum semispan of triangular wing
t	time in $\bar{x},\bar{y},\bar{z}$ system
te	trailing edge
$\bar{u}, \bar{v}, \bar{w}$	velocities along $\bar{x}$, $\bar{y}$, and $\bar{z}$ axes
v, w	velocities along y and z axes
v_1, w_1	velocities along ξ and η axes
v_r, v_θ	radial and tangential velocities in y,z plane
V_0	free-stream velocity
W	complex potential in $\mathfrak{z}$ plane
W_1	complex potential in σ plane
x, y, z	body axes for triangular and rectangular wings
x_{cp}	streamwise distance to wing center of pressure
$\bar{x}, \bar{y}, \bar{z}$	Fig. 2-1
$\mathfrak{z}$	complex variable, $y + iz$
α	angle of attack
α_n	angle of attack in plane normal to leading edge
γ	ratio of specific heats
δ	half angle of wedge
θ	polar angle in y,z plane
Λ	sweep angle of leading edge
ν	$\tan^{-1}(y/x)$
ρ	mass density of fluid
ρ_0	mass density of free-stream fluid in $\bar{x},\bar{y},\bar{z}$ system
ρ_∞	mass density of fluid at infinity in ξ,η,ζ system
ξ,η,ζ	Fig. 2-1
σ	complex variable of physical plane, $\xi + i\eta$
τ	time in ξ,η,ζ system

ϕ velocity potential in $\bar{x},\bar{y},\bar{z}$ system; also amplitude of elliptic integral

Φ velocity potential in ξ,η,ζ system

χ frequency parameter

ψ stream function

ω semiapex angle of triangular wing

REFERENCES

1. Liepmann, H., and A. Puckett: "Aerodynamics of a Compressible Fluid," sec. 7-3, John Wiley & Sons, Inc., New York, 1947.

2. Garrick, I. E., and S. I. Rubinow: Theoretical Study of Air Forces on an Oscillating or Steady Thin Wing in a Supersonic Main Stream, *NACA Tech. Repts.* 872, 1947.

3. Staff of the Ames 1- by 3-foot Supersonic Wind-tunnel Section: Notes and Tables for Use in the Analysis of Supersonic Flow, *NACA Tech. Notes* 1428, December, 1947.

4. Ames Research Staff: Equations, Tables, and Charts for Compressible Flow, *NACA Tech. Repts.* 1135, 1953.

5. Sauer, R.: "Theoretische Einführung in die Gasdynamik," Edwards Bros., Inc., Ann Arbor, Mich., 1947.

6. Jones, R. T.: Effects of Sweepback on Boundary Layer and Separation, *NACA Tech. Repts.* 884, 1947.

7. Jones, Robert T., and Doris Cohen: Aerodynamics of Wings at High Speeds, sec. A in "Aerodynamic Components of Aircraft at High Speeds," vol. VII of "High-speed Aerodynamics and Jet Propulsion," Princeton University Press, Princeton, 1957.

8. Lagerstrom, P. A.: Linearized Supersonic Theory of Conical Wings, *NACA Tech. Notes* 1685, 1950.

9. Nielsen, Jack N.: Quasi-cylindrical Theory of Wing-Body Interference at Supersonic Speeds and Comparison with Experiment, *NACA Tech. Repts.* 1252, 1955.

10. Jones, R. T.: Thin Oblique Airfoils at Supersonic Speed, *NACA Tech. Notes* 1107, 1946.

11. Stewart, H. J.: The Lift of a Delta Wing at Supersonic Speeds, *Quart. Appl. Math.*, vol. 4, no. 3, pp. 246–254, 1946.

12. Rogers, A. W.: Application of Two-dimensional Vortex Theory to the Prediction of Flow Fields behind Wing-Body Combinations at Subsonic and Supersonic Speeds, *NACA Tech. Notes* 3227, 1954.

13. Busemann, A.: Infinitesimal Conical Supersonic Flow, *NACA Tech. Mem.* 1100, 1947.

14. Milne-Thompson, L. M.: "Theoretical Hydrodynamics," 2d ed., pp. 136–140, The Macmillan Company, New York, 1950.

15. Byrd, P. F., and M. D. Friedman: "Handbook of Elliptic Integrals for Engineers and Physicists," "Grundlehren der Mathematischen Wissenschaften," Band LXVIII, Springer-Verlag, Berlin, 1953.

CHAPTER 3

SLENDER-BODY THEORY AT SUPERSONIC AND SUBSONIC SPEEDS

The principal purpose of this chapter is to derive a number of general formulas for slender bodies at subsonic and supersonic speeds having application to a wide range of slender missiles. The formulas yield pressure coefficients, forces including drag, and moments for such configurations as slender bodies of revolution, bodies of noncircular cross section, wing-body combinations, and wing-body-tail combinations. The basic results of this chapter are applied to nonslender missiles in subsequent chapters.

Slender-body theory is greatly simplified if only bodies of revolution are taken into consideration. Then the mathematical analysis can proceed along the intuitive lines of sources and doublets. The first part of this chapter including Secs. 3-1 and 3-2 considers the problems of determining the potentials for slender bodies of revolution. It also serves as an introduction to the theory for bodies of noncircular section, the analysis of which is not so direct. The second part of the chapter, Secs. 3-4 to 3-11, is concerned with the more general analysis based principally on the methods of G. N. Ward.[1] The analysis for bodies of revolution suggests certain procedures used in the general analysis. The third part of the chapter is concerned with slender configurations at subsonic speeds. No results for specific configurations are considered here, but this subject is reserved for later chapters. The emphasis is on the mathematical methods and general formulas. Therefore, the reader who would avail himself of specific results can pass lightly over the mathematics herein, particularly the Laplace and Fourier transform theories. The theory of this chapter is limited in application to that range of angle of attack of a slender missile over which its aerodynamic characteristics are essentially linear. It is further limited to steady flow in the missile reference system.

SLENDER BODIES OF REVOLUTION

3-1. Slender Bodies of Revolution at Zero Angle of Attack at Supersonic Speeds; Sources

In the study of bodies of revolution let us denote the potential at zero angle of attack, the thickness potential, by ϕ_t and that due to angle of

34

attack ϕ_α. To obtain a solution for the potential ϕ_t of a slender body of revolution, it is convenient first to set up the potential to the full accuracy of linear theory, and then to specialize the general results to slender bodies of revolution. The basis for the linear theory potential is Eq. (2-18) for steady flow expressed in cylindrical coordinates (Fig. 3-1)

$$(M_0{}^2 - 1)\frac{\partial^2\phi}{\partial x^2} - \left(\frac{\partial^2\phi}{\partial r^2} + \frac{1}{r}\frac{\partial\phi}{\partial r} + \frac{1}{r^2}\frac{\partial^2\phi}{\partial\theta^2}\right) = 0 \tag{3-1}$$

wherein $B^2 = M_0{}^2 - 1$. The potential for a body of revolution at zero angle of attack is constructed from axially symmetric solutions of Eq. (3-1), solutions not dependent on θ. Some axially symmetric solutions of Eq. (3-1) are

$$\phi_{s_1} = \mathbf{RP}\cosh^{-1}\frac{x}{Br}$$

$$\phi_{s_2} = \mathbf{RP}\frac{1}{(x^2 - B^2r^2)^{1/2}} \tag{3-2}$$

as may be verified directly by differentiation. The second solution is the x derivative of the first solution. It is easy to see that ϕ_x and ϕ_θ also satisfy Eq. (3-1) so that x and θ derivatives of solutions are also solutions. The solution ϕ_{s_2} is sometimes termed the supersonic source with center at the origin because of its obvious similarity to the potential for an incompressible source, $1/(x^2 + r^2)^{1/2}$.

It is intuitively obvious that a body of revolution in a uniform flow can be constructed by adding sources and sinks in just the right strengths along the axis of the body. Let the source strength per unit length along the x axis be $f(\xi)$. The continuous distribution of sources (and sinks) represented by $f(\xi)$ can be summed by integration to yield their combined potentials.

$$\phi_t = \int_{x_l}^{x_u}\frac{f(\xi)\,d\xi}{[(x - \xi)^2 - B^2r^2]^{1/2}} \tag{3-3}$$

The sources used are of the ϕ_{s_2} type, and the limits of integration are purposely not specified. The limits are established on the basis of certain arguments explainable with the help of Fig. 3-1. The Mach cone from point P will intersect the x axis at a distance $x - Br$ downstream from the origin. Downstream of this intersection no source can influence point P since the region of influence of a source is confined to its downstream Mach cone. The upper limit is therefore $x - Br$. The sources start at $x = 0$ in the present case, and $f(\xi) = 0$ if $\xi \leq 0$. Therefore any lower limit equal to zero or less is possible. We therefore write

$$\phi_t = \int_0^{x - Br}\frac{f(\xi)\,d\xi}{[(x - \xi)^2 - B^2r^2]^{1/2}} \tag{3-4}$$

It is to be noted that a potential V_0x due to the uniform flow is additive to ϕ_t to obtain the total potential.

The source strength distribution $f(\xi)$ must be determined from the shape of the body.   To the accuracy required here the boundary condition yields with reference to Fig. 3-1

$$\frac{\partial \phi_t / \partial r}{V_0} \equiv \frac{v_r}{V_0} \approx \frac{dr_0}{dx} = \frac{S'(x)}{2\pi r_0} \tag{3-5}$$

The quantity $S(x)$ is the cross-sectional area of the body of revolution. To utilize this boundary condition we must determine $\partial \phi_t / \partial r$ from Eq.

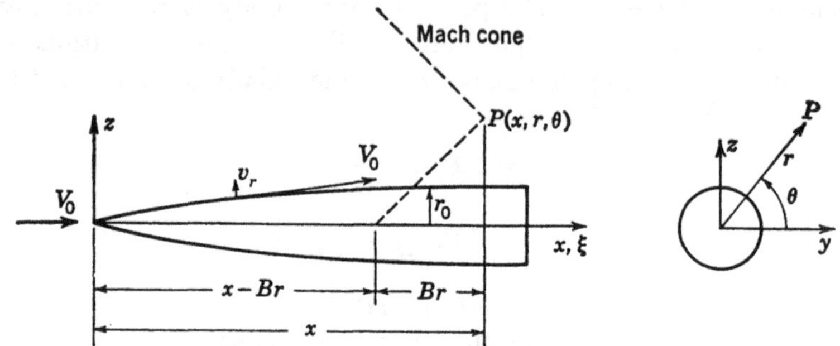

Fig. 3-1. Axes and notation for body of revolution at zero incidence.

(3-4).   Assume that $f(0)$ is zero, and rewrite Eq. (3-4) as

$$\phi_t = \frac{\partial}{\partial x} \int_0^{x-Br} f(\xi) \, \cosh^{-1} \frac{x - \xi}{Br} \, d\xi \tag{3-6}$$

For a "slender body," the body radius is small compared to x, and the quantity $(x - \xi)/Br$ is large except for a limited interval near the upper limit, which we can neglect.   The inverse hyperbolic cosine can then be expanded

$$\cosh^{-1} \frac{x - \xi}{Br} = \log \frac{2(x - \xi)}{Br} + \cdots \tag{3-7}$$

For a slender body, Eq. (3-6) therefore assumes the form

$$\phi_t = \frac{\partial}{\partial x} \int_0^x f(\xi) \log (x - \xi) \, d\xi - \frac{\partial}{\partial x} \int_0^x f(\xi) \log \frac{Br}{2} \, d\xi \tag{3-8}$$

from which
$$\frac{\partial \phi_t}{\partial r} = - \frac{f(x)}{r} \tag{3-9}$$

From Eq. (3-5) the source strength is directly related to the body shape

$$f(x) = - \frac{V_0 S'(x)}{2\pi} \tag{3-10}$$

and the potential from Eq. (3-8) is then

$$\frac{\phi_t}{V_0} = + \frac{S'(x)}{2\pi} \log \frac{Br}{2} - \frac{1}{2\pi} \frac{\partial}{\partial x} \int_0^x S'(\xi) \log (x - \xi) \, d\xi \tag{3-11}$$

For purposes of physical interpretation, separate ϕ_t into a part dependent on r and a part independent of r.

$$\frac{\phi_t}{V_0} = + \frac{S'(x)}{2\pi} \log r + g(x)$$

$$g(x) = + \frac{S'(x)}{2\pi} \log \frac{B}{2} - \frac{1}{2\pi} \frac{\partial}{\partial x} \int_0^x S''(\xi) \log (x - \xi) \, d\xi \qquad (3\text{-}12)$$

With reference to Fig. 3-2 the thickness potential is the sum of a part which depends on the position in the crossflow plane AA, and a part which has the same value for every point in the plane. The part of ϕ_t depending on r is precisely the potential function for an incompressible source flow in the crossflow plane. The flow velocities in the crossflow plane depend only on this term since $g(x)$ has the same value all over the

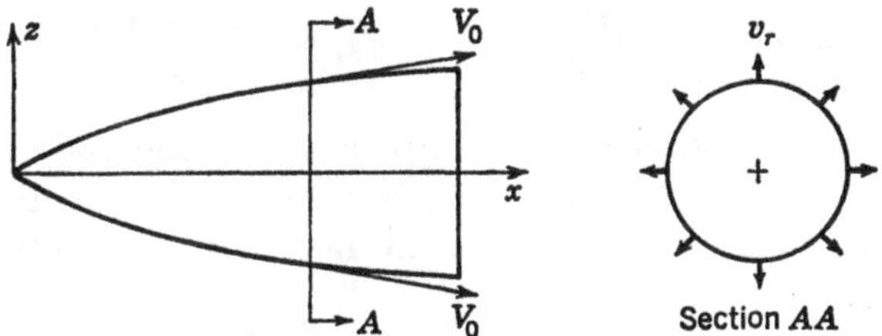

Fig. 3-2. Source flow in crossflow plane of body of revolution.

plane. The $g(x)$ term can, however, influence the pressure coefficient, which depends principally on $\partial \phi_t / \partial x$. To obtain the function $g(x)$, it was necessary to specialize the full linear theory potential to a slender body of revolution. We will consider next the effect of angle of attack, which is additive to that of thickness in a simple way. The question of pressure coefficients and forces is left until later.

3-2. Slender Bodies of Revolution at Angle of Attack at Supersonic Speed; Doublets

The axis system and the body of revolution at angle of attack are oriented with respect to the uniform flows as shown in Fig. 3-3. The component of velocity $V_0 \cos \alpha_c$ along x causes ϕ_t as discussed in Sec. 3-1,

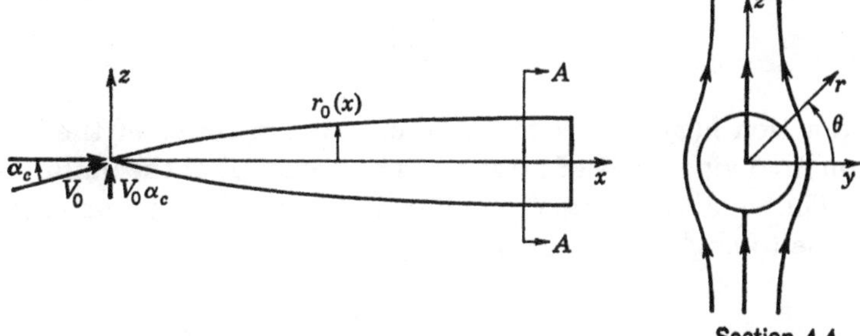

Fig. 3-3. Incompressible crossflow around body of revolution at angle of attack.

and the component of the flow velocity $V_0\alpha_c$ along z causes the potential component ϕ_α now to be evaluated. Just as ϕ_t was constructed by distributing sources along the body axis, so ϕ_α is constructed by superimposing dipoles along the axis. First consider the dipoles formed from the axially symmetric solutions of Eq. (3-2). The dipole is formed by placing a sink directly above a source of equal magnitude and letting the source approach the sink, while keeping the product of the magnitude and distance between source and sink a constant. This physical process is mathematically equivalent to taking the derivative of the source solutions with respect to z. The constant multiplying the solution, the so-called *dipole strength*, is of no concern at this point; only the analytical form of the dipoles is of interest: For the two source solutions of Eq. (3-2) we have the two corresponding dipole solutions:

$$\phi_{d_1} = \text{RP} \frac{x \sin \theta}{r(x^2 - B^2 r^2)^{\frac{1}{2}}} \qquad \phi_{d_2} = \text{RP} \frac{r \sin \theta}{(x^2 - B^2 r^2)^{\frac{3}{2}}} \qquad (3\text{-}13)$$

Consider now a superposition of dipoles along the body axis. If $d(\xi)$ were the dipole strength per unit length, we could form a dipole potential similarly as the source potential was formed from ϕ_{s_2} solutions.

$$\phi_\alpha = r \sin \theta \int_0^{x - Br} \frac{d(\xi)\, d\xi}{[(x - \xi)^2 - B^2 r^2]^{\frac{3}{2}}} \qquad (3\text{-}14)$$

Unfortunately this integral is infinite because of the 3/2 power infinity at the upper limit. Though the singularity is mathematically tractable by the use of the concept of the finite part of an integral,[2] we will avoid the singularity by other means. Specifically we will obtain a potential by superimposing dipoles of the ϕ_{d_1} type in strength $h(\xi)$ for unit length along the body axis, and then taking the x derivative of the sum which is itself a dipole-type solution.

$$\phi_\alpha = \frac{\sin \theta}{r} \frac{\partial}{\partial x} \int_0^{x - Br} \frac{h(\xi)(x - \xi)\, d\xi}{[(x - \xi)^2 - B^2 r^2]^{\frac{1}{2}}} \qquad (3\text{-}15)$$

For a slender body of revolution, $x \gg Br$, and Eq. (3-15) takes on the simple form

$$\phi_\alpha = \frac{\sin \theta}{r} h(x) \qquad (3\text{-}16)$$

The function $h(x)$ is now to be determined in terms of the boundary condition involving angle of attack. The potential of the uniform flow is $V_0\alpha_c z$. The condition of no radial flow in the crossflow plane at the body surface due to angle of attack yields

$$\frac{\partial}{\partial r}(\phi_\alpha + V_0\alpha_c z) = 0 \qquad (3\text{-}17)$$

with the result that

$$h(x) = V_0 \alpha_c r_0{}^2 \tag{3-18}$$

The potential of a slender body of revolution due to angle of attack is thus simply

$$\phi_\alpha = V_0 \alpha_c r_0{}^2 \frac{\sin \theta}{r} \tag{3-19}$$

The physical interpretation of the potential ϕ_α is that of an incompressible two-dimensional doublet in the crossflow plane. There is no additive function such as $g(x)$ in Eq. (3-12) for the potential due to thickness. The entire potential due to angle of attack could have been constructed by considering the flow in each crossflow plane to be incompressible. In fact, a simplified slender-body theory based on this procedure is described in the next section.

3-3. Slender-body Theory for Angle of Attack

The distinguishing characteristics of flow about slender bodies was discussed by Munk in his early work on the aerodynamics of airship hulls.[3] In this work he laid down the basis of Munk's airship theory which has subsequently been extended into what is now known as slender-body

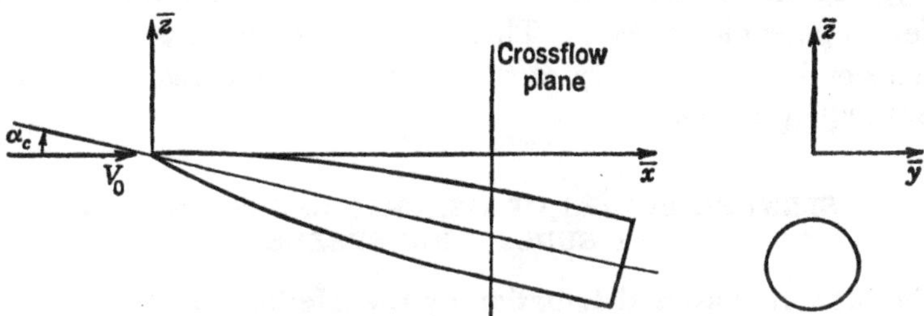

FIG. 3-4. Axes used in slender-body theory.

theory. Consider a slender body, not necessarily a body of revolution, flying through still air at a speed V_0, at Mach number M_0, and at an angle of attack α, and passing through a plane fixed in the fluid. The flow as viewed in the plane is nonsteady as the body passes through it. If, however, the plane is fixed in the missile, the flow will appear steady. Let the $\bar{x},\bar{y},\bar{z}$ axis system be oriented as shown in Fig. 3-4 with the $\bar{x}$ axis parallel to V_0, and let the crossflow plane correspond to $\bar{x}$ equal a constant. The flow about the missile is governed by Eq. (2-31) for linear theory

$$B^2 \frac{\partial^2 \phi}{\partial \bar{x}^2} - \frac{\partial^2 \phi}{\partial \bar{y}^2} - \frac{\partial^2 \phi}{\partial \bar{z}^2} = 0 \tag{3-20}$$

If the body is sufficiently slender (or if the Mach number is close to unity), the first term of the equation is negligible, so that we have

$$\phi_{\bar{y}\bar{y}} + \phi_{\bar{z}\bar{z}} = 0 \tag{3-21}$$

For an observer fixed with respect to the body, the flow in any plane normal to the $\bar{x}$ axis is thus the steady incompressible flow based on boundary conditions in that plane. It is independent of the crossflow in all other planes. An example of the incompressible flow in a normal plane is shown in Fig. 3-3. The *normal plane* will be defined as the plane normal to the body axis. The planes $\bar{x} = $ constant are *crossflow planes*. For small angles of attack, the flow patterns in the normal plane and crossflow plane can be considered identical for slender bodies.

The foregoing simplified analysis of the flow about a slender body is generally applicable to the calculation of the potential due to angle of attack as we have seen in the preceding section. However, it is not adequate for obtaining the potential due to thickness existing at zero angle of attack. The mathematical reason for this inadequacy is readily apparent. In descending from three variables in Eq. (3-20) to two variables in Eq. (3-21), we eliminated the possibility of determining explicitly the dependence of the potential on $\bar{x}$. For instance, any solution to Eq. (3-21) is still a solution if a function of $\bar{x}$ is added to it. Furthermore the addition of a function $\bar{x}$ will not change the velocities in the crossflow plane. Such a function of $\bar{x}$ does, however, change the axial velocity, and therefore the pressure coefficient, which depends principally on this velocity. It turns out that the function of $\bar{x}$ is different for subsonic and for supersonic speeds. This feature is the essential difference between slender-body theory at subsonic and at supersonic speeds, as we shall subsequently see.

SLENDER BODIES OF GENERAL CROSS SECTION AT SUPERSONIC SPEEDS

3-4. Solution of Potential Equation by the Method of Ward

In the ensuing sections it is our purpose to derive the principal formulas of slender-body theory for supersonic speeds following the method of Ward.[1] Some attempt will be made to maintain mathematical rigor and to carry order-of-magnitude estimations of the terms neglected in the analysis. The essential method of the analysis is to find a general solution for the wave equation of linear theory, and to select those terms out of the general solution that remain under the assumption of a slender body. In this way all terms that should appear in slender-body theory are found explicitly.

The body is assumed pointed at the front end, and is either pointed or blunt at the rear end. The body length is taken to be unity, and the maximum radial dimension is t. The angle between the free-stream direction and planes tangent to the body should be small, as well as the rate of change of this angle with streamwise distance. The assumptions assure that there are no discontinuities in the streamwise slope, and hence no singularities in ϕ. If one requires no singularity in local pressure, which depends principally on the axial derivative of ϕ, he must impose the additional requirement of no discontinuities in streamwise curvature. If d is the maximum diameter of any cross section, the curvature in the crossflow plane at any point on the body where it convexes outward should be $O(1/d)$. No such restriction is necessary for points where the body is convex inward as in a wing-body juncture. The foregoing restrictions simply assure that the perturbation velocities due to the body are small compared to the free stream. At a sharp leading edge, the curvature is convex outward and certainly of much greater order than $1/d$. The slender-body theory gives infinite perturbation velocities and pressure coefficients at such points so that the estimates of the orders of magnitude of the terms neglected in slender-body theory become invalid. Certain gross terms such as lift and possibly drag may, nevertheless, be accurate to the order of magnitude indicated in the following formulas. Thus, while the local pressure coefficient is physically untenable, it is confined to a restricted region, and its net influence on gross forces can be negligible. The order of magnitude of the remainder terms in the equations for the physical quantities due to the approximations of the analysis will be given in terms of the maximum radial dimension, which is assumed small compared to unity ($l = 1$), and which is designated t.

Let ϕ be the perturbation velocity potential for unit free-stream velocity with the system of axes shown in Fig. 3-4. The perturbation velocities are then

$$\bar{u} = \frac{\partial \phi}{\partial \bar{x}} \qquad \bar{v} = \frac{\partial \phi}{\partial \bar{y}} \qquad \bar{w} = \frac{\partial \phi}{\partial \bar{z}} \tag{3-22}$$

Equation (2-18) specialized to steady flow forms the basis of the present analysis:

$$\phi_{\bar{y}\bar{y}} + \phi_{\bar{z}\bar{z}} - B^2\phi_{\bar{x}\bar{x}} = 0 \qquad B^2 = M_0^2 - 1 \tag{3-23}$$

In the analysis which follows we seek a general solution of Eq. (3-23), and then pick out the terms of an expansion of the general solution appropriate to a slender body. For supersonic flow the mathematical tool convenient for doing this is the Laplace transform theory. Let us first rewrite Eq. (3-23) in terms of cylindrical coordinates

$$\phi_{rr} + \frac{1}{r}\phi_r + \frac{1}{r^2}\phi_{\theta\theta} - B^2\phi_{\bar{x}\bar{x}} = 0 \tag{3-24}$$

The transformation we will use converts the potential $\phi(\bar{x},r,\theta)$ into a transformed potential $\Phi(p,r,\theta)$ by means of the Laplace operator L.

$$L[\phi(\bar{x},r,\theta)] \equiv \int_0^\infty e^{-p\bar{x}}\phi(\bar{x},r,\theta)\,d\bar{x} \tag{3-25}$$

With reference to Churchill,[4]

$$
\begin{aligned}
L[\phi_{rr}] &= \Phi_{rr} \\
L[\phi_{r}] &= \Phi_{r} \\
L[\phi_{\theta\theta}] &= \Phi_{\theta\theta} \\
L[\phi_{\bar{x}\bar{x}}] &= p^2\Phi - p\phi(0^+,r,\theta) - \phi_{\bar{x}}(0^+,r,\theta)
\end{aligned}
\qquad (3\text{-}26)
$$

Since in supersonic flow there is no influence of the body for $\bar{x}$ less than zero and since ϕ is continuous, we have that $\Phi(0^+,r,\theta)$ is zero. Also we may assume that $\dfrac{\partial \phi}{\partial \bar{x}}(0^+,r,\theta)$ is zero on the basis of the following physical argument. If $\dfrac{\partial \phi}{\partial \bar{x}}(0^+,r,\theta)$ jumps discontinuously crossing $\bar{x} = 0$, we can make it continuous by an infinitesimal fairing of the body without significantly influencing the jump. On this basis we take $\dfrac{\partial \phi}{\partial \bar{x}}(0^+,r,\theta)$ as zero. The physical argument is not actually required, and the mathematical treatment with $\dfrac{\partial \phi}{\partial \bar{x}}(0^+,r,\theta)$ not zero will give the same final results as proved by Fraenkel.[5] The transformation of Eq. (3-24) is thus

$$
\Phi_{rr} + \frac{1}{r}\,\Phi_{r} + \frac{1}{r^2}\,\Phi_{\theta\theta} = B^2 p^2 \Phi
\qquad (3\text{-}27)
$$

A solution of Eq. (3-27) can easily be found by the method of the separation of variables in the form

$$
\Phi = \sum_{n=0}^{\infty} [C_n(p)\sin n\theta + D_n(p)\cos n\theta]I_n(Bpr)
$$

$$
+ \sum_{u=0}^{\infty} [E_n(p)\sin n\theta + F_n(p)\cos n\theta]K_n(Bpr)
\qquad (3\text{-}28)
$$

The functions $C_n(p)$, $D_n(p)$, etc., can be considered constant so far as Eq. (3-27) is concerned. Actually they are arbitrary functions of p chosen so as to satisfy the boundary conditions. The functions I_n and K_n are modified Bessel functions of the first and second kinds, respectively. For large arguments they have the following asymptotic behaviors:

$$
I_n(Bpr) \sim \frac{e^{Bpr}}{(2\pi Bpr)^{1/2}}
\qquad (3\text{-}29)
$$

$$
K_n(Bpr) \sim \left(\frac{\pi}{2Bpr}\right)^{1/2} e^{-Bpr}
\qquad (3\text{-}30)
$$

The dominant term of the inverse transform of the I_n function represents upstream waves increasing exponentially in strength along upstream Mach waves, $\bar{x} + Br = $ constant. The $K_n(Bpr)$ functions, on the other hand, represent downstream waves attenuating exponentially along the downstream Mach waves, $\bar{x} - Br = $ constant. It is clear that we are thus interested only in the $K_n(Bpr)$ functions except for possible rare cases.   For only the $K_n(Bpr)$ function Eq. (3-28) can be written into the following compact form by combining the sin $n\theta$ and cos $n\theta$ by means of arbitrary phase angles $\delta_n(p)$:

$$\Phi = A_0(p)K_0(Bpr) + \sum_{n=1}^{\infty} A_n(p)K_n(Bpr) \cos [n\theta + \delta_n(p)] \quad (3\text{-}31)$$

The next step in the analysis is to find the special form of Eq. (3-31) appropriate to slender configurations.   A slender configuration is one characterized by the fact that its r dimensions are small compared to its $\bar{x}$ dimensions.   We therefore seek a form of Eq. (3-31) valid for small values of r.   The questions which then arise are: In what region will the new form be valid, and how large an error occurs in Φ as compared with Φ from Eq. (3-31)?   To obtain the form of Eq. (3-31) for small r, we note the following expansions of the Bessel functions for small values of r (γ is Euler's constant):

$$K_0(Bpr) = -\left(\gamma + \log \frac{Bpr}{2}\right)[1 + O(r^2)]$$

$$K_1(Bpr) = \frac{1}{2}\frac{2}{Bpr}[1 + O(r^2 \log r)] \qquad (3\text{-}32)$$

$$K_n(Bpr) = \frac{(n-1)!}{2}\left(\frac{2}{Bpr}\right)^n [1 + O(r^2)]$$

The dominant terms in r therefore yield

$$\Phi_0 = -\left(\gamma + \log \frac{Bpr}{2}\right) A_0(p)$$

$$+ \frac{1}{2}\sum_{n=1}^{\infty} (n-1)!\left(\frac{2}{Bp}\right)^n A_n(p)r^{-n} \cos (n\theta + \delta_n) \quad (3\text{-}33)$$

We use the subscript zero to denote the value of Φ for small r.   The *fractional error* in Φ_0 is at most $O(r^2 \log r)$ and, if the K_1 term is missing, then the error is $O(r^2)$.   Inspection of Eq. (3-33) shows that the series converges if r is greater than some value r_1.   The series converges external to a cylinder enclosing the body as shown in Fig. 3-5.   The series will usually not converge inside this cylinder and does not represent the solution at the body surface.   It must be continued inside the cylinder by the process of analytical continuation.   Since r_1 is some dimension of the same order of magnitude as t, the fractional error in Φ_0 is $O(t^2 \log t)$.

To establish the potential ϕ_0 in the $\mathfrak{z}$ plane, we must take the inverse transform of Eq. (3-33) term by term.   For this purpose denote the inverse transforms of the various terms in the equation as

$$a_0(\bar{x}) = L^{-1}[-A_0(p)]$$

$$b_0(\bar{x}) = L^{-1}\left[-\left(\gamma + \log \frac{Bp}{2}\right) A_0(p)\right]$$

$$a_n = a_n{}^*(\bar{x}) - ib_n{}^*(\bar{x}) = L^{-1}\left[\frac{(n-1)!}{2}\left(\frac{2}{Bp}\right)^n A_n(p)e^{i\delta_n}\right] \qquad (3\text{-}34)$$

The inverse transform of Eq. (3-33) is then

$$\phi_0 = a_0 \log r + b_0 + \sum_{n=1}^{\infty} \frac{a_n{}^* \cos n\theta + b_n{}^* \sin n\theta}{r^n} \qquad (3\text{-}35)$$

It is clear that ϕ_0 is the real part of the function $W(\mathfrak{z})$ of a complex variable of $\mathfrak{z} = re^{i\theta}$

$$\phi_0 = \mathbf{RP}\ W(\mathfrak{z}) \qquad \mathfrak{z} = re^{i\theta} \qquad (3\text{-}36)$$

$$W(\mathfrak{z}) = a_0 \log \mathfrak{z} + b_0 + \sum_{n=1}^{\infty} \frac{a_n}{\mathfrak{z}^n} \qquad (3\text{-}37)$$

What we have shown is that ϕ_0 is a solution of Laplace's equation in the crossflow planes, $\bar{x}$ equal to a constant. The function $b_0(\bar{x})$ is the

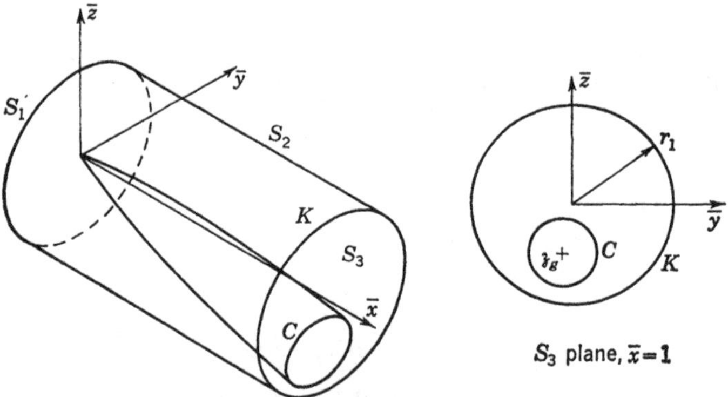

FIG. 3-5. Cylindrical control surface enclosing slender body.

function left indeterminate in the simplified treatment of slender-body theory in Sec. 3-3. In the form of Eq. (3-35) the series converges outside a cylinder of radius r_1 enclosing the body. Although the series converges for large values of r, it does not follow that it represents the flow about a slender body for large values of r. This is apparent when we recall that Eq. (3-33) was established by extracting from Eq. (3-31) for the full linearized theory those terms dominant for small values of r. Slender-body theory is accurate, therefore, only in the field near the body. To obtain solutions for slender bodies for distances far from the body it is necessary to retain the full linearized equation.   The slender-body potential ϕ_0 has the fractional error in the form $(\phi - \phi_0)/\phi$ of order $t^2 \log t$.

3-5. Boundary Conditions; Accuracy of Velocity Components

First let us consider the matter of boundary conditions and then turn our attention to the accuracy of the velocity components compared with those for the full linearized theory.   Consider contours C_1 and C_2 in the crossflow planes corresponding to $\bar{x}$ and $\bar{x} + d\bar{x}$ with the body as shown in Fig. 3-6.   Let the normal and the tangent to the contour in the crossflow plane be ν and τ, respectively.   Consider a streamwise plane containing ν

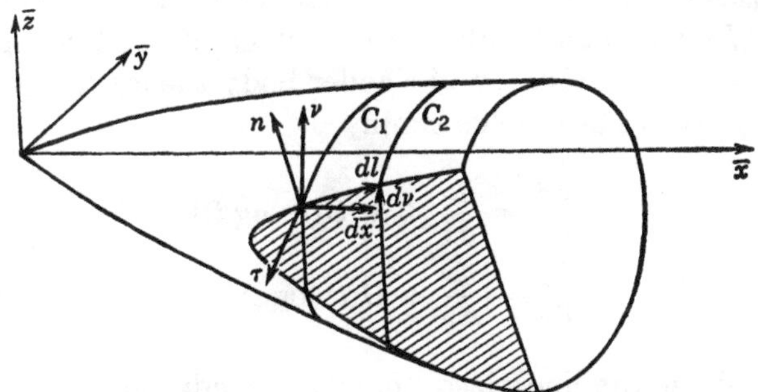

Fig. 3-6. Boundary conditions in streamwise plane through slender body.

shown with section lining in Fig. 3-6.   The plane is normal to τ and intersects an element of length dl of the body surface between C_1 and C_2.   Let n be the outward normal to τ and dl.   Thus n is the normal to the surface and lies in the streamwise plane.   In the streamwise plane the exact condition of flow tangency is

$$\frac{V_0(1 + \partial\phi/\partial\bar{x})}{d\bar{x}} = \frac{V_0\,\partial\phi/\partial\nu}{d\nu}$$

or
$$\frac{\partial\phi}{\partial\nu} = \left(1 + \frac{\partial\phi}{\partial\bar{x}}\right)\frac{d\nu}{d\bar{x}} \tag{3-38}$$

The quantity $d\nu/d\bar{x}$ is the streamwise slope of the body surface.   Since by hypothesis $d\nu/d\bar{x}$ is $O(t)$, we have from Eq. (3-38) that the velocity normal to the body in the crossflow plane is $O(t)$.   With the assumptions regarding the streamwise body slopes and the curvature of the body cross sections, the velocity at the body surface will not deviate from the free-stream direction by an angle greater than $O(t)$.   Since the magnitude of the velocity is of the order unity by hypothesis, the velocities in the crossflow plane normal to and tangential to the body contour are both $O(t)$.   Therefore in Eq. (3-37) $|dW/d\bar{z}|$ is $O(t)$.   If each term in the $dW/d\bar{z}$ equation is of like magnitude, then

$$a_0 = O(t^2) \qquad a_n = O(t^{n+2})$$

Since differentiation with respect to $\bar{x}$ does not change order of magnitude, we have from Eq. (3-35) for the order of magnitude of the perturbation velocities (assuming b_0' is not a dominant term in $\partial\phi/\partial\bar{x}$)

$$\frac{\partial\phi}{\partial\bar{x}} = O(t^2 \log t) \qquad \frac{\partial\phi}{\partial\bar{y}} = O(t) \qquad \frac{\partial\phi}{\partial\bar{z}} = O(t) \tag{3-39}$$

and for the potential

$$\phi = O(t^2 \log t)$$

Let us now linearize the boundary condition, Eq. (3-38), and estimate the order of the error introduced thereby. Let us also consider the errors due to the use of ϕ_0 rather than ϕ associated with the complete linearized equation. We will then be able to tell which of the two simplifications actually controls the accuracy of slender-body theory.

Linearizing Eq. (3-38) yields simply

$$\frac{\partial\phi}{\partial\nu} = \frac{d\nu}{d\bar{x}} [1 + O(t^2 \log t)]$$
$$\frac{\partial\phi}{\partial\nu} = \frac{d\nu}{d\bar{x}} + O(t^3 \log t) \tag{3-40}$$

For ϕ_0 we will use the linearized boundary condition

$$\frac{\partial\phi_0}{\partial\nu} = \frac{d\nu}{d\bar{x}} \tag{3-41}$$

Now the error due to use of ϕ_0 for ϕ was shown to be

$$\phi = \phi_0 + O(t^4 \log^2 t) \tag{3-42}$$

The error in axial velocity is the same since derivation by $\bar{x}$ does not change the order of magnitude

$$\frac{\partial\phi}{\partial\bar{x}} = \frac{\partial\phi_0}{\partial\bar{x}} + O(t^4 \log^2 t)$$

The error in the crossflow velocity components is

$$\frac{\partial\phi}{\partial\bar{y}} = \frac{\partial\phi_0}{\partial\bar{y}} + O(t^3 \log t)$$
$$\frac{\partial\phi}{\partial\bar{z}} = \frac{\partial\phi_0}{\partial\bar{z}} + O(t^3 \log t) \tag{3-43}$$

since the fractional error in the velocity components is the same as the fractional error in ϕ_0 (as Ward has proved). It is seen that the error due to linearizing the boundary condition equation, Eq. (3-40), is the same as the error due to the use of ϕ_0 for ϕ, Eq. (3-43), so that the two simplifications are compatible.

3-6. Determination of $a_0(\bar{x})$ and $b_0(\bar{x})$

It is possible to obtain the values of $a_0(\bar{x})$ and $b_0(\bar{x})$ in Eq. (3-35) from the distribution of the body cross-sectional area along the body axis regardless of the cross-sectional shape.   The higher-order coefficients $a_n(\bar{x})$ depend on the shape.   The $a_0 \log r$ term corresponds to source flow in the crossflow plane and is zero if the body cross section is not changing size. To evaluate a_0 consider the contour K shown in Fig. 3-7.   From the

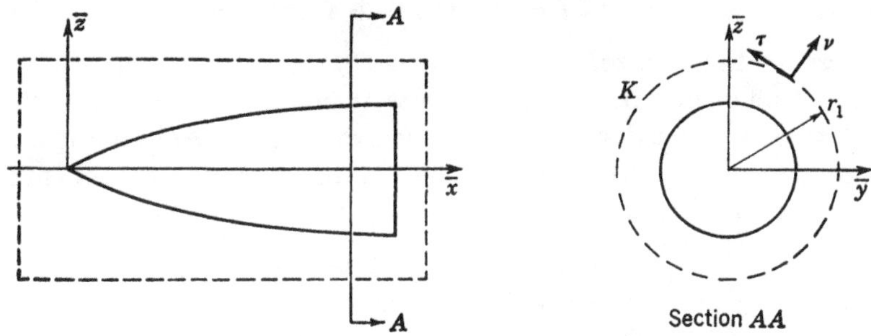

FIG. 3-7. Contour for evaluation of $a_0(\bar{x})$.

integral of the outflow across K

$$\oint_K \frac{\partial \phi_0}{\partial \nu} \, d\tau = \int_0^{2\pi} \frac{\partial \phi_0}{\partial r} r_1 \, d\theta = \int_0^{2\pi} \left(\frac{a_0}{r_1} - \sum_{n=1}^{\infty} \frac{n a_n}{r^{n+1}} \cos n\theta \right) r_1 \, d\theta$$

$$= 2\pi a_0 \tag{3-44}$$

Now invoke the linearized boundary condition (Eq. 3-41) and reevaluate the contour integral about K:

$$\oint_K \frac{\partial \phi_0}{\partial \nu} \, d\tau = \oint_K \frac{\partial \nu}{\partial \bar{x}} \, d\tau = \frac{d}{d\bar{x}} \oint_K d\nu \, d\tau = S'(\bar{x}) \tag{3-45}$$

Here $S'(\bar{x})$ is the cross-sectional area in the crossflow plane.   The final result for $a_0(\bar{x})$ is

$$a_0(\bar{x}) = \frac{S'(\bar{x})}{2\pi} \tag{3-46}$$

The function $b_0(\bar{x})$ is uniform in any cross-sectional plane and yields nothing to the crossflow velocity components.   It does contribute to the pressure coefficient but not to the loading.   Now from Eq. (3-34)

$$b_0(\bar{x}) = -L^{-1}[\gamma A_0(p)] - L^{-1}\left[A_0(p) \log \frac{Bp}{2} \right] \tag{3-47}$$

The use of the convolution theorem[4] yields

$$b_0(\bar{x}) = \frac{1}{2\pi}\left[S'(\bar{x}) \log \frac{B}{2} - \int_0^{\bar{x}} \log(\bar{x} - \xi) S''(\xi) \, d\xi \right] \tag{3-48}$$

where we have made use of the condition for a pointed body $S'(0) = 0$. It is seen that $b_0(\bar{x})$ depends on Mach number.

3-7. Pressure Coefficient

With the magnitudes of the velocity components known, it is a simple matter to formulate the pressure coefficient from Eq. (2-29). For unit V_0, the pressure coefficient, including quadratic terms in the perturbation velocity components, is

$$P = -2\bar{u} - (\bar{v}^2 + \bar{w}^2) + \bar{u}^2(M_0^2 - 1) \tag{3-49}$$

If we ignore the last term, the error is $O(t^4 \log^2 t)$. Let us express the pressure coefficient with respect to velocity components u, v, and w along the body axes x, y, z displaced from the axes $\bar{x}$, $\bar{y}$, $\bar{z}$ by pitch angle α_c and bank angle φ (Fig. 1-2). From Table 1-1 the velocity components $\bar{u}$, $\bar{v}$, and $\bar{w}$ are related to u, v, w by

$$\begin{aligned}
\bar{u} &= u - v\alpha_c \sin \varphi + w\alpha_c \cos \varphi \\
\bar{v} &= v \cos \varphi + w \sin \varphi \\
\bar{w} &= -u\alpha_c - v \sin \varphi + w \cos \varphi
\end{aligned} \tag{3-50}$$

Direct substitution into Eq. (3-49) yields

$$P = -2(u - v\alpha_c \sin \varphi + w\alpha_c \cos \varphi) - (v^2 + w^2) + O(t^4 \log^2 t) \tag{3-51}$$

where we have discarded the terms $u^2\alpha_c^2$, $uv\alpha_c \sin \varphi$, and $uw\alpha_c \cos \varphi$ as terms of higher order than $t^4 \log^2 t$ if α_c is $O(t)$. In terms of the angle of attack α and the angle of sideslip β, we have

$$P = -2(u - v\beta + w\alpha) - (v^2 + w^2) \tag{3-52}$$

It is probably important to note that the superposition principle does not necessarily hold for pressure coefficient in slender-body theory. The principle of superposition has been retained for the potential, however.

3-8. Lift, Sideforce, Pitching Moment, and Yawing Moment

The lift $\bar{Z}$ and sideforce $\bar{Y}$ can be evaluated by taking the rate of change of momentum in the $\bar{z}$ and $\bar{y}$ directions through a control surface of the type shown in Fig. 3-5. The net transfer of momentum in the vertical direction through the cylinder $r = r_1$ and the base plane area S_3 is

$$\bar{Z} = - \int_{S_2} \left(\rho V_0^2 \frac{\partial \phi}{\partial r} \frac{\partial \phi}{\partial \bar{z}} + p \sin \theta \right) dS_2 - \int_{S_3} \rho V_0^2 \left(1 + \frac{\partial \phi}{\partial \bar{x}} \right) \frac{\partial \phi}{\partial \bar{z}} dS_3 \tag{3-53}$$

and in the lateral direction

$$\bar{Y} = - \int_{S_2} \left(\rho V_0^2 \frac{\partial \phi}{\partial r} \frac{\partial \phi}{\partial \bar{y}} + p \cos \theta \right) dS_2$$

$$- \int_{S_3} \rho V_0^2 \left(1 + \frac{\partial \phi}{\partial \bar{x}} \right) \frac{\partial \phi}{\partial \bar{y}} dS_3 \tag{3-54}$$

It is convenient to form the complex force

$$F = \bar{Y} + i\bar{Z}$$

so that

$$\frac{F}{\frac{1}{2}\rho_0 V_0{}^2} = -2 \int_{S_2} \left[\frac{\rho}{\rho_0} \frac{\partial \phi}{\partial r}\left(\frac{\partial \phi}{\partial \bar{y}} + i\frac{\partial \phi}{\partial \bar{z}}\right) + \left(\frac{p - p_0}{2q_0}\right) e^{i\theta} \right] dS_2$$

$$- 2 \int_{S_2} \frac{\rho}{\rho_0}\left(1 + \frac{\partial \phi}{\partial \bar{x}}\right)\left(\frac{\partial \phi}{\partial \bar{y}} + i\frac{\partial \phi}{\partial \bar{z}}\right) dS_3 \quad (3\text{-}55)$$

The pressure coefficient is equal in magnitude to the percentage change in absolute pressure, namely, $O(t^2 \log t)$. By the isentropic relationship, the percentage change in density is the same magnitude so that

$$\frac{\rho}{\rho_0} = 1 + O(t^2 \log t) \quad (3\text{-}56)$$

Also we have the relationship

$$\frac{dW}{d\mathfrak{z}} = \frac{\partial \phi}{\partial \bar{y}} - i\frac{\partial \phi}{\partial \bar{z}}$$

$$\frac{d\bar{W}}{d\bar{z}} = \frac{\partial \phi}{\partial \bar{y}} + i\frac{\partial \phi}{\partial \bar{z}} \quad (3\text{-}57)$$

These relationships reduce Eq. (3-55) to

$$\frac{F}{q_0} = -2 \int_{S_2} \frac{\partial \phi_0}{\partial r} \frac{d\bar{W}}{d\bar{z}} dS_2 + \int_{S_2} \left(2\frac{\partial \phi_0}{\partial \bar{x}} + \frac{dW}{d\mathfrak{z}}\frac{d\bar{W}}{d\bar{z}}\right) e^{i\theta} dS_2$$

$$- 2 \int_{S_3} \left(\frac{\partial \phi_0}{\partial \bar{y}} + i\frac{\partial \phi_0}{\partial \bar{z}}\right) dS_3 + O(t^5 \log^2 t) \quad (3\text{-}58)$$

It is interesting to note that the variation of density does not enter into Eq. (3-58) since the variation represents in part the error term. The first term is simply handled by Stokes's theorem. The contours C and K are shown in Fig. 3-5.

$$- \int_{S_3} \left(\frac{\partial \phi_0}{\partial \bar{y}} + i\frac{\partial \phi_0}{\partial \bar{z}}\right) dS_3 = -i \oint_C \phi_0(d\bar{y} + i\,d\bar{z}) + i \oint_K \phi_0(d\bar{y} + i\,d\bar{z})$$

$$= -i \oint_C \phi_0\,d\mathfrak{z} - \int_0^{2\pi} \phi_0 r_1 e^{i\theta}\,d\theta$$

$$= -i \oint_C \phi_0\,d\mathfrak{z} - \int_0^{2\pi} r_1 e^{i\theta}\,d\theta \int_0^1 \frac{\partial \phi_0}{\partial \bar{x}}\,d\bar{x} \quad (3\text{-}59)$$

If we substitute Eq. (3-59) into Eq. (3-58), we have

$$\frac{F}{q_0} = -2i \oint_C \phi_0\,d\mathfrak{z} - \int_0^1 \int_0^{2\pi} \left(2\frac{\partial \phi_0}{\partial r}\frac{d\bar{W}}{d\bar{z}} - \frac{dW}{d\mathfrak{z}}\frac{d\bar{W}}{d\bar{z}}\,e^{i\theta}\right) r_1\,d\theta\,d\bar{x}$$

$$+ O(t^5 \log^2 t) \quad (3\text{-}60)$$

The terms $\dfrac{\partial \phi_0}{\partial r}$, $\dfrac{d\bar{W}}{d\bar{z}}$, $\dfrac{dW}{d\hat{z}}$ are all velocities in the crossflow plane, which decrease as $1/r$ or faster as r approaches ∞. As a result the double integral is zero in the limit. Thus

$$\frac{F}{q_0} = -2i \oint_C \phi_0 \, d\hat{z} + O(t^5 \log^2 t) \tag{3-61}$$

The contour C is the outline of the base intended to pass around any singular points that may occur on the body surface. It is noted that the force depends only on the line integral of the potential around the base. Since ϕ_0 depends (except for a constant) only on the base configuration and angle of attack, we have the simple result that the force depends only on the base characteristics and is independent of the forward shape of the body. The formation of vortices behind the position of maximum span can modify this result for wing-body combinations.

If $\hat{z}_o$ is the center of area of the base, the complex force F can be expressed as in the following form (derived in Appendix A at the end of the chapter):

$$\frac{F}{q_0} = 4\pi a_1 + 2S'(1)\hat{z}_o(1) + 2S(1)\hat{z}_o{}'(1) + O(t^5 \log^2 t) \tag{3-62}$$

If
$$\hat{z}_o = \bar{y}_o + i\bar{z}_o \tag{3-63}$$

the forces become

$$\frac{\bar{Y}}{q_0} = 4\pi \, \mathbf{RP} \, a_1 + 2S'(1)\bar{y}_o + 2S(1) \frac{d\bar{y}_o}{d\bar{x}} + O(t^5 \log^2 t)$$

$$\frac{\bar{Z}}{q_0} = 4\pi \, \mathbf{IP} \, a_1 + 2S'(1)\bar{z}_o + 2S(1) \frac{d\bar{z}_o}{d\bar{x}} + O(t^5 \log^2 t) \tag{3-64}$$

The quantity $\bar{Y}$ is the sideforce, and $\bar{Z}$ is the lift.

To obtain the moment we can write Eq. (3-62) for the force at any axial distance and integrate the local loading times $\bar{x}$ to obtain the moments. Thus, if $M_{\bar{x}}$ is the moment about the $\bar{y}$ axis (positive when $\bar{z}$ moves toward $\bar{x}$) and $M_{\bar{z}}$ is the moment about the $\bar{z}$ axis (positive when $\bar{x}$ moves toward $\bar{y}$), we have

$$M = M_{\bar{y}} + iM_{\bar{z}} = i \int_0^{\bar{x}} \bar{x} F'(\bar{x}) \, d\bar{x}$$

$$= i\bar{x}F(\bar{x}) - i \int_0^{\bar{x}} F'(\xi) \, d\xi \tag{3-65}$$

$$\frac{M(1)}{q_0} = 4\pi i a_1(1) - 4\pi i \int_0^1 a_1 \, d\bar{x}$$

$$+ 2i[S'(1)\hat{z}_o(1) + S(1)\hat{z}_o{}'(1) - S(1)\hat{z}_o(1)] + O(t^5 \log^2 t) \tag{3-66}$$

Thus for the entire slender body

$$M_{\bar{y}} = -4\pi\,\mathrm{IP}\,(a_1) + 4\pi\,\mathrm{IP}\int_0^1 a_1\,d\bar{x}$$
$$- 2\,\mathrm{IP}\,[S'(1)\mathfrak{z}_\sigma(1) + S(1)\mathfrak{z}_\sigma'(1) - S(1)\mathfrak{z}_\sigma(1)]$$
$$M_{\bar{z}} = 4\pi\,\mathrm{RP}\,(a_1) - 4\pi\,\mathrm{RP}\int_0^1 a_1\,d\bar{x}$$
$$+ 2\,\mathrm{RP}\,[S'(1)\mathfrak{z}_\sigma(1) + S(1)\mathfrak{z}_\sigma'(1) - S(1)\mathfrak{z}_\sigma(1)] \tag{3-67}$$

The quantity $M_{\bar{y}}$ is the pitching moment, and $M_{\bar{z}}$ is minus the yawing moment.

3-9. Drag Force

The drag formula of slender-body theory is a widely used result which for special types of slender bodies exhibits elegant mathematical properties. In the derivation of the drag formula, use is made of a cylindrical control surface as shown in Fig. 3-5. It is easy to set up the drag force in terms of pressure and momentum transfer.

$$D = \int_{S_1} (p_0 + \rho_0 V_0{}^2)\,dS_1 - \int_{S_2} \rho V_0{}^2\left(1 + \frac{\partial\phi}{\partial\bar{x}}\right)\frac{\partial\phi}{\partial r}\,dS_2$$
$$- \int_{S_3}\left[p + \rho V_0{}^2\left(1 + \frac{\partial\phi}{\partial\bar{x}}\right)^2\right]dS_3 - p_B S(1) \tag{3-68}$$

The symbol p_B stands for the static pressure acting on the base. To simplify Eq. (3-68) we introduce the conservation of mass.

$$\int_{S_1} \rho_0 V_0\,dS - \int_{S_2} \rho V_0\frac{\partial\phi}{\partial r}\,dS_2 - \int_{S_3} \rho V_0\left(1 + \frac{\partial\phi}{\partial\bar{x}}\right)dS_3 = 0 \tag{3-69}$$

By multiplying Eq. (3-69) by V_0 and subtracting it from Eq. (3-68) some simplification is achieved within the framework of exactness.

$$D = \int_{S_1} p_0\,dS_1 - \int_{S_2} \rho V_0{}^2\frac{\partial\phi}{\partial\bar{x}}\frac{\partial\phi}{\partial r}\,dS_2$$
$$- \int_{S_3}\left[p + \rho V_0{}^2\frac{\partial\phi}{\partial\bar{x}}\left(1 + \frac{\partial\phi}{\partial\bar{x}}\right)\right]dS_3 - p_B S(1) \tag{3-70}$$

This result is now further simplified by assuming that the density is uniform, and the resulting error is recorded in the error term with the help of Eq. (3-56). Also, the static pressure is eliminated by means of Bernoulli's equation, Eq. (3-49), which in cylindrical coordinates is

$$\frac{p - p_0}{q_0} = P = -2\frac{\partial\phi_0}{\partial\bar{x}} - \left[\left(\frac{\partial\phi_0}{\partial r}\right)^2 + \frac{1}{r^2}\left(\frac{\partial\phi_0}{\partial\theta}\right)^2\right] + O(t^4\log^2 t) \tag{3-71}$$

With these approximations, Eq. (3-70) now becomes

$$\frac{D}{q_0} = -2\int_{S_2}\frac{\partial\phi_0}{\partial r}\frac{\partial\phi_0}{\partial\bar{x}}\,dS_2 + \int_{S_3}\left[\left(\frac{\partial\phi_0}{\partial r}\right)^2 + \frac{1}{r^2}\left(\frac{\partial\phi_0}{\partial\theta}\right)^2\right]dS_3$$
$$- P_B S(1) + O(t^6\log^2 t) \tag{3-72}$$

where P_B is the base-pressure coefficient. From the results of Appendix B at the end of the chapter, Eq. (3-72) takes the form

$$\frac{D}{q_0} = -4\pi \int_0^1 a_0 b_0' \, d\bar{x} + 2\pi a_0(1)b_0(1) - \oint_C \phi_0 \frac{\partial \phi_0}{\partial \nu} \, d\tau$$
$$- P_B S(1) + O(t^6 \log^2 t)$$
$$= 4\pi \int_0^1 a_0' b_0 \, d\bar{x} - 2\pi a_0(1)b_0(1) - \oint_C \phi_0 \frac{\partial \phi_0}{\partial \nu} \, d\tau$$
$$- P_B S(1) + O(t^6 \log^2 t) \quad (3\text{-}73)$$

The drag can now be evaluated since the values of a_0 and b_0 are given explicitly by Eqs. (3-46) and (3-48). We obtain

$$\frac{D}{q_0} = \frac{1}{2\pi} \int_0^1 \int_0^1 \log \frac{1}{|s - \xi|} S''(s)S''(\xi) \, d\xi \, ds - \frac{S'(1)}{2\pi} \int_0^1 \log \frac{1}{1 - \xi} S''(\xi) \, d\xi$$
$$- \oint_C \phi_0 \frac{\partial \phi_0}{\partial \nu} \, d\tau - P_B S(1) + O(t^6 \log^2 t) \quad (3\text{-}74)$$

This is the Ward drag formula for a slender body. It is interesting to note that the drag represented by the first two terms depends only on the axial distribution of the body cross-sectional area and is independent of cross-sectional shape. The second two terms depend on the slope of the body cross section at the base only. We will investigate the various terms of this drag formula at considerable length in Sec. 9-3.

Two important classes of slender bodies result in considerable simplification of Eq. (3-74). These classes occur when the base is pointed or when the body is tangent to the cylindrical extension of its base. In both instances the drag formula reduces to the symmetrical form

$$\frac{D}{q_0} = \frac{1}{2\pi} \int_0^1 \int_0^1 \log \frac{1}{|s - \xi|} S''(s)S'(\xi) \, d\xi \, ds + O(t^6 \log^2 t) - P_B S(1) \quad (3\text{-}75)$$

Minimum drag bodies are derived on the basis of this result in Sec. 9-5.

3-10. Drag Due to Lift

The following treatment is good not only for supersonic speeds but also for subsonic speeds, as we will subsequently show. When a body develops lift, it develops a wake of one kind or another. A lifting surface usually develops a well-defined vortex wake. In this case the contour C must be enlarged to enclose the vortices as shown in Fig. 3-8b. Another kind of wake arises when flow separates from a surface under angle of attack as shown in Fig. 3-8a. We imagine a dead water region to form in the separation region which is then enclosed by vortex sheets. The wake can then be considered a solid body extension and the contour deformed to enclose the dead water region. The force acting on the body enlarged to include the dead water will be the same as that on the solid boundaries of the body since the resultant force on the dead water region is zero.

An inspection of Eq. (3-74) reveals that the drag represented by the first two terms is independent of the lift, depending as it does only on the axial distribution of body cross-sectional area.   Thus the drag due to lift is to be found in the integral about C, neglecting any changes in the base-pressure coefficient due to changes in angle of attack.   To evaluate the drag due to lift we must inspect this integral under the conditions of no

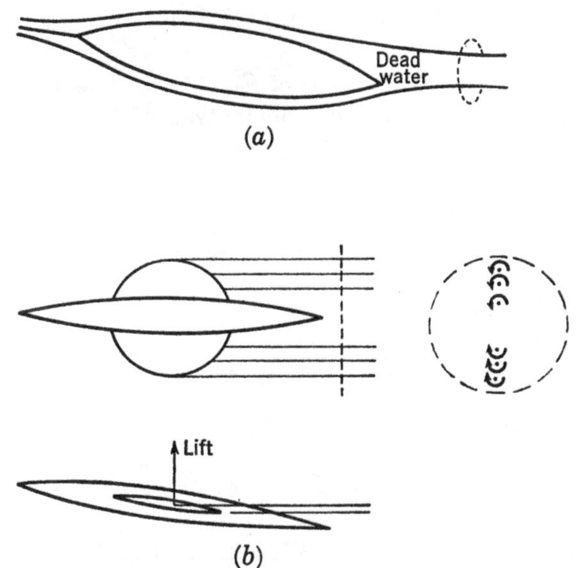

FIG. 3-8. Separated or vortex flows requiring distortion of contour of integration. (a) Body; (b) Wing-body combination.

lift and of lift.   Let the potential ϕ_0 be composed of a part ϕ_{00} at zero lift and a part ϕ_{01} due to lift

$$\phi_0 = \phi_{00} + \phi_{01} \tag{3-76}$$

The integral about the contour C of the base becomes

$$\oint_C \phi_0 \frac{\partial \phi_0}{\partial \nu}\, d\tau = \oint_C \phi_{00} \frac{\partial \phi_{00}}{\partial \nu}\, d\tau + \oint_C \phi_{00} \frac{\partial \phi_{01}}{\partial \nu}\, d\tau +$$
$$\oint_C \phi_{01} \frac{\partial \phi_{00}}{\partial \nu}\, d\tau + \oint_C \phi_{01} \frac{\partial \phi_{01}}{\partial \nu}\, d\tau \tag{3-77}$$

The first integral is not part of the drag due to lift.   The second and third integrals are *coupling terms* between the potential at lift and zero lift, while the fourth integral is a "pure" lifting effect.   The derivative $\partial \phi_{01}/\partial \nu$ in accordance with the boundary condition, Eq. (3-41), represents the change in the streamwise slope of the body surface due to angle of attack α if the angle of attack is arbitrarily taken to be zero at zero lift.

With reference to Fig. 3-6 the potential ϕ_{01} must produce a velocity normal to the body sufficient to offset the component of the free-stream velocity normal to the body. Thus

$$\frac{\partial \phi_{01}}{\partial \nu} = -\alpha \cos(\nu,\bar{z}) \tag{3-78}$$

where $\cos(\nu,\bar{z})$ is the direction cosine of ν with respect to the $\bar{z}$ axis. The second integral then becomes

$$\oint_C \phi_{00} \frac{\partial \phi_{01}}{\partial \nu} d\tau = -\alpha \oint_C \phi_{00} \cos(\nu,\bar{z}) d\tau = \alpha \oint_C \phi_{00} d\bar{y} = 0 \tag{3-79}$$

The second integral is zero because of Eq. (3-61) since the lift is zero for ϕ_{00}. The third integral is zero by the analysis of Appendix C at the end of the chapter. The drag due to lift is therefore all due to the fourth integral, which by Eq. (3-78) becomes

$$\oint_C \phi_{01} \frac{\partial \phi_{01}}{\partial \nu} d\tau = \alpha \oint_C \phi_{01} d\bar{y} \tag{3-80}$$

Again Eq. (3-61) shows that the lift is

$$\bar{Z} = -2 \oint_C \phi_{01} d\bar{y} \tag{3-81}$$

The drag due to lift for constant base pressure is now

$$D - D_0 = \frac{\alpha}{2} \bar{Z} = -\oint_C \phi_{01} \frac{\partial \phi_{01}}{\partial \nu} d\tau \tag{3-82}$$

where D_0 is the drag at zero lift. We can put Eq. (3-74) into the following form for lift

$$D = D_0 + \frac{\alpha}{2} \bar{Z} + \Delta P_B S(1) + O(t^6 \log^2 t) \tag{3-83}$$

where ΔP_B is the change in base-pressure coefficient due to angle of attack.

The physical significance of Eq. (3-82) is that the lift creates a drag $\alpha \bar{Z}/2$ rather than $\alpha \bar{Z}$, which would be expected for a flat plate. Thus the resultant force on a slender configuration due to angle of attack is inclined backward at an angle $\alpha/2$ from the normal to the free-stream direction.

3-11. Formula Explicitly Exhibiting Dependence of Drag on Mach Number

Let us divide the drag given by Eq. (3-74) into parts dependent on and independent of Mach number. Examination of Eq. (3-74) shows directly that any part of the drag dependent on M_0 must occur as a result of the contour integral about C. To obtain this part, let us write

$$\phi_0 - b_0 = \phi^* \tag{3-84}$$

The discussion following Eq. (3-37) and Eqs. (3-46) and (3-48) shows that all the dependence of ϕ_0 on M_0 enters through b_0, so that ϕ^* is independent of M_0. Also b_0 is uniform in the plane of C. Thus $\partial b_0/\partial \nu$ is zero, and the contour integral can be written

$$\oint_C \phi_0 \frac{\partial \phi_0}{\partial \nu} \, d\tau = b_0 \oint_C \frac{\partial \phi_0}{\partial \nu} \, d\tau + \oint_C \phi^* \frac{\partial \phi^*}{\partial \nu} \, d\tau \tag{3-85}$$

The first integral is readily evaluated by means of Eq. (3-45)

$$b_0 \oint_C \frac{\partial \phi_0}{\partial \nu} \, d\tau = 2\pi a_0(1) b_0(1)$$

Introducing these relationships into Eq. (3-74) yields the desired drag equation.

$$\frac{D}{q_0} = \frac{-[S'(1)]^2}{2} \log \frac{B}{2} - P_B S(1)$$

$$+ \frac{1}{2\pi} \int_0^1 \int_0^1 \log \frac{1}{|s - \xi|} S''(s) S''(\xi) \, d\xi \, ds$$

$$- \frac{S'(1)}{\pi} \int_0^1 \log \frac{1}{1 - \xi} S''(\xi) \, d\xi - \oint_C \phi^* \frac{\partial \phi^*}{\partial \nu} \, d\tau + O(t^6 \log^2 t) \tag{3-86}$$

The first two terms depend on Mach number. If the base is pointed, $S'(1)$ is zero, and the first term is zero as well as the base pressure. The drag on the basis of slender-body theory is then independent of Mach number (neglecting separation over the base). If the base is tangent to its own cylindrical extension, $S'(1)$ is zero, and the only effect of Mach number on drag is through its influence on base pressure. The potential ϕ^* in the equation is just that potential which would be obtained by applying Laplace's equation to the flow in the crossflow plane as described in Sec. 3-3.

**SLENDER BODIES OF GENERAL CROSS SECTION
AT SUBSONIC SPEEDS**

3-12. Solution of the Potential Equation

The treatment by Ward of supersonic slender-body theory has its counterpart for subsonic flow. Mathematically, the difference is one of using Fourier transforms instead of Laplace transforms. Actually, the entire difference between the subsonic and supersonic cases enters through the b_0 term. Thus all results derived for the supersonic case not depending explicitly on b_0 are unchanged for subsonic speeds. Let us now find the operational solution to the potential equation, Eq. (3-24), on the basis of Fourier transforms. Consider the Fourier operator F and the inverse Fourier operator F^{-1} defined by the following pair of reciprocal relationships:

$$F[\phi(\bar{x},r,\theta)] = \int_{-\infty}^{+\infty} e^{i\bar{x}\omega} \phi(\bar{x},r,\theta) \, d\bar{x} = \Phi(\omega,r,\theta)$$

$$F^{-1}[\Phi(\omega,r,\theta)] = \frac{1}{2\pi} \int_{-\infty}^{+\infty} e^{-i\bar{x}\omega}\Phi(\omega,r,\theta) \, d\omega = \phi(\bar{x},r,\theta) \qquad (3\text{-}87)$$

Note the use of the complex Fourier transform and the placement of the 2π factor. Integration by parts establishes the following transforms for $\bar{x}$ derivatives of ϕ:

$$F\left[\frac{\partial\phi}{\partial\bar{x}}\right] = e^{i\bar{x}\omega}\phi\Big|_{-\infty}^{\infty} - i\omega\Phi$$

$$F\left[\frac{\partial^2\phi}{\partial\bar{x}^2}\right] = e^{i\bar{x}\omega}\frac{\partial\phi}{\partial\bar{x}}\Big|_{-\infty}^{+\infty} - i\omega e^{i\bar{x}\omega}\phi\Big|_{-\infty}^{+\infty} - \omega^2\Phi \qquad (3\text{-}88)$$

If we can invoke the boundary conditions

$$\phi(+\infty) = \phi(-\infty) = 0$$

$$\frac{\partial\phi}{\partial\bar{x}}(+\infty) = \frac{\partial\phi}{\partial\bar{x}}(-\infty) = 0 \qquad (3\text{-}89)$$

Eq. (3-24) becomes

$$\frac{\partial^2\Phi}{\partial r^2} + \frac{1}{r}\frac{\partial\Phi}{\partial r} + \frac{1}{r^2}\frac{\partial^2\Phi}{\partial\theta^2} = B_0^2\omega^2\Phi$$

$$B_0^2 = 1 - M_0^2 \qquad (3\text{-}90)$$

Again, as in the case of supersonic flow, a suitable general solution of Eq. (3-90) for the present purpose can be obtained by separation of variables. In fact, the solution is of the following form in complete analogy to Eq. (3-28):

$$\Phi = \sum_{n=0}^{\infty} K_n(B_0\omega r)[E_n(\omega)\sin n\theta + F_n(\omega)\cos n\theta]$$

$$+ I_n(B_0\omega r)[C_n(\omega)\sin n\theta + D_n(\omega)\cos n\theta] \qquad (3\text{-}91)$$

The value of ω ranges from $-\infty$ to $+\infty$, and the arbitrary functions, $C_n(\omega)$, $D_n(\omega)$, etc., are to be suitably chosen so that (1) the behavior of ϕ is not divergent as $r \to \infty$, and (2) ϕ is real. The requirement that ϕ is not divergent as $r \to \infty$ transforms to the requirement that Φ not be divergent as $r \to \infty$, since the transformation does not involve r. We must discuss separately the cases for positive ω and negative ω. For positive ω, we have already seen that $I_n(B_0\omega r)$ varies as e^r, and is not admissible on account of the first condition. Thus

$$C_n(\omega) = D_n(\omega) = 0 \qquad \omega \geq 0 \qquad (3\text{-}92)$$

For negative values of ω, we must make use of the relationship between Bessel functions of negative and positive arguments.

$$K_n(-x) = (-1)^n K_n(x) - \pi i I_n(x)$$

$$I_n(-x) = (-1)^n I_n(x) \qquad (3\text{-}93)$$

The $\sin n\theta$ part of the solution then becomes

$$E_n(-\omega)K_n(-B_0\omega r) + C_n(-\omega)I_n(-B_0\omega r)$$
$$= E_n(-\omega)[(-1)^nK_n(B_0\omega r) - \pi i I_n(B_0\omega r)]$$
$$+ C_n(-\omega)(-1)^nI_n(B_0\omega r) \quad (3\text{-}94)$$

The $I_n(B_0\omega r)$ terms must have zero coefficient if the behavior as $r \to \infty$ is not to be divergent.   This behavior is assured if

$$-\pi i E_n(-\omega) + (-1)^nC_n(-\omega) = 0 \quad (3\text{-}95)$$

We are then left with the solution

$$E_n(-\omega)K_n(-B_0\omega r) + C_n(-\omega)I_n(-B_0\omega r)$$
$$= E_n(-\omega)(-1)^nK_n(B_0\omega r) \quad (3\text{-}96)$$

If the coefficients of $K_n(B_0\omega r)$ are chosen to be new functions as follows,

$$\begin{aligned}
B_n(\omega) &= E_n(\omega) & \omega &> 0 \\
B_n(\omega) &= (-1)^nE_n(-\omega) & \omega &< 0 \\
A_n(\omega) &= F_n(\omega) & \omega &> 0 \\
A_n(\omega) &= (-1)^nF_n(-\omega) & \omega &< 0
\end{aligned} \quad (3\text{-}97)$$

the general solution of Eq. (3-91) with the correct behavior can be expressed as

$$\Phi = \sum_{n=0}^{\infty} K_n(B_0|\omega|r)[B_n(\omega) \sin n\theta + A_n(\omega) \cos n\theta] \quad (3\text{-}98)$$

It should be noted that Eq. (3-97) does not place any condition on $A_n(\omega)$ and $B_n(\omega)$ since $E_n(\omega)$ and $F_n(\omega)$ are quite arbitrary.   The second condition that ϕ be real can be simply satisfied by choosing

$$\begin{aligned}
A_n(-\omega) &= \bar{A}_n(\omega) \\
B_n(-\omega) &= \bar{B}_n(\omega)
\end{aligned} \quad (3\text{-}99)$$

Equation (3-98) is the solution in the transformed plane of the full linearized equation which is appropriate for subsonic speeds.   The value of ϕ it gives will become small as $r \to \infty$ and will extend upstream and downstream.   The problem now is to extract from the full linearized solution that special solution suitable for slender configurations.   The problem is solved in exactly the same manner as for the supersonic case: by expanding Φ as given by Eq. (3-98) in a series valid for small r, and retaining the dominant terms.   In fact, the expansions for ϕ are identical in form

$$\phi = a_0(\bar{x}) \log r + b_0(\bar{x}) + \sum_{n=1}^{\infty} \frac{a_n^*(\bar{x}) \cos n\theta + b_n^*(\bar{x}) \sin n\theta}{r^n} \quad (3\text{-}100)$$

but the coefficients are now determined as inverse Fourier transforms rather than inverse Laplace transforms.

$$a_0(\bar{x}) = -F^{-1}[A_0(\omega)]$$
$$b_0(\bar{x}) = -F^{-1}\left[\gamma + \log \frac{B_0|\omega|}{2}\right] A_0(\omega) \tag{3-101}$$

The only term that can differ from that for the supersonic case is $b_0(\bar{x})$. The rest of the terms in ϕ are solutions to Laplace's equation in the crossflow plane and are uniquely determined by the boundary condition in the crossflow plane regardless of the Mach number.

3-13. Determination of $a_0(\bar{x})$ and $b_0(\bar{x})$

The value of $a_0(\bar{x})$ in this case is precisely the same as for supersonic speeds since the part of ϕ involving a_0 is independent of Mach number. Thus

$$a_0(\bar{x}) = \frac{S'(\bar{x})}{2\pi} \tag{3-102}$$

The function $b_0(\bar{x})$ is obtained from Eq. (3-101) on a purely operational basis

$$b_0(\bar{x}) = +\left(\gamma + \log \frac{B_0}{2}\right) a_0(\bar{x}) - F^{-1}[A_0(\omega) \log |\omega|] \tag{3-103}$$

The inverse transform of a product of transforms can be obtained by means of the convolution integral

$$F^{-1}[G(\omega)H(\omega)] = \int_{-\infty}^{+\infty} g(\xi)h(\bar{x} - \xi)\, ds \tag{3-104}$$

To insure the existence of the separate transforms let

$$G(\omega) = \omega A_0(\omega)$$
$$H(\omega) = \frac{\log |\omega|}{\omega} \tag{3-105}$$

so that
$$g(\bar{x}) = -ia_0'(\bar{x})$$
and
$$h(\bar{x}) = \frac{1}{2\pi} \int_{-\infty}^{+\infty} \frac{\log |\omega|}{\omega} e^{-i\omega\bar{x}}\, d\omega$$
$$= -\frac{i}{\pi} \int_0^{\infty} \frac{\log |\omega|}{\omega} \sin \omega\bar{x}\, d\omega \tag{3-106}$$

With the help of Erdélyi et al.[6]

$$h(\bar{x}) = \frac{i}{2}(\gamma + \log \bar{x}) \qquad \bar{x} > 0$$
$$h(\bar{x}) = -\frac{i}{2}(\gamma + \log |\bar{x}|) \qquad \bar{x} < 0 \tag{3-107}$$

Finally with the help of Eq. (3-94)

$$b_0(\bar{x}) = a_0(\bar{x}) \log \frac{B_0}{2} - \frac{1}{2} \int_{0^+}^{\bar{x}} a_0'(\xi) \log (\bar{x} - \xi) \, d\xi$$

$$+ \frac{1}{2} \int_{\bar{x}}^{1^-} a_0'(\xi) \log (\xi - \bar{x}) \, d\xi - \tfrac{1}{2} a_0(0^+) \log \bar{x}$$

$$- \tfrac{1}{2} a_0(1^-) \log (1 - \bar{x}) \quad (3\text{-}108)$$

3-14. Drag Formula for Subsonic Speeds; d'Alembert's Paradox

The drag force for subsonic speeds will be developed from Eq. (3-73). Though we have developed the formulas for subsonic slender-body theory on the basis of a possible blunt base, $S(1) \neq 0$, such a body will not fulfill the requirements of slenderness. A blunt base in subsonic flow can send strong upstream signals, which it cannot do in supersonic flow because of the rule of forbidden signals. As a consequence we must now assume that $S'(1)$ and $S(1)$ are both zero; that is, the base is pointed. With $S'(1) = 0$, $b_0(\bar{x})$ becomes [for $S(\bar{x})$ continuous and $S'(0) = 0$]

$$b_0(\bar{x}) = a_0(\bar{x}) \log \frac{B_0}{2} - \frac{1}{2} \int_0^{\bar{x}} a_0'(\xi) \log (\bar{x} - \xi) \, d\xi$$

$$+ \frac{1}{2} \int_{\bar{x}}^1 a_0'(\xi) \log (\xi - \bar{x}) \, d\xi \quad (3\text{-}109)$$

By Eq. (3-73) the drag is then

$$\frac{D}{q_0} = \frac{1}{2\pi} \log \frac{B_0}{2} \{ [S'(1)]^2 - [S'(0)]^2 \}$$

$$- \frac{1}{2\pi} \int_0^1 S''(\bar{x}) \int_0^{\bar{x}} S''(\xi) \log (\bar{x} - \xi) \, d\xi \, d\bar{x}$$

$$+ \frac{1}{2\pi} \int_0^1 S''(\bar{x}) \int_{\bar{x}}^1 S''(\xi) \log (\xi - \bar{x}) \, d\xi \, d\bar{x} - \oint_c \phi_0 \frac{\partial \phi_0}{\partial \nu} \, d\tau$$

$$- P_B S(1) + O(t^6 \log^2 t) \quad (3\text{-}110)$$

or

$$\frac{D}{q_0} = - \oint_c \phi_0 \frac{\partial \phi_0}{\partial \nu} \, d\tau + O(t^6 \log^2 t) \quad (3\text{-}111)$$

Noting the next to last equation of Appendix B which follows, we have with $a_0(1)$ equal to zero

$$\frac{D}{q_0} = 0$$

Slender-body theory thus yields d'Alembert's paradox in subsonic flow as it should.

SYMBOLS

$a_0(\bar{x})$ coefficient of log term in expansion for ϕ_0

$A_n(p)$ arbitrary function of p

$a_n(\bar{x})$ coefficients in expansion for $W(\mathfrak{z})$

$A_n(\omega)$ arbitrary function of ω

$b_0(\bar{x})$ coefficient in expansion for ϕ_0

B $(M_0^2 - 1)^{1/2}$

B_0 $(1 - M_0^2)^{1/2}$

$B_n(\omega)$ arbitrary function of ω

D drag force

D_0 drag force at zero lift

$f(\xi)$ source strength per unit ξ distance

F $\bar{Y} + i\bar{Z}$

F, F^{-1} Fourier transform operator, and inverse Fourier transform operator

$h(\xi)$ dipole strength per unit ξ distance

I_n, K_n modified Bessel functions of first and second kinds

L, L^{-1} Laplace transform operator, and inverse transform operator

M $M_{\bar{y}} + iM_{\bar{z}}$

M_0 free-stream Mach number

$M_{\bar{y}}$ moment about $\bar{y}$ axis, pitching moment

$M_{\bar{z}}$ moment about $\bar{z}$ axis, negative yawing moment

p static pressure, variable of Laplace transform

p_0 free-stream static pressure

p_B base static pressure

P pressure coefficient, $(p - p_0)/q_0$

P_B base-pressure coefficient

q_0 free-stream dynamic pressure

r radius vector in $\bar{y}, \bar{z}$ plane (also in y,z plane in Secs. 3-1 and 3-2)

r_0 local body radius

r_1 radius of cylindrical control surface, Fig. 3-5

RP   real part of a complex function

$S(\bar{x})$ area of slender configuration in crossflow plane

t maximum radial dimension of slender configuration

u, v, w perturbation velocity components along x, y, z

$\bar{u}, \bar{v}, \bar{w}$ perturbation velocity components along $\bar{x}, \bar{y}, \bar{z}$

v_r radial velocity component

V_0 free-stream velocity

$W(\mathfrak{z})$ complex potential, $\phi + i\psi$

$\bar{W}(\mathfrak{z})$ conjugate complex potential, $\phi - i\psi$

x, y, z principal body axes, Fig. 3-1

$\bar{x}, \bar{y}, \bar{z}$ body axes for $\alpha_c = 0$ and $\varphi = 0$, Fig. 3-4

$\bar{y}_g$ $\bar{y}$ coordinate of centroid of $S(\bar{x})$

$\bar{Y}$ sideforce along $\bar{y}$ axis

$\bar{z}_g$ $\bar{z}$ coordinate of centroid of $S(\bar{x})$

$\bar{Z}$ force along $\bar{Z}$ axis, lift

$\mathfrak{z}$ $\bar{y} + i\bar{z}$

$\mathfrak{z}_g$ $\bar{y}_g + i\bar{z}_g$

α	angle of attack, $\alpha_c \cos \varphi$
α_c	included angle between body axis and free-stream velocity
β	angle of sideslip, $\alpha_c \sin \varphi$
γ	Euler's constant, 0.5772
$\delta_n(p)$	phase angle
θ	polar angle in $\bar{x}$, $\bar{y}$, $\bar{z}$ coordinates
ν	normal to body contour in crossflow plane
ξ	variable of integration
ρ	local mass density
ρ_0	free-stream mass density
τ	tangent to body contour in crossflow plane
ϕ	general potential solution of Eq. (3-24)
φ	angle of bank
ϕ_0	approximation to ϕ valid for slender configurations
ϕ_d	potential for a doublet
ϕ_s	potential for a source, an axially symmetric potential
$\Phi(p,r,\theta)$	Laplace transform of ϕ
$\Phi(\omega,r,\theta)$	Fourier exponential transform of ϕ
Φ_0	transform of ϕ_0
ω	variable of the Fourier transform

REFERENCES

1. Ward, G. N.: Supersonic Flow Past Slender Pointed Bodies, *Quart. J. Mech. and Appl. Math.*, vol. 2, part I, p. 94, 1949.

2. Heaslet, Max A., and Harvard Lomax: Supersonic and Transonic Small Perturbation Theory, sec. D in "General Theory of High-speed Aerodynamics," vol. VI of "High-speed Aerodynamics and Jet Propulsion," Princeton University Press, Princeton, 1954.

3. Munk, Max M.: The Aerodynamic Forces on Airship Hulls, *NACA Tech. Repts.* 184, 1924.

4. Churchill, Ruel V.: "Operational Mathematics," 2d ed., McGraw-Hill Book Company, Inc., New York, 1958.

5. Fraenkel, L. E.: On the Operational Form of the Linearized Equation of Supersonic Flow, *J. Aeronaut. Sci.*, vol. 20, no. 9, pp. 647–648, Readers' Forum, 1953.

6. Erdélyi, A. (ed.): "Tables of Integral Transforms," vol. I, McGraw-Hill Book Company, Inc., New York, 1954.

APPENDIX 3A

In Sec. 3-8, the complex force, $F = \bar{Y} + i\bar{Z}$, was put into the following form:

$$\frac{F}{q_0} = -2i \oint_C \phi_0 \, d\bar{z} + O(t^5 \log^2 t) \qquad (3A\text{-}1)$$

It is possible to find a somewhat more appropriate form for calculative purposes by replacing ϕ_0 by $W - i\psi_0$

$$\frac{F}{q_0} = -2i \oint_C W \, d\mathfrak{z} - 2 \oint_C \psi_0 \, d\mathfrak{z} \tag{3A-2}$$

Some care must be taken in connection with Eq. (3A-2) because the expansion for $W(\mathfrak{z})$, Eq. (3-37), contains a logarithmic term which is not single-valued

$$W(\mathfrak{z}) = a_0 \log \mathfrak{z} + b_0 + \sum_{n=1}^{\infty} a_n \mathfrak{z}^{-n} \tag{3A-3}$$

To make the $W(\mathfrak{z})$ function single-valued, we put a cut in the $\mathfrak{z}$ plane from $\mathfrak{z}_0$ to ∞ as shown in Fig. 3-9, and the argument of the logarithm increases

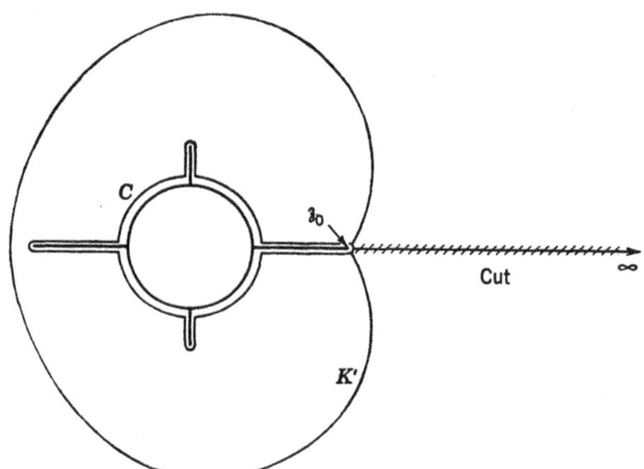

FIG. 3-9. Distortion of contour in cut $\mathfrak{z}$ plane.

by $2\pi i$ every time $\mathfrak{z}$ crosses the cut. Now the contour C encloses the cross section of the body base but indents any singular points of $W(\mathfrak{z})$ as shown. The nature of the series in the expansion for $W(\mathfrak{z})$ is such that it converges if $|\mathfrak{z}|$ is greater than the largest value associated with any singular point. The series for $W(\mathfrak{z})$ will not converge on all of C, and so we expand the contour to K' on which Eq. (3A-3) is convergent. Then

$$\oint_C W \, d\mathfrak{z} = \oint_{K'} W \, d\mathfrak{z} = \oint_{K'} \left(a_0 \log \mathfrak{z} + b_0 + \sum_{n=1}^{\infty} \frac{a_n}{\mathfrak{z}^n} \right) d\mathfrak{z}$$
$$= a_0(2\pi i \mathfrak{z}_0) + 2\pi i a_1 \tag{3A-4}$$

Note that the value of the integral depends on where the cut starts. We will get a compensating term from the other integral.

For this second integral, integration by parts yields

$$\oint_C \psi_0 \, d\mathfrak{z} = [\psi_0 \mathfrak{z}]_C - \oint_C \mathfrak{z} \, d\psi_0$$

$$= \mathfrak{z}_0 [\psi_0]_C - \oint \mathfrak{z} \frac{d\nu}{d\bar{x}} \, d\tau \tag{3A-5}$$

since
$$\frac{\partial \psi_0}{\partial \tau} = \frac{\partial \phi_0}{\partial \nu} = \frac{d\nu}{d\bar{x}} \tag{3A-6}$$

We have the geometric integrals giving area and moment of area

$$S'(1) = \oint_C \frac{d\nu}{d\bar{x}} \, d\tau = [\psi_0]_C \tag{3A-7}$$

$$\frac{d}{d\bar{x}} [\mathfrak{z}_0 S(1)] = \oint_C \mathfrak{z} \frac{d\nu}{d\bar{x}} \, d\tau \tag{3A-8}$$

The integral thus becomes

$$\oint_C \psi_0 \, d\mathfrak{z} = S'(1)\mathfrak{z}_0 - S'(1)\mathfrak{z}_0(1) - \mathfrak{z}_0'(1)S(1) \tag{3A-9}$$

Finally from Eq. (3A-2) there is obtained

$$\frac{F}{q_0} = 4\pi a_1(1) + 2S'(1)\mathfrak{z}_0(1) + 2\mathfrak{z}_0'(1)S(1) \tag{3A-10}$$

APPENDIX 3B

We now evaluate the two integrals of Eq. (3-72)

$$I_1 = \int_{S_2} \frac{\partial \phi_0}{\partial r} \frac{\partial \phi_0}{\partial \bar{x}} \, dS_2$$

$$I_2 = \int_{S_3} \left[\left(\frac{\partial \phi_0}{\partial r} \right)^2 + \frac{1}{r^2} \left(\frac{\partial \phi_0}{\partial \theta} \right)^2 \right] dS_3 \tag{3B-1}$$

With reference to Fig. 3-5

$$I = \int_0^1 d\bar{x} \int_0^{2\pi} r_1 \frac{\partial \phi_0}{\partial r} \frac{\partial \phi_0}{\partial \bar{x}} \, d\theta \tag{3B-2}$$

Now
$$\phi = a_0 \log r + b_0 + \sum_{n=1}^{\infty} \frac{a_n{}^* \cos n\theta + b_n{}^* \sin n\theta}{r^n} \tag{3B-3}$$

so that (neglecting the sin $n\theta$ terms)

$$I_1 = \int_0^1 d\bar{x} \int_0^{2\pi} r_1 \left(\frac{a_0}{r_1} - \sum_{m=1}^{\infty} \frac{m a_m{}^* \cos m\theta}{r_1{}^{m+1}} \right) \left(a_0' \log r_1 + b_0' \right.$$

$$\left. + \sum_{n=1}^{\infty} \frac{a_n{}'^* \cos n\theta}{r_1{}^n} \right) d\theta$$

$$= 2\pi \int_0^1 (a_0 a_0' \log r_1 + a_0 b_0') \, d\bar{x} \tag{3B-4}$$

The same result is obtained with the $\sin n\theta$ terms. Let us operate on I_2 by Stokes's theorem to convert it from an integral over S_3 to integrals about C, its inner boundary, and K, its outer boundary.

$$I_2 = \int_{S_3}\left[\left(\frac{\partial \phi_0}{\partial r}\right)^2 + \frac{1}{r^2}\left(\frac{\partial \phi}{\partial \theta}\right)^2\right] dS_3$$

$$= -\oint_C \phi_0 \frac{\partial \phi_0}{\partial \nu}\, d\tau + \int_0^{2\pi} \phi_0 \frac{\partial \phi_0}{\partial r}\, r_1\, d\theta \qquad (3B\text{-}5)$$

$$\int_0^{2\pi} \phi_0 \frac{\partial \phi_0}{\partial r}\, r_1\, d\theta = \int_0^{2\pi} r_1 \left(a_0 \log r_1 + b_0 \right.$$

$$\left. + \sum_{n=1}^{\infty} \frac{a_n{}^* \cos n\theta}{r_1{}^n}\right)\left(\frac{a_0}{r_1} - \sum_{m=1}^{\infty} \frac{m a_m{}^* \cos m\theta}{r_1{}^{m+1}}\right) d\theta$$

$$= 2\pi(a_0{}^2 \log r_1 + a_0 b_0)$$

Thus

$$I_2 = 2\pi(a_0{}^2 \log r_1 + a_0 b_0) - \oint_C \phi_0 \frac{\partial \phi_0}{\partial \nu}\, d\tau \qquad (3B\text{-}6)$$

APPENDIX 3C

The integral to evaluate is that of Eq. (3-77)

$$I = \oint_C \phi_{01} \frac{\partial \phi_{00}}{\partial \nu}\, d\tau \qquad (3C\text{-}1)$$

where the contour C is shown in Fig. 3-5. We can subtract the integral of Eq. (3-79) from I since it is zero

$$I = \oint_C \phi_{01} \frac{\partial \phi_{00}}{\partial \nu}\, d\tau - \oint_C \phi_{00} \frac{\partial \phi_{01}}{\partial \nu}\, d\tau \qquad (3C\text{-}2)$$

Consider the contours C and K enclosing S_3, and apply Green's theorem to area S_3,

$$I = \oint_K \left(\phi_{01} \frac{\partial \phi_{00}}{\partial \nu} - \phi_{00} \frac{\partial \phi_{01}}{\partial \nu}\right) d\tau$$

$$+ \int_{S_3} (\phi_{01}\nabla^2 \phi_{00} - \phi_{00}\nabla^2 \phi_{01})\, dS_3 \qquad (3C\text{-}3)$$

Here ∇^2 is the Laplace operator, and ϕ_{00} and ϕ_{01} are solutions of Laplace's equation. Hence,

$$I = \oint_K \left(\phi_{01} \frac{\partial \phi_{00}}{\partial \nu} - \phi_{00} \frac{\partial \phi_{01}}{\partial \nu}\right) d\tau \qquad (3C\text{-}4)$$

Since the integral has been transposed to the contour K, we can use the expansion for ϕ_{00} and ϕ_{01} which converge on K but not on C. On K the expansions for ϕ_{00} and ϕ_{01} have the form (neglecting $\sin n\theta$ terms)

$$\phi_{00} = c \log r + c_0 + \sum_{n=1}^{\infty} \frac{c_n \cos n\theta}{r^n}$$

$$\phi_{01} = \sum_{n=1}^{\infty} \frac{d_n \cos n\theta}{r^n} \tag{3C-5}$$

where c, c_0, c_n, and d_n are constants.   The form of the integrand then becomes

$$\phi_{01} \frac{\partial \phi_{00}}{\partial \nu} - \phi_{00} \frac{\partial \phi_{01}}{\partial \nu} = \frac{cd_1 + c_0 d_1}{r_1^2} \cos \theta$$

$$+ cd_1 \frac{\log r_1}{r_1^2} \cos \theta + O\left(\frac{1}{n^3}\right) + O\left(\frac{\log r_1}{n^3}\right) \tag{3C-6}$$

$$I = \int_0^{2\pi} r_1 \left(\phi_{01} \frac{\partial \phi_{00}}{\partial \nu} - \phi_{00} \frac{\partial \phi_{01}}{\partial \nu} \right) d\theta = O\left(\frac{\log r_1}{r_1}\right) \tag{3C-7}$$

Since I does not depend on r_1 because the drag cannot depend on the radius r_1 of the control surface, we can let r_1 approach ∞.

$$I = O\left(\frac{\log r_1}{r_1}\right)$$

$$= O \qquad \text{as } r_1 \to \infty \tag{3C-8}$$

CHAPTER 4

AERODYNAMICS OF BODIES; VORTICES

In the present and ensuing chapters we will be concerned with application of the general results of the preceding chapter to various types of configurations such as bodies, wing-body combinations, and wing-body-tail combinations. Concurrently, it will be our purpose to investigate how departures from slenderness modify the slender-body results, as well as how viscosity introduces additional effects, some of which can be treated by extensions of slender-body theory. In the first half of the chapter inviscid slender-body theory is applied to bodies of circular and elliptical cross section. Also, the theory of quasi-cylindrical bodies of nearly circular cross section is treated. No discussion is included of non-linear theory or of nonslender bodies, since for zero angle of attack these subjects are considered in Sec. 9-4 in connection with drag.

The appearance at high angles of attack of vortices on the leeward side of slender bodies constitutes one of the most important single causes of the breakdown of inviscid slender-body theory. However, in one sense the slender-body theory has not failed at all, but rather the slender-body model must be generalized. In fact, if discrete vortices are introduced into the slender-body model to account for the effects of viscosity, it is not difficult to extend slender-body theory to include the vortex effects. This is the principal purpose of the second half of the chapter. Results will be obtained for slender configurations with panels present.

INVISCID FLOW

4-1. Lift and Moment of Slender Bodies of Revolution

In Secs. 3-1 and 3-2, the potentials were derived for slender bodies of revolution at zero angle of attack, and at angle of attack by introducing the assumption of slenderness into the solutions based on the full linearized theory of supersonic flow. The potential for a slender body of revolution due to angle of attack, Eq. (3-19), is independent of Mach number, so that the distributions of lift and sideforce along the body are also independent of Mach number. Let us use the general formula, Eq. (3-62), to calculate the forces and moments on a slender body of revolution. The body is taken oriented with respect to the $\bar{x}$, $\bar{y}$, $\bar{z}$ axes as in Fig. 3-4 at angle of attack α_c. Its centroid then lies along the line $\mathfrak{z}_g = -i\alpha_c\bar{x}$. Since $S(\bar{x})$ and $\mathfrak{z}_g(\bar{x})$ are known, it remains only to determine a_1 in Eq. (3-62) to obtain the forces. Although only the potential due to angle of attack creates lift or sideforce for a body of revolution, the coefficient a_1 arises as a result of both angle of attack and thickness because of compensating terms in Eq. (3-62). We can, however, ignore $b_0(\bar{x})$ in Eq. (3-47) since it has no contribution to a_1.

The complex potential for a slender body of revolution consists of the part $W_t(\mathfrak{z})$ existing at zero angle of attack plus a part $W_\alpha(\mathfrak{z})$ due to angle of attack. The part at zero angle of attack is the sum of $b_0(\bar{x})$ and a logarithmic term proportional to the rate of body expansion. With reference to Table 2-3, the equation for $W_t(\mathfrak{z})$ with due regard for shift in origin is

$$\begin{aligned}
W_t(\mathfrak{z}) &= b_0(\bar{x}) + r_s v_r \log(\mathfrak{z} - \mathfrak{z}_g) \\
&= b_0(\bar{x}) + \frac{S'(\bar{x})}{2\pi} \log(\mathfrak{z} - \mathfrak{z}_g)
\end{aligned} \tag{4-1}$$

where r_s is the local body radius. From Table 2-3 the complex potential for angle of attack suitably modified for shift in origin is

$$W_\alpha(\mathfrak{z}) = -iV_0\alpha_c\left[(\mathfrak{z} - \mathfrak{z}_g) - \frac{r_s^2}{\mathfrak{z} - \mathfrak{z}_g}\right] \tag{4-2}$$

The entire complex potential with $V_0 = 1$ is

$$\begin{aligned}
W(\mathfrak{z}) = b_0(\bar{x}) + \frac{S'(\bar{x})}{2\pi}&\left[\log\mathfrak{z} + \log\left(1 - \frac{\mathfrak{z}_g}{\mathfrak{z}}\right)\right] \\
&- i\alpha_c\left[(\mathfrak{z} - \mathfrak{z}_g) - \frac{r_s^2}{\mathfrak{z}(1 - \mathfrak{z}_g/\mathfrak{z})}\right]
\end{aligned}$$

Expansion of this equation yields the coefficient $a_1(\bar{x})$ of the $\mathfrak{z}^{-1}$ term:

$$\begin{aligned}
a_1 &= -\frac{\mathfrak{z}_g S'(\bar{x})}{2\pi} + ir_s^2\alpha_c \\
&= -\mathfrak{z}_g\frac{S'(\bar{x})}{2\pi} - \mathfrak{z}_g'\frac{S(\bar{x})}{\pi}
\end{aligned} \tag{4-3}$$

All the necessary quantities are now at hand for evaluating $\bar{Y}$ and $\bar{Z}$ from Eq. (3-62):

$$\begin{aligned}
\frac{\bar{Y}}{q_0} + i\frac{\bar{Z}}{q_0} &= -2S(\bar{x})\mathfrak{z}_g'(\bar{x}) = 2i\alpha_c S(\bar{x}) \\
\frac{\bar{Y}}{q_0} &= 0 \qquad \frac{\bar{Z}}{q_0} = 2\alpha_c S(\bar{x})
\end{aligned} \tag{4-4}$$

The lift per unit axial distance along the span of a cone-cylinder has been calculated by Eq. (4-4) and is shown in Fig. 4-1a. A similar calculation has been made for a parabolic-arc body and is shown in Fig. 4-1b. Since $S'(\bar{x})$ is linear in $\bar{x}$ for a cone, the lift distribution is linear as shown.

Behind the shoulder of the body where $S'(\bar{x})$ falls discontinuously to zero, the lift distribution also falls to zero on the basis of theory. Unless the body is very slender, some measurable lift would intuitively be expected to be carried over past the shoulder, and in practice such is the case. The second example exhibits equal areas of positive and negative lift. The net lift on the basis of Eq. (4-4) is zero for this case in inviscid flow, since the base area is zero. The body boundary layer will usually not

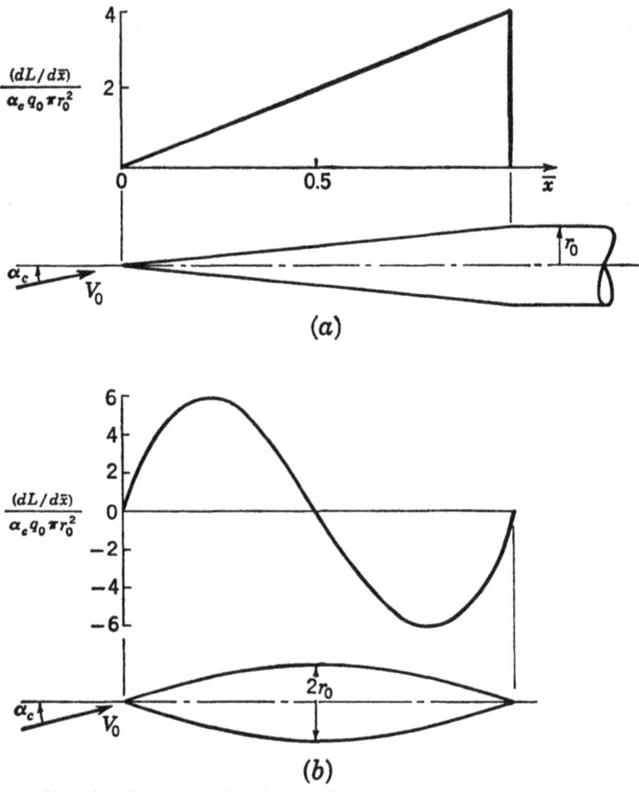

Fig. 4-1. Lift distributions for slender bodies of revolution. (a) Cone-cylinder; (b) parabolic body.

stay attached up to the point for a body of revolution at angle of attack, and, if it did, it would give the point an effective base area by virtue of the displacement thickness of the boundary layer. In either circumstance some lift would be expected from the body.

The lift coefficient with the base area as reference area for a body of unit length is

$$C_L = \frac{\bar{Z}(1)}{q_0 S(1)} = 2\alpha_c \tag{4-5}$$

Slender-body theory thus yields the simple result that the lift-curve slope of a slender body of revolution is two based on its base area.

Consider now the center of pressure of a slender body of revolution. Since the lift per unit length is proportional to $S'(\bar{x})$, the center of pressure is a distance $\bar{x}_{cp}$ behind the body vertex as given by

$$\bar{x}_{cp} = \frac{\int_0^1 \bar{x} S'(\bar{x})\, d\bar{x}}{\int_0^1 S'(\bar{x})\, d\bar{x}} = 1 - \frac{\text{Vol.}}{\pi r_B{}^2} \qquad (4\text{-}6)$$

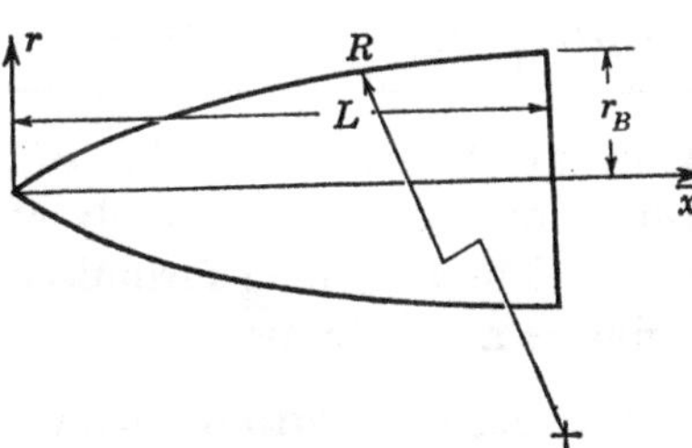

Fig. 4-2. Coordinates of tangent ogive.

Vol. is the volume of the body of unit length, and r_B is the radius of its base.

Illustrative Example

Calculate the center of pressure for a tangent ogive as a function of its caliber, the length of the ogive in diameters of the base.

With reference to Fig. 4-2, the equation for the local radius of the ogive comes from the equation for a circle

$$[r + (R - r_B)]^2 + (L - \bar{x})^2 = R^2$$

with

$$R = \frac{L^2 + r_B{}^2}{2r_B}$$

Introduce the nondimensional parameters

$$r^* = \frac{r}{r_B} \qquad x^* = \frac{\bar{x}}{L} \qquad K = \frac{L}{2r_B} \qquad \text{calibers}$$

and express the value of r^* in terms of these parameters

$$r^* = \frac{-(4K^2 - 1) + [(4K^2 - 1)^2 + 16K^2 x^*(2 - x^*)]^{1/2}}{2} \qquad (4\text{-}7)$$

The volume of the ogive is

$$\text{Vol.} = \pi r_B{}^2 L \int_0^1 r^{*2}\, dx^* \qquad (4\text{-}8)$$

If we introduce the value of r^* from Eq. (4-7) into Eq. (4-8), carry out the integration, and substitute into Eq. (4-6), we obtain the desired result

$$\frac{\bar{x}_{cp}}{L} = 1 - \left[\frac{8K^2}{3} + \frac{(4K^2 - 1)^2}{4} - \frac{(4K^2 - 1)(4K^2 + 1)^2}{16K} \sin^{-1} \frac{4K}{4K^2 + 1} \right] \qquad (4\text{-}9)$$

The center-of-pressure positions in decimal parts of the total body length calculated by this formula are presented in Table 4-1. According to the slender-body theory, there is relatively little shift in center of pressure with change in caliber K for a tangent ogive. The lower value of K to which the theory is valid depends to some extent on the Mach number.

TABLE 4-1. CENTER OF PRESSURE OF TANGENT OGIVE FROM SLENDER-BODY THEORY

K	2	3	3.5	4	6	8	10
$\bar{x}_{cp}/L$	0.457	0.462	0.464	0.464	0.465	0.466	0.466

Only in the limiting value of a hemisphere, for which $K = \frac{1}{2}$, is the center of pressure invariably at the geometric center for all Mach numbers. The foregoing formula does not, of course, apply to such a blunt ogive as a hemisphere.

4-2. Pressure Distribution and Loading of Slender Bodies of Revolution; Circular Cones

In the foregoing section on gross forces of bodies of revolution, the $\bar{x}$, $\bar{y}$, $\bar{z}$ axes were used, but in this section on pressure coefficients it is more

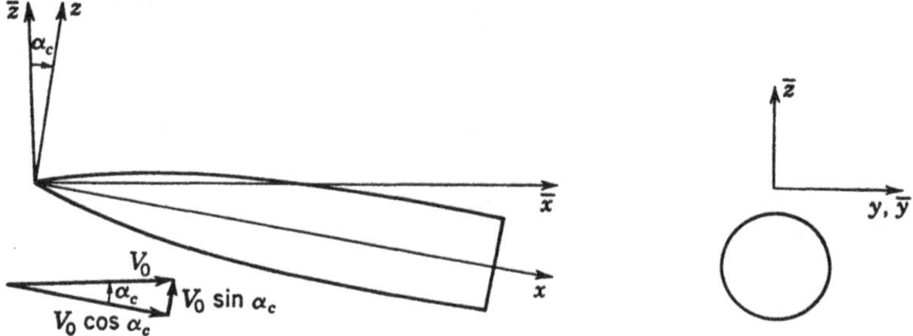

FIG. 4-3. Axis conventions.

convenient to use the x, y, z body axes. In the crossflow planes the body cross sections of a body of revolution at angle of attack are ellipses rather than circles as in the preceding section. The fractional change in the major and minor axes of the ellipse from the radius of the circle is $O(\alpha_c^2)$ and can be ignored in the computation of forces and moments. However, the pressure coefficient involves quadratic terms in the velocity components, and it is not clear that the difference between the circles and ellipses can be ignored. However, this question is circumvented because it is convenient to use normal planes rather than crossflow planes, so that we have simple potential problems for circular boundaries. To obtain the potential we can use the principle of superposition. With reference to Fig. 4-3 we can calculate the perturbation potential ϕ_t due to the component of velocity $V_0 \cos \alpha_c$ along the body axis and a potential ϕ_α due to $V_0 \sin \alpha_c$ normal to the body axis, and thereby obtain the total perturbation potential

$$\phi = \phi_t + \phi_\alpha \tag{4-10}$$

No question of bank angle arises for bodies of revolution, and the pressure coefficient depends only on α_c, which we will simply write as α with the bank angle zero.

To obtain the pressure coefficient we can utilize Eq. (3-52) in terms of velocity components u, v, and w along the body axes. Both ϕ_t and ϕ_α produce velocity components which are linearly superposable. However, the corresponding pressure coefficients P_t and P_α are not generally superposable because of the quadratic terms in the velocity components. However, for bodies of revolution the superposition of P_t and P_α can be demonstrated, as will now be done. Let u_t, v_t, and w_t be the perturbation velocity components associated with ϕ_t, and let u_α, v_α, and w_α be those associated with ϕ_α. If the superscript plus refers to the lower surface and minus to the upper, then from Eq. (3-52) we have for ϕ_t

$$P_t^+ = -2u_t^+ - [(v_t^+)^2 + (w_t^+)^2]$$
$$P_t^- = -2u_t^- - [(v_t^-)^2 + (w_t^-)^2] \tag{4-11}$$

For ϕ_α we have a comparable set of pressure coefficients

$$P_\alpha^+ = -2(u_\alpha^+ + \alpha w_\alpha^+) - [(v_\alpha^+)^2 + (w_\alpha^+)^2]$$
$$P_\alpha^- = -2(u_\alpha^- + \alpha w_\alpha^-) - [(v_\alpha^-)^2 + (w_\alpha^-)^2] \tag{4-12}$$

For the combined effects of ϕ_t and ϕ_α with the velocity components additive, the pressure coefficients are

$$P_{t+\alpha}^+ = P_t^+ + P_\alpha^+ - 2(v_\alpha^+ v_t^+ + w_\alpha^+ w_t^+ + \alpha w_t^+)$$
$$P_{t+\alpha}^- = P_t^- + P_\alpha^- - 2(v_\alpha^- v_t^- + w_\alpha^- w_t^- + \alpha w_t^-) \tag{4-13}$$

The last term in each instance can be considered as arising from coupling between ϕ_t and ϕ_α. For a body of revolution the velocity vector $v_t + iw_t$ in the normal plane is normal to the body surface. The perturbation velocity in the normal plane due to ϕ_α is $v_\alpha + iw_\alpha$, while the total velocity tangential to the body is $v_\alpha + iw_\alpha + i\alpha$. Since the two velocities are perpendicular to each other, their dot product is zero.

$$(v_t + iw_t) \cdot (v_\alpha + iw_\alpha + i\alpha) = v_t v_\alpha + w_t w_\alpha + \alpha w_t = 0 \tag{4-14}$$

This proves that the coupling terms in Eq. (4-13) are zero. The pressure coefficients P_t and P_α can therefore be separately calculated and then added.

The absence of coupling terms in the loading coefficient can also be demonstrated. The loading coefficient is

$$\Delta P = P_{t+\alpha}^+ - P_{t+\alpha}^-$$
$$= P_\alpha^+ - P_\alpha^-$$
$$- 2(w_\alpha^+ w_t^+ - w_\alpha^- w_t^- + v_\alpha^+ v_t^+ - v_\alpha^- v_t^- + \alpha w_t^+ - \alpha w_t^-) \tag{4-15}$$

The coupling term in this instance can be simplified by the obvious symmetry properties resulting from the horizontal plane of symmetry of the body of revolution.

$$u_t{}^+ = u_t{}^- \qquad v_t{}^+ = v_t{}^- \qquad w_t{}^+ = -w_t{}^-$$
$$u_\alpha{}^+ = -u_\alpha{}^- \qquad v_\alpha{}^+ = -v_\alpha{}^- \qquad w_\alpha{}^+ = w_\alpha{}^- \qquad (4\text{-}16)$$

The coupling term becomes

$$(w_\alpha{}^+w_t{}^+ - w_\alpha{}^-w_t{}^- + v_\alpha{}^+v_t{}^+ - v_\alpha{}^-v_t{}^- + \alpha w_t{}^+ - \alpha w_t{}^-)$$
$$= 2(w_\alpha{}^+w_t{}^+ + v_\alpha{}^+v_t{}^+ + \alpha w_t{}^+) \qquad (4\text{-}17)$$

Since this is the same coupling term that occurred for the pressure coefficient, it is zero. Thus to obtain the loading coefficient we need only calculate $(P_\alpha{}^+ - P_\alpha{}^-)$.

Having proved the principle of superposition for the pressure coefficient and loading coefficient, let us consider calculating ϕ_t and P_t first. The function ϕ_t is found as the real part of $W_t(\mathfrak{z})$ obtained from Eqs. (3-37), (3-46), and (3-48) for the free-stream velocity $V_0 \cos \alpha$ along the body axis (taken as unity).

$$W_t(\mathfrak{z}) = b_0(\bar{x}) + a_0(\bar{x}) \log \mathfrak{z}$$
$$a_0 = \frac{S'(\bar{x})}{2\pi} \qquad (4\text{-}18)$$
$$b_0 = \frac{1}{2\pi}\left[S(\bar{x}) \log \frac{B}{2} - \int_0^{\bar{x}} \log (\bar{x} - \xi)S''(\xi)\, d\xi \right]$$

For a circular cone (Fig. 4-4) the values of a_0 and b_0 are

$$a_0 = \omega^2 \bar{x}$$
$$b_0 = \omega^2 \bar{x} \log \frac{B}{2} - \omega^2 \bar{x}(\log \bar{x} - 1) \qquad (4\text{-}19)$$

The real part of $W_t(\mathfrak{z})$ then gives the potential

$$\phi_t = \omega^2 \bar{x} \log \frac{Br}{2} - \omega^2 x(\log \bar{x} - 1) \qquad (4\text{-}20)$$

with velocity components

$$u_t = \omega^2 \log \frac{Br}{2\bar{x}}$$
$$v_t = r_s\omega \frac{\cos \theta}{r} \qquad (4\text{-}21)$$
$$w_t = r_s\omega \frac{\sin \theta}{r}$$

The pressure coefficient from Eq. (4-11) is

$$P_t = -2\omega^2 \log \frac{B\omega}{2} - \omega^2 \qquad (4\text{-}22)$$

The thickness pressure coefficients for circular cones as determined from this equation are compared with the values in Fig. 4-4. The Taylor and Maccoll[1] values are accurate to the order of the full nonlinear potential theory. Fortunately, the range of validity of slender-body theory is broader in other cases than in the present connection. For bodies of revolution a large number of approximate methods of varying degrees of accuracy are available for calculating P_t. These methods are discussed in Sec. 9-4.

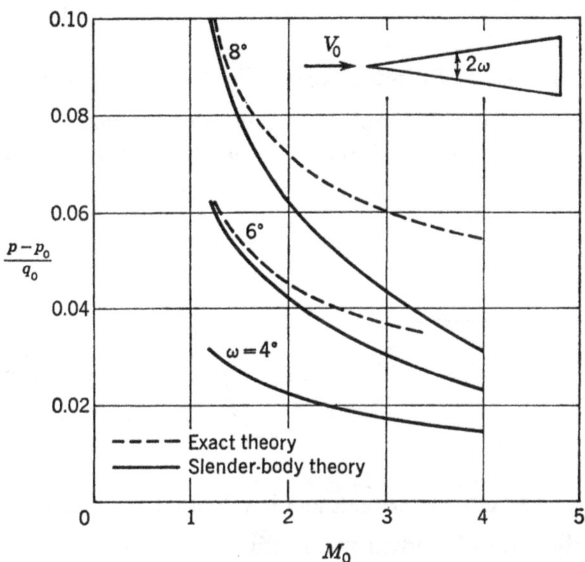

FIG. 4-4. Thickness pressure coefficients for circular cones.

The separate effects of angle of attack are now considered. The perturbation velocity potential ϕ_α due to the crossflow velocity $V_0 \sin \alpha$ has already been given in Sec. 3-2 for $V_0 = 1$.

$$\phi_\alpha = \alpha r_s^2 \frac{\sin \theta}{r} \tag{4-23}$$

The velocity components associated with ϕ_α are

$$u_\alpha = 2\alpha r_s \frac{dr_s}{dx} \frac{\sin \theta}{r}$$

$$v_\alpha = -\alpha r_s^2 \frac{\sin 2\theta}{r^2} \tag{4-24}$$

$$w_\alpha = \alpha r_s^2 \frac{\cos 2\theta}{r^2}$$

On the body surface, $r = r_s$, the pressure coefficient by Eq. (4-12) is

$$P_\alpha = -4\alpha \frac{dr_s}{dx} \sin \theta - \alpha^2 (2 \cos 2\theta + 1) \tag{4-25}$$

The body of revolution need not be a cone. An examination of the pressure coefficient shows in the first place a term proportional to the product of angle of attack and rate of body expansion. The term is odd in θ, thereby producing a body loading. In the second place we have a term proportional to α^2, and even in θ. Though this term influences the pressure coefficient, it produces no body loading. A body loading is thus developed only under the combined action of body expansion and angle attack. For a cone of semiapex angle of the same order of magnitude as α, the contributions of both terms to the pressure coefficient P_α are significant.

4-3. Slender Bodies of Elliptical Cross Section; Elliptical Cones

As an example in the application of slender-body theory to noncircular bodies, consider the forces, moments, velocity components, and

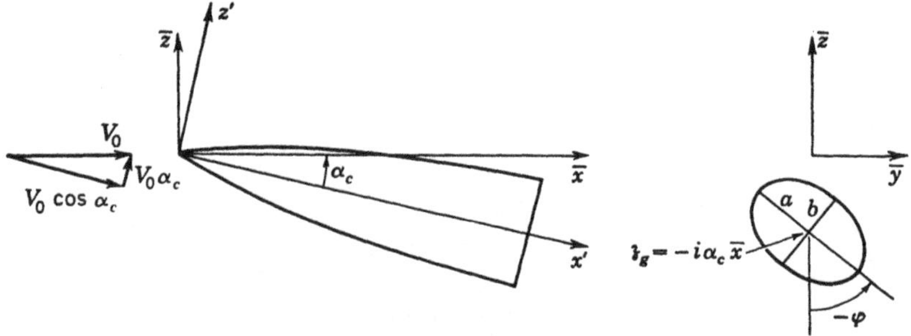

Fig. 4-5. Axis conventions and notation for elliptical bodies.

pressure coefficients of bodies with elliptical cross sections. First, consider gross forces and moments, and then velocity components and pressure coefficients. The aerodynamic characteristics of noncircular bodies depend on two independent variables: the included angle α_c, and the bank angle φ, as shown in Fig. 4-5. Zero angle of bank is taken to correspond to a vertical position of the major axis. Slender bodies of elliptical cross section have been treated by Kahane and Solarski[2] and by Fraenkel.[3]

The general analysis is made for the body and axes as shown in Fig. 4-5. The x, y, z and x', y', z' standards conform to those of Fig. 1-2. The complex variable $\mathfrak{z}$ is $y' + iz'$ in this section. Resolve the velocity V_0 into a component $V_0 \cos \alpha_c$ parallel to x' and a component $V_0 \sin \alpha_c$ normal to x'. Let the complex potentials for the perturbation velocities be $W_t(\mathfrak{z})$ and $W_\alpha(\mathfrak{z})$. Then the total complex potential is

$$W(\mathfrak{z}) = W_t(\mathfrak{z}) + W_\alpha(\mathfrak{z})$$

The question of coupling between thickness and angle of attack will not be examined in detail for elliptical bodies as it was for bodies of revolution. Instead we proceed directly to the calculation of the gross forces and moments depending on $W_\alpha(\mathfrak{z})$.

The complex force $\bar{Y} + i\bar{Z}$ given by Eq. (3-62) requires a knowledge of $W_\alpha(\mathfrak{z})$, the perturbation complex potential for the flow about a banked ellipse due to angle of attack. With reference to Table 2-3, the complex potential is

$$W_\alpha(\mathfrak{z}) = \frac{-iV_0\alpha_c}{2}\left[(\xi^2 + c^2)^{1/2} - \xi - \frac{(a + b)^2}{\xi + (\xi^2 + c^2)^{1/2}}\right] \qquad (4\text{-}26)$$

$$\xi = \mathfrak{z} - \mathfrak{z}_0 \qquad \mathfrak{z}_0 = -i\alpha_c\bar{x} \qquad c^2 = (a^2 - b^2)e^{-2i\varphi}$$

Expansion of $W_\alpha(\mathfrak{z})$ readily yields a_1, the coefficient of the $\mathfrak{z}^{-1}$ term

$$(a_1)_\alpha = \frac{-iV_0\alpha_c}{4}[c^2 - (a + b)^2] \qquad (4\text{-}27)$$

Now Eq. (3-62) has a term depending on $S'(1)\mathfrak{z}_0(1)$ which for bodies of revolution is canceled by the contribution of $W_t(\mathfrak{z})$ to a_1. For elliptical bodies with $S'(1) = 0$ or with constant a/b ratio approaching the base, it will be shown that similar cancellation occurs. Therefore, in these cases we have

$$\frac{F(1)}{q_0} = \frac{\bar{Y} + i\bar{Z}}{q_0} = i\pi\alpha_c[-(a^2 - b^2)e^{-2i\varphi} + (a + b)^2 - 2ab] \qquad (4\text{-}28)$$

The lift $\bar{Z}$ and sideforce $\bar{Y}$ are then given by

$$\frac{\bar{Y}}{q_0} = -2\pi\alpha_c(a^2 - b^2)\sin\varphi\cos\varphi$$

$$\frac{\bar{Z}}{q_0} = 2\pi\alpha_c(a^2\sin^2\varphi + b^2\cos^2\varphi) \qquad (4\text{-}29)$$

To show the cancellation of the $S'(1)\mathfrak{z}_0(1)$ term when $S'(1)$ is not equal to zero but a/b is uniform approaching the base, consider the case of an expanding ellipse of constant a/b ratio for $\alpha_c = 0$ and $\varphi = 0$ as given in Table 2-3.

$$W_t(\mathfrak{z}) = b_0(x') + \frac{S'(x')}{2\pi}\log\frac{\mathfrak{z} + (\mathfrak{z}^2 + a^2 - b^2)^{1/2}}{2} \qquad (4\text{-}30)$$

The major axis is vertical, as shown in Fig. 4-6. To convert this complex potential to the case α_c not equal to zero, we must substitute $\mathfrak{z} - \mathfrak{z}_0$ for $\mathfrak{z}$. To take into account the effect of bank angle, we must then substitute $(\mathfrak{z} - \mathfrak{z}_0)e^{i\varphi}$ for $(\mathfrak{z} - \mathfrak{z}_0)$. Thus for α_c and φ both not zero

$$W_t(\mathfrak{z}) = b_0(x') + \frac{S'(x')}{2\pi}\log\left\{\frac{e^{i\varphi}}{2}(\mathfrak{z} - \mathfrak{z}_0) + \frac{e^{i\varphi}}{2}[(\mathfrak{z} - \mathfrak{z}_0)^2 \right.$$
$$\left. + (a^2 - b^2)e^{-2i\varphi}]^{1/2}\right\} \qquad (4\text{-}31)$$

Expansion of $W_t(\mathfrak{z})$ yields the coefficient of the $\mathfrak{z}^{-1}$ term

$$(a_1)_t = \frac{-S'(x')\mathfrak{z}_0}{2\pi} \qquad (4\text{-}32)$$

This term cancels the $S'(1)\dot{\delta}_\theta(1)$ term of Eq. (3-62), so that Eqs. (4-28) and (4-29) are valid for elliptical bodies of uniform a/b ratio approaching the base.

We can derive simple results for the moments of slender bodies of elliptical cross section under the restriction that the body cross sections are all of uniform a/b ratio. To do this we use Eq. (3-66) and carry out

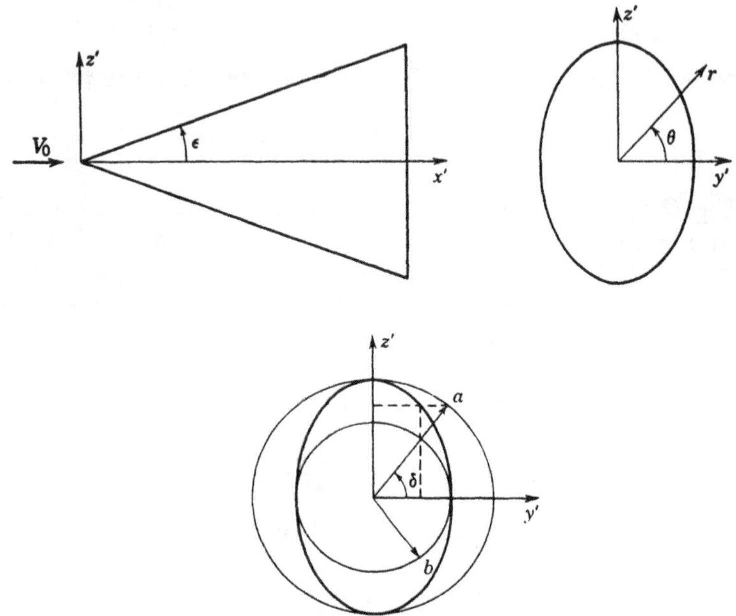

FIG. 4-6. Elliptical cone at zero angle of attack.

the integration. The uniform a/b ratio permits us to write

$$a_1(x') = (a_1)_\alpha + (a_1)_t = \frac{-i\alpha_c}{4\pi}\frac{a}{b}\left[\left(1 - \frac{b^2}{a^2}\right)e^{-2i\varphi}\right.$$
$$\left. - \left(1 + \frac{b}{a}\right)^2\right]S(x') + \frac{S'(x')}{2\pi}(+i\alpha_c x') \quad (4\text{-}33)$$

The integral with respect to x' of a_1 then yields simple integrals in terms of the body volume, regardless of the body shape, because a/b is constant. The moments are given then simply as

$$M_y = -2\pi\alpha_c(a^2\sin^2\varphi + b^2\cos^2\varphi)\left[1 - \frac{\text{Vol.}}{S(1)}\right] \quad (4\text{-}34)$$

$$M_z = -2\pi\alpha_c(a^2 - b^2)\sin\varphi\cos\varphi\left[1 - \frac{\text{Vol.}}{S(1)}\right] \quad (4\text{-}35)$$

Here a and b are the semiaxes of the base section for a body of unit length with base area $S(1)$. It is noted that the pitching moment is M_y and the yawing moment is $-M_z$.

The pressure coefficients P_t and P_α will be separately evaluated without regard for possible coupling effects. Since the angle of bank will not influence the value of P_t, let the angle of bank be zero as in Fig. 4-6.   The velocities v_t' and w_t' along the y' and z' axis can be evaluated directly from Eq. (4-30), but the velocity u_t' along x' requires a knowledge of $b_0(x')$. From Eq. (3-48), $b_0(x')$ for an elliptical cone is

$$b_0(x') = b\epsilon \log \frac{B}{2} - \frac{b}{a}\,\epsilon^2 \int_0^{x'} \log \xi \, d\xi \qquad (4\text{-}36)$$

Here ϵ is the semiapex angle of the cone in the plane of the major axis. Actual evaluation of $b_0(x')$ is unnecessary since it will be differentiated by x' to obtain u_t'.   The velocity perturbation components are

$$u_t' = \mathbf{RP}\,\frac{\partial}{\partial x'}\,[W_t(\mathfrak{z})]$$
$$v_t' - iw_t' = \frac{d}{dz}\,[W_t(\mathfrak{z})] \qquad (4\text{-}37)$$

The carrying out of the integrations yields the velocity components, which can conveniently be expressed in terms of the angle δ illustrated in Fig. 4-6.   It is to be noted that the angle δ is not the polar angle of the ellipse.

$$v_t' = \frac{ab\epsilon \cos \delta}{a^2 \cos^2 \delta + b^2 \sin^2 \delta}$$
$$w_t' = \frac{b^2\epsilon \sin \delta}{a^2 \cos^2 \delta + b^2 \sin^2 \delta} \qquad (4\text{-}38)$$
$$u_t' = \frac{b}{a}\,\epsilon^2 \log \frac{B(a+b)}{4x'} + \frac{(b/a)\epsilon^2(a/b - 1)[(a/b)\cos^2 \delta - \sin^2 \delta]}{(a/b)^2 \cos^2 \delta + \sin^2 \delta}$$

The angle δ is such that

$$\begin{aligned} y' &= b \cos \delta \\ z' &= a \sin \delta \end{aligned} \qquad (4\text{-}39)$$

The pressure coefficient due to thickness is easily found since the velocity components are specified

$$P_t = -2\frac{b}{a}\,\epsilon^2 \log \frac{B(a+b)}{4x'} - 2\epsilon^2 \frac{b}{a} + \frac{\epsilon^2(b/a)^2}{(b/a)^2 \sin^2 \delta + \cos^2 \delta} \qquad (4\text{-}40)$$

Let us now turn our attention to the determination of the velocity components and pressure coefficient due to angle of attack. For this purpose the complex potential of Eq. (4-26) is to be used, and neither α_c nor φ is taken as zero as it was for the thickness calculations.   We will be concerned with the velocity components along the x', y', z' axes due to angle of attack, namely, u_α', v_α', and w_α'.   It is convenient to use those axes because the body cross sections are true ellipses normal to the x' axes

(Fig. 4-5). Then we do not have to be concerned with any distortions of the body cross sections from true elliptical shapes. The appropriate form of Bernoulli's equation for the computation of the pressure coefficient for the x', y', z' system is

$$P_\alpha = -2(u_\alpha' + \alpha_c w_\alpha') - [(v_\alpha')^2 + (w_\alpha')^2] \tag{4-41}$$

The lifting pressure coefficient depends on the rate of expansion of the major and minor axes. In this development, the case of constant ratio of minor to major axis will be taken. Some latitude in the shape of the body is retained in that it need not be conical.

The velocity components have been calculated from Eq. (4-26) by means of the formulas

$$u_\alpha' = \mathrm{RP}\, \frac{\partial W_\alpha(\mathfrak{z})}{\partial x'}$$
$$v_\alpha' - i w_\alpha' = \frac{dW(\mathfrak{z})}{d\mathfrak{z}} \tag{4-42}$$

The actual operations are lengthy, and the velocity components come out in somewhat cumbersome form.

$$u_\alpha' = \alpha_c \left(1 + \frac{b}{a}\right) \frac{da}{dx'} \left(\frac{-a^2 \sin \varphi \cos \delta + b^2 \cos \varphi \sin \delta}{a^2 \cos^2 \delta + b^2 \sin^2 \delta}\right) \tag{4-43}$$

$$v_\alpha' = \frac{\alpha_c(a + b)(a \sin^2 \varphi - b \cos^2 \varphi) \sin \delta \cos \delta}{a^2 \cos^2 \delta + b^2 \sin^2 \delta}$$
$$+ \frac{\alpha_c(a + b)(a \cos^2 \delta - b \sin^2 \delta) \sin \varphi \cos \varphi}{a^2 \cos^2 \delta + b^2 \sin^2 \delta} \tag{4-44}$$

$$w_\alpha' = \frac{-\alpha_c(a \sin^2 \varphi - b \cos^2 \varphi)(a \cos^2 \delta - b \sin^2 \delta)}{a^2 \cos^2 \delta + b^2 \sin^2 \delta}$$
$$+ \frac{\alpha_c(a + b)^2 \sin \varphi \cos \varphi \sin \delta \cos \delta}{a^2 \cos^2 \delta + b^2 \sin^2 \delta} \tag{4-45}$$

The pressure coefficient due to angle of attack calculated from Eq. (4-41) is also cumbersome:

$$P_\alpha = \frac{2\alpha_c \left(1 + \dfrac{b}{a}\right) \dfrac{da}{dx'} (a^2 \sin \varphi \cos \delta - b^2 \cos \varphi \sin \delta)}{a^2 \cos^2 \delta + b^2 \sin^2 \delta}$$
$$+ \frac{\alpha_c^2[(a^2 - b^2) \cos 2\delta - (a + b)^2 \cos (2\varphi - 2\delta) - 2ab]}{2(a^2 \cos^2 \delta + b^2 \sin^2 \delta)} \tag{4-46}$$

The angle δ is measured from the minor axis, as shown in Fig. 4-6, even when the ellipse is banked.

We have neglected any questions of coupling between P_α and P_t for elliptical bodies. Just as in the case of bodies of revolution, this coupling is zero for elliptical cones. The thickness pressure distribution for an elliptical cone is shown in Fig. 4-7. The pressure coefficient is greatest at the end of the major axis, as would be expected.

As a final subject under the general heading of elliptical bodies, the drag of an elliptical cone will be evaluated at zero angle of attack by direct integration of the pressure over the surface area. With reference to Fig. 4-8, the drag D_e of an elliptical cone can be written

$$\frac{D_e}{q_0} = \oint P_t \frac{\lambda}{2}\, d\tau \qquad (4\text{-}47)$$

where P_t is given by Eq. (4-40).

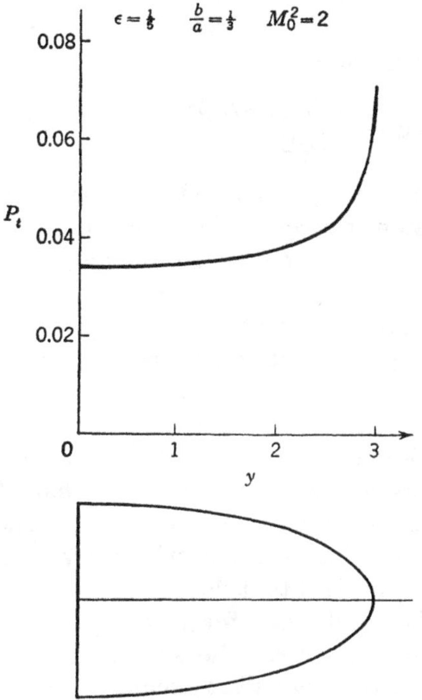

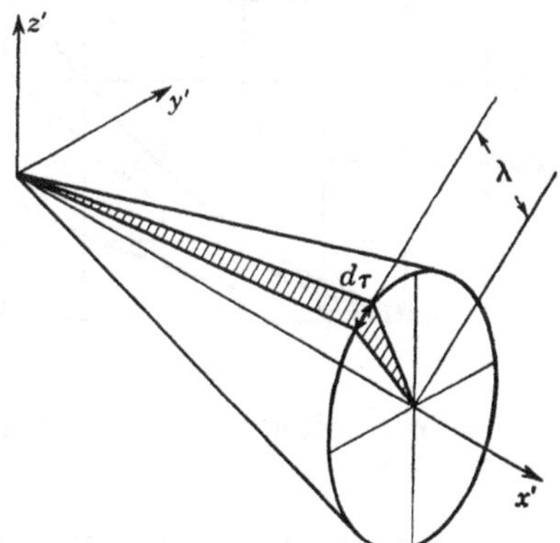

FIG. 4-7. Thickness pressure distribution for elliptical cone.

FIG. 4-8. Notation for elliptical cone.

The values of λ and $d\tau$ are given in terms of δ as illustrated in Fig. 4-6.

$$\lambda = \frac{ab}{(b^2 \sin^2 \delta + a^2 \cos^2 \delta)^{\frac{1}{2}}} \qquad (4\text{-}48)$$
$$d\tau = (b^2 \sin^2 \delta + a^2 \sin^2 \delta)^{\frac{1}{2}}\, d\delta$$

Evaluation of the drag integral yields the drag coefficient $(C_D)_e$ for the elliptical cone with the base areas as reference area:

$$(C_D)_e = \frac{D_e}{q_0 \pi ab} = -2 \frac{b}{a} \epsilon^2 \log \frac{B\epsilon(1 + b/a)}{4} - \frac{b}{a} \epsilon^2 \qquad (4\text{-}49)$$

The angle ϵ is the semiapex angle in the plane of the major axis.

Illustrative Example:

Compare the drag coefficient of circular and elliptical cones of equal base area and length. Let the semiapex angle of the circular cone be ω, and let its drag coefficient be $(C_D)_c$. Its drag coefficient from Eq. (4-49) is

$$(C_D)_c = -2\omega^2 \log \frac{B\omega}{2} - \omega^2 \qquad (4\text{-}50)$$

The angles ω and ϵ are related for equal base area and length

$$\omega^2 = \frac{b}{a}\,\epsilon^2$$

The difference in drag coefficients then becomes

$$(C_D)_c - (C_D)_e = 2\omega^2 \log \frac{(1 + b/a)(a/b)^{\frac{1}{2}}}{2} \tag{4-51}$$

Slender-body theory thus yields the interesting result that the drag of an

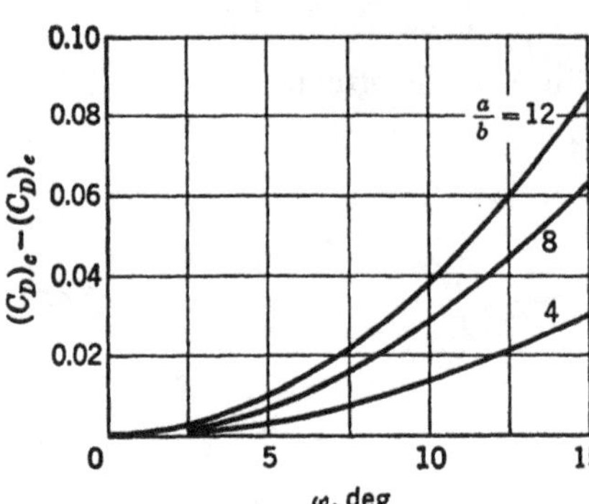

FIG. 4-9. Wave drag of circular cones versus elliptical cones.

elliptical cone is less than that of a circular cone by an amount independent of Mach number. The drag coefficient increment given by Eq. (4-51) is shown in Fig. 4-9 for various values of ω and a/b. As a/b increases, the elliptical cone becomes more winglike and has lower drag compared with that of the equivalent circular cone. The foregoing results must be interpreted in the light that slender-body theory is valid only for small semiapex angles. Also, the surface area of the elliptical cone is greater than that of the equivalent circular cone and therefore causes greater skin friction.

4-4. Quasi-cylindrical Bodies

One class of bodies not generally included within the scope of slender-body theory is that class of bodies the surfaces of which lie everywhere

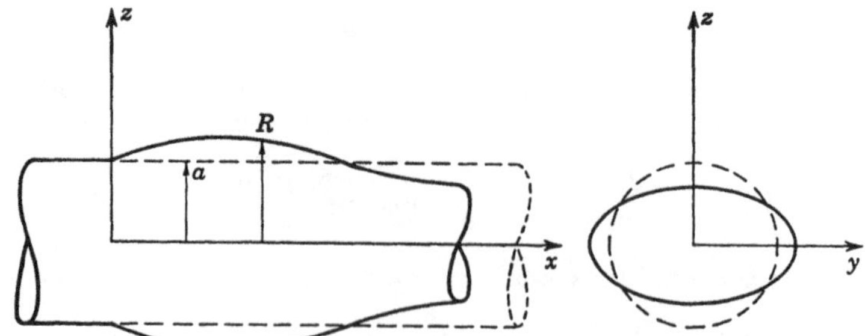

FIG. 4-10. Axes and notation for quasi-cylindrical bodies.

close to a cylinder. The cylinder need not be circular in the general case of a quasi-cylindrical body. We will, however, confine our attention to quasi-cylinders, which lie everywhere close to a circular cylinder as in Fig. 4-10. Let the surface be defined by the local radius R which is both a function of x and θ

$$R = R(x,\theta)$$

We shall be concerned with the slope of the surface in the streamwise direction dR/dx, which can be expanded in a Fourier series as follows, with coefficients which are functions of x:

$$\frac{dR}{dx} = \sum_{n=0}^{\infty} f_n(x) \cos n\theta \tag{4-52}$$

Only cosine terms are retained on the assumption of a horizontal plane of symmetry. The mean surface to which the quasi-cylinder is close is the surface $r = a$, as shown. The problem to be solved is to calculate the pressure coefficient in the flow due to the quasi-cylinder. For this purpose the full linearized theory of supersonic flow is used, and the boundary condition represented by Eq. (4-52) is applied on the $r = a$ cylinder.

The method of solution follows, in principle, the Laplace transform treatment of the wave equation, which is the basis of the slender-body theory of Ward (Sec. 3-4). The potential ϕ is the solution of the wave equation in cylindrical coordinates (we assume that $M_0{}^2 = 2$ during the derivation)

$$\phi_{rr} + \frac{1}{r}\,\phi_r + \frac{1}{r^2}\,\phi_{\theta\theta} - \phi_{xx} = 0 \tag{4-53}$$

Under the Laplace transform operation

$$L(\phi) = \Phi = \int_0^{\infty} e^{-px}\phi(x,r,\theta)\,dx \tag{4-54}$$

Eq. (4-53) becomes

$$\Phi_{rr} + \frac{1}{r}\,\Phi_r + \frac{1}{r^2}\,\Phi_{\theta\theta} = p^2\Phi \tag{4-55}$$

on the assumption that

$$\phi(0^+,r,\theta) = 0$$
$$\frac{\partial \phi}{\partial x}\,(0^+,r,\theta) = 0 \tag{4-56}$$

The solution to Eq. (4-55) of interest here is that given by Eq. (3-31) containing Bessel functions. However, the Bessel functions $K_n(pr)$ are the only ones that should be retained, as discussed in Sec. 3-4.

The solution of Eq. (4-55) then is

$$\Phi = \sum_{n=0}^{\infty} C_n(p)K_n(pr) \cos n\theta \tag{4-57}$$

where $C_n(p)$ are functions of p to be chosen to satisfy the boundary conditions. The boundary condition to be satisfied is that the flow velocity at the quasi-cylinder be tangent to its surface.   The actual calculation is made at the cylinder $r = a$.   Thus

$$\frac{(\partial\phi/\partial r)_{r=a}}{V_0} = \frac{dR}{dx} = \sum_{n=0}^{\infty} f_n(x)\cos n\theta \qquad (4\text{-}58)$$

Let
$$L[f_n(x)] = F_n(p) \qquad (4\text{-}59)$$

and evaluate $C_n(p)$ in Eq. (4-57) by means of the boundary condition to obtain

$$C_n(p) = \frac{V_0 F_n(p)}{pK_n{}'(pa)} \qquad (4\text{-}60)$$

so that
$$\Phi = V_0 \sum_{n=0}^{\infty} \cos n\theta\, F_n(p)\, \frac{K_n(pr)}{pK_n{}'(pa)} \qquad (4\text{-}61)$$

We are interested in obtaining the pressure coefficient as follows,

$$P = \frac{-2(\partial\phi/\partial x)}{V_0} = \frac{-2}{V_0} L^{-1}(p\Phi) \qquad (4\text{-}62)$$

where L^{-1} denotes taking the inverse Laplace transform.   Before taking the inverse transform, let us write Eq. (4-61) as

$$p\Phi = V_0 \sum_{n=0}^{\infty} \cos n\theta\, F_n(p)e^{-p(r-1)} \left[\left(\frac{e^{p(r-1)}K_n(pr)}{K_n{}'(p)} + \frac{1}{r^{1/2}}\right) - \frac{1}{r^{1/2}}\right] \qquad (4\text{-}63)$$

where we have let $a = 1$ without any loss in generality.   The technique now employed is to split the expression into two parts, one dependent of the boundary conditions as represented by $F_n(p)$, and the other independent of the boundary conditions, as follows:

$$L[f_n(x - r + 1)] = F_n(p)e^{-p(r-1)} \qquad (4\text{-}64)$$

$$L[W_n(x,r)] = e^{p(r-1)}\frac{K_n(pr)}{K_n{}'(p)} + \frac{1}{r^{1/2}} \qquad (4\text{-}65)$$

The part independent of the boundary conditions represented by Eq. (4-65) has been made the basis for the definition of a set of characteristic functions $W_n(x,r)$.   Assuming that these functions are known, we can write the inversion of Eq. (4-63) by the convolution theorem that gives the inverse transform of a product of transforms.   We thereby obtain the pressure coefficient from Eq. (4-62) as

$$P = \frac{2}{B} \sum_{n=0}^{\infty} \left\{\frac{f_n(x - Br + Ba)}{(r/a)^{1/2}} - \frac{1}{Ba}\int_0^{x-Br+Ba} f_n(\xi)W_n\left(\frac{x}{Ba} - \frac{r}{a}\right.\right.$$
$$\left.\left. + 1 - \frac{\xi}{Ba}, \frac{r}{a}\right)d\xi\right\} \cos n\theta \qquad (4\text{-}66)$$

This result has been written for cylinders of any average radius for any supersonic Mach number. The equation can be used to calculate the pressure coefficient of any quasi-cylinder of nearly circular cross section as specified by Eq. (4-52). The $W_n(x,r)$ functions required for the calculation have been tabulated elsewhere.[4] The calculation is made by numerical or graphical integration. In the reference the physical significance of the $W_n(x,r)$ functions is discussed. They represent downstream pressure waves associated with a sudden ramp on the body surface.

Illustrative Example

To show how Eq. (4-66) might be used, let us calculate the pressure distribution on an axially symmetric bump on a circular cylinder as

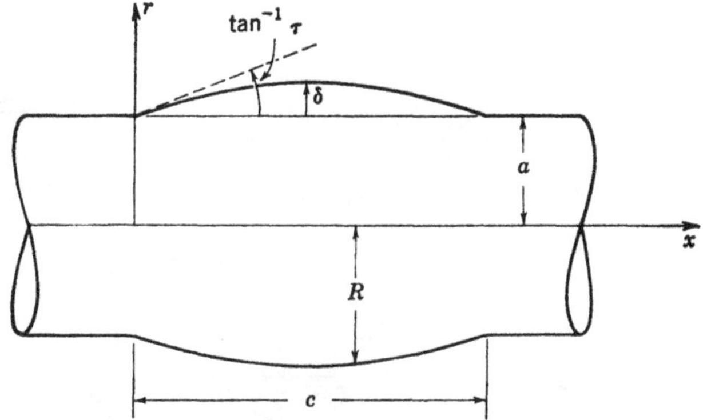

FIG. 4-11. Circular cylinder with axially symmetric bump.

shown in Fig. 4-11. The equation of the bump is taken to be

$$R = a + 4\delta \frac{x}{c}\left(1 - \frac{x}{c}\right)$$

$$\frac{dR}{dx} = 4\frac{\delta}{c}\left(1 - \frac{2x}{c}\right) \qquad 0 \le x \le c$$

$$= 0 \qquad c < x \tag{4-67}$$

From Eq. (4-52) the $f_n(x)$ functions are

$$f_0(x) = 4\frac{\delta}{c}\left(1 - \frac{2x}{c}\right) \qquad 0 \le x \le c$$

$$= 0 \qquad c < x \tag{4-68}$$

$$f_n(x) = 0 \qquad n > 0$$

Only one term remains in the summation of Eq. (4-66) for the pressure coefficient:

$$P = \frac{2}{B}\left\{ \frac{4\dfrac{\delta}{c}\left[1 - \dfrac{2(x - Br + Ba)}{c}\right]}{(r/a)^{\frac{1}{2}}} \right.$$

$$\left. - \frac{1}{Ba}\int_0^{x-Br+Ba} 4\frac{\delta}{c}\left(1 - \frac{2\xi}{c}\right) W_0\left(\frac{x}{Ba} - \frac{r}{a} + 1 - \frac{\xi}{Ba}, \frac{r}{a}\right) d\xi \right\}$$

$$0 \leq x \leq c \quad (4\text{-}69)$$

The pressure coefficient on the body has been calculated for several values of B, and the results are expressed in the form of BP/τ in Fig. 4-12. The symbol τ indicates the initial ramp angle as shown in Fig. 4-11. Of

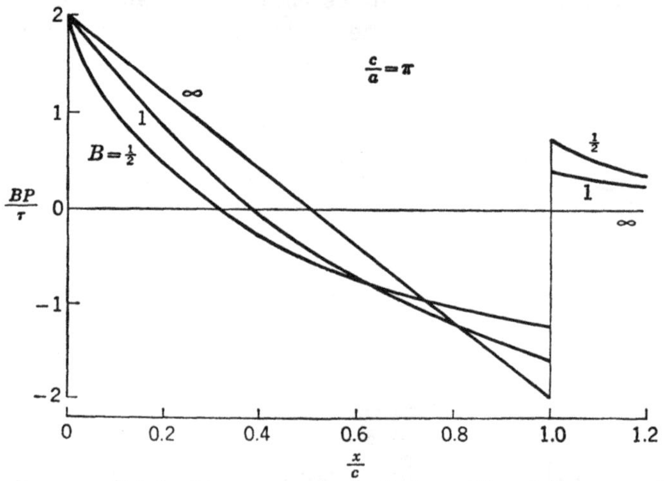

FIG. 4-12. Pressure distribution on circular cylinder with axially symmetric bump.

interest is the fact that pressure coefficient always starts off with a value of $2\tau/B$. Such a value corresponds precisely to the Ackeret value, the value to be expected for the full linearized theory and two-dimensional flow. Since the flow is essentially two-dimensional to start, the result is to be expected. However, as the flow continues downstream, it sees part of the bump in its forward Mach cone as curved rather than on a flat surface. If the bump had remained flat, we would continue to have only the first term of Eq. (4-69). The second term thus represents the influence of the curvature of the surface on which the bump is fitted. In this sense the second term represents three-dimensional influences. If M_0 is large, the second term is small. Such a result is in accordance with the fact that the upstream Mach cone has a narrow field of view and cannot "see" much curvature of the body. As B approaches infinity, the upstream Mach cone "sees" only a planar strip of body so that the calculated pressure coefficient has the local two-dimensional value everywhere.

VORTICES

4-5. Positions and Strengths of Body Vortices

The subject of this second half of the chapter is body vortices. The appearance of vortices in the flow can cause significant departures between experiment and inviscid slender-body theory. One of the most direct ways of illustrating the effects of vortices is to examine the pressure distribution around a body of revolution at high angles of attack. Such

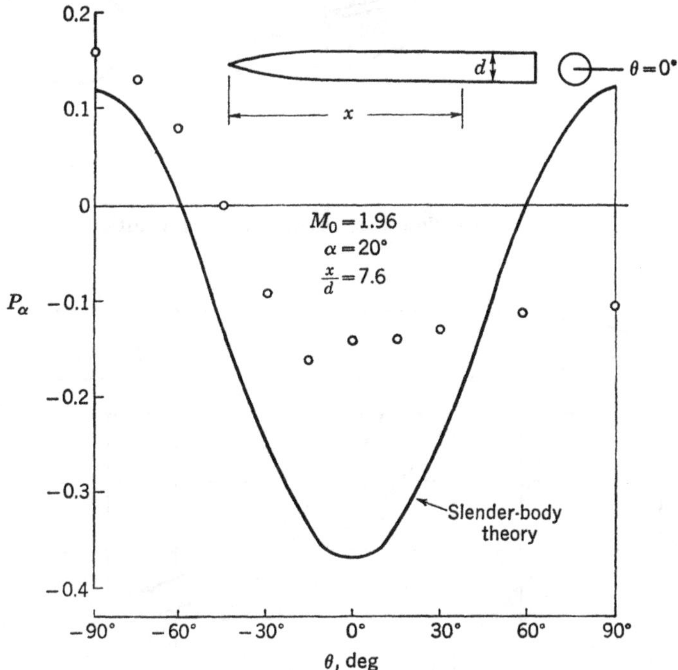

FIG. 4-13. Pressure distributions around body of revolution; comparison of theory and experiment.

a pressure distribution taken from Perkins and Jorgensen[5] is shown in Fig. 4-13. In this figure the experimental pressure distribution is compared with the theoretical distribution predicted by inviscid slender-body theory, Eq. (4-25). According to slender-body theory, the pressure distribution on a nonexpanding body section is symmetric above and below the horizontal plane of symmetry; that is, the positive pressure existing on the windward face of the body is also recovered on the leeward face of the body. An examination of the data points reveals that no such pressure recovery appears. In fact, somewhere near the side edge of the body the pressure change ceases, and a fairly uniform pressure level exists over the top of the body. The lack of pressure recovery is ascribed to the body boundary layer, which separates from the body with the resultant formation of a "dead water region" of more or less uniform pressure on the leeward side of the body. The boundary layer itself rolls up into vortices. Let us now examine the vortex formation in greater detail.

The general features of flow separation on bodies of revolution at supersonic speed have been studied by Jorgensen and Perkins,[6] Raney,[7]

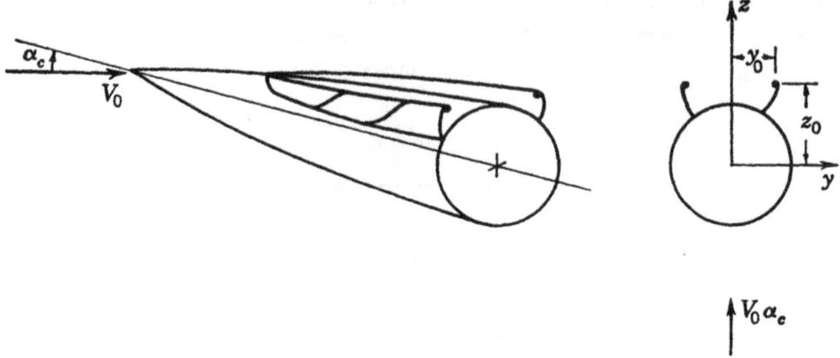

FIG. 4-14. Crossflow vortices of body of revolution.

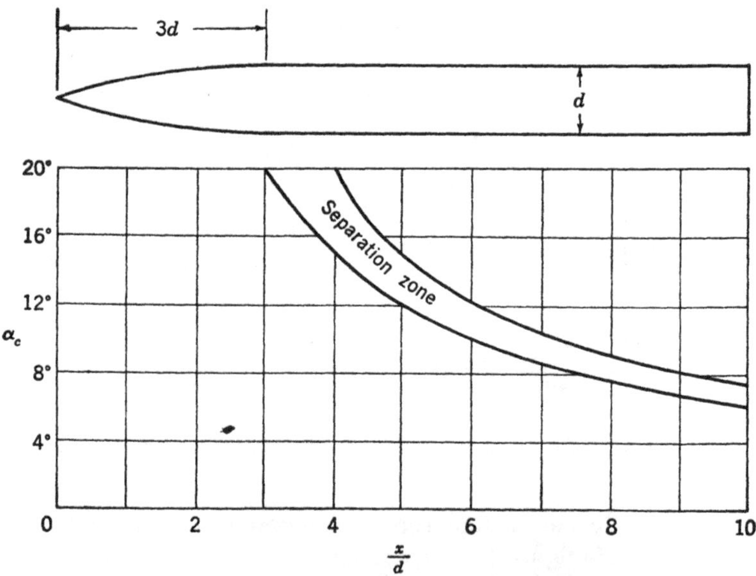

FIG. 4-15. Location of vortex separation for body of revolution.

and others. These features are illustrated in Fig. 4-14. As the boundary layer flows from the underside of the body around to the leeward side, it separates along a line of separation shown on the body. After separating, the boundary layer continues as vortex filaments, which rise above the body and curl up into strong body vortices on each side of the body. As the body vortices proceed downstream, more vortex filaments originating at the separation lines feed into the cores and increase their strengths.

One of the pertinent questions is: At what distance x_S behind the body apex do the body vortices first form? The distance will depend strongly on the angle of attack; but, since the controlling phenomenon is boundary-layer separation under pressure gradients, the Reynolds number and Mach number are also involved, as indeed is the shape of the body itself. Some data exist[6] for the dependence of x_S on α_c. These data are reproduced in Fig. 4-15 for an ogive-cylinder combination at a Mach number of 2. At the higher angles of attack, the vortices tend to originate at the body shoulder. This is reasonable, since the expansion of the body in front of the shoulder tends to thin out the boundary layer and inhibit

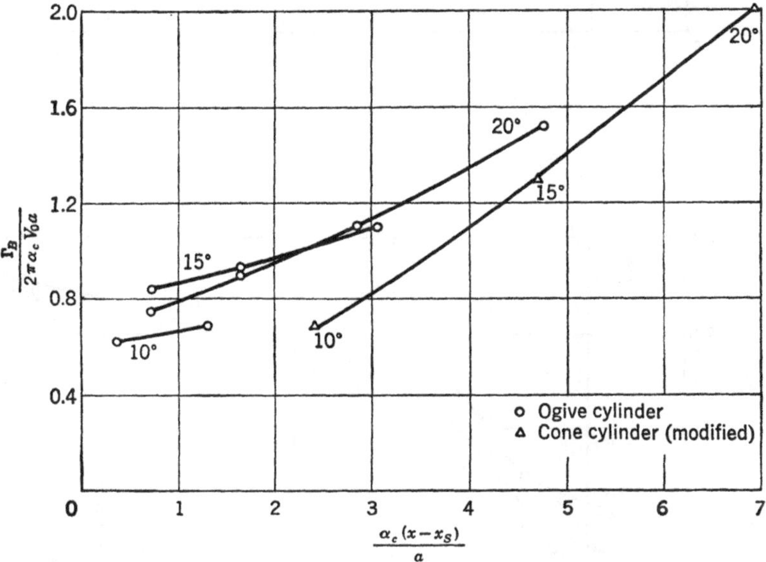

FIG. 4-16. Nondimensional vortex strengths for bodies of revolution.

separation. The precise location of vortex formation could not be ascertained, but rather a region of vortex formation was obtained.

It is possible to obtain a nondimensional correlation of the strength and position of the body vortex cores as a function of x and α_c on the basis of certain plausible arguments. Consider the body vortices as seen in planes normal to the body axis. Assume that the change in the pattern of the flow with changes in x is analogous to the change in the flow pattern about a two-dimensional cylinder with time if it is impulsively moved normal to itself at velocity V_n. If zero time corresponds to the distance x_S, then time and distance are related by

$$x - x_S = V_0 t = \frac{V_n t}{\alpha_c} \tag{4-70}$$

The nondimensional parameter which characterizes the impulsive flow N

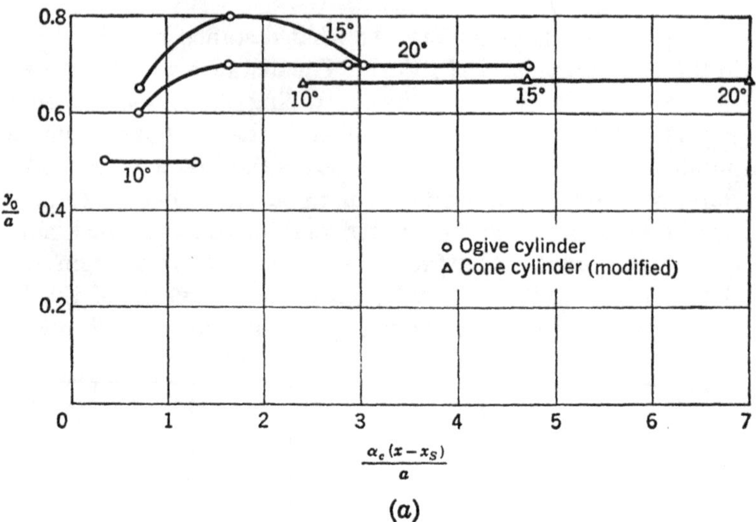

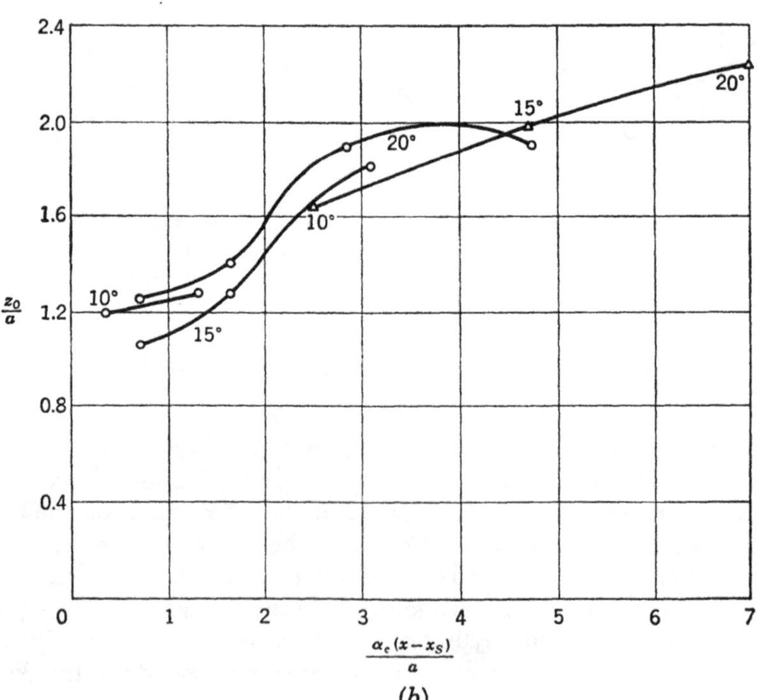

Fig. 4-17. Vortex positions for bodies of revolution. (a) Lateral location; (b) vertical location.

is

$$N = \frac{V_n t}{a} \tag{4-71}$$

By analogy the corresponding dimensionless number for our case is

$$N = \frac{\alpha_c(x - x_S)}{a} \tag{4-72}$$

If the analogy is correct, then the vortex strengths and positions in non-dimensional form should correlate on the basis of N alone for different values of x and α_c.

The analogy has been tested,[8] using data from Jorgensen and Perkins,[6] and Raney.[7] The measure of the nondimensional vortex strength is $\Gamma/2\pi V_0 a\alpha_c$. This parameter is shown as a function of N in Fig. 4-16. A rough correlation exists. It must be remembered that correlation is hampered by experimental difficulties of measuring Γ. The vortex positions are simply specified by the nondimensional quantities y_0/a and z_0/a. These quantities are correlated as functions of N in Fig. 4-17, and the correlation is considered fairly good.

4-6. Forces and Moments Due to Body Vortices; Allen's Crossflow Theory

Since the body vortices can significantly influence the pressure distribution, they will have large effects on the body forces and moments in certain cases. It is our purpose now to present the theory of Allen for such effects. The theory is based on the concept of the crossflow drag coefficient $(c_d)_c$. If dN_v/dx is the normal force per unit length (viscous crossforce per unit length) developed normal to an infinite cylinder of radius a at angle of attack α_c, then the crossflow drag coefficient is so defined that

$$\frac{dN_v}{dx} = (c_d)_c(2a)q_0\alpha_c^2 \tag{4-73}$$

The crossflow drag coefficients of a number of different cylinders have been measured and are reported by Lindsey.[9]

By adding the viscous crossforce N_v directly to the lift developed by a slender body on the basis of slender-body theory, one has the basic results of Allen's crossflow theory.[10] The total normal force or lift, since no distinction will be made between lift and normal force here, is then given per unit length by

$$\frac{dN}{dx} = 2q_0\alpha_c\frac{dS}{dx} + (c_d)_c q_0 2a\alpha_c^2 \tag{4-74}$$

where S is the body cross-sectional area. Integration then gives the total body normal force

$$N = 2q_0\alpha_c S_B + (\bar{c}_d)_c q_0 \alpha_c^2 S_C \qquad (4\text{-}75)$$

where S_B is the body base area, and S_C the body planform area subject to viscous crossflow. The area S_C is behind the body cross section corresponding to x_S (as given by Fig. 4-15, for instance). The tacit assumption in the integration of Eq. (4-74) is that $(c_d)_c$ is uniform along the body length. There is some evidence that $(c_d)_c$ is not uniform,[11] but an average value of $(c_d)_c$ has been assumed. It is clear that the pitching moment can easily be calculated since Eq. (4-74) gives the body normal force distribution.

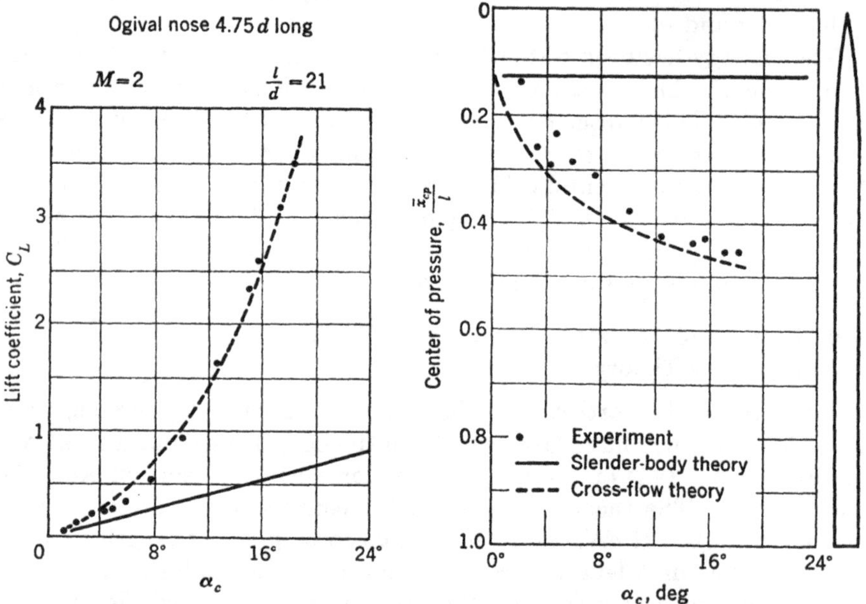

FIG. 4-18. Comparison of measured and predicted body aerodynamic characteristics.

The lift coefficient and center of pressure of a body of revolution have been calculated on the basis of slender-body theory and of Allen's crossflow theory. The calculated values are compared to experimental values in Fig. 4-18. The actual body is of very high fineness ratio, and the viscous crossforce for such a body is much greater than the lift predicted by slender-body theory. The large rearward shift of the center of pressure with increase in angle of attack is noteworthy. Generally speaking, the lift predicted by slender-body theory acts on the expanding sections of the body in front of the vortex separation region, and the viscous crossforce acts behind the region of vortex separation. As the angle of attack increases, the viscous crossforce increases approximately as α_c^2, while the slender-body lift increases as α_c. The rearward shift of the center of pressure is the result.

4-7. Motion of Symmetrical Pair of Crossflow Vortices in Presence of Circular Cylinder

Many problems of interest in missile aerodynamics require a detailed knowledge of the vortex flow due to bodies or lifting surfaces. In this section we will explore the behavior of a symmetrical vortex pattern of two vortices in the presence of a circular cylinder. As pictured in Fig. 4-14, the vorticity is moving along the feeding sheets into the cores at all times. If we neglect any influence of the feeding sheets in comparison with that of the cores, then we can idealize the flow model as shown in Fig. 4-19. Two external vortices occur with equal vortex strength but opposite rotation, and with the vortex strengths changing with time. Inside the body are located two image vortices to insure that the body surface is a streamline. The right image vortex has the opposite sense of rotation of the right external vortex but the same magnitude; with a similar result for the left vortices. If the external right vortex has position ζ_0, then the image vortex must be located by the method of reciprocal radii, namely, so that

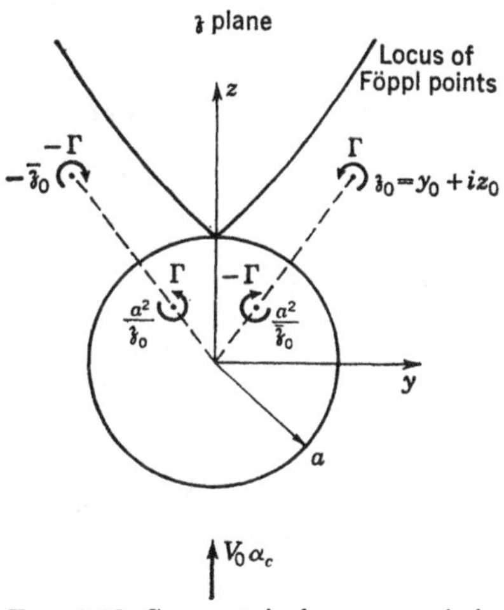

FIG. 4-19. Symmetrical vortex pair in presence of circular cylinder.

$$\zeta_i = \frac{a^2}{\bar{\zeta}_0} \tag{4-76}$$

where ζ_i is the coordinate of the image vortex. The complex potential for a vortex of strength Γ counterclockwise at position ζ_0 is

$$\frac{-i\Gamma}{2\pi} \log (\zeta - \zeta_0)$$

The complex potential for the model of Fig. 4-19, including potential crossflow and four vortices, is

$$W(\zeta) = \phi + i\psi$$
$$= -iV_0\left(\zeta - \frac{a^2}{\zeta}\right)\alpha_c - \frac{i\Gamma}{2\pi}\log\left[\frac{(\zeta - \zeta_0)(\zeta + a^2/\bar{\zeta}_0)}{(\zeta + \bar{\zeta}_0)(\zeta - a^2/\zeta_0)}\right] \tag{4-77}$$

A number of interesting special cases of the general case will now be explored.

One question which might be asked is whether there exist combinations of vortex positions and strengths for which the resultant velocities at the external vortices are zero. Such a question was studied by Föppl.[12] The velocity $v_0 - iw_0$ of the right vortex is given by

$$
\begin{aligned}
v_0 - iw_0 &= \lim_{\mathfrak{z}\to\mathfrak{z}_0} \frac{d}{d\mathfrak{z}}\left[W(\mathfrak{z}) + \frac{i\Gamma}{2\pi}\log(\mathfrak{z} - \mathfrak{z}_0)\right] \\
&= \lim_{\mathfrak{z}\to\mathfrak{z}_0}\left[-iV_0\alpha_c\left(1 + \frac{a^2}{\mathfrak{z}^2}\right) + \frac{i\Gamma}{2\pi}\left(\frac{1}{\mathfrak{z} + \bar{\mathfrak{z}}_0}\right.\right. \\
&\qquad\qquad\qquad\left.\left. + \frac{1}{\mathfrak{z} - a^2/\bar{\mathfrak{z}}_0} - \frac{1}{\mathfrak{z} + a^2/\mathfrak{z}_0}\right)\right]
\end{aligned}
\tag{4-78}
$$

The resultant velocities at the vortex are

$$
v_0 = \frac{-2a^2 V_0\alpha_c y_0 z_0}{r_0^4} + \frac{\Gamma}{2\pi}\left[\frac{z_0}{r_0^2 - a^2} - \frac{z_0(r_0^2 - a^2)}{(r_0^2 - a^2)^2 + 4a^2 y_0^2}\right]
$$

$$
w_0 = \alpha_c V_0\left[1 + \frac{a^2(y_0^2 - z_0^2)}{r_0^4}\right]
\tag{4-79}
$$

$$
\qquad + \frac{\Gamma}{2\pi}\left[-\frac{1}{2y_0} - \frac{y_0}{r_0^2 - a^2} + \frac{y_0(r_0^2 + a^2)}{(r_0^2 - a^2)^2 + 4a^2 y_0^2}\right]
$$

where
$$
\mathfrak{z}_0 = y_0 + iz_0 = r_0 e^{i\theta_0}
\tag{4-80}
$$

The condition that $v_0 - iw_0$ be zero leads to the condition, after eliminating the vortex strength,

$$
(\mathfrak{z}_0\bar{\mathfrak{z}}_0 - a^2)^2 = \mathfrak{z}_0\bar{\mathfrak{z}}_0(\mathfrak{z}_0 + \bar{\mathfrak{z}}_0)^2
\tag{4-81}
$$

After reduction to polar form, this equality yields

$$
r_f - \frac{a^2}{r_f} = 2r_f\cos\theta_f
\tag{4-82}
$$

The subscript f has been used to denote the equilibrium or Föppl positions and strengths. The vortex strength Γ_f corresponding to r_f is

$$
\frac{\Gamma_f}{2\pi V_0\alpha_c} = \frac{(r_f^2 - a^2)^2(r_f^2 + a^2)}{r_f^5}
\tag{4-83}
$$

See Milne-Thompson[14] for details of the derivation. The locus of the equilibrium positions given by Eq. (4-82) is shown in Fig. 4-19. For equilibrium positions far from the body the vortex strength is large, the strength increasing in accordance with Eq. (4-83). One thing to remember is that, though the equilibrium positions are points of zero flow velocity, they are not stagnation points of the crossflow in the usual sense, since the flow velocity changes discontinuously from infinity to zero as the points are approached from any direction.

Another relationship of interest is that between the vortex strength and vortex position when there is to be a stagnation point in the crossflow on the body at the point specified by $\mathfrak{z}_s = ae^{i\theta_s}$. The total velocity in the crossflow plane is

$$\frac{dW}{d\mathfrak{z}} = -i\alpha_c V_0\left(1 + \frac{a^2}{\mathfrak{z}^2}\right) + \frac{i\Gamma}{2\pi}\left(\frac{1}{\mathfrak{z} + \bar{\mathfrak{z}}_0} + \frac{1}{\mathfrak{z} - a^2/\bar{\mathfrak{z}}_0} - \frac{1}{\mathfrak{z} - \mathfrak{z}_0} - \frac{1}{\mathfrak{z} + a^2/\mathfrak{z}_0}\right) \quad (4\text{-}84)$$

For zero velocity on the body at $\mathfrak{z}_s$

$$\frac{\Gamma}{2\pi V_0\alpha_c} = \frac{1 + a^2/\mathfrak{z}_s^2}{\dfrac{1}{\mathfrak{z}_s + \bar{\mathfrak{z}}_0} + \dfrac{1}{\mathfrak{z}_s - a^2/\bar{\mathfrak{z}}_0} - \dfrac{1}{\mathfrak{z}_s - \mathfrak{z}_0} - \dfrac{1}{\mathfrak{z}_s + a^2/\mathfrak{z}_0}} \quad (4\text{-}85)$$

Manipulation of Eq. (4-85) and the requirement that Γ is real yields

$$\frac{\Gamma}{2\pi V_0\alpha_c}$$
$$= \frac{[\tfrac{1}{2}(r_0 + a^2/r_0) - a\cos(\theta_0 - \theta_s)][\tfrac{1}{2}(r_0 + a^2/r_0) + a\cos(\theta + \theta_s)]}{\tfrac{1}{2}(r_0 - a^2/r_0)\cos\theta_0}$$
$$(4\text{-}86)$$

For a given vortex strength Γ and stagnation point, Eq. (4-86) will yield a curve on which the vortex must be located.

The actual streamlines in the crossflow plane of the vortices depend on how the vortex strength varies with time. Actually to consider variable strength of the vortices without including the feeding sheet leads to a physically inconsistent model. One important case for which the vortex streamlines can be found analytically is that for constant vortex strength. If the function ψ_v is the stream function of the vortex streamline, then

$$d\psi_v = \frac{\partial\psi_v}{\partial y_0}dy_0 + \frac{\partial\psi_v}{\partial z_0}dz_0$$
$$= v_0\,dz_0 - w_0\,dy_0$$
$$= \mathbf{IP}\,(v_0 - iw_0)\,d\mathfrak{z}_0$$

Thus
$$\psi_v = \mathbf{IP}\int(v_0 - iw_0)\,d\mathfrak{z}_0 \quad (4\text{-}87)$$

The integration with the aid of Eq. (4-79) yields

$$\psi_v = \frac{2y_0}{a}\left(1 - \frac{a^2}{r_0^2}\right) - \frac{\Gamma}{4\pi V_0\alpha_c a}\log\left[\frac{y_0^2(r_0^2 - a^2)^2}{(r_0^2 - a^2)^2 + 4a^2 y_0^2}\right] \quad (4\text{-}88)$$
$$= \text{constant}$$

The constant is to be evaluated from the knowledge of one point on a particular vortex path. A different set of streamlines occurs for each value of the nondimensional vortex strength $\Gamma/4\pi V_0 a\alpha_c$. For a value of this parameter of unity, the vortex streamlines have the general pattern shown in Fig. 4-20. Vortices near the body move downward against the flow, and those far from the body move with the flow. The Föppl position for the given value of the nondimensional vortex strength is included in the figure. A region of circulatory flow exists about the Föppl point. The asymptotic lateral positions of the vortices at infinity y_∞ shown in Fig. 4-20 can be obtained implicitly from Eq. (4-88) as follows:

$$\frac{2y_\infty}{a} - \frac{\Gamma}{2\pi V_0 a} \log y_\infty = \frac{2y_i}{a}\left(1 - \frac{a^2}{y_i^2}\right) - \frac{\Gamma}{2\pi V_0 a} \log \frac{y_i(y_i^2 - a^2)}{y_i^2 + a^2} \quad (4\text{-}89)$$

For the general case in which the vortex strength is changing with time, an analytical solution for the vortex path seems not to be generally possible. In fact, a stream function for the vortex path in the usual sense does not exist for this case. To obtain the paths we must integrate Eq. (4-79) numerically, using small time increments. Another problem which is also analytically intractible except in special cases is the determination of the positions y_0 and z_0 as functions of the time. To obtain such relationships the following equations must be solved.

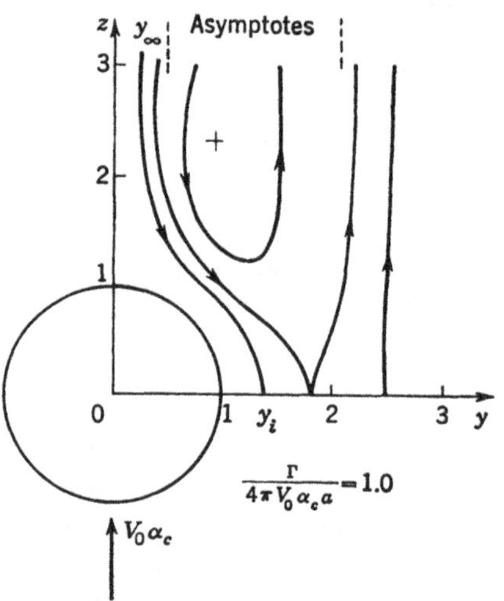

Fig. 4-20. Paths of symmetrical vortex pair in presence of circular cylinder.

$$t = \int \frac{dy_0}{v_0(y_0,z_0)} = \int \frac{dz_0}{w_0(y_0,z_0)} \quad (4\text{-}90)$$

The functions v_0 and w_0 are to be taken from Eq. (4-79). For the special case of $V_0 = 0$ and two symmetrical vortices as shown in Fig. 4-19, Sacks,[13] has determined the time explicitly from Eq. (4-90).

4-8. Motion of Vortices in Presence of a Noncircular Slender Configuration

Let us consider a pair of vortices not necessarily of equal strength in the presence of a noncircular slender configuration as shown in Fig. 4-21. The number of vortices considered is of no importance since the method is valid for any number of vortices. The external vortices induce velocities normal to the body and panels. Single image vortices of the type considered in connection with circular cross sections will not be adequate in this case. In fact, a complicated image system is required. For this reason it is easier to transform the body cross section into a circular one for which the image system is known, and then to relate the vortex velocity in the δ plane to that in the σ plane (Fig. 4-21).

Let $W(\sigma)$ be the complex potential for the complete flow in the σ plane.

With reference to Table 2-3 we have

$$* \quad W(\sigma) = -iV_0\alpha_c e^{-i\theta}\left(\sigma - \frac{r_e^2}{\sigma}e^{2i\theta}\right) - \frac{i\Gamma_1}{2\pi}\log\frac{\sigma - \sigma_1}{\sigma - r_e^2/\bar{\sigma}_1}$$

$$- \frac{i\Gamma_2}{2\pi}\log\frac{\sigma - \sigma_2}{\sigma - r_e^2/\bar{\sigma}_2} \quad (4\text{-}91)$$

Let the transformation equations between the $\mathfrak{z}$ and σ planes leaving the flow at infinity unaltered be

$$\sigma = \sigma(\mathfrak{z}) \qquad \mathfrak{z} = \mathfrak{z}(\sigma) \tag{4-92}$$

The complex potential for the flow in the physical plane is now $W(\sigma(\mathfrak{z}))$.

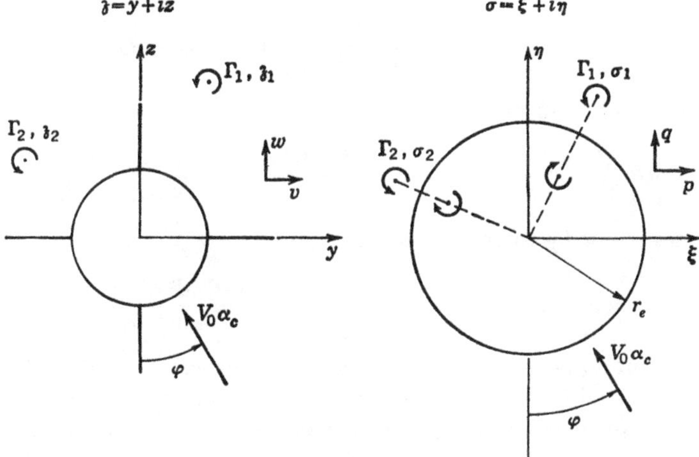

FIG. 4-21. Transformation of missile cross section into circle.

The vortices are transformed as vortices. Look now at the velocity $v_1 - iw_1$ of the vortex at $\mathfrak{z}_1$ in the physical plane.

$$v_1 - iw_1 = \lim_{\mathfrak{z}\to\mathfrak{z}_1}\frac{d}{d\mathfrak{z}}\left[W(\sigma(\mathfrak{z})) + \frac{i\Gamma_1}{2\pi}\log(\mathfrak{z} - \mathfrak{z}_1)\right] \tag{4-93}$$

The velocity of the vortex in the σ plane is denoted by $p_1 - iq_1$

$$p_1 - iq_1 = \lim_{\sigma\to\sigma_1}\frac{d}{d\sigma}\left[W(\sigma) + \frac{i\Gamma_1}{2\pi}\log(\sigma - \sigma_1)\right] \tag{4-94}$$

If we were concerned with the flow velocities at any point other than the vortices, the velocities would be related simply by $d\sigma/d\mathfrak{z}$ in the usual fashion of potential theory. However, the fact that the complex potentials for the vortices in the $\mathfrak{z}$ and σ planes do not transform the same way as $W(\sigma)$ modifies the usual rule.

To relate the vortex velocities in the two planes let us rewrite Eq.

(4-93) as

$$v_1 - iw_1 = \lim_{\mathfrak{z}\to\mathfrak{z}_1} \frac{d}{d\sigma}\left[W(\sigma(\mathfrak{z})) + \frac{i\Gamma_1}{2\pi}\log(\sigma - \sigma_1) \right]\frac{d\sigma}{d\mathfrak{z}}$$

$$+ \lim_{\mathfrak{z}\to\mathfrak{z}_1} \frac{d}{d\mathfrak{z}}\left[\frac{i\Gamma_1}{2\pi}\log(\mathfrak{z} - \mathfrak{z}_1) - \frac{i\Gamma_1}{2\pi}\log(\sigma - \sigma_1) \right] \quad (4\text{-}95)$$

or $\qquad v_1 - iw_1 = (p_1 - iq_1)\left(\frac{d\sigma}{d\mathfrak{z}}\right)_{\mathfrak{z}=\mathfrak{z}_1} + \lim_{\mathfrak{z}\to\mathfrak{z}_1} \frac{d}{d\mathfrak{z}}\left(\frac{i\Gamma_1}{2\pi}\log\frac{\mathfrak{z}-\mathfrak{z}_1}{\sigma-\sigma_1}\right) \quad (4\text{-}96)$

The logarithmic term can be evaluated by differentiating and using the Taylor expansion.

$$\sigma = \sigma_1 + \sigma_1'(\mathfrak{z} - \mathfrak{z}_1) + \frac{\sigma_1''}{2}(\mathfrak{z} - \mathfrak{z}_1)^2 + O(\mathfrak{z} - \mathfrak{z}_1)^3 \quad (4\text{-}97)$$

The Taylor expansion required is

$$\frac{\sigma - \sigma_1}{\mathfrak{z} - \mathfrak{z}_1} = \left(\frac{d\sigma}{d\mathfrak{z}}\right)_{\mathfrak{z}=\mathfrak{z}_1} + \frac{1}{2}\left(\frac{d^2\sigma}{d\mathfrak{z}^2}\right)_{\mathfrak{z}=\mathfrak{z}_1}(\mathfrak{z} - \mathfrak{z}_1) + O(\mathfrak{z} - \mathfrak{z}_1)^2 \quad (4\text{-}98)$$

The limit is then simply

$$\lim_{\mathfrak{z}\to\mathfrak{z}_1} \frac{d}{d\mathfrak{z}}\log\frac{\mathfrak{z}-\mathfrak{z}_1}{\sigma-\sigma_1} = -\frac{1}{2}\frac{d^2\sigma/d\mathfrak{z}^2}{d\sigma/d\mathfrak{z}}\bigg|_{\mathfrak{z}=\mathfrak{z}_1} = \frac{1}{2}\frac{d^2\mathfrak{z}/d\sigma^2}{(d\mathfrak{z}/d\sigma)^2}\bigg|_{\sigma=\sigma_1} \quad (4\text{-}99)$$

The vortex velocity in the physical plane is now

$$v_1 - iw_1 = (p_1 - iq_1)\frac{d\sigma}{d\mathfrak{z}}\bigg|_{\mathfrak{z}=\mathfrak{z}_1} - \frac{i\Gamma_1}{4\pi}\frac{d^2\sigma/d\mathfrak{z}^2}{d\sigma/d\mathfrak{z}}\bigg|_{\mathfrak{z}=\mathfrak{z}_1} \quad (4\text{-}100)$$

The term involving the second derivative arises as an addition to the first term which would be anticipated if the vortex velocities transformed in the same manner as ordinary flow velocities.

The calculation of $v_1 - iw_1$ for the vortices in the presence of a general cross section will usually proceed streamwise step by step in a numerical solution. The initial vortex positions and strengths Γ_1, $\mathfrak{z}_1$, Γ_2, and $\mathfrak{z}_2$ are given. The positions $\mathfrak{z}_1$ and $\mathfrak{z}_2$ are transformed into σ_1 and σ_2. Then the velocity of the vortex in the transformed plane, $p_1 - iq_1$, is computed by the method of Sec. 4-7. The vortex velocity in the physical plane is calculated from Eq. (4-100). The change in vortex position is then obtained by assuming the vortex velocities uniform over the time or distance interval chosen for the calculation. The cycle is repeated in a step-by-step calculation to establish the vortex paths. The vortex path in the σ plane is not the transformation of that in the $\mathfrak{z}$ plane. Variations in body cross section and in vortex strength are easily accounted for in a step-by-step calculation.

4-9. Lift and Sideforce on Slender Configuration Due to Free Vortices

If free vortices follow their natural streamlines in flowing past a slender configuration, the lift and sideforce due to the vortices can be established simply in terms of the vortex strengths and positions. Since a method for calculating vortex paths was described in the previous section for a slender configuration of general cross section, the possibility is at hand of determining the lift and sideforce distributions along such a missile. It is the purpose of this section to derive the necessary formulas in terms of vortex strengths and positions. Consider a single free vortex of strength Γ_1 developed by a vortex generator (Fig. 4-22), or any other means such as body vortex separation. The vortex is free to follow the general flow past the winged part of the configuration. Before starting the derivation

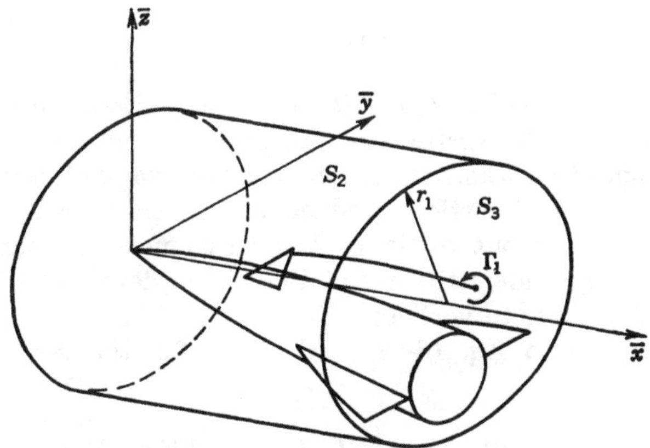

Fig. 4-22. Control areas for calculating forces and moments due to free vortices.

of the formula, it is desirable to determine the magnitudes of the lateral velocities due to the vortex, and then to compare them with the magnitude of the velocities without vortices.

The complex potential due to a vortex of strength Γ_1 at $\mathfrak{z}_1$ is

$$W_1(\mathfrak{z}) = \frac{-i\Gamma_1}{2\pi} \log\left(\mathfrak{z} - \mathfrak{z}_1\right) \qquad (4\text{-}101)$$

and the lateral velocity components are given by

$$\bar{v}_1 - i\bar{w}_1 = \frac{dW_1}{d\mathfrak{z}} = \frac{-i\Gamma_1}{2\pi(\mathfrak{z} - \mathfrak{z}_1)} \qquad (4\text{-}102)$$

The bars on $\bar{v}_1$ and $\bar{w}_1$ indicate that the velocity components are along the $\bar{y}$ and $\bar{z}$ axes. Equation (4-102) will yield the magnitude of the lateral velocities if the magnitude of Γ is known. In this matter we must distinguish between wing-induced vortices and body-induced vortices. If the vortex is body-induced, then with reference to Fig. 4-16

$$\frac{\Gamma_B}{2\pi V_0 a\alpha_c} = O\left(\frac{\alpha_c l}{a}\right) \qquad (4\text{-}103)$$

where l is the body length.   For unit body length

$$\frac{\Gamma_B}{V_0} = O(\alpha_c{}^2) \tag{4-104}$$

Since the angle of attack is $O(t)$, and the lateral dimensions such as $(\mathfrak{z} - \mathfrak{z}_1)$ are also $O(t)$, we find that $\bar{v}_1 - i\bar{w}_1$ is $O(t)$ for $V_0 = 1$.   Here t is the maximum radial dimension of the slender configuration of unit length. For a vortex induced by a wing of semispan s_m at angle of attack α_c, Eq. (6-21) gives

$$\frac{\Gamma_0}{V_0} = 2\alpha_c s_m \tag{4-105}$$

Since the body is slender, s_m is $O(t)$ just as α_c.   Equations (4-104) and (4-105) show that the vortex strength is of the same magnitude for a slender configuration whether body-induced or wing-induced.   Thus, for $V_0 = 1$ both types of vortices produce lateral velocities $O(t)$ just as the lateral velocities without vortices.   What this means is that we can use the order-of-magnitude estimates of Chap. 3 in developing formulas for lift and sideforce due to vortices.

With reference to Eq. (3-58) and Fig. 4-22, the generalized force $\bar{Y} + i\bar{Z}$ is

$$\frac{\bar{Y} + i\bar{Z}}{q_0} = -2 \int_{S_2} \phi_r \frac{d\bar{W}}{d\bar{\mathfrak{z}}}\, dS_2 + \int_{S_2} \left(2\phi_x + \frac{dW}{d\mathfrak{z}} \frac{d\bar{W}}{d\bar{\mathfrak{z}}}\right) e^{i\theta}\, dS_2$$
$$- 2\int_{S_1} \frac{d\bar{W}}{d\bar{\mathfrak{z}}}\, dS_3 + O(t^5 \log^2 t) \tag{4-106}$$

To evaluate the forces requires a knowledge of the complex potential W_0 without vortices and W_1 due to the vortices.   The complex potential has the general form

$$W_0 = a_0 \log \mathfrak{z} + b_0 + \sum_{n=1}^{\infty} \frac{b_n}{\mathfrak{z}^n} \tag{4-107}$$

and the complex potential due to the vortex plus its image is

$$W_1 = \frac{-i\Gamma_1}{2\pi} \log (\mathfrak{z} - \mathfrak{z}_1) + W_i(\mathfrak{z}) \tag{4-108}$$

Actually, the precise form of $W_i(\mathfrak{z})$ is hard to write down in the plane unless the cross section is some simple shape like a circle.   It is easier to transform the missile cross section into a circle of radius r_e in the σ plane, while leaving the field at infinity undisturbed.   The transformations between $\mathfrak{z}$ and σ under these circumstances have the forms

$$\sigma = \mathfrak{z} + \sum_{m=1}^{\infty} \frac{c_m}{\mathfrak{z}^m} \qquad \mathfrak{z} = \sigma + \sum_{n=1}^{\infty} \frac{d_n}{\sigma^n} \qquad (4\text{-}109)$$

We can now write the complex potential W_1 in the σ plane explicitly

$$W_1(\mathfrak{z}(\sigma)) = \frac{-i\Gamma}{2\pi} \log \frac{\sigma - \sigma_1}{\sigma - \sigma_i} \qquad (4\text{-}110)$$

See Fig. 4-23.

To make the complex potential single-valued, we must put cuts into the planes. First, in $W_0(\mathfrak{z})$ there is the log $\mathfrak{z}$ term which is indeterminate

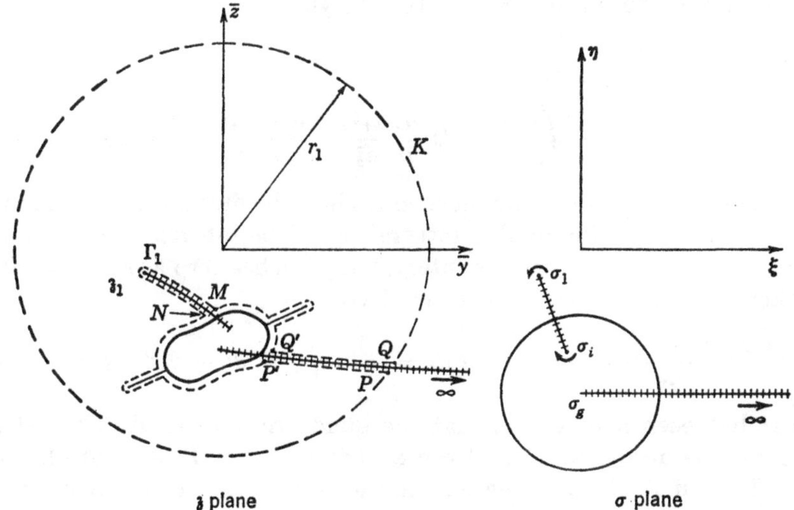

Fig. 4-23. Cuts and contours of integration.

to multiples of $2\pi i$. The logarithm term arises because of sources within the body cross section. Thus a *source cut* must extend from some point within the body to infinity as shown in Fig. 4-23. So long as no path crosses over the cut, the $W_0(\mathfrak{z})$ function will be single-valued. If any path crosses the cut, then $W_0(\mathfrak{z})$ must be increased or decreased by $2\pi i a_0$, depending on which direction the cut is crossed. If $S'(\bar{x})$ is zero, no logarithmic term occurs in $W_0(\mathfrak{z})$. Two logarithms appear in the term $W_1(\mathfrak{z})$. Actually, a *vortex cut* from σ_i to σ_1 will render $W_1(\mathfrak{z})$ single-valued. It can easily be shown that ψ_1 is continuous crossing the cut but that ϕ_1 has the value $-\Gamma/2$ on the right side of the cut and $\Gamma/2$ on the left side.

Examine now the integral over S_3 given in Eq. (4-106). The area S_3 is enclosed by the contour $QQ'MNP'PQ$, which has been chosen to cross over no cuts. Let the contour K be the outer circle of radius r_1, let C_0 be that part of the contour next to the body, let C_1 be the contour consisting of segments QQ' and $P'P$, and let C_2 be the contour segment MN. The whole internal contour from Q to P is denoted by C, and

$$C = C_0 + C_1 + C_2 \tag{4-111}$$

Applying Green's theorem to the area S_3 yields

$$\int_{S_3} \frac{d\bar{W}}{d\bar{\mathfrak{z}}}\, dS_3 = i \oint_C \phi\, d\mathfrak{z} - i \oint_K \phi\, d\mathfrak{z} \tag{4-112}$$

The contour integral about K can be simply rewritten as

$$\oint_K \phi\, d\mathfrak{z} = i \int_0^1 \oint_0^{2\pi} \frac{\partial \phi}{\partial \bar{x}}\, r_1 e^{i\theta}\, d\theta\, d\bar{x} \tag{4-113}$$

and then introduced into Eq. (4-106) to yield

$$\frac{\bar{Y} + i\bar{Z}}{q_0} = -2i \oint_C \phi\, d\mathfrak{z}$$
$$- \int_0^1 \int_0^{2\pi} \left(2 \frac{\partial \phi}{\partial r} \frac{d\bar{W}}{d\bar{\mathfrak{z}}} - \frac{dW}{d\mathfrak{z}} \frac{d\bar{W}}{d\bar{\mathfrak{z}}}\, e^{i\theta}\right) r_1\, d\theta\, d\bar{x} \tag{4-114}$$

Now $\bar{Y} + i\bar{Z}$ cannot depend on r_1, and, since the first integral is independent of r_1, so must the double integral be. The integrand of the double integral is $O(1/r_1)$ so that the integral approaches zero as $r_1 \rightarrow \infty$ with or without vortices present. We now have

$$\frac{\bar{Y} + i\bar{Z}}{q_0} = -2i \oint_C \phi\, d\mathfrak{z} = -2i \oint_C W\, d\mathfrak{z} - 2 \oint_C \psi\, d\mathfrak{z} \tag{4-115}$$

What has been achieved is that the quadratic integrand of the double integral has disappeared, and the contributions to $\bar{Y} + i\bar{Z}$ are linear in W. Thus, if $\bar{Y}_v + i\bar{Z}_v$ is the contribution due to the vortex, an expression for this quantity can be written down immediately

$$\frac{\bar{Y}_v + i\bar{Z}_v}{q_0} = -2i \oint_C W_1\, d\mathfrak{z} - 2 \oint_C \psi_1\, d\mathfrak{z} \tag{4-116}$$

The integral around C of W_1 can be distorted to K since $W_1(\mathfrak{z})$ is an analytic function in S_3, and K can be transformed into K_σ in the σ plane.

$$\oint_C W_1(\mathfrak{z})\, d\mathfrak{z} = \oint_K W_1(\mathfrak{z})\, d\mathfrak{z} = \oint_{K_\sigma} W_1(\mathfrak{z}(\sigma)) \frac{d\mathfrak{z}}{d\sigma}\, d\sigma \tag{4-117}$$

where K_σ is in a large contour into which K is transformed in the σ plane. The expansions

$$\frac{d\mathfrak{z}}{d\sigma} = 1 + O\left(\frac{1}{\sigma^2}\right)$$
$$W_1(\mathfrak{z}(\sigma)) = \frac{i\Gamma_1}{2\pi} (\sigma_1 - \sigma_i) \left(\frac{1}{\sigma}\right) + O\left(\frac{1}{\sigma^2}\right) \tag{4-118}$$

permit the contour integral easily to be evaluated by the residue theorem

$$\oint_C W_1(\mathfrak{z})\, d\mathfrak{z} = 2\pi i \left(\frac{i\Gamma_1}{2\pi}\right)(\sigma_1 - \sigma_i) = -\Gamma_1(\sigma_1 - \sigma_i) \qquad (4\text{-}119)$$

The integral around C of ψ_1 is zero because ψ_1 is constant on C_0 and is continuous across the cuts bracketed by the contours C_1 and C_2.

Equations (4-116) and (4-119) thus yield the final result

$$\frac{\bar{Y}_v + i\bar{Z}_v}{q_0} = 2i\Gamma_1(\sigma_1 - \sigma_i) \qquad (4\text{-}120)$$

If $\sigma_1 - \sigma_0$ is the point $\gamma_1 e^{i\delta_1}$ in the σ plane, the sideforce and lift are then

$$\frac{\bar{Y}_v}{q_0} = -2\Gamma_1\gamma_1\left(1 - \frac{r_e^2}{\gamma_1^2}\right)\sin \delta_1$$

$$\frac{\bar{Z}_v}{q_0} = 2\Gamma_1\gamma_1\left(1 - \frac{r_e^2}{\gamma_1^2}\right)\cos \delta_1 \qquad (4\text{-}121)$$

These simple formulas provide a means of calculating the forces due to the vortices up to any axial position in terms of the vortex positions and strengths. However, their use presupposes a knowledge of the vortex positions. Such knowledge is obtained by a step-by-step calculation of the type described in the previous section. The effects of many vortices may be found from Eq. (4-121) by superposition. Any coupling between the vortex effects enters through mutual interference between vortex paths. It is interesting to note that, if the contribution to a_1 of $W_1(\mathfrak{z})$ had been introduced into Eq. (3-62) derived on the basis of no vortices, exactly the vortex contributions found here would have arisen. Sacks[16] makes an equivalent statement. Also, Eq. (4-121) is obviously applicable to the determination of the force between any two crossflow planes due to one or more vortices, whether they originate on the missile or not.

4-10. Rolling Moment of Slender Configuration Due to Free Vortices

It is possible to derive a formula for the rolling moment developed by free vortices passing a slender configuration in terms of quantities in the plane of the base analogous to the lift and sideforce formulas of the preceding section. For convenience consider the same circumstances as those prevailing in Figs. 4-22 and 4-23, except that in Fig. 4-23 transform the body cross section so that the center of the circle falls on the origin in the σ plane. The pressure forces on control surfaces S_2 and S_3 do not contribute to the rolling moment. Only the transport of tangential momentum across areas S_2 and S_3 can cause rolling moment, and, of these, it turns out that only S_3 has a contribution. The rolling moment L' is

$$L' = +V_0^2 \int_{S_2} \rho\phi_\theta\phi_r\, dS_2 + V_0^2 \int_{S_3} \rho(1 + \phi_{\mathfrak{x}})\phi_\theta\, dS_3 \qquad (4\text{-}122)$$

with positive L' taken in the negative θ sense, $\bar{z} \to \bar{y}$.   Division by q_0 and the use of the density relationship, Eq. (3-56), yields

$$\frac{L'}{q_0} = +2 \int_{S_2} \phi_\theta \phi_r \, dS_2 + 2 \int_{S_3} \phi_\theta \, dS_3 + O(t^4 \log t) \qquad (4\text{-}123)$$

To show that the integral over S_2 contributes nothing to the rolling moment, rewrite the integral as

$$\int_{S_2} \phi_\theta \phi_r \, dS_2 = \int_0^1 dx \int_0^{2\pi} r_1 \phi_\theta \phi_r \, d\theta \qquad (4\text{-}124)$$

The general form of the potential function including vortex effects can be written in the following form convergent on a contour K enclosing the vortices

$$\phi = a_0 \log r + b_0 + \sum_{n=1}^{\infty} \frac{a_n \cos n\theta + b_n \sin n\theta}{r^n} \qquad (4\text{-}125)$$

The source cut in this case is of no importance since ϕ is continuous across the source cut.   The vortex cut is important for that part of ϕ due to vortices.   On S_2 the values of ϕ_θ and ϕ_r can be calculated by differentiation of Eq. (4-125).   If the values of ϕ_θ and ϕ_r are substituted into Eq. (4-124) and the integrations carried out, it is found that the integral is zero.

Consider now the contribution of the area S_3.   At this point let us confine our attention only to that part of the rolling moment due to the vortex.   This is now possible because the remaining integral in Eq. (4-123) is linear in ϕ.   While the rolling moment due to the vortex can be evaluated in terms of the vortex position in the base plane, all components of the flow will influence this position.   The surface integral over S_3 is taken over the area within the dashed contour in Fig. 4-23. The area integral is converted to contour integrals by means of Green's theorem

$$\int_{S_3} \phi_\theta \, dS_3 = \frac{1}{2} \int_C \phi \, d(r^2) - \frac{1}{2} \int_K \phi \, d(r^2) \qquad (4\text{-}126)$$

The contour C is composed of the part C_0 in proximity with the body, the part C_1 composed of segments PP' and $Q'Q$ about the source cut, and the part C_2 comprising segment MN about the vortex cut.   The integral around K is zero since r is a constant.   The stream function ψ_1 due to the vortex and its image has a constant value on C_0 and is continuous across C_1 and C_2, so that

$$\oint_C \psi_1 d(r^2) = 0 \qquad (4\text{-}127)$$

Thus, the integral over S_3 for the part of ϕ_θ due to the vortex can be written

$$\int_{S_1} \frac{\partial \phi_1}{\partial \theta}\, dS_3 = \frac{1}{2} \oint_C (\phi_1 + i\psi_1)d(r^2) = \frac{1}{2} \oint_C W_1(\mathfrak{z})d(\mathfrak{z}\bar{\mathfrak{z}}) \qquad (4\text{-}128)$$

The evaluation of the contour integral cannot be made directly by the residue calculus because the integrand is not analytic.   Let us transform the contour C in the $\mathfrak{z}$ plane into a circle of radius r_e in the σ plane with $\mathfrak{z}_0$ at the origin.

$$\mathfrak{z} - \mathfrak{z}_0 = \sigma + \sum_{n=1}^{\infty} \frac{d_n}{\sigma^n} = f(\sigma) \qquad (4\text{-}129)$$

The field at infinity suffers a finite translation only.   The coefficients d_n are usually complex, and the function $f(\sigma)$ can usually be written in finite form for most cases of interest.

The integral about C can be broken up into two convenient parts with the aid of the following identity:

$$\mathfrak{z}\bar{\mathfrak{z}} = (\mathfrak{z} - \mathfrak{z}_0)(\bar{\mathfrak{z}} - \bar{\mathfrak{z}}_0) + (\mathfrak{z}_0\bar{\mathfrak{z}} + \bar{\mathfrak{z}}_0\mathfrak{z}) - \mathfrak{z}_0\bar{\mathfrak{z}}_0 \qquad (4\text{-}130)$$

With the following notation

$$I_1 = \frac{1}{2} \oint_C W_1(\mathfrak{z})d(\mathfrak{z}_0\bar{\mathfrak{z}} + \bar{\mathfrak{z}}_0\mathfrak{z})$$

$$I_2 = \frac{1}{2} \oint_C W_1(\mathfrak{z})d(\mathfrak{z} - \mathfrak{z}_0)(\bar{\mathfrak{z}} - \bar{\mathfrak{z}}_0)$$

$$(4\text{-}131)$$

we see that I_2 is the contribution when $\mathfrak{z}_0 = 0$, and I_2 is the additional contribution when $\mathfrak{z}_0$ is not equal to zero.

$$\int \frac{\partial \phi_1}{\partial \theta}\, dS_3 = I_1 + I_2 \qquad (4\text{-}132)$$

Confine the analysis to the evaluation of I_1 for the present.   The integral I_1 can be written

$$I_1 = y_0 \oint_C W_1(\mathfrak{z})\, dy + z_0 \oint_C W_1(\mathfrak{z})\, dz \qquad (4\text{-}133)$$

Also, since ψ_1 is constant on C_0 and continuous across the cuts

$$\oint_C \psi_1\, dy = \oint_C \psi_1\, dz = 0 \qquad (4\text{-}134)$$

and I_1 can be written

$$I_1 = y_0\, \mathrm{RP} \oint_C W_1(\mathfrak{z})\, d\mathfrak{z} + z_0\, \mathrm{IP} \oint_C W_1(\mathfrak{z})\, d\mathfrak{z} \qquad (4\text{-}135)$$

The contour C can be transformed into the σ plane and then enlarged into a large circular contour D, centered on the origin and enclosing the body and vortex cut.   We can then expand the integrand in a series in σ and integrate term by term. In the σ plane the complex potential is

$$W_1(\mathfrak{z}(\sigma)) = \frac{-i\Gamma}{2\pi} [\log(\sigma - \sigma_1) - \log(\sigma - \sigma_i)] \tag{4-136}$$

$$\sigma_1\sigma_i = r_e^2$$

The expansion valid on D is

$$W_1(\mathfrak{z}(\sigma)) = \frac{i\Gamma}{2\pi} \left[\sum_{n=1}^{\infty} \frac{1}{n} \left(\frac{\sigma_1}{\sigma}\right)^n - \sum_{n=1}^{\infty} \frac{1}{n} \left(\frac{\sigma_i}{\sigma}\right)^n \right] \tag{4-137}$$

Thus

$$\oint_D W_1(\mathfrak{z}(\sigma))\, d\mathfrak{z} = \oint_D \frac{i\Gamma}{2\pi} \left[\sum_{n=1}^{\infty} \frac{1}{n} \left(\frac{\sigma_1}{\sigma}\right)^n - \sum_{n=1}^{\infty} \frac{1}{n} \left(\frac{\sigma_i}{\sigma}\right)^n \right]$$

$$\left(1 - \sum_{m=1}^{\infty} \frac{m d_m}{\sigma^{m+1}} \right) d\sigma \tag{4-138}$$

Since only the σ^{-1} term contributes to the integral

$$\oint_D W_1(\mathfrak{z}(\sigma))\, d\mathfrak{z} = -\Gamma(\sigma_1 - \sigma_i) \tag{4-139}$$

The value I_1 is thus

$$I_1 = -\Gamma[y_g\, \mathbf{RP}\,(\sigma_1 - \sigma_i) + z_g\, \mathbf{IP}\,(\sigma_1 - \sigma_i)] \tag{4-140}$$

From Eq. (4-120)

$$\sigma_1 - \sigma_i = \frac{\bar{Z}_v}{2q_0\Gamma} - i\frac{\bar{Y}_v}{2q_0\Gamma} \tag{4-141}$$

so that

$$I_1 = -\left(\frac{\bar{Z}_v y_g}{2q_0} - \frac{\bar{Y}_v z_g}{2q_0} \right) \tag{4-142}$$

The evaluation of I_2 requires different treatment from that of I_1. It is first decomposed into integrals over C_0 and C_2 since the source at C_1 is of no concern here.

$$I_2 = \frac{1}{2}\int_{C_0} W_1(\mathfrak{z})d(\mathfrak{z} - \mathfrak{z}_g)(\bar{\mathfrak{z}} - \bar{\mathfrak{z}}_g) + \frac{1}{2}\int_{C_2} W_1(\mathfrak{z})d(\mathfrak{z} - \mathfrak{z}_g)(\bar{\mathfrak{z}} - \bar{\mathfrak{z}}_g) \tag{4-143}$$

The integral along the vortex cut is easily evaluated since $\phi = -\Gamma/2$ on the right side and $\Gamma/2$ on the left.

$$\frac{1}{2}\int_{C_2} W_1(\mathfrak{z})d(\mathfrak{z} - \mathfrak{z}_g)(\bar{\mathfrak{z}} - \bar{\mathfrak{z}}_g) = -\frac{\Gamma}{2}\left(|\mathfrak{z}_1 - \mathfrak{z}_g|^2 - |\mathfrak{z}_M - \mathfrak{z}_g|^2 \right)$$

$$= -\frac{\Gamma}{2}\left(\lambda_1^2 - \lambda_M^2 \right) \tag{4-144}$$

The integral around C_0 is transformed into an integral about C_0', $|\sigma| = r_e$, in the σ plane.   In the σ plane generally

$$(\mathfrak{z} - \mathfrak{z}_0)(\bar{\mathfrak{z}} - \bar{\mathfrak{z}}_0) = f(\sigma)\bar{f}(\sigma) \tag{4-145}$$

and on C_0' in particular

$$(\mathfrak{z} - \mathfrak{z}_0)(\bar{\mathfrak{z}} - \bar{\mathfrak{z}}_0) = f(\sigma)\bar{f}\left(\frac{r_e{}^2}{\bar{\sigma}}\right) \tag{4-146}$$

By using the series of Eq. (4-129) we can expand the product in a Laurent series

$$(\mathfrak{z} - \mathfrak{z}_0)(\bar{\mathfrak{z}} - \bar{\mathfrak{z}}_0) = \left(\sigma + \sum_{n=1}^{\infty} \frac{d_n}{\sigma^n}\right)\left(\frac{r_e{}^2}{\sigma} + \sum_{m=1}^{\infty} \frac{\bar{d}_m \sigma^m}{r_e{}^{2m}}\right)$$

$$= \sum_{n=-\infty}^{\infty} \frac{k_n}{\sigma^n} \qquad |\sigma| = r_e \tag{4-147}$$

The coefficients k_n turn out to be

$$k_n = \sum_{m=-1}^{\infty} \frac{\bar{d}_m d_{n+m}}{r_e{}^{2m}} \qquad n \text{ positive}$$

$$\tag{4-148}$$

$$k_{-n} = \sum_{m=-1}^{\infty} \frac{d_m \bar{d}_{m-n}}{r_e{}^{2(m-n)}} \qquad n \text{ negative}$$

with $\qquad d_{-1} = 1 \qquad d_0 = 0 \qquad d_m = 0 \qquad m < 1 \tag{4-149}$

We shall confine our attention to those cases wherein the series converges on C_0', although its derivatives are of no concern.

The integral around C_0' now becomes on integration by parts

$$\frac{1}{2}\oint_{C_0'} W_1(\mathfrak{z})d(\mathfrak{z} - \mathfrak{z}_0)(\bar{\mathfrak{z}} - \bar{\mathfrak{z}}_0)$$

$$= -\frac{\Gamma}{2}\lambda_M{}^2 - \frac{1}{2}\oint_{C_0'} (\mathfrak{z} - \mathfrak{z}_0)(\bar{\mathfrak{z}} - \bar{\mathfrak{z}}_0)\, dW_1 \tag{4-150}$$

From the series expansion for $(\mathfrak{z} - \mathfrak{z}_0)(\bar{\mathfrak{z}} - \bar{\mathfrak{z}}_0)$ and that for $dW_1/d\sigma$

$$\frac{dW_1}{d\sigma} = \frac{i\Gamma}{2\pi}\left(\frac{1}{\sigma_1} + \frac{1}{\sigma} + \frac{1}{\sigma_1}\sum_{n=1}^{\infty}\left(\frac{\sigma}{\sigma_1}\right)^n + \frac{1}{\sigma}\sum_{n=1}^{\infty}\left(\frac{\sigma_i}{\sigma}\right)^n\right) \tag{4-151}$$

direct integration yields

$$\frac{1}{2}\int_{C_0'} W_1(\mathfrak{z})d(\mathfrak{z} - \mathfrak{z}_0)(\bar{\mathfrak{z}} - \bar{\mathfrak{z}}_0) = -\frac{\Gamma}{2}\lambda_M{}^2$$

$$+ \frac{\Gamma}{2}\left(\frac{k_1}{\sigma_1} + k_0 + \frac{1}{\sigma_1}\sum_{n=1}^{\infty}\frac{k_{n+1}}{\sigma_1{}^n} + \sum_{n=1}^{\infty}k_{-n}\sigma_i{}^n\right) \tag{4-152}$$

and

$$I_2 = -\frac{\Gamma}{2}(\lambda_1{}^2 - \lambda_M{}^2) + \frac{\Gamma}{2}\left[(k_0 - \lambda_M{}^2) + \sum_{n=1}^{\infty}\frac{k_n}{\sigma_1{}^n} + \sum_{n=1}^{\infty}\frac{\bar{k}_n}{\bar{\sigma}_1{}^n}\right] \qquad (4\text{-}153)$$

wherein we have made use of the relationship

$$k_{-n} = \bar{k}_n r_e{}^{-2|n|} \qquad (4\text{-}154)$$

The final result for the rolling moment is

$$\frac{L'}{q_0} = +2(I_1 + I_2)$$

$$= -\frac{\bar{Z}_v y_g - \bar{Y}_v z_g}{q_0} + \frac{\Gamma}{V_0}(k_0 - \lambda_1{}^2) + \frac{\Gamma}{V_0}\sum_{n=1}^{\infty}\left(\frac{k_n}{\sigma_1{}^n} + \frac{\bar{k}_n}{\bar{\sigma}_1{}^n}\right) \qquad (4\text{-}155)$$

where V_0 is no longer unity. It should be remembered that this result contains any moment due to the vortex generator (Fig. 4-22). The rolling moment between two crossflow planes can be found by differencing as shown in the following example.

Illustrative Example

Calculate the rolling moment due to a free vortex of strength Γ_1 as it passes a triangular wing as shown in Fig. 4-24.

This example is a case wherein the series are finite. The rolling

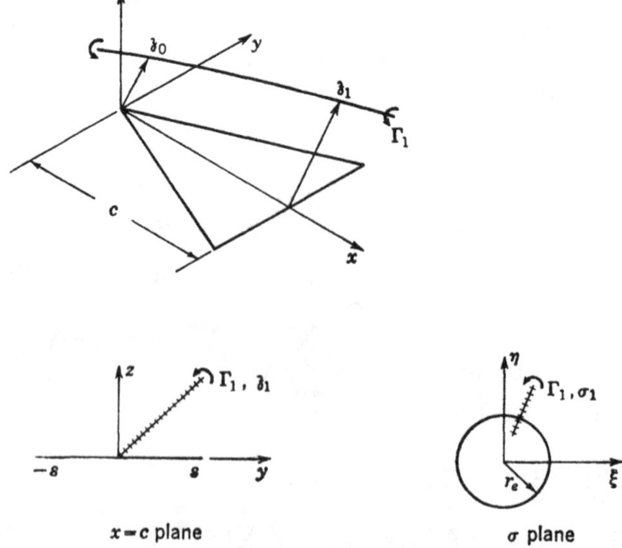

Fig. 4-24. Free vortex passing triangular wing.

moment is given from Eq. (4-155) as

$$\frac{L'}{q_0} = \left(\frac{L'}{q_0}\right)_{x=c} - \left(\frac{L'}{q_0}\right)_{x=0}$$

The transformation taking the wing cross section into a circle with center at the origin is

$$\text{\textzh} = \sigma + \frac{r_e^2}{\sigma} \qquad r_e = \frac{s}{2}$$

The values of the coefficients d_n not identically zero are

$$d_{-1} = 1 \qquad d_1 = r_e^2$$

From Eq. (4-148) the values of k_n not zero are

$$k_{-2} = 1 \qquad k_0 = 2r_e^2 \qquad k_2 = r_e^4$$

As a result the rolling moment is

$$\frac{L'}{q_0} = \frac{\Gamma_1}{V_0}\left(2r_e^2 - |\text{\textzh}_1|^2\right) + \frac{\Gamma_1 r_e^4}{V_0}\left(\frac{1}{\sigma_1^2} + \frac{1}{\bar{\sigma}_1^2}\right) + \frac{\Gamma_1}{V_0}|\text{\textzh}_0|^2$$

The quantity $\text{\textzh}_1$ is not independent of $\text{\textzh}_0$. In fact, $\text{\textzh}_1$ is determined from the initial position $\text{\textzh}_0$ by a step-by-step calculation of the vortex path.

SYMBOLS

a	mean radius of quasi-cylinder
a, b	major and minor axes of ellipse
$a_0(x)$	coefficient of log term in ϕ expansion
a_1	coefficient of r^{-1} term in ϕ expansion
$(a_1)_\alpha$	part of a_1 due to angle of attack
a_n	coefficient of r^{-n} term in ϕ expansion
$(a_1)_t$	part of a_1 due to thickness
b_0	additive function of $\bar{x}$ in ϕ expansion
B	$(M_0^2 - 1)^{1/2}$
c	length of bump in circular body
$(c_d)_c$	crossflow drag coefficient
c_m	coefficients in expansion for σ
c^2	$(a^2 - b^2)e^{-2i\varphi}$
$(C_D)_c$	drag coefficient of circular cone
$(C_D)_e$	drag coefficient of elliptical cone
C_L	lift coefficient
$C_n(p)$	functions of p
d	body diameter
d_n	coefficient in expansion for $\text{\textzh}$
D_e	drag of elliptical cone

$f_n(x)$	functions of x specifying shape of quasi-cylindrical body
$F_n(p)$	Laplace transform of $f_n(x)$
IP	imaginary part
K	caliber of tangent ogive
K_n	modified Bessel function of second kind
l	length of slender configuration
L	length of tangent ogive
L, L^{-1}	Laplace transform operator, and inverse transform operator
L'	rolling moment about $\bar{x}$ axis
M_0	free-stream Mach number
$M_{\bar{y}}$	moment about $\bar{y}$ axis, pitching moment
$M_{\bar{z}}$	moment about $\bar{z}$ axis, yawing moment
n	number of Fourier component of quasi-cylindrical body
n,m	summation indices
N	dimensionless number for viscous crossflow; also normal force
N_v	normal force due to viscous crossflow
$O(t)$	order of magnitude of t in physical sense
p	variable of plane of Laplace transform; local static pressure
p_0	free-stream static pressure
$p_1 - iq_1$	complex conjugate velocity of vortex Γ_1 in σ plane
P	pressure coefficient, $(p - p_0)/q_0$
P_α	pressure coefficient due to angle of attack
P_t	pressure coefficient due to thickness ($\alpha_c = 0$)
P^+, P^-	pressure coefficients on impact and leeward surfaces
ΔP	loading coefficient, $P^+ - P^-$
q_0	free-stream dynamic pressure
r, θ	polar coordinates
r_0	radius position of right external vortex of a symmetrical pair
r_1	radius of control surface
r_B	radius of base of body of revolution
r_e	radius of circle in σ plane
r_f, θ_f	vortex polar coordinates in Föppl equilibrium condition
r_s	local radius of body of revolution
r^*	r/r_B for tangent ogive
R	radius of curvature of tangent ogive; local radius of quasi-cylindrical body
s_m	maximum semispan of wing panel
S	body cross-sectional area
S_2, S_3	control surfaces, Fig. 4-22
S_B	base area of slender body
S_C	body planform area subject to viscous crossflow
t	maximum lateral dimension of slender configuration for unit length; time
u, v, w	perturbation velocity components along x, y, and z

u_t, v_t, w_t	perturbation velocity components due to thickness
$u_\alpha, v_\alpha, w_\alpha$	perturbation velocity components due to angle of attack
u', v', w'	perturbation velocity components along x', y', and z'
$v_0 - iw_0$	complex conjugate velocity of right external vortex of a symmetrical pair
$v_1 - iw_1$	complex conjugate velocity of vortex Γ_1 in $\mathfrak{z}$ plane
V_0	free-stream velocity
V_n	velocity of flow normal to cylinder
Vol.	volume
W	complex potential, $\phi + i\psi$
W_1	complex potential due to vortex Γ_1
W_i	complex potential due to image system of vortex Γ_1
W_t	complex potential at zero angle of attack
W_α	complex potential due to angle of attack
x, y, z	axis systems described in Sec. 1-3
x_S	axial distance to vortex separation points of body
x', y', z'	axis systems described in Sec. 1-3
x^*	$\bar{x}/L$
$\bar{x}, \bar{y}, \bar{z}$	axis systems described in Sec. 1-3
$\bar{x}_{cp}$	axial distance to center of pressure
$y_0 + iz_0$	position of right external vortex of a symmetrical pair
y_i	value of y_0 when $z_0 = 0$
y_∞	value of y_0 when $z_0 = \infty$
Y, Z	forces along y and z
Y_v, Z_v	forces due to vortex
$\bar{Y}, \bar{Z}$	forces along $\bar{y}$ and $\bar{z}$
$\mathfrak{z}$	$y + iz$
$\mathfrak{z}_0$	external position of right vortex of a symmetrical pair
$\mathfrak{z}_1$	position of vortex Γ_1
$\mathfrak{z}_g$	position of centroid of body cross section
$\mathfrak{z}_i$	internal position of right image vortex of a symmetrical pair
$\mathfrak{z}_S$	position of separation point on body surface
α_c	included angle between free-stream direction and body axis
γ_1	radial distance to vortex Γ_1 in σ plane
Γ	vortex strength
Γ_0	strength of wing circulation at root chord
Γ_1, Γ_2	strength of vortices
Γ_f	vortex strength of Föppl equilibrium position
Γ_B	strength of body vortices
δ	polar angle in construction of ellipse, Fig. 4-6; also height of bump on cylinder, Fig. 4-11
δ_1	polar angle of vortex Γ_1 in σ plane
ϵ	semiapex angle of elliptical cone in plane of major axis
ζ	variable of integration; also $z - z_g$

ρ_0	free-stream density
θ	polar angle in z plane
θ_0	polar angle of right external vortex of symmetrical pair
θ_S	polar angle of stagnation point on body
λ	an elliptical distance, Fig. 4-8
λ_1	$\|z_1 - z_0\|$
λ_M	$\|z_M - z_0\|$, Fig. 4-23
σ	variable of transformed plane
σ_1, σ_2	positions of vortices Γ_1 and Γ_2 in σ plane
σ_i	position of image vortex for Γ_1
τ	distance along tangent direction to body cross section; also ramp angle
ϕ	velocity potential
Φ	Laplace transform of ϕ
ψ	stream function for complete flow
ψ_v	stream function for vortex path
ω	semiapex angle of circular cone

REFERENCES

1. Taylor, G. I., and J. W. Maccoll: The Air Pressure on a Cone Moving at High Speeds, *Proc. Roy. Soc. London A*, vol. 139, pp. 278–311, 1933.

2. Kahane, A., and A. Solarski: Supersonic Flow about Slender Bodies of Elliptic Cross Section, *J. Aeronaut. Sci.*, vol. 20, no. 8, pp. 513–524, 1953.

3. Fraenkel, L. E.: Supersonic Flow past Slender Bodies of Elliptic Cross Section, *Brit. ARC R & M* 2954, 1955.

4. Nielsen, Jack N.: Tables of Characteristic Functions for Solving Boundary-value Problems of the Wave Equation with Application to Supersonic Interference, *NACA Tech. Notes* 3873, February, 1957.

5. Perkins, Edward W., and Leland H. Jorgensen: Comparison of Experimental and Theoretical Normal-force Distributions (including Reynolds Number Effects) on an Ogive-cylinder Body at Mach Number 1.98, *NACA Tech. Notes* 3716, May, 1956.

6. Jorgensen, Leland H., and Edward W. Perkins: Investigation of Some Wake Vortex Characteristics of an Inclined Ogive-cylinder Body at Mach Number 1.98, *NACA Research Mem.* A55E31, August, 1955.

7. Raney, D. J.: Measurement of the Cross Flow around an Inclined Body at Mach Number of 1.91, *RAE Tech. Note Aero.* 2357, January, 1955.

8. Nielsen, Jack N., and George E. Kaattari: The Effects of Vortex and Shock-expansion Fields on Pitch and Yaw Instabilities of Supersonic Airplanes, *Inst. Aeronaut. Sci. Preprint* 743, 1957.

9. Lindsey, W. F.: Drag of Cylinders of Simple Shapes, *NACA Tech. Repts.* 619, 1938.

10. Allen, H. J., and E. W. Perkins: A Study of Effects of Viscosity on Flow over Slender Inclined Bodies of Revolution, *NACA Tech. Repts.* 1048, 1951.

11. Goldstein, S.: "Modern Developments in Fluid Dynamics," vol. II, pp. 418–421, Clarendon Press, Oxford, 1938.

12. Föppl, L.: Wirbelbewegung hinter einen Kreiszylinder, *Sitzber. bayer Akad. Wiss.*, 1913.

13. Sacks, Alvin H.: Theoretical Lift Due to Wing Incidence of Slender Wing-Body-Tail Combinations at Zero Angle of Attack, *NACA Tech. Notes* 3796, 1956.

14. Milne-Thompson, L. M.: "Theoretical Hydrodynamics," 2d ed., pp. 331–332, The Macmillan Company, New York, 1950.

15. Bryson, Arthur E., Jr.: Evaluation of the Inertia Coefficients of the Cross Section of a Slender Body, *J. Aeronaut. Sci.*, vol. 21, no. 6, Readers' Forum, pp. 424–427, 1954.

16. Sacks, Alvin H.: Vortex Interference on Slender Airplanes, *NACA Tech. Notes* 3525, November, 1955.

17. Lin, C. C.: On the Motion of Vortices in Two Dimensions, University of Toronto Press, Toronto, 1943.

WING-BODY INTERFERENCE

The purpose of this chapter is to present methods for predicting the aerodynamic characteristics of configurations formed by the addition of lifting surfaces to a body. The lifting surfaces can be wing panels, empennage panels, etc., and will be termed panels for short. The primary focus here is on planar and cruciform wing-body combinations. By a *planar wing-body combination* we mean one with two wing panels, usually of the same shape and size, symmetrically disposed to the left and right sides of the missile. By a *cruciform combination*, we mean one with four panels of equal size and shape, disposed around the missile 90° apart. Configurations built up by the addition of panels of unequal size as in an empennage are treated in Chap. 10. Traditionally in airplane design the aerodynamic characteristics of the wing-body combination have been viewed as dominated by the wing as though the body were not there. For subsonic air frames where wing spans are usually large compared to the body diameter, the traditional assumption can be defended. However, the use of very small wings in comparison to the body diameter, which characterizes many missile designs, requires a different approach. The point of view is taken that neither the panels nor the body necessarily have a dominant influence on the aerodynamic characteristics of the wing-body combination. Rather, the over-all characteristics result from the body and wing acting together with mutual interference between each other.

The chapter starts in Sec. 5-1 with an enumeration of the various definitions and notations, and then in Secs. 5-2 and 5-3 takes up the subject of planar wing-body combinations for zero bank angle. The loadings, lifts, and centers of pressure are determined for the pressure fields acting on the panel and body. In Sec. 5-4 the characteristics of banked cruciform combinations are investigated. The influence of the angle of bank of the interference between panels is treated in Sec. 5-5 for both planar and cruciform configurations. In Sec. 5-6 the results are summarized for a complete wing-body configuration. The question of the application of these results to nonslender configurations and a calculative example illustrating the theoretical methods are the subjects of Sec. 5-7. Finally, the chapter concludes with a discussion of a simplified vortex model of a wing-body combination useful for such purposes as calculating the flow field about the wing-body combination.

5-1. Definitions; Notation

For purposes of wing-body interference, the *wing alone* will be taken to be the exposed wing panels joined together so that no part of the wing alone is blanketed by the body.   Thus, when the exposed wing panels disappear, so does the wing alone.   The *body alone* is the wing-body combination less the wing panels.   Actually the *precise* definitions would

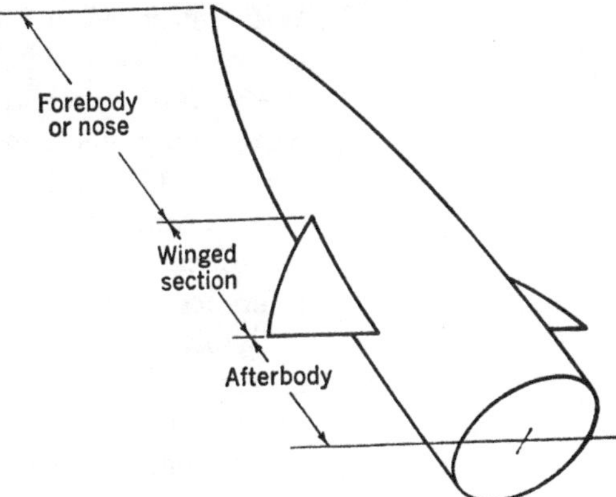

FIG. 5-1. Sections of wing-body combination.

require a specification of how the panels are parted from the body, but we will forego this refinement.   The interference can be specified once the wing-alone and body-alone definitions are specified.   The interference for any quantity is the difference between the quantity for the complete wing-body combination less the sum of the quantities for the wing alone and the body alone.   For instance—the interference potential would be

$$\phi_i = \phi_C - (\phi_W + \phi_B) \tag{5-1}$$

where the subscripts i, C, W, and B refer, respectively, to interference, combination, wing alone, and body alone.   If the wing-alone definition is changed, it is clear that the interference will change since the characteristics of the complete combination are independent of how the wing alone is defined.   The interference potential can influence part or all of the body or wing.   The values of ϕ_i at the body surface account for the effect of the wing on the body, and the values of ϕ_i at the wing surface account for the effect of the body on the wing.

The various sections of a wing-body combination are illustrated in Fig. 5-1.   For convenience, the various sections of the body are subdivided into the *forebody* in front of the wing panels, the *winged section* of the body with the wing panels, and the *afterbody* behind the trailing edge of the wing panels.

Two sets of axes are of importance in so far as forces and moments are concerned.   The axes x', y', z' correspond to the principal body axes of symmetry for $\varphi = 0$ but α_c not equal to zero.   The axes x, y, z are the principal body axes under all combinations of φ and α_c.   The forces on the body due to the wing or on the complete configuration will generally be referred to the x',y',z' system.   The force along z' is the *lift L*; the force C along y' is the *cross-wind force;* and the moment about y' is the *pitching moment*.   We will also be interested in the panel forces which, for φ not equal to zero, are not conveniently specified with respect to x', y', z' axes. With reference to Fig. 5-2, the panel normal force coefficient is denoted $(C_Z)_P$ (no panel deflection).   The panel hinge-moment coefficient is denoted C_h, and the hinge line is taken normal to the body axis at the same location as the pitching-moment reference axis.

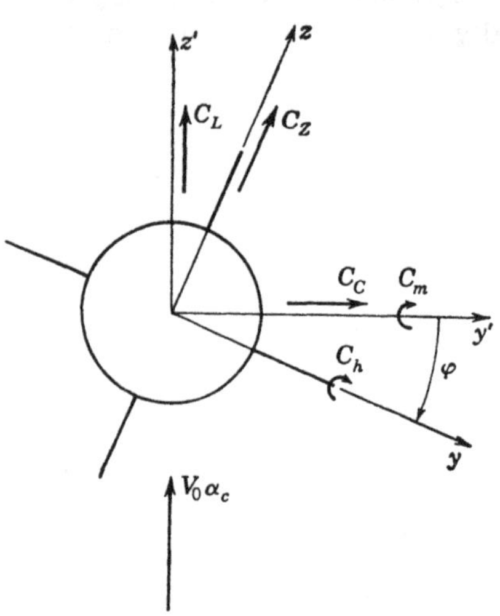

FIG. 5-2. Force and moment coefficients for panels and complete configuration.

Before consideration of the application of slender-body theory to wing-body interference, it is probably well to mention that wing-body interference problems can in certain instances be solved by full linear theory.   For rectangular wings and circular bodies, for instance, the formal boundary-value problem presented by the full linear theory has been solved.[1]   Also, another solution for part of the interference field is given by Morikawa.[2]   However, these methods are generally too complex for actual engineering use, but they do serve as useful yardsticks for evaluating more approximate but simpler engineering methods.   One such method is the essential subject matter of this chapter.   A general survey of the subject of wing-body interference has been presented by Lawrence and Flax.[3]

5-2. Planar Wing and Body Interference

The utility of slender-body theory is never better exemplified than in its application to wing-body interference.   From it we can derive the loading coefficients, span-load distribution, lift, and moment of a slender wing-body combination, as well as the components of these quantities acting on the panel and the body.   Consider a planar wing and body combination at zero angle of bank as shown in Fig. 5-3, for which the perturbation complex potential will now be constructed.   Let the body radius a and the semispan s be functions of x.   The complex potential can be separated into two parts: $W_t(\mathfrak{z})$ due to thickness, which exists at zero angle of attack, and $W_\alpha(\mathfrak{z})$ due to angle of attack. The part of the perturbation complex potential due to angle of attack is easily found from Table 2-3.

$$W_\alpha(\mathfrak{z}) = -iV_0\alpha_c \left\{\left[\left(\mathfrak{z} + \frac{a^2}{\mathfrak{z}}\right)^2 - \left(s + \frac{a^2}{s}\right)^2\right]^{\frac12} - \mathfrak{z}\right\} \qquad (5\text{-}2)$$

The complex potential due to thickness is precisely that due to the body of revolution taken to be the body alone. Thus the entire perturbation complex potential for unit V_0 is

$$W(\mathfrak{z}) = b_0(x) + a\frac{da}{dx}\log\mathfrak{z} - i\alpha_c \left\{\left[\left(\mathfrak{z} + \frac{a^2}{\mathfrak{z}}\right)^2 - \left(s + \frac{a^2}{s}\right)^2\right]^{\frac12} - \mathfrak{z}\right\} \qquad (5\text{-}3)$$

Since the wing panels have no thickness, they have no contribution to $W_t(\mathfrak{z})$.

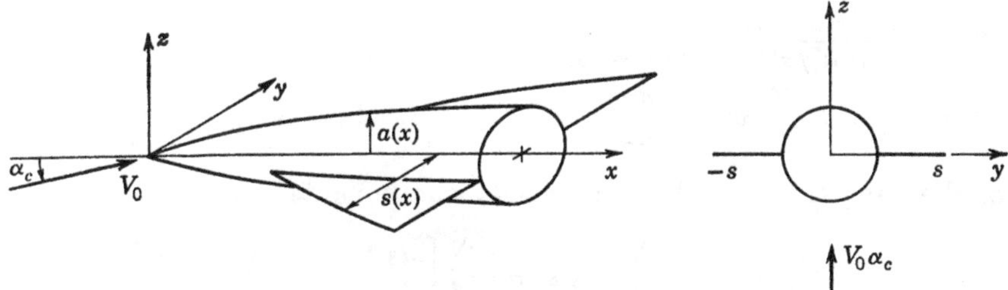

FIG. 5-3. Planar wing and body combination at zero angle of bank.

The velocity components entering the loading coefficients differ for the wing and body. The velocity components u_α, v_α, w_α are those due to $W_\alpha(\mathfrak{z})$ with V_0 of unity and α of unity. Correspondingly we have u_t, v_t, w_t due to $W_t(\mathfrak{z})$. The superscript $+$ indicates the lower impact surface, and $-$ the upper suction surface. The loading on the body is not influenced by thickness effects as discussed in connection with Eq. (4-15). Thus from Eq. (4-12)

$$\begin{aligned}(\Delta P)_{B(W)} &= (P^+ - P^-)_{B(W)} \\ &= -2\alpha(u_\alpha{}^+ + \alpha w_\alpha{}^+ - u_\alpha{}^- - \alpha w_\alpha{}^-)\end{aligned} \qquad (5\text{-}4)$$

The symmetry properties of the missile yield

$$u_\alpha{}^+ = -u_\alpha{}^- \qquad w_\alpha{}^+ = w_\alpha{}^-$$

so that
$$(\Delta P)_{B(W)} = -4\alpha u_\alpha{}^+ \qquad (5\text{-}5)$$

For the wing panel in the presence of the body, we have from Eq. (3-52)

$$\begin{aligned}(\Delta P)_{W(B)} &= (P^+_{t+\alpha} + P^-_{t+\alpha})_{W(B)} \\ &= -2[\alpha u_\alpha{}^+ + u_t{}^+ + (\alpha w_\alpha{}^+ + w_t{}^+)\alpha] \\ &\quad + 2[\alpha u_\alpha{}^- + u_t{}^- + (\alpha w_\alpha{}^- + w_t{}^-)\alpha] \\ &\quad - [(v_t{}^+ + \alpha v_\alpha{}^+)^2 + (w_t{}^+ + \alpha w_\alpha{}^+)^2] \\ &\quad + [(v_t{}^- + \alpha v_\alpha{}^-)^2 + (w_t{}^- + \alpha w_\alpha{}^-)^2]\end{aligned} \qquad (5\text{-}6)$$

which for the following symmetries in the velocity components for the wing

$$u_t{}^+ = u_t{}^- \qquad v_t{}^+ = v_t{}^- \qquad w_t{}^+ = -w_t{}^-$$
$$u_\alpha{}^+ = -u_\alpha{}^- \qquad v_\alpha{}^+ = -v_\alpha{}^- \qquad w_\alpha{}^+ = w_\alpha{}^- = -1 \tag{5-7}$$

yields
$$(\Delta P)_{W(B)} = -4\alpha u_\alpha{}^+ - 4\alpha v_\alpha{}^+ v_t{}^+ \tag{5-8}$$

We note that the wing loading has a quadratic form while the body loading does not.

The velocity components needed to obtain the loading coefficients can be obtained simply from Eq. (5-3). For the body we obtain

$$u_\alpha{}^+ = -\frac{-2a\dfrac{da}{dx}(1 + \cos 2\theta) + \left(s + \dfrac{a^2}{s}\right)\left[\dfrac{ds}{dx}\left(1 - \dfrac{a^2}{s^2}\right) + 2\dfrac{a}{s}\dfrac{da}{dx}\right]}{[(s + a^2/s)^2 - 4a^2\cos^2\theta]^{1/2}}$$

$$v_\alpha{}^+ = \frac{2a\sin 2\theta \sin\theta}{[(s + a^2/s)^2 - 4a^2\cos^2\theta]^{1/2}} \tag{5-9}$$

$$w_\alpha{}^+ = \frac{-2a\sin 2\theta \cos\theta}{[(s + a^2/s)^2 - 4a^2\cos^2\theta]^{1/2}} - 1$$

$$v_t{}^+ = \frac{da}{dx}\cos\theta \qquad w_t{}^+ = \frac{da}{dx}\sin\theta$$

For the wing the perturbation velocity components are

$$u_\alpha{}^+ = -\frac{-2a\dfrac{da}{dx}\left(1 + \dfrac{a^2}{y^2}\right) + \left(s + \dfrac{a^2}{s}\right)\left[\dfrac{ds}{dx}\left(1 - \dfrac{a^2}{s^2}\right) + 2\dfrac{a}{s}\dfrac{da}{dx}\right]}{[(s + a^2/s)^2 - y^2(1 + a^2/y^2)^2]^{1/2}}$$

$$v_\alpha{}^+ = \frac{y(1 - a^4/y^4)}{[(s + a^2/s)^2 - y^2(1 + a^2/y^2)^2]^{1/2}} \tag{5-10}$$

$$w_\alpha{}^+ = -1 \qquad v_t{}^+ = \frac{a}{y}\frac{da}{dx} \qquad w_t{}^+ = 0$$

where we have assumed that the wing has no thickness in calculating the thickness velocity components. The loadings as obtained for the velocity components are

$$(\Delta P)_{B(W)} = \frac{4\alpha}{[(1 + a^2/s^2)^2 - 4y^2/s^2]^{1/2}}$$
$$\left[\left(1 - \frac{a^4}{s^4}\right)\frac{ds}{dx} + 2\frac{a}{s}\left(1 + \frac{a^2}{s^2} - 2\frac{y^2}{a^2}\right)\frac{da}{dx}\right] \tag{5-11}$$

$$(\Delta P)_{W(B)} = \frac{4\alpha}{[(1 + a^4/s^4) - (y^2/s^2)(1 + a^4/y^4)]^{1/2}}$$
$$\left\{\left(1 - \frac{a^4}{s^4}\right)\frac{ds}{dx} + \frac{a}{s}\frac{da}{dx}\left[2\left(\frac{a^2}{s^2} - 1\right) + \left(1 - \frac{a^2}{y^2}\right)^2\right]\right\} \tag{5-12}$$

It is noted that the loadings on both wing and body depend on the expansion rate of both wing semispan and body radius. It is interesting to compare the loading for body cross sections of identical shape but for

$da/dx = 0$ and $da/dx \neq 0$. Such a comparison is made in Fig. 5-4, which shows loadings on a combination of a triangular wing and a circular cylinder, and a combination of a triangular wing and a cone. The influence of body expansion on the shape of the loadings is not important in this case.

These loadings with $da/dx = 0$ are the same as those obtained by Lennertz as a solution to a problem of minimum induced drag. The problem, one of subsonic flow, is based on Trefftz plane methods. The vortex wake is assumed to retain the general shape of the body in end

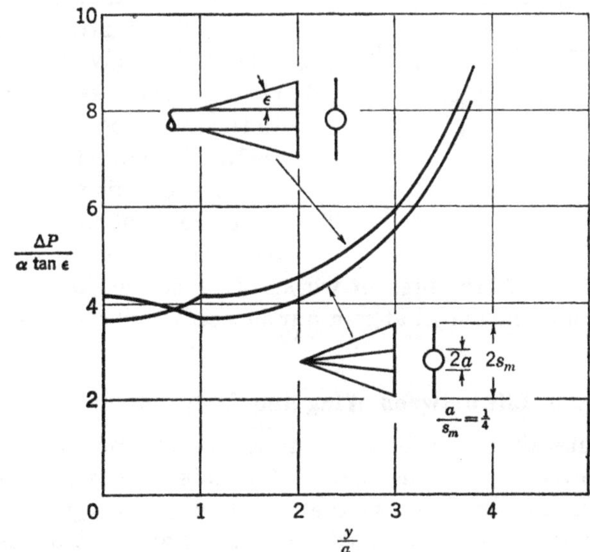

FIG. 5-4. Pressure differences at trailing edges of slender wing-body combinations.

view in moving backward to the so-called Trefftz plane. Here the criterion of minimum drag is that the vortex wake move downward undistorted. Mathematically the problem is to solve the Laplace equation for the cross section of the wake moving downward with uniform speed. It is mathematically equivalent to the present problem with no body expansion. The details of the solution are given by Durand.[4]

Consider now the total lift of the wing-body combination as given by Eq. (3-64). Let s_m be the maximum span of the combination, and let $\bar{a}$ be the accompanying body radius. Then the lift up to this axial station comes out to be

$$\frac{L_C}{q_0} = 2\pi\alpha s_m{}^2 \left(1 - \frac{\bar{a}^2}{s_m{}^2} + \frac{\bar{a}^4}{s_m{}^4}\right) \tag{5-13}$$

The lift includes that developed by the missile forebody. Actually, the total lift of the combination is given by Eq. (5-13), independent of the shape of the combination in front of the axial position for s_m, or of the shape of the wing panels behind this axial position.   The loadings given by Eqs. (5-11) and (5-12) do depend on the planform through ds/dx, but the total integrated lift does not.   If the trailing edge of the panel is normal to the flow at the axial position for s_m, then no question of lift due to additional wing area behind this position arises.   However, even if wing area with $s < s_m$ does occur behind this position, no increased lift occurs on the basis of slender-body theory.   The reason for this behavior

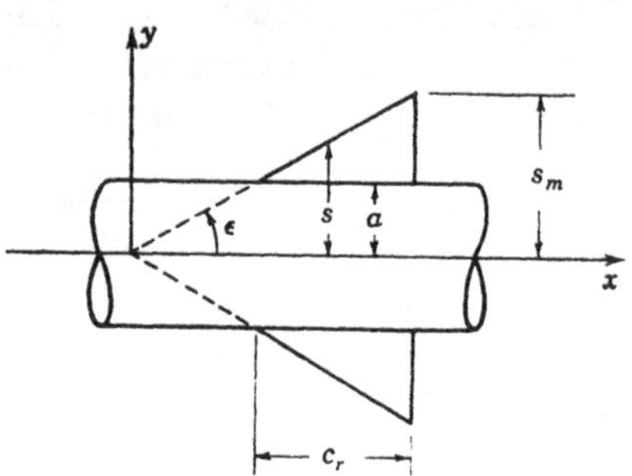

FIG. 5-5. Triangular wing and body combination.

is discussed in Sec. 7-1.   Actually, the trailing-vortex system from the panel trailing edge induces downwash on the area, which just offsets the angle of attack.   The precise role of the body expansion is not so clear.   If the body is expanding up to the axial position for s_m, then Eq. (5-13) is correct.   Body contraction aft of this position may influence the total combination lift, but a consideration of this problem is beyond the scope of the present work.   In fact, we shall *assume* that the afterbody is a circular cylinder in our succeeding discussion of wing-body interference.

5-3. Division of Lift between Wing and Body; Panel Center of Pressure

It is of interest to see how the total combination lift is distributed between the panels and the body.   For this purpose, assume that the body is a circular cylinder so that we have no body expansion term. Also, for purposes of definiteness, assume that the wing is triangular, although this assumption will shortly be relaxed.   With reference to Fig. 5-5 the lift on the panel is

$$\frac{L_{W(B)}}{q_0} = 4\alpha \tan \epsilon \int_a^{s_m} dy \int_{y/\tan\epsilon}^{s_m/\tan\epsilon} \frac{(1 - a^4/s^4)\, dx}{[(1 + a^4/s^4) - y^2/s^2(1 + a^4/y^4)]^{1/2}}$$

$$(5\text{-}14)$$

One integration yields (one panel)

$$\frac{L_{W(B)}}{q_0} = 4\alpha \int_a^{s_m} \left[\left(s_m + \frac{a^2}{s_m}\right)^2 - \left(y + \frac{a^2}{y}\right)^2\right]^{1/2} dy \qquad (5\text{-}15)$$

The integrand gives the shape of the span loading.   The span loading is very closely elliptical, as discussed in connection with Table 6-1.   Though the integration has been carried out for a panel of triangular planform, the span loading is independent of the exact shape of the panel for a slender configuration.   What follows is therefore valid for panels of general planform.   The total lift on the wing panels is conveniently expressed as a fraction of the lift developed by the wing alone L_W:

$$\frac{L_W}{q_0} = 2\pi\alpha(s_m - a)^2 \tag{5-16}$$

The lift ratio is denoted by K_W, and the value as found from Eqs. (5-15) and (5-16) is

$$K_W \equiv \frac{L_{W(B)}}{L_W}$$

$$= \frac{1}{\pi(\lambda - 1)^2}\left[\frac{\pi}{2}\left(\frac{\lambda^2 - 1}{\lambda}\right)^2 + \left(\frac{\lambda^2 + 1}{\lambda}\right)^2 \sin^{-1}\left(\frac{\lambda^2 - 1}{\lambda^2 + 1}\right)\right.$$

$$\left. - \frac{2(\lambda^2 - 1)}{\lambda}\right]$$

$$\lambda = \frac{s_m}{a} \tag{5-17}$$

The lift ratio is a function solely of a/s_m.

TABLE 5-1. SLENDER-BODY PARAMETERS FOR LOADING DUE TO PITCH*

$\dfrac{a}{s_m}$	K_W	K_B	$\left(\dfrac{\bar{x}_\alpha}{c_r}\right)_{W(B)}$ †	$\left(\dfrac{\bar{x}_\alpha}{c_r}\right)_{B(W)}$	$\dfrac{\bar{y}_\alpha - a}{s_m - a}$
0	1.000	0	0.667(⅔)	0.500(½)	0.424(4/3π)
0.1	1.077	0.133	0.657	0.521	0.421
0.2	1.162	0.278	0.650	0.542	0.419
0.3	1.253	0.437	0.647	0.563	0.418
0.4	1.349	0.611	0.646	0.581	0.417
0.5	1.450	0.800	0.647	0.598	0.417
0.6	1.555	1.005	0.650	0.613	0.416
0.7	1.663	1.227	0.654	0.628	0.418
0.8	1.774	1.467	0.658	0.641	0.420
0.9	1.887	1.725	0.662	0.654	0.422
1.0	2.000	2.000	0.667(⅔)	0.667(⅔)	0.424(4/3π)

* The accuracy of the tabulated results is estimated to be ± 0.002.
† Triangular panel.

An analogous lift ratio to K_W also serves to specify the lift on the body due to the wing:

$$K_B = \frac{L_{B(W)}}{L_W} \tag{5-18}$$

The lift on the body due to the wing is easily evaluated since

$$L_{B(W)} = L_C - L_{W(B)} - L_N \tag{5-19}$$

where the lift on the missile nose L_N is given by

$$\frac{L_N}{q_0} = 2\pi\alpha a^2 \tag{5-20}$$

The value of K_B turns out to be

$$K_B = \left(1 + \frac{a}{s_m}\right)^2 - K_W \qquad (5\text{-}21)$$

so that K_B and K_W are both functions solely of a/s_m. They are given in Table 5-1, and plotted versus a/s_m in Fig. 5-6.

The values of K_B and K_W shown in Fig. 5-6 reveal some of the salient gross facts about wing-body interference. At a value of a/s_m of zero, the value of K_W is unity because of the way in which K_W has been defined, and K_B is zero because there is no body. However, at the upper limit of a/s_m of unity, the panels are very small and are effectively mounted on an infinite reflection plane. From the potential ϕ_α given by Eq. (3-19) it is easy to see that the body produces a local angle of attack along its side edge of 2α, since the velocity here is twice the velocity of the main flow normal to the body. The wing panels therefore develop twice as much lift as they would at angle of attack α so that K_W is 2. Thus, interference of the body on the wing through upwash has increased the panel lift to twice its usual value. As a rough rule of thumb, Fig. 5-6 shows that the fractional increase in wing panel lift due to body upwash is a/s_m.

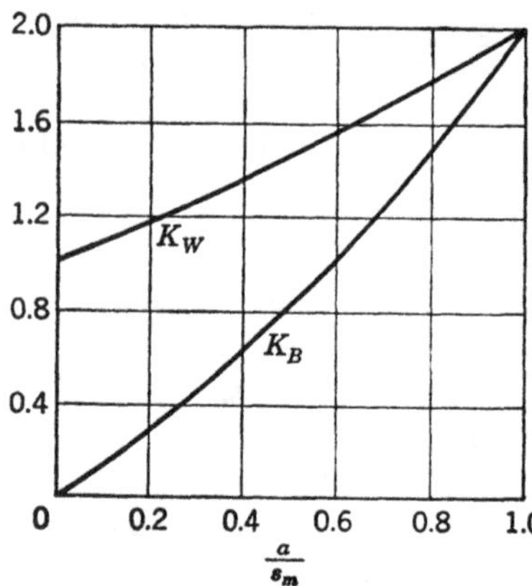

FIG. 5-6. Interference lift ratios for lift associated with pitch.

The parameter a/s_m is thus the primary measure of the importance of interference on lift.

The nature of the lift on the body due to the wing panels represented by K_B is of interest. Actually, the lift is entirely transferred or "carried over" onto the body from the wing. The wing is the primary generator of the lift, and certain of the lift is carried over onto the body because of its proximity to the wing panel. For a very small panel and a very large body that prevails as a/s_m approaches unity, there is a large expanse of body to "catch" the lift generated by the wing. This area accounts for the fact that the body "catches" as much lift as acts on the wing panels themselves, as a/s_m becomes unity. The application of the ratios K_W and K_B to nonslender configuration is shown in Sec. 5-7.

In addition to the division of lift between body and panels, the center of pressure of the panel is of some interest. The center of pressure of the lift on the body due to the wing is significantly influenced by afterbody length and is discussed in Sec. 5-6 where afterbody effects are considered. The lateral center of pressure is denoted by $(\bar{y}_\alpha)_{W(B)}$, and the longitudinal position by $(\bar{x}_\alpha)_{W(B)}$ measured behind the leading edge of the wing-body juncture. Since the shape of the span loading is given by the integrand of Eq. (5-15), it is easy to write down the expression for $(\bar{y}_\alpha)_{W(B)}$:

$$(\bar{y}_\alpha)_{W(B)} = \frac{\int_a^{s_m} [(s_m + a^2/s_m)^2 - (y + a^2/y)^2]^{1/2}\, y\, dy}{\int_a^{s_m} [(s_m + a^2/s_m)^2 - (y + a^2/y)^2]^{1/2}\, dy} \qquad (5\text{-}22)$$

yielding

$$\left(\frac{\bar{y}_\alpha}{a}\right)_{W(B)} = \frac{2}{3\pi K_W \lambda^3(\lambda - 1)^2} [4\lambda(\lambda^2 + 1)(\lambda - \lambda^2 - 1)K(k)$$
$$+ (\lambda + 1)^2(\lambda^4 + 1)E(k) + (\lambda^2 - 1)^3] \quad (5\text{-}23)$$

where $K(k)$ and $E(k)$ are complete elliptic integrals of modulus k.

$$k = \left(\frac{\lambda - 1}{\lambda + 1}\right)^2 \qquad (5\text{-}24)$$

The values of $(\bar{y}_\alpha - a)_{W(B)}/(s_m - a)$ depend solely on a/s_m and are given in Table 5-1. The lateral center of pressure does not depart significantly from the value of $4/3\pi$ that is obtained for an elliptical span loading. This result, independent of wing planform, really shows that wing-body interference does not influence the lateral center of pressure appreciably.

It can easily be shown that the streamwise center-of-pressure position is definitely not independent of panel planform, as is the lateral position. For instance, to the extent that slender-body theory can be applied to a rectangular wing panel, slender-body theory would place its center of pressure on the leading edge. It is worthwhile calculating the center-of-pressure location for a triangular wing to see what effect interference has on the location as far as its axial position is concerned. The values of $(\bar{x}_\alpha/c_r)_{W(B)}$ have been calculated from the loading of the panel as given by Eq. (5-12). The calculation is not reproduced here, but the values are given in Table 5-1. Actually, the variation in the value of $(\bar{x}_\alpha/c_r)_{W(B)}$ from the value of two-thirds for the wing alone is very small for triangular panels. In fact, the effect of interference on both the lateral and longitudinal center-of-pressure positions can be neglected for most purposes on the basis of slender-body theory.

5-4. Cruciform Wing and Body Interference

The load distribution and the lift and cross-wind forces will be calculated for a cruciform wing-body combination formed of a flat-plate wing and a circular body. Actually, the vertical panels can possess a semispan $t(x)$ different from the horizontal panels, which have semispan $s(x)$. As shown in Fig. 5-7, the configuration is pitched through α_c and banked by angle φ, so that the combination is at angle of attack $\alpha = \alpha_c \cos \varphi$ and at angle of sideslip $\beta = \alpha_c \sin \varphi$. The fact we shall use to establish the flow is that the flow field due to α will be unaltered by the presence of the vertical panels, and that due to β will be unaltered by the horizontal panels. As a result, we need only compound two flow fields for a planar wing-body combination at right angles to obtain that for a cruciform configuration. This addition follows from the fact that potential functions and flows can be added linearly in slender-body theory. We must, however, perform an analysis to see what happens to the pressure coefficient under these circumstances.

To study the pressure coefficient, let us consider the total potential function for the perturbation velocities to be composed as follows,

$$\phi = \phi_t + \alpha\phi_\alpha + \beta\phi_\beta$$

where ϕ_α and ϕ_β are the perturbation potentials for unit velocities in the α and β direction of Fig. 5-7. Then the perturbation velocities are of the form

$$u = u_t + \alpha u_\alpha + \beta u_\beta \quad (5\text{-}25)$$

The form of the pressure coefficient equation appropriate to the present problem is

$$P = -2(u + \alpha w - \beta v) - (v^2 + w^2) \quad (5\text{-}26)$$

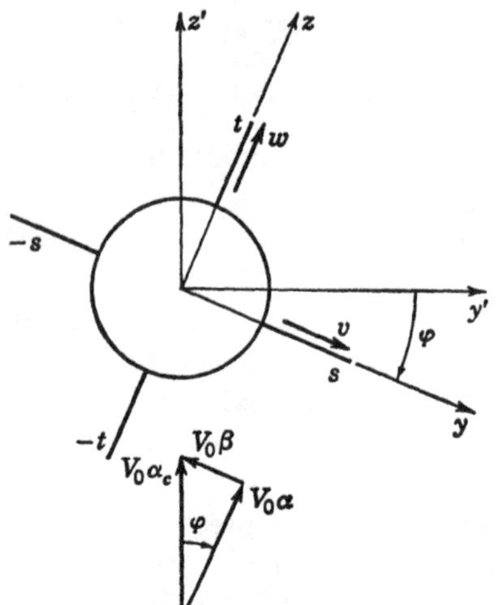

Fig. 5-7. Axis systems for cruciform missile under combined pitch and sideslip.

where the velocity perturbation components are along the x, y, z axes. Let us now apply Eq. (5-26) to calculating the loading on the right horizontal panel given by $P^+ - P^-$ or $(\Delta P)_P$. The perturbation velocity components possess the following symmetry properties and boundary values for the wing panel (with panels of no thickness):

$$
\begin{array}{lll}
u_t{}^+ = u_t{}^- & v_t{}^+ = v_t{}^- & w_t{}^+ = -w_t{}^- = 0 \\
u_\alpha{}^+ = -u_\alpha{}^- & v_\alpha{}^+ = -v_\alpha{}^- & w_\alpha{}^+ = w_\alpha{}^- = -1 \\
u_\beta{}^+ = u_\beta{}^- & v_\beta{}^+ = v_\beta{}^- & w_\beta{}^+ = -w_\beta{}^- = 0
\end{array}
\quad (5\text{-}27)
$$

From these values it is easily established that

$$(\Delta P)_P = -4\alpha u_\alpha{}^+ - 4\alpha v_\alpha{}^+ v_t{}^+ + 4\alpha\beta v_\alpha{}^+(1 - v_\beta{}^+) \quad (5\text{-}28)$$

The first two terms correspond precisely to those for a planar configuration as given by Eq. (5-8) for $\varphi = 0$. The loading of the cruciform configuration is thus the same as the planar configuration for $\varphi = 0$. For φ not equal to zero an additional term arises: a term proportional to $\alpha\beta$. This third term represents the effect of bank angle on the panel loading. Its nature is discussed in the next section for both plane and cruciform configurations, and it is termed the $\alpha\beta$ *coupling* term. Note that it is asymmetrical from left to right.

The velocity components involved in the panel loading, u_α^+, v_α^+, and v_t^+, have already been given in Eqs. (5-9) and (5-10). The remaining velocity component v_β^+ can be obtained from Eq. (5-3) by appropriately interchanging v and w.

$$v_\beta^+ = \frac{-(y/t)(1 - a^4/y^4)}{[(1 + a^4/t^4) + (y^2/t^2)(1 + a^4/y^4)]^{1/2}} + 1 \tag{5-29}$$

The complete loading for the right horizontal panel is now

$$(\Delta P)_P = \frac{4\alpha_c \cos\varphi \left\{ \left(1 - \frac{a^4}{s^4}\right)\frac{ds}{dx} + \frac{a}{s}\frac{da}{dx}\left[2\left(\frac{a^2}{s^2} - 1\right) + \left(1 - \frac{a^2}{y^2}\right)^2 \right] \right\}}{[(1 + a^4/s^4) - (y^2/s^2)(1 + a^4/y^4)]^{1/2}}$$
$$+ \frac{4(y/s)(y/t)(1 - a^4/y^4)^2\alpha_c^2 \sin\varphi \cos\varphi}{\left[\left(1 + \frac{a^4}{s^4}\right) - \frac{y^2}{s^2}\left(1 + \frac{a^4}{y^4}\right) \right]^{1/2} \left[\left(1 + \frac{a^4}{t^4}\right) + \frac{y^2}{t^2}\left(1 + \frac{a^4}{y^4}\right) \right]^{1/2}} \tag{5-30}$$

Because of the second term the normal force on the right panel is increased as it moves downward with positive φ, and the upgoing left panel loses the same amount of normal force. It is clear that the loading can be obtained on any panel from Eq. (5-30) by changing bank angle or interchanging s and t.

A similar analysis of the loading can be carried out for the body. However, the boundary conditions for the body result in different relationships for the velocity components than for the wing panels. With the symmetry properties of Eq. (5-27) (but not the boundary values), the loading on the body becomes

$$(\Delta P)_{B(W)} = -4\alpha u^+ - 4\alpha(w_t^+ + v_\alpha^+ v_t^+ + w_\alpha^+ w_t^+)$$
$$- 4\alpha\beta(w_\beta^+ - v_\alpha^+ + v_\alpha^+ v_\beta^+ + w_\alpha^+ w_\beta^+) \tag{5-31}$$

The second term is zero since the velocity in the crossflow plane due to thickness $v_t + iw_t$ is perpendicular to the velocity due to α, $[\alpha v_\alpha^+ + i\alpha(1 + w_\alpha^+)]$. The first term exists at all bank angles and is the same term given by Eq. (5-5) for a planar wing. The third term is the coupling term for the body loading analogous to that for the panels. The velocity components v_α^+ and w_α^+ occurring in the coupling term are given by Eq. (5-9) as for a planar configuration. The velocity components v_β^+ and w_β^+ are easily obtained from symmetry considerations from the results for v_α^+ and w_α^+.

$$v_\beta^+ = -\frac{2a \sin\theta \sin 2\theta}{[(t + a^2/t)^2 - 4a^2 \sin^2\theta]^{1/2}} + 1$$
$$w_\beta^+ = \frac{2a \sin 2\theta \cos\theta}{[(t + a^2/t)^2 - 4a^2 \sin^2\theta]^{1/2}} \tag{5-32}$$

The body loading is

$$(\Delta P)_{B(W)} = \frac{4\alpha_c \cos\varphi \left[\left(1 - \frac{a^4}{s^4}\right)\frac{ds}{dx} + 2\frac{a}{s}\left(1 + \frac{a^2}{s^2} - 2\frac{y^2}{a^2}\right)\frac{da}{dx}\right]}{[(1 + a^2/s^2)^2 - 4y^2/s^2]^{1/2}}$$
$$+ \frac{64(y/s)(y/t)(1 - y^2/a^2)\alpha_c{}^2 \cos\varphi \sin\varphi}{\left[\left(1 + \frac{a^2}{s^2}\right)^2 - 4\frac{y^2}{s^2}\right]^{1/2}\left[\left(1 - \frac{a^2}{t^2}\right)^2 + 4\frac{y^2}{t^2}\right]} \qquad (5\text{-}33)$$

The body loading contains in the first term a part proportional to rate of body expansion and a part proportional to rate of change of wing semispan. However, neither of these quantities influences the loading associated with combined pitch and sideslip.

With regard to the total forces on a cruciform configuration, it has been noted that the coupling term proportional to $\alpha\beta$ caused as much increase in loading on the right panel for positive φ as it caused decrease on the left panel. Likewise, the coupling term in Eq. (5-33) causes a similar situation with respect to the right and left halves of the body. In consequence the coupling terms produce no net lift but only cause an asymmetrical loading. The total force on the configuration can be calculated by adding the forces due to two planar configurations at right angles just as the flow can be similarly constructed. This is true since the gross forces are independent of the coupling terms. Thus, the force Z along the z axis is from Eq. (5-13)

$$Z = 2\pi\alpha s_m{}^2\left(1 - \frac{a^2}{s_m{}^2} + \frac{a^4}{s_m{}^4}\right)$$
$$Y = -2\pi\beta t_m{}^2\left(1 - \frac{a^2}{t_m{}^2} + \frac{a^4}{t_m{}^4}\right) \qquad (5\text{-}34)$$

The lift in the z' direction (Fig. 5-7) is

$$L = Z \cos\varphi - Y \sin\varphi$$

which for a true cruciform configuration, $s_m = t_m$, becomes

$$L = 2\pi\alpha_c s_m{}^2\left(1 - \frac{a^2}{s_m} + \frac{a^4}{s_m{}^4}\right) \qquad (5\text{-}35)$$

and the cross-wind force along the y' axis is zero. It is seen that changing the bank orientation for a constant value of α_c does not change the lift force, nor does it develop any cross-wind force. Thus, as the missile rolls, it will continue to develop lift in the plane defined by the relative wind and the missile axis. This characteristic, an important property of the cruciform configuration, is also true of the triform configuration.

The aerodynamics of slender cruciform configurations have been studied by Spreiter and Sacks.[5]

5-5. Effect of Angle of Bank on Triangular Panel Characteristics; Panel-Panel Interference

For a cruciform or planar wing-body combination, banking the missile in a positive direction introduces an additional loading on the right panel proportional to $\alpha\beta$, and subtracts a like loading from the left panel. The amount of asymmetric panel loading so produced is given for planar and cruciform configurations by the third term of Eq. (5-28). For cruciform configurations, the loading is given explicitly by the second term of Eq. (5-30). It is the purpose of this section to evaluate the asymmetrical panel normal forces due to bank angle. The difference in the results for the planar and cruciform configurations is an illustration of panel-panel interference.

Consider now the second term in Eq. (5-30) for the loading and designate it by P_φ.

$$P_\varphi = \frac{4\alpha_c{}^2 \sin\varphi \cos\varphi(\delta^2 - 1)^2\tau}{\delta^2(\tau^2 - \delta^2)^{\frac{1}{2}}(\delta^2\tau^2 - 1)^{\frac{1}{2}}} \tag{5-36}$$

with

$$\tau = \frac{s^2}{a^2} \qquad \delta = \frac{y^2}{a^2} \tag{5-37}$$

With the notation of Fig. 5-5, the total normal force on the right panel due to P_φ is

$$\begin{aligned}
\frac{\Delta Z_P}{q_0} &= \frac{1}{\tan\epsilon} \int_a^{s_m} dy \int_y^{s_m} P_\varphi \, ds \\
&= \frac{a^2\alpha_c{}^2 \sin\varphi \cos\varphi}{\tan\epsilon} \int_1^{(s_m/a)^2} \frac{(\delta^2 - 1)^2}{\delta^{\frac{5}{2}}} \, d\delta \\
&\qquad\qquad \int_\delta^{(s_m/a)^2} \frac{\tau^{\frac{1}{2}} \, d\tau}{(\tau^2 - \delta^2)^{\frac{1}{2}}(\delta^2\tau^2 - 1)^{\frac{1}{2}}}
\end{aligned} \tag{5-38}$$

The integration with respect to τ yields elliptic integrals

$$\frac{\Delta Z_P}{q_0} = \frac{a^2\alpha_c{}^2 \sin\varphi \cos\varphi}{\tan\epsilon} \int_1^{(s_m/a)^2} \frac{(\delta^2 - 1)^{\frac{3}{2}}}{2^{\frac{1}{2}}\delta^3} [F(\psi_1,k) + F(\psi_2,k')] \, d\delta \tag{5-39}$$

wherein

$$\begin{aligned}
\cos\psi_1 &= \frac{(s_m/a)(\delta - 1)}{(s_m{}^2/a^2 - 1)\delta^{\frac{1}{2}}} & k^2 &= \frac{(\delta + 1)^2}{2(\delta^2 + 1)} \\
\cos\psi_2 &= \frac{(s_m/a)(\delta + 1)}{(s_m{}^2/a^2 + 1)\delta^{\frac{1}{2}}} & k'^2 &= \frac{(\delta - 1)^2}{2(\delta^2 + 1)}
\end{aligned} \tag{5-40}$$

A further integration to obtain the panel normal force appears formidable, and the evaluation has actually been performed numerically. The normal force Z_P is conveniently made nondimensional in such a way that it is specified by a lift ratio K_φ depending only on s_m/a.

$$K_\varphi = \frac{\Delta Z_P}{Z_P} \frac{\tan\epsilon}{\beta} \tag{5-41}$$

Here Z_P is the normal force on the panel as a part of the wing alone at angle of attack α.

$$\frac{Z_P}{q_0} = \pi(s_m - a)^2 \alpha \tag{5-42}$$

Equations (5-38), (5-41), and (5-42) give

$$K_\varphi = \frac{1}{\pi(s_m/a - 1)^2} \int_1^{(s_m/a)^2} \frac{(\delta^2 - 1)^{3/2}}{2^{1/2}\delta^3} [F(\psi_1,k) + F(\psi_2,k')] \, d\delta \tag{5-43}$$

for a cruciform wing-body combination.

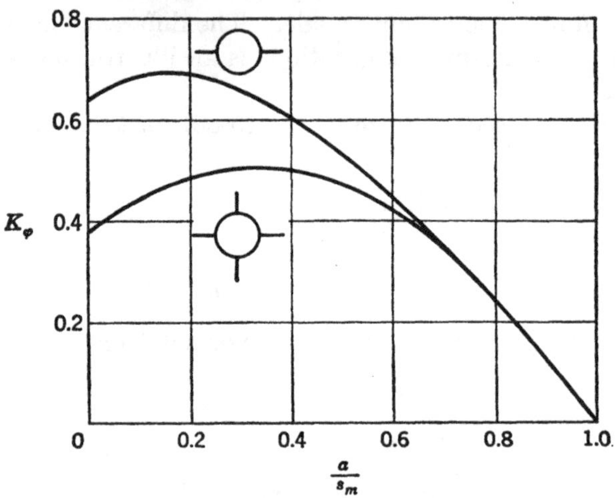

FIG. 5-8. Interference lift ratios for loading associated with bank angle.

For a planar wing-body combination the expression for K_φ can be determined in the same fashion as for a cruciform combination. The loading coefficient due to $\alpha\beta$ is obtained from the second term of Eq. (5-30) with $t = a$. The equation for K_φ is

$$K_\varphi = \frac{2}{\pi(s_m/a - 1)^2} \int_1^{(s_m/a)^2} \frac{(\delta - 1)(\delta^2 - 1)}{\delta^3}$$
$$\tanh^{-1} \left(\frac{s_m{}^2/a^2 - \delta}{s_m{}^2/a^2 - 1/\delta} \right)^{1/2} d\delta \tag{5-44}$$

The values of K_φ are tabulated in Table 5-2 for ready use and are plotted in Fig. 5-8. The difference in K_φ between the planar and cruciform cases is associated with a form of panel-panel interference. In Eq. (5-28) it is seen that the force associated with K_φ depends on a coupling between the sidewash velocities v_α and v_β due to angles of attack and sideslip. The presence of the vertical panels between the horizontal panels in the cruciform apparently has the gross effect of diminishing the coupling and reducing the value of K_φ. In the illustrative example which follows, the nature of K_φ for a triangular wing will be explained, but first let us consider the center of pressure of the loading associated with K_φ.

The centers of pressure associated with the K_φ loadings of planar and cruciform combinations with triangular wings have been calculated numerically and are listed in Table 5-2. The centers of pressure so

TABLE 5-2. SLENDER-BODY PARAMETERS FOR LOADING DUE TO BANK; TRIANGULAR PANELS*

$\dfrac{a}{s_m}$	Planar			Cruciform		
	K_φ	$\left(\dfrac{\bar{x}_\varphi}{c_r}\right)_{W(B)}$	$\dfrac{\bar{y}_\varphi - a}{s_m - a}$	K_φ	$\left(\dfrac{\bar{x}_\varphi}{c_r}\right)_{W(B)}$	$\dfrac{\bar{y}_\varphi - a}{s_m - a}$
0	0.637(2/π)	0.667(⅔)	0.524(π/6)	0.382	0.667(⅔)	0.556
0.1	0.687	0.667	0.518	0.447	0.654	0.532
0.2	0.681	0.677	0.531	0.490	0.660	0.530
0.3	0.649	0.688	0.546	0.508	0.673	0.540
0.4	0.597	0.699	0.560	0.502	0.687	0.554
0.5	0.529	0.709	0.575	0.471	0.700	0.569
0.6	0.447	0.719	0.588	0.417	0.714	0.585
0.7	0.352	0.729	0.601	0.342	0.725	0.598
0.8	0.246	0.736	0.614	0.244	0.734	0.612
0.9	0.128	0.744	0.616	0.127	0.743	0.625
1.0	0	0.750(¾)	0.637(2/π)	0	0.750(¾)	0.637(2/π)

* The accuracy of the tabulated results is estimated to be ± 0.002.

calculated are useful for predicting the variation with bank angle of the rolling moments and root bending moments as well as the panel hinge moments. Comparison of Tables 5-1 and 5-2 shows that the migration of panel center-of-pressure position with a/s_m is much greater for the K_φ panel force than for the K_W panel force. No integrated results are presented for the loading on the body which is asymmetrical with respect to the vertical plane of symmetry. It is clear that the body loading has no net effect on body normal force, rolling moment, or pitching moment.

Illustrative Example

Let us examine the variation of the force on the panel of a triangular wing as it sideslips at constant angle of attack. Calculate the fractional change in the panel force, and compare it with the change computed on the basis of slender-body theory using K_φ.

With regard to Fig. 5-9 consider a triangular wing with no thickness of semiapex angle ϵ, at angle of attack α and angle of sideslip β. The pressure of distribution on the wing is conical with respect to the apex, and the loading of the right panel is greater on the average than that of the left panel for positive sideslip. The change in the panel force with sideslip can be calculated on the basis of linear theory from the results of

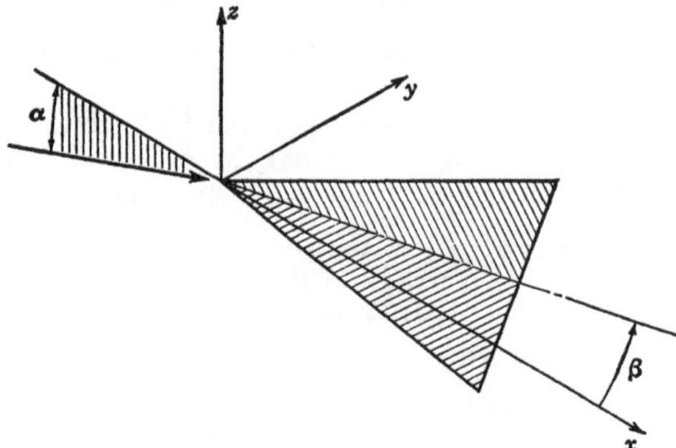

Fɪɢ. 5-9. Triangular lifting surface at combined pitch and sideslip.

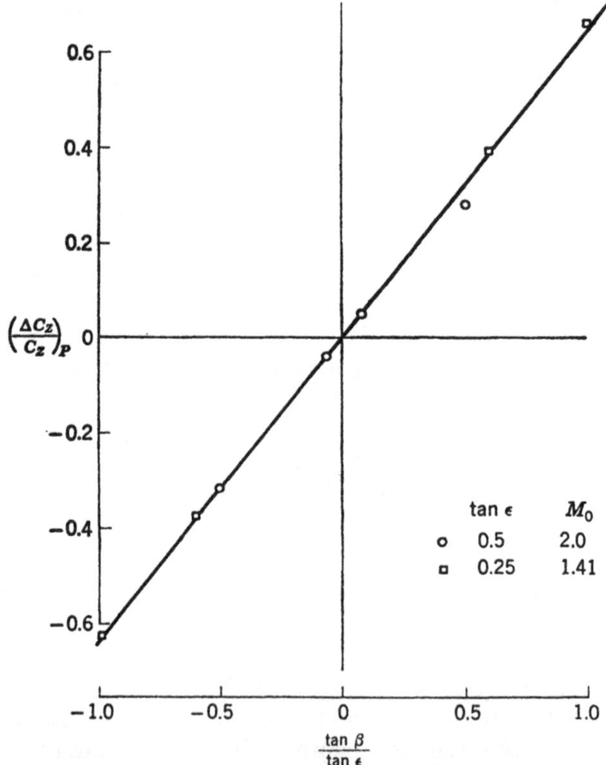

Fɪɢ. 5-10. Change in loading of panel of triangular wing due to sideslip.

A. L. Jones and A. Alksne.[9]   The results for the pressure distribution
have been integrated to obtain the panel normal force coefficient $(C_Z)_P$.
Let $(\Delta C_Z)_P$ be the change in force coefficient due to changing the angle of
sideslip for 0 to β.   Then $(\Delta C_Z/C_Z)_P$ is the fractional change in panel
force due to sideslip.   Normalize the sideslip angle by forming the param-

eter $\tan \beta / \tan \epsilon$. Then, for a value of the parameter of unity, the left side edge is streamwise. The values for the triangular wings are shown in Fig. 5-10 for two different conditions. For $\tan \epsilon = 0.5$ and $M_0 = 2.0$, the right edge becomes supersonic for a few degrees of sideslip. Actually the force gained by the left panel is not precisely counteracted by the force lost by the left panel, but the balance is nearly precise.

Let us now apply the K_φ factor to calculate the panel force on the basis of slender-body theory. From Eq. (5-41) the force coefficient ratio is

$$\left(\frac{\Delta C_Z}{C_Z}\right)_P = \frac{\beta K_\varphi}{\tan \epsilon} \tag{5-45}$$

Let us substitute $\tan \beta$ for β so that

$$\left(\frac{\Delta C_Z}{C_Z}\right)_P = K_\varphi \frac{\tan \beta}{\tan \epsilon} \tag{5-46}$$

The meaning of K_φ is now clear since it is the slope of the curve shown in Fig. 5-10. For a planar wing Table 5-2 yields a value of K_φ of $2/\pi$. The straight line shown in this figure has this slope and therefore represents slender-body theory. It is surprising that slender-body theory fits the results of linear theory so well when the large semiapex angles and angles of sideslip are considered.

5-6. Summary of Results; Afterbody Effects

The previous results apply as derived to slender planar and cruciform wing-body combinations. It is the purpose of this section to gather together the results into a compact form for application to actual missiles. The formulas are illustrated by application to a cruciform missile under a banked condition in the next section. Before summarizing the results, let us note that the panel forces and moments are not all referred to the same axes as the forces and moments of the other components. The two axis systems and the corresponding notations are given in Fig. 5-2. For simplicity, the hinge axis of the panel is assumed to have the same longitudinal position as the center of moments. Transfer of hinge moments to any other axis can easily be made. The results for the right panel apply to all panels since the bank angle is arbitrary.

Planar Configuration

Forces and moments of right panel:

$$(C_Z)_P = K_W \left(\frac{1}{2}\frac{dC_L}{d\alpha}\right)_W \alpha_c \cos\varphi + \frac{K_\varphi}{\tan\epsilon}\left(\frac{1}{2}\frac{dC_L}{d\alpha}\right)_W \alpha_c^2 \sin\varphi \cos\varphi$$

$$(C_h)_P = -K_W \left(\frac{1}{2}\frac{dC_L}{d\alpha}\right)_W \frac{(\bar{x}_\alpha)_{W(B)}}{l_r} \alpha_c \cos\varphi$$

$$- \frac{K_\varphi}{\tan\epsilon}\left(\frac{1}{2}\frac{dC_L}{d\alpha}\right)_W \frac{(\bar{x}_\varphi)_{W(B)}}{l_r} \alpha_c^2 \sin\varphi \cos\varphi \tag{5-47}$$

$$(C_l)_P = -K_W \left(\frac{1}{2}\frac{dC_L}{d\alpha}\right)_W \frac{(\bar{y}_\alpha)_{W(B)}}{l_r} \alpha_c \cos \varphi$$

$$- \frac{K_\varphi}{\tan \epsilon}\left(\frac{1}{2}\frac{dC_L}{d\alpha}\right)_W \frac{(\bar{y}_\varphi)_{W(B)}}{l_r} \alpha_c{}^2 \sin \varphi \cos \varphi$$

Forces and moments on body due to wing:

$$(C_L)_{B(W)} = K_B \left(\frac{dC_L}{d\alpha}\right)_W \alpha_c \cos^2 \varphi$$

$$(C_C)_{B(W)} = K_B \left(\frac{dC_L}{d\alpha}\right)_W \alpha_c \sin \varphi \cos \varphi \tag{5-48}$$

$$(C_m)_{B(W)} = -K_B \left(\frac{dC_L}{d\alpha}\right)_W \frac{(\bar{x}_\alpha)_{B(W)}}{l_r} \alpha_c \cos^2 \varphi$$

$$(C_l)_{B(W)} = 0$$

Forces and moments of complete configuration:

$$(C_L)_C = (C_L)_N + (C_L)_{B(W)} + K_W \left(\frac{dC_L}{d\alpha}\right)_W \alpha_c \cos^2 \varphi$$

$$(C_C)_C = (C_C)_N + (C_C)_{B(W)} + K_W \left(\frac{dC_L}{d\alpha}\right)_W \alpha_c \sin \varphi \cos \varphi \tag{5-49}$$

$$(C_m)_C = (C_m)_N + (C_m)_{B(W)} - K_W \left(\frac{dC_L}{d\alpha}\right)_W \frac{(\bar{x}_\alpha)_{W(B)}}{l_r} \alpha_c \cos^2 \varphi$$

$$(C_l)_C = -K_W \left(\frac{dC_L}{d\alpha}\right)_W \frac{(\bar{y}_\alpha)_{W(B)}}{l_r} \alpha_c \cos \varphi$$

Cruciform Configuration

Forces and moments on right panel:

$$(C_Z)_P$$
$$(C_h)_P$$
$$(C_l)_P \qquad \text{Same analytical form as Eq. (5-47)}$$

Forces and moments on body due to wing:

$$(C_L)_{B(W)} = K_B \left(\frac{dC_L}{d\alpha}\right)_W \alpha_c$$

$$(C_C)_{B(W)} = 0$$

$$(C_m)_{B(W)} = -K_B \left(\frac{dC_L}{d\alpha}\right)_W \frac{(\bar{x}_\alpha)_{B(W)}}{l_r} \alpha_c \tag{5-50}$$

$$(C_l)_{B(W)} = 0$$

Forces and moments on complete configuration:

$$(C_L)_C = (C_L)_N + (K_B + K_W)\left(\frac{dC_L}{d\alpha}\right)_W \alpha_c$$

$$(C_C)_C = 0$$

$$(C_m)_C = (C_m)_N - \frac{[K_B(\bar{x}_\alpha)_{B(W)} + K_W(\bar{x}_\alpha)_{W(B)}]}{l_r}\left(\frac{dC_L}{d\alpha}\right)_W \alpha_c \tag{5-51}$$

$$(C_l)_C = 0$$

The quantities due to the missile nose can be calculated by any method applicable to bodies alone.

The results for the over-all forces and moments on the cruciform missile show several interesting properties. First, the resultant force is independent of bank angle in magnitude and direction, being always in the plane of α_c. Second, the rolling moments of the individual panels add up to zero. These two factors produce an air frame, the characteristics of which are independent of bank attitude in contrast to a planar wing-body combination. The technological importance of the cruciform configuration is associated in part with this result.

Before discussing the application of the foregoing formulas to an actual nonslender case, let us be concerned with the values of the lift ratios and centers of pressure to be used in the theory as given in Tables 5-1 and 5-2. Actually three lift ratios are concerned: K_φ, K_W, and K_B. The values of K_W and K_B as derived do not depend on the wing planform although K_φ does. Nevertheless, as a first approximation to the $\alpha\beta$ coupling term, it is believed that K_φ can be used for panels other than triangular. With

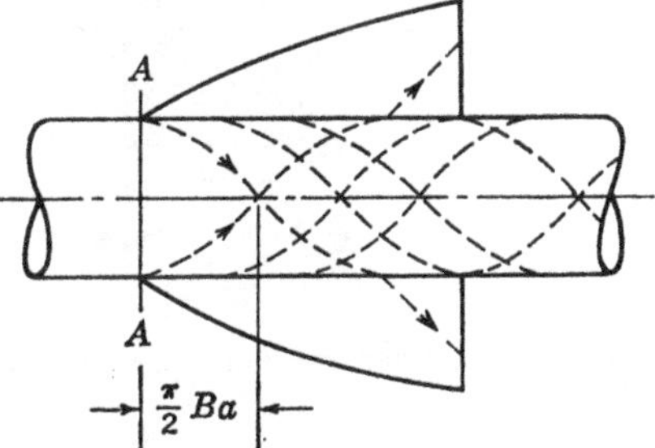

FIG. 5-11. Transference of lift from wing to body.

regard to the panel center of pressure, the values of $\bar{x}_\alpha$ and $\bar{y}_\alpha$ are very close to the values for the wing alone for the triangular panels considered in the derivation. Actually, the value of $\bar{y}_\alpha$ does not depend on the planform and should apply to panels other than triangular. For panels other than triangular, it is recommended that the center of pressure of the wing alone be used for $\bar{x}_\alpha$, since wing-body interference has little effect on panel center of pressure for a triangular panel. For rectangular panels some linear theory calculations are given by Pitts et al.[6] to show the effect of wing-body interference on x_α for a rectangular panel. At worst, interference causes a few per cent forward shift. The values of the center of lift on the body due to the wing are open to some criticism when calculated by slender-body theory in certain instances. Let us now consider afterbody effects from the point of view of K_B and $(\bar{x}_\alpha)_{B(W)}$.

For slender configurations, the length of the body behind the wing should not have an important effect on the body lift or center of pressure. However, for nonslender configurations, the existence of an afterbody can have a large influence on the values of K_B or $(\bar{x}_\alpha)_{B(W)}$. In slender-body theory it is assumed that the loading on the body due to the wing carries straight across the body diameter along AA as shown in Fig. 5-11. Actually the pressure waves travel around the body and follow the helices intersecting the parallel generators of the body at the Mach angle.

The pressures on the body are thus transferred a distance downstream anywhere from zero at the juncture to $\pi Ba/2$ on the top of the body. The importance of this effect depends among other things on afterbody length and Mach number. Behind the Mach helices from the wing trailing edge, the wave system from the trailing edge causes a decrease in afterbody loading. A swept trailing edge further complicates the problem. An approximate model for calculating the loading and center of pressure on the body is shown in Fig. 5-12. The body is assumed to be planar and to act at zero angle of attack to "catch" the lift developed by

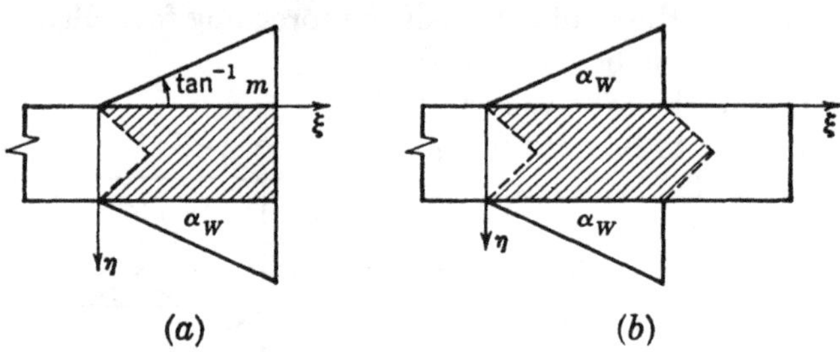

Fig. 5-12. Planar models for calculating afterbody effects. (a) No afterbody; (b) afterbody.

the wing. If no afterbody is present, the loading on the body is integrated only over the region in front of the trailing edge. However, if the trailing-edge Mach waves intersect on the afterbody, the region in front of the waves is considered to be effective in lift.

The pressure field due to either panel is considered to be the pressure field of the isolated panel. With reference to the coordinate system of Fig. 5-12, the pressure field for a supersonic edge is (Eq. 2-33)

$$P = \frac{2\alpha_W m}{\pi(m^2 B^2 - 1)^{1/2}} \cos^{-1} \frac{\xi/B + mB\eta}{\eta + m\xi} \tag{5-52}$$

and for a subsonic edge[7] is

$$P = \frac{4\alpha_W (Bm)^{3/4}}{\pi B(mB + 1)} \left(\frac{\xi/B - \eta}{m\xi + \eta} \right)^{1/2} \tag{5-53}$$

In the application of these fields to the wing-body combination, it has been assumed that the Mach wave from the leading edge of the wing tip falls behind the trailing edge of the wing-body juncture. This assumption, which insures that no tip effects fall on the lift-catching areas, leads to the condition

$$BA(1 + \lambda)\left(1 + \frac{1}{mB}\right) \geq 4$$

The values of K_B and $(\bar{x}_\alpha/c_r)_{B(W)}$ calculated on the basis of the planar models are shown in Figs. 5-13 and 5-14. It is apparent that the effect

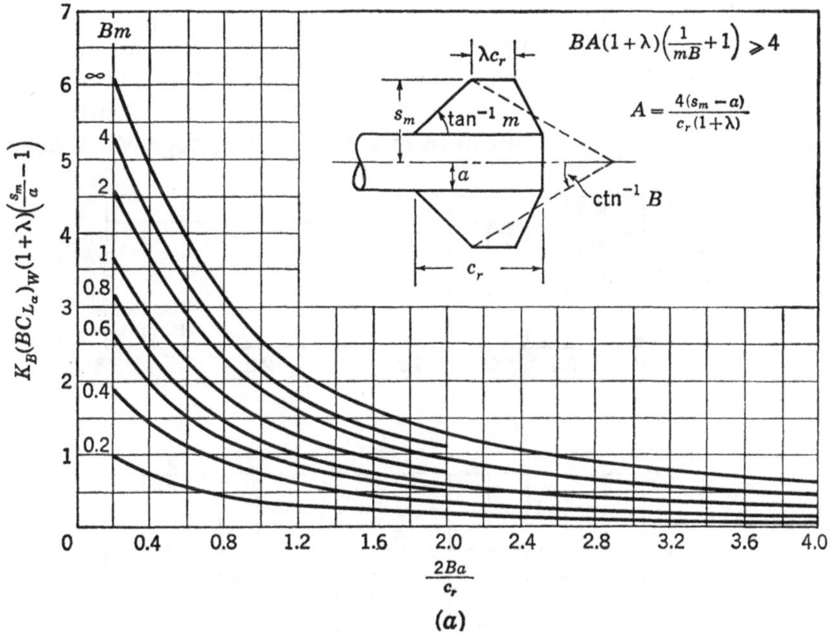

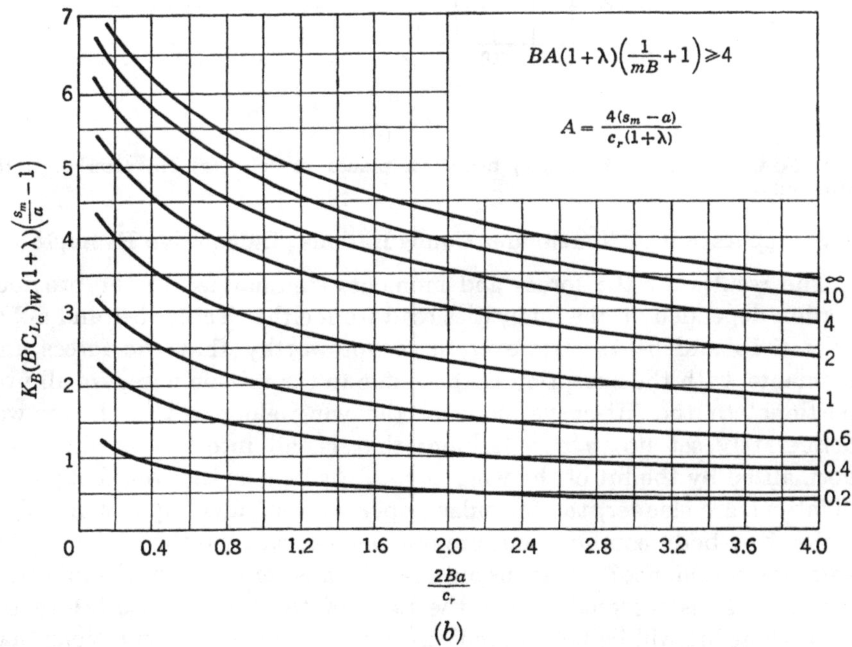

FIG. 5-13. Values of K_B based on planar model. (a) No afterbody; (b) afterbody

of the afterbody depends principally on the value of the parameter $2Ba/c_r$. For large values of the parameter the large lifting-catching area behind the wing trailing edge causes larger values of K_B and more rearward positions of the center of pressure. The importance of afterbody effects increases with Mach number if the afterbody is sufficiently long so that more afterbody area falls in front of the trailing-edge Mach helices.

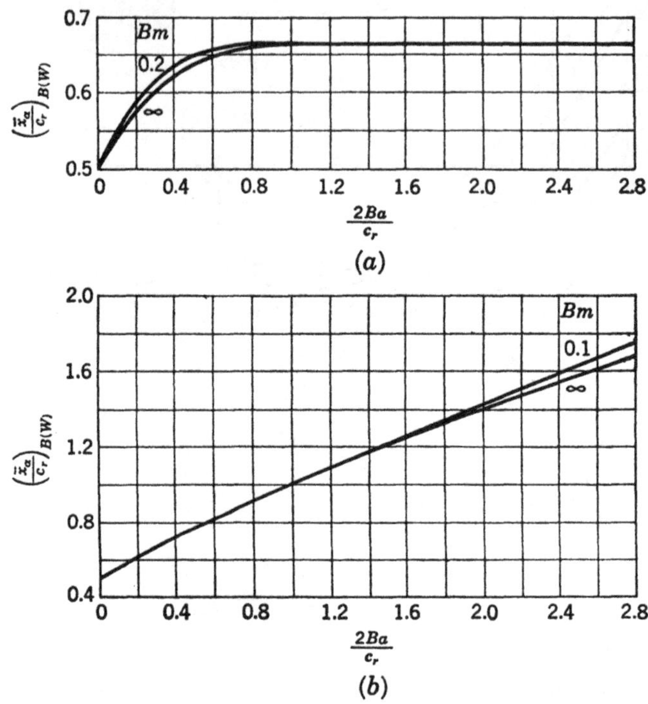

Fig. 5-14. Values of $(\bar{x}_\alpha/c_r)_{B(W)}$ based on planar model. (a) No afterbody; (b) afterbody.

5-7. Application to Nonslender Configurations; Calculative Example

The results for the forces and moments summarized in the previous section depended on the quantities read from either Tables 5-1 and 5-2 or Figs. 5-13 and 5-14. However, it is noteworthy that the forces and moments, with the exception of those due to the missile nose, are all proportional to the lift-curve slope of the wing alone. The theory was deliberately set up in this fashion; that is, all interference lifts were normalized by the lift of the wing alone. As long as the wing-body combinations are slender, the formulas apply without much question. But, if the wing-body combinations are not slender, can the theory be applied with any confidence? It turns out that the answer is yes for the following reasons. It is reasonable that the ratio of the interference lift to the wing-alone lift will be better predicted for nonslender configurations than

the absolute magnitude of the interference lift itself. In fact, if it is
assumed that *lift ratios* and *centers of pressure* are accurately predicted by
slender-body theory, then the foregoing formulas apply directly to nonslender configurations, provided an accurate value of the lift-curve slope

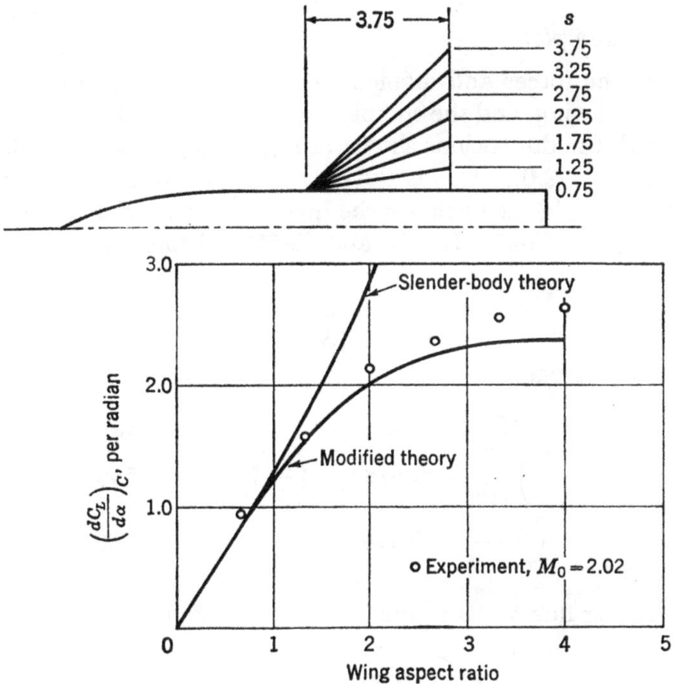

Fig. 5-15. Comparison of theory and experiment for triangular wing and body
combinations.

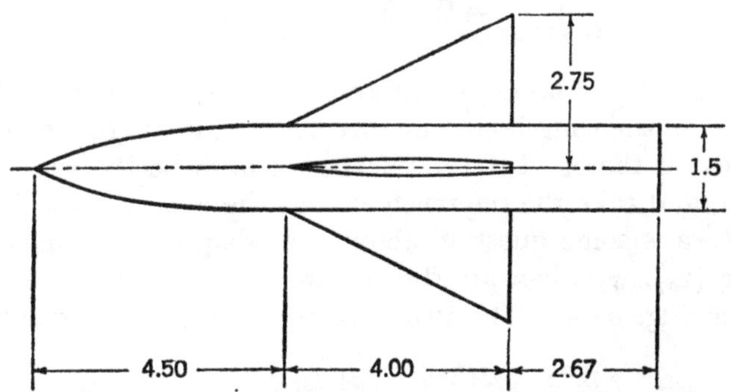

Fig. 5-16. Dimensions of model used in calculative example.

of the wing alone for the nonslender wing is used. (For this lift-curve
slope, either the value from the linear theory or an experimental value
will do.) The proof of the assumption lies in being able to predict
accurately the measured lift and moments of wing-body combinations by
the method. Actually, the method has been tested successfully for large

numbers of wing-body combinations at subsonic and supersonic speeds.[6] In Fig. 5-15 a favorable comparison is made between the predictions of the formulas and the measured characteristics of a systematic series of triangular wing-body combinations varying from slender to nonslender. These data are those of Nielsen, Katzen, and Tang.[8]

Calculative Example:

Calculate the forces and moments of the right panel, the body in the presence of the wing, and the complete configuration for a wing-body combination with the dimensions shown in Fig. 5-16. Take $\alpha_c = 0.3$ radian, $\varphi = 22.5°$, $M_0 = 2.0$. This is the configuration studied by Spahr.[10]

As a first step, let us evaluate the quantities occurring in the formulas as given by Eqs. (5-47), (5-50), and (5-51). From the dimensions, we have

$$\frac{a}{s_m} = \frac{0.75}{2.75} = 0.27$$

Table 5-1 then gives

$$K_W = 1.23 \qquad\qquad K_B = 0.39$$

$$\left(\frac{\bar{x}_\alpha}{c_r}\right)_{W(B)} = 0.648 \qquad \left(\frac{\bar{x}_\alpha}{c_r}\right)_{B(W)} = 0.556$$

$$\frac{\bar{y}_\alpha - a}{s_m - a} = 0.418 \qquad\qquad \bar{y}_\alpha = 1.586$$

For the loading due to $\alpha\beta$ coupling, Table 5-2 gives

$$K_\varphi = 0.50 \qquad \frac{\bar{y}_\varphi - a}{s_m - a} = 0.537$$

$$\left(\frac{\bar{x}_\varphi}{c_r}\right)_{W(B)} = 0.669 \qquad \bar{y}_\varphi = 1.914$$

Since the centers of pressure are given already in distance behind the leading edge of the wing-body juncture, let the pitching-moment reference axis be located there. Let the reference length be the wing-body juncture chord, and take the reference area as the wing-alone area.

Since there is some question about the adequacy of the slender-body theory for $(\bar{x}_\alpha)_{B(W)}$ when afterbodies are present, let us determine this factor from Fig. 5-14. The two parameters required for doing this are

$$\frac{2Ba}{c_r} = \frac{2(3)^{1/2}(0.75)}{4} = 0.65$$

$$Bm = 3^{1/2} \tan \epsilon = 3^{1/2}(0.5) = 0.866$$

The figure then gives

$$\left(\frac{\bar{x}_\alpha}{c_r}\right)_{B(W)} = 0.85$$

a value considerably greater than the slender-body value of 0.556.   The final quantity required to evaluate the forces and moments is the lift-curve slope of the wing alone.   From Eq. (2-36)

$$\left(\frac{dC_L}{d\alpha}\right)_W = \frac{2\pi \tan \epsilon}{E[(1 - B^2m^2)^{\frac{1}{2}}]}$$
$$(1 - B^2m^2)^{\frac{1}{2}} = 0.5 \qquad \sin^{-1} 0.5 = 30°$$
$$E(30°) = 1.4675$$
$$\left(\frac{dC_L}{d\alpha}\right)_W = \frac{2\pi(0.5)}{1.4675} = 2.14 \text{ per radian}$$

Let us now evaluate the force and moment coefficients for the right wing panel as given by Eq. (5-47).

$$(C_Z)_P = (1.23)\frac{2.14}{2}(0.3)(0.924)$$
$$+ \frac{0.50}{0.50}\left(\frac{2.14}{2}\right)(0.3)^2(0.924)(0.383) = 0.399$$
$$(C_h)_P = -(1.23)\frac{2.14}{2}(0.648)(0.3)(0.924)$$
$$- \frac{0.50}{0.50}\left(\frac{2.14}{2}\right)(0.669)(0.3)^2(0.924)(0.383) = -0.258$$
$$(C_l)_P = -(1.23)\frac{2.14}{2}\left(\frac{1.586}{4}\right)(0.3)(0.924)$$
$$- \frac{0.50}{0.50}\left(\frac{2.14}{2}\right)\frac{1.914}{4}(0.3)^2(0.924)(0.383) = -0.160$$

The coefficients for any other panel can be calculated as if the right-hand panel had been rotated by angle φ into its position.

The force and moments for the body in the presence of the wing are given by Eq. (5-50).

$$(C_L)_{B(W)} = 0.39(2.14)(0.3)$$
$$= 0.25$$
$$(C_M)_{B(W)} = -0.39(2.14)(0.85)(0.3)$$
$$= -0.212$$
$$(C_C)_{B(W)} = (C_l)_{B(W)} = 0$$

The forces on the complete wing-body combination are given by Eq. (5-51).

$$(C_L)_C = (C_L)_N + (0.39 + 1.23)(2.14)(0.3)$$
$$= (C_L)_N + 1.04$$
$$(C_C)_C = 0$$
$$(C_m)_C = (C_m)_N - [0.39(0.85) + 1.23(0.648)](2.14)(0.3) = (C_m)_N - 0.725$$
$$(C_l)_C = 0$$

5-8. Simplified Vortex Model of Wing-Body Combination

A simplified vortex model of a wing-body combination is useful for many purposes, and such a model is illustrated by Fig. 5-17. Consider the circulation distribution across the wing panels shown in the figure. The actual shape of the distribution is given by the integrand of the integral in Eq. (5-15). If Γ_0 is the circulation at the wing-body juncture,

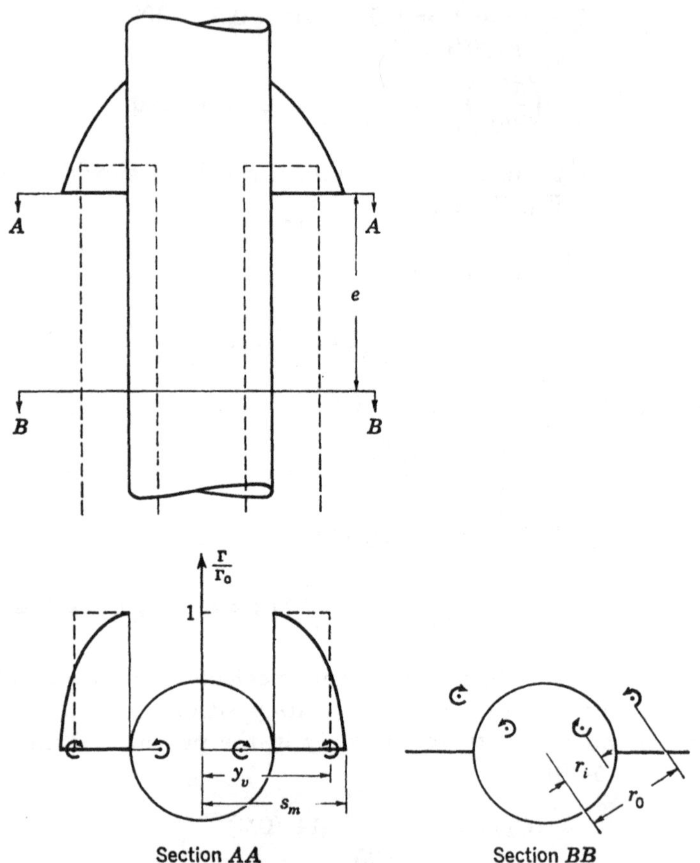

Fig. 5-17. Simplified vortex model of slender wing-body combination.

then

$$\frac{\Gamma}{\Gamma_0} = \frac{[(s_m^2 y^2 - a^4)(s_m^2 - y^2)]^{1/2}}{y(s_m^2 - a^2)} \tag{5-54}$$

The trailing vorticity is proportional to the slope of the circulation distribution curve and is distributed continuously across the wing span, being concentrated toward the wing tips. According to the discussion of Sec. 6-2, the trailing vorticity soon rolls up into a concentrated vortex near the center of gravity of the vortex sheet. The center of vorticity for the present circulation distribution, which is nearly elliptical, lies very close to $\pi/4$ of the panel semispan from the wing-body juncture. See Table 6-1. Assume therefore that the external wing panels are replaced by a bound vortex in the panel plus a trailing vortex on each side as shown in Fig. 5-17. The presence of the circular afterbody requires an image vortex system to cancel the velocity normal to the body induced by the external vortices. Or, from another point of view, the bound vortex in the wing has to be terminated inside the body in some fashion. In so far as the flow in each crossflow plane can be considered independent of that in other crossflow planes, as in slender-body theory, we can satisfy the body boundary condition by the introduction of the image trailing vortices as shown. The image vortices must be so located that

$$r_i = \frac{a^2}{r_0} \tag{5-55}$$

It is possible to complete the vortices by extending them forward to form horseshoe vortices as shown in the figure.

It is to be pointed out that the foregoing model is not accurate in the immediate neighborhood of the wing because many vortex lines lie on the wing surface. Nevertheless, the model accurately predicted the division of lift between wing and body. Since we have replaced the wing-body combination by a pair of horseshoe vortices, we have a uniform loading along the part of the vortex normal to the stream, the so-called lifting line. The load per unit spanwise distance of a lifting line in $\rho_0 V_0 \Gamma_0$, and the lift on the body is represented by the length of the line inside the body, and similarly for the lift on the wing. Thus

$$\begin{aligned}
L_C &= 2\rho_0 V_0 \Gamma_0 \left(y_v - \frac{a^2}{y_v} \right) \\
&= q_0 2\pi\alpha \left(1 - \frac{a^2}{s_m{}^2} + \frac{a^4}{s_m{}^4} \right) s_m{}^2
\end{aligned} \tag{5-56}$$

It will be recognized that this equation is a special case of Eq. (4-121). The vortex strength is

$$\Gamma_0 = \frac{(\pi/2)\alpha V_0 s_m{}^2 (1 - a^2/s_m{}^2 + a^4/s_m{}^4)}{y_v - a^2/y_v} \tag{5-57}$$

The ratio of the lift on the body to that on the wing panel is

$$\begin{aligned}
\frac{L_{B(W)}}{L_{W(B)}} &= \frac{K_B}{K_W} \\
&= \frac{a - a^2/y_v}{y_v - a}
\end{aligned} \tag{5-58}$$

The values of K_B/K_W obtained from the simplified model are compared with the corresponding values from slender-body theory for several values of the radius-semispan ratio in Table 5-3. These values are based on the value of $(y_v - a)/(s_m - a)$ of $\pi/4$. It is interesting that the approximate model predicts a division of lift between body and wing very close to that of slender-body theory.

Behind the trailing edge, vortices roll up and follow the streamline given implicitly by Eq. (4-88). Actually there is developed a load on the afterbody because of the motions of the vortices. The actual load can be calculated by Eq. (4-121). As the vortices pass along the body in the

TABLE 5-3. VALUES OF K_B/K_W

a/s_m	0	0.2	0.4	0.6	0.8	1.0
Slender-body theory	0	0.239	0.454	0.646	0.826	1.0
Vortex model	0	0.242	0.459	0.656	0.837	1.0

downstream direction, their lateral spacing decreases. It can readily be seen from Eq. (4-121) that the afterbody loading is then downward, that is, negative. The problem of afterbody loading for a symmetrical vortex pair in the presence of a circular cylinder was studied by Lagerstrom and Graham.[11]

SYMBOLS

a	body radius
$\bar{a}$	body radius occurring with s_m
A	aspect ratio of wing alone
$b_0(x)$	function of x occurring in complex potential
B	$(M_0^2 - 1)^{1/2}$
c_r	chord at wing-body juncture
C	cross-wind force, Fig. 5-2
C_C	cross-wind force coefficient, Fig. 5-2
C_h	hinge-moment coefficient of wing panel, Fig. 5-2
C_l	rolling-moment coefficient
$dC_L/d\alpha$	lift-curve slope per radian
C_L	lift coefficient, Fig. 5-2
C_m	pitching-moment coefficient, Fig. 5-2
C_Z	Z force coefficient, Fig. 5-2
E	complete elliptic integral of second kind
k	modulus of elliptic integral
k'	complementary modulus, $(1 - k^2)^{1/2}$
K	complete elliptic integral of first kind
K_B	ratio of lift on body in presence of wing to lift of wing alone, $\varphi = 0$
K_W	ratio of lift of wing panels in presence of body to lift of wing alone, $\varphi = 0$

K_φ lift ratio specifying additional wing load due to sideslip at constant angle of attack

l_r reference length

L lift force in plane of V_0 and missile longitudinal axis

m tangent of wing semiapex angle

M_0 free-stream Mach number

p local static pressure

P pressure coefficient, $(p - p_0)/q_0$

P_φ additional pressure coefficient due to sideslip at constant angle of attack

P^+ pressure on impact surface (positive α)

P^- pressure on suction surface (positive α)

ΔP $P^+ - P^-$

q_0 free-stream dynamic pressure

r_0 radial distance to external vortex

r_i radial distance to image vortex

s local semispan of right wing panel

s_m maximum semispan of right wing panel

t local semispan of vertical panel

t_m maximum semispan of vertical panel

u, v, w perturbation velocity components along x, y, and z, respectively, for unit V_0

u_t, v_t, w_t perturbation velocity component at $\alpha = \beta = 0$ for unit V_0

$u_\alpha, v_\alpha, w_\alpha$ perturbation velocity components due to angle of attack for unit V_0 and unit α

$u_\beta, v_\beta, w_\beta$ perturbation velocity components due to angle of sideslip for unit V_0 and unit β

V_0 free-stream velocity

W_t complex potential at $\alpha = 0$

W_α complex potential due to angle of attack

x, y, z missile axes of symmetry

x', y', z' missile axes of symmetry for angle of attack with $\varphi = 0$, Fig. 5-2

$\bar{x}_\alpha, \bar{y}_\alpha$ coordinates of center of pressure for loading due to angle of attack

$\bar{x}_\varphi, \bar{y}_\varphi$ coordinates of center of pressure of additional loading due to sideslip at constant angle of attack

y_v lateral position of concentrated vortex

Y, Z forces along y and z axes

$\mathfrak{z}$ $y + iz$

α angle of attack, $\alpha_c \cos \varphi$

α_c included angle between V_0 and missile longitudinal axis

α_W wing angle of attack

β angle of sideslip, $\alpha_c \sin \varphi$

Γ_0	value of Γ at wing-body juncture
$\Gamma(y)$	circulation distribution
δ	y^2/a^2
ϵ	semiapex angle of wing alone
θ	polar angle
λ	s_m/a; also panel taper ratio
τ	s^2/a^2
ξ, η	Fig. 5-12
ϕ_i	interference potential
ϕ_t	potential due to thickness, $\alpha = 0$
ϕ_α	potential due to angle of attack
ϕ_β	potential due to angle of sideslip
φ	angle of bank
ψ_1, ψ_2	Eq. (5-40)

Subscripts:

B	body alone
$B(W)$	body in presence of wing panels
C	complete configuration
N	missile nose or forebody
P	wing panel
W	wing alone formed by joining exposed wing panels together
$W(B)$	wing panels in presence of body

REFERENCES

1. Nielsen, Jack N.: Quasi-cylindrical Theory of Wing-Body Interference at Supersonic Speeds and Comparison with Experiment, *NACA Tech. Rept.* 1252, 1956.

2. Morikawa, G. K.: The Wing-Body Problem for Linearized Supersonic Flow, doctoral thesis, California Institute of Technology, Pasadena, 1949.

3. Lawrence, H. R., and A. H. Flax: Wing-Body Interference at Subsonic and Supersonic Speeds: Survey and new developments, *J. Aeronaut. Sci.*, vol. 21, no. 5, 1954.

4. Durand, William Frederick: "Aerodynamic Theory," vol. IV, pp. 152–157, Durand Reprinting Committee, California Institute of Technology, Pasadena, 1943.

5. Spreiter, John R., and Alvin H. Sacks: A Theoretical Study of the Aerodynamics of Slender Cruciform-wing Arrangements and Their Wakes, *NACA Tech. Repts.* 1296, 1957.

6. Pitts, William C., Jack N. Nielsen, and George E. Kaattari: Lift and Center of Pressure of Wing-Body-Tail Combinations at Subsonic, Transonic, and Supersonic Speeds, *NACA Tech. Repts.* 1307, 1957.

7. Lagerstrom, P. A.: Linearized Theory of Conical Wings, *NACA Tech. Notes* 1685, 1948.

8. Nielsen, Jack N., Elliott D. Katzen, and Kenneth K. Tang: Lift and Pitching-moment Interference between a Pointed Cylindrical Body and Triangular Wings of Various Aspect Ratios at Mach Numbers of 1.50 and 2.02, *NACA Tech. Notes* 3795, 1956.

9. Jones, A. L., and Alberta Y. Alksne: The Load Distribution Due to Sideslip on Triangular, Trapezoidal, and Related Planforms in Supersonic Flow, *NACA Tech. Notes* 2007, January, 1950.

10. Spahr, J. Richard: Contribution of the Wing Panels to the Forces and Moments of Supersonic Wing-Body Combinations at Combined Angles, *NACA Tech. Notes* 4146, January, 1958.

11. Lagerstrom, P. A., and M. E. Graham: Aerodynamic Interference in Supersonic Missiles, *Douglas Aircraft Co. Rept.* SM-13743, 1950.

CHAPTER 6

DOWNWASH, SIDEWASH, AND THE WAKE

In this chapter we will be concerned with methods for predicting the streamline directions behind a lifting surface, alone or in combination with a body. This knowledge is necessary for the determination of the aerodynamic characteristics of any aerodynamic shape, such as a tail, immersed in the flow. For this purpose the direction of the streamlines

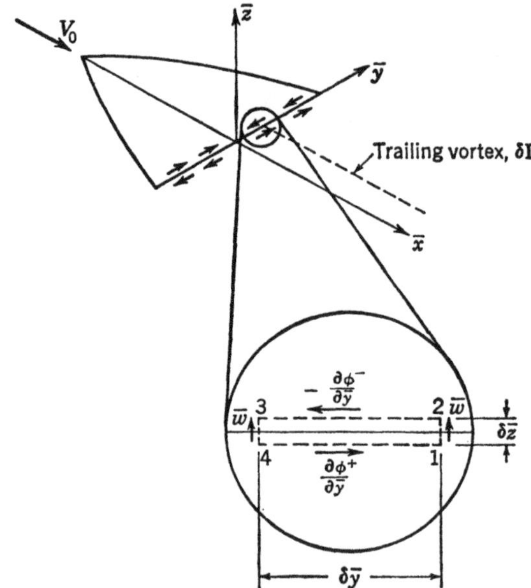

Fig. 6-1. Wind axes and sidewash at the trailing edge of the wing.

will be specified with respect to the system of wind axes shown in Fig. 6-1. Let the components of the streamline velocity V *with respect to the missile* be $\bar{u}$, $\bar{v}$, and $\bar{w}$ along the positive axes of $\bar{x}$, $\bar{y}$, and $\bar{z}$, respectively. Then the *downwash angle* ϵ and the *sidewash angle* σ are defined to be

$$\epsilon = -\arctan\frac{\bar{w}}{\bar{u}} \approx \frac{-\bar{w}}{V}$$

$$\sigma = \arcsin\frac{\bar{v}}{V} \approx \frac{\bar{v}}{V}$$

(6-1)

These definitions based on the local streamline velocity are to be compared with the tangent definition of the angle of attack and the sine

144

definition of the angle of sideslip in Sec. 1-4, based on the free-stream velocity. Thus the angles of downwash and sidewash for the streamline velocity have the opposite sign conventions of the angles of attack and sideslip for V_0. The term *wake* is used in reference to the regions of vorticity or the vortex sheet associated with flow behind an aerodynamic shape.

6-1. Vortex Model Representing Slender Wing with Trailing Edge Normal to Flow

Consider the sidewash velocities at the trailing edge of a slender wing, as shown in Fig. 6-1. The $\bar{x}$ axis is aligned in the V_0 direction, and the wing as drawn shows no angle of attack because α is assumed small. Let the potential on the bottom surface be ϕ^+ and that on the top surface be ϕ^- so that the positive sidewash velocities on the bottom and top are $\partial\phi^+/\partial\bar{y}$ and $\partial\phi^-/\partial\bar{y}$. Consider an enlarged section of the trailing edge. The *circulation* Γ around the contour is defined to be

$$\Gamma = \oint q_t\, ds \tag{6-2}$$

where q_t is the velocity component tangent to the contour, and the line integral is taken in the counterclockwise sense. The quantity Γ is then taken as the measure of the strength of all vortex lines threading through the contour. In evaluating the circulation, let us for the moment ignore the presence of any shock waves. Then we can evaluate the circulation around the contour as follows:

$$\delta\Gamma_{12} = \bar{w}\,\delta\bar{z} \qquad \delta\Gamma_{23} = -\frac{\partial\phi^-}{\partial\bar{y}}\,\delta\bar{y}$$

$$\delta\Gamma_{34} = -\bar{w}\,\delta\bar{z} \qquad \delta\Gamma_{41} = \frac{\partial\phi^+}{\partial\bar{y}}\,\delta\bar{y} \tag{6-3}$$

Thus
$$\delta\Gamma_{1234} = \left(\frac{\partial\phi^+}{\partial\bar{y}} - \frac{\partial\phi^-}{\partial\bar{y}}\right)\delta\bar{y} \tag{6-4}$$

Since $\partial\phi^+/\partial\bar{y}$ is positive as shown, and $\partial\phi^-/\partial\bar{y}$ is negative, the circulation will be positive corresponding to a counterclockwise vortex. Let us define the *potential difference* as the positive quantity

$$\Delta\phi_{te} = \phi^- - \phi^+ \tag{6-5}$$

so that
$$\frac{\partial\Gamma_{1234}}{\partial\bar{y}} = -\frac{d(\Delta\phi)_{te}}{d\bar{y}} \tag{6-6}$$

or the *trailing-vortex strength per unit span* is the negative slope of the potential-difference curve. Alternatively,

$$\delta\Gamma = -\delta(\Delta\phi)_{te} \tag{6-7}$$

or the total vortex strength trailing back from the trailing edge between any two spanwise points is equal to the negative of the change in potential

difference across the trailing edge between the two points. Thus, from
a knowledge of the velocity potential at the wing trailing edge, the
strength of the vortex lines leaving the edge can be directly calculated.

From the simple preceding result the vortex model of the flow at the
wing trailing edge can be constructed, and such a vortex model is shown
in Fig. 6-2. The potential difference at the trailing edge produces a
trailing-vortex sheet, the strength of which is $d\Gamma/dy$ per unit span given
by Eq. (6-6). The tendency of the vortex strength per unit span to
approach infinity at the side edges of the sheet is noteworthy. The vortex lines do not terminate at the wing trailing edge but can be considered

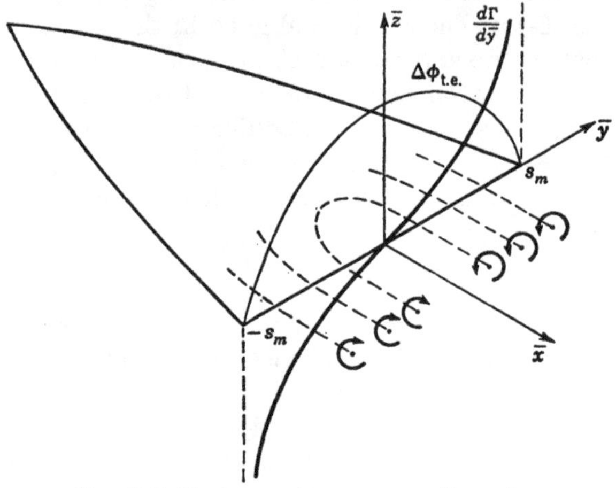

Fig. 6-2. Vortex system representing wing.

to lie in the wing surface as shown. These *bound vortex lines* can be
shown to lie along contours of constant potential difference. The fact
that these lines do not lie along the quarter-chord line is the only difference between the foregoing model and that of simple lifting-line theory.
Modifications of simple lifting-line theory to account in part for this difference have, of course, been made in an effort to adapt lifting-line theory to
lower aspect ratios.[8,9] The exact positions of the vortex lines on the
planform of the wing will have an influence on the downwash and sidewash fields right behind the trailing edge, but their influence is appreciable only a short distance downstream, as will subsequently be shown.

Consider now a mathematical determination of the vortex strengths at
the trailing edge of the wing in Fig. 6-2. The complex potential for the
slender flat-plate wing on the basis of Table 2-3 is

$$W(\tilde{s}) = \phi + i\psi = b_0(\bar{x}) - i\alpha V_0(\tilde{s}^2 - s_0^2)^{1/2} \qquad (6\text{-}8)$$

in any crossflow plane with the local semispan equal to s_0. The potential
at the trailing edge on the upper surface is

$$\phi^- = b_0(0) + \alpha V_0(s_m^2 - \bar{y}^2)^{1/2} \qquad (6\text{-}9)$$

and on the lower surface

$$\phi^+ = b_0(0) - \alpha V_0(s_m{}^2 - \bar{y}^2)^{1/2} \tag{6-10}$$

so that

$$\Delta\phi_{te} = 2\alpha V_0(s_m{}^2 - \bar{y}^2)^{1/2} \tag{6-11}$$

The strength of the trailing-vortex sheet per unit span is

$$\frac{d\Gamma}{d\bar{y}} = - \frac{\partial(\Delta\phi)_{te}}{\partial\bar{y}} = 2\alpha V_0 \frac{\bar{y}}{(s_m{}^2 - \bar{y}^2)^{1/2}} \tag{6-12}$$

The vortex strength per unit span exhibits square-root singularities at the side edges of the vortex sheet. Only the part of ϕ asymmetrical with respect to $\bar{z}$ can contribute to the potential difference at the trailing edge: that is, the part due to angle of attack or camber. Within the framework

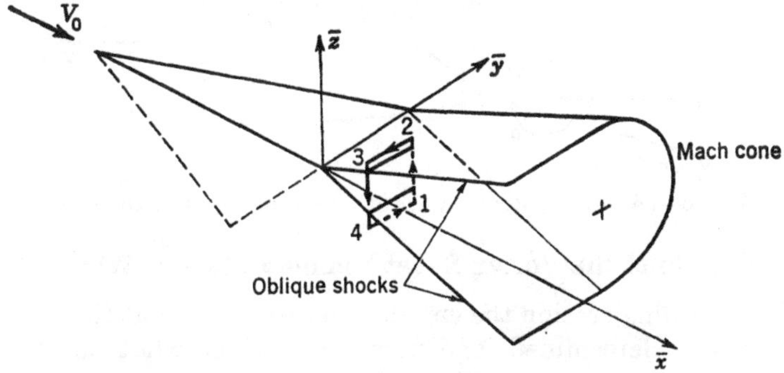

Fig. 6-3. Circulation contour with trailing shock waves.

of slender-body theory the shape of the planform does not affect the potential difference at a trailing edge normal to the stream, and therefore has no effect on the trailing-vortex strengths. It is to be noted that the potential difference can differ from the span loading if the square terms of Bernoulli's equation contribute to the span loading.

No particular attention has been paid to the shock-wave system at the trailing edge of the wing. Figure 6-3 illustrates the state of affairs for an edge normal to the air stream. The contour of integration 1234 of Fig. 6-1 for evaluating the circulation is repeated. Although the contour straddles the two plane shock waves as shown, the contributions of sides 12 and 34 to the circulation still vanish as in the original derivation. Also, if the sides 23 and 41 are brought down between the shock waves, the circulation will still be the same, since the velocities along 23 and 41, being tangential to the shock fronts, will remain unaltered passing through them.

Although the simple case of a trailing edge normal to the flow was assumed in the derivation, this restriction can be relaxed. Consider the trailing edge at an angle of sideslip β as in Fig. 6-4. The velocity can be broken down into a component $V_0 \cos \beta$ perpendicular to the trailing edge, which produces a potential ϕ_α and a component $V_0 \sin \beta$ parallel to the trailing edge, which produces a potential ϕ_β. It is clear that ϕ_β will produce a potential which has the same value at corresponding points on the top and bottom surfaces and which therefore adds nothing to $\Delta\phi_{te}$. However, in the calculation of $\Delta\phi_{te}$ the appropriate free-stream velocity and angle of attack normal to the trailing edge must be used. The sidewash velocity $V_0 \sin \beta$, when superimposed on the flow due to ϕ_α, will straighten out the vortex lines in the free-stream direction, as shown in Fig. 6-4.

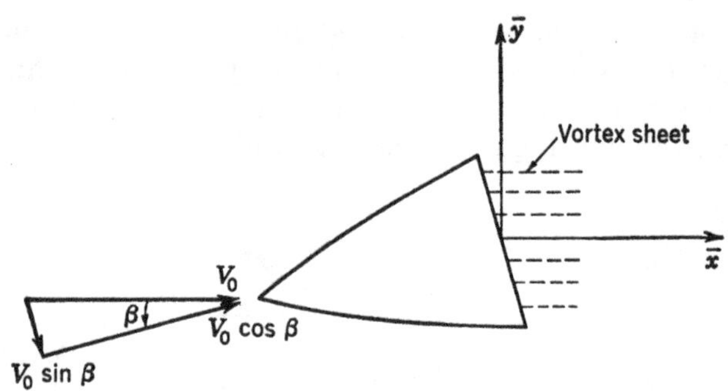

Fig. 6-4. Vortex system representing slender wing with sideslip.

6-2. Rolling Up of the Vortex Sheet behind a Slender Wing

In the preceding section the circulation distribution at the trailing edge of a wing was determined, and now we consider what happens as the vortex sheet leaves the trailing edge and moves downstream. Two slender-body solutions exist for the shape of the downstream sheet. The first solution is that proposed by Jones,[14] and subsequently treated also by Ward (Ref. 1 of Chap. 3); the second solution is that of Westwater.[2] In the Jones-Ward solution two linearized conditions are used: first, that the velocity is tangential to the vortex sheet on both sides, and second, that the pressure is continuous through the sheet, as calculated by the linearized Bernoulli equation. The consequence of these two assumptions is that the vortex lines are straight and parallel, as in lifting-line theory. If the two conditions above are not linearized, then the Jones-Ward solution is modified in two aspects. In the first place, the vortex lines are no longer straight and parallel, but a more serious difficulty arises. The infinite velocities at the outer edges of the vortex sheets give a finite force tending to tear the sheet apart, whereas no such force arose with the linearized Bernoulli equation. As a consequence of this force, the sheet, instead of tearing apart immediately, starts to roll up at the edge. A more detailed discussion of this phenomenon is given by Ward.[15]

Let us now turn to the work of Westwater. As the vortices stream backward, they induce velocities on each other in such a manner that the center of the sheet is depressed relative to the outer edges which roll up. If the usual assumption of slender-body theory is made that the flow in each crossflow plane is independent of that in others, a simple calculation can be performed to see how the sheet rolls up.   With reference to Fig. 6-5, the magnitude of the velocity induced on one vortex by another, say the velocity induced on vortex 2 by vortex 1, is

$$v_{2(1)} = \frac{\Gamma_1}{2\pi r} \qquad (6\text{-}13)$$

the velocity acts normal to the radius vector joining the vortices.   Westwater[2] has calculated the rolling up of the vortex sheet due to an elliptical potential difference at the wing trailing edge.   In his calculations, Westwater replaced the continuous vortex sheet by 20 vortices of equal strength, and computed their mutual interactions by means of Eq. (6-13).   Having calculated the velocities of the vortices in a given crossflow plane, he was able to determine their new positions in a crossflow plane a short distance downstream.   By continuing this step-by-step process, he was able to calculate the rolling up of the vortex sheet for the elliptical case.

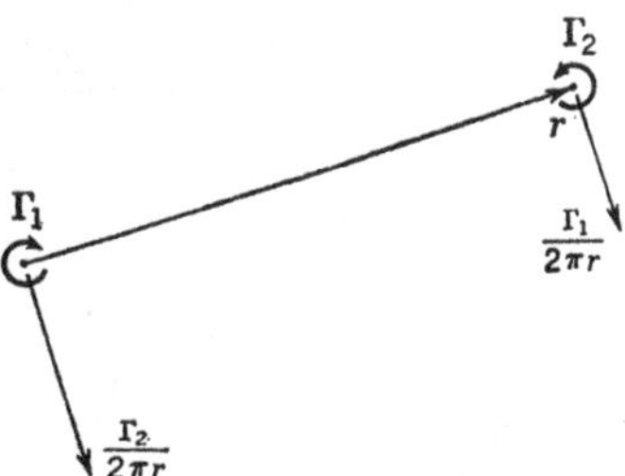

FIG. 6-5. Mutual induction between pair of two-dimensional vortices.

The results of Westwater's calculations are illustrated in Fig. 6-6.

The edge of the vortex sheet starts to curl up by virtue of vortices moving along the sheet toward the edges on each side.   At the same time the center of the sheet moves downward.   The vortex sheet tends to roll up into a concentrated vortex on each side, with a lateral spacing between vortices somewhat less than the wing span.   For an elliptical loading, the vortex sheet can thus be approximately represented by a pair of concentrated vortices for sufficiently large distances behind the wing trailing edge.   It should be borne in mind, however, that a potential difference at the trailing edge of other than elliptical shape can produce a different type of vortex system.   See Fig. 6-21.

It is desirable to know at what distance behind the wing trailing edge the vortex sheet is "essentially rolled up."   Mathematically, the vortex sheet approached a completely rolled-up condition only in an asymptotic sense and never achieves it.   Thus, some arbitrary criterion must be specified to indicate when the sheet can be said to be rolled up.   Kaden,[1] using a particular model and a particular mathematical criterion, has established the following distance for the sheet to roll up for elliptical distributions,

$$e = 0.28 \frac{A}{C_L} b \qquad (6\text{-}14)$$

where b is the wing span and A the aspect ratio. The form of this equation can be established on the basis of similarity arguments. The distance to roll up e, or any other significant downstream distance, is directly proportional to some linear dimension of the wing and to the free-stream velocity, and is inversely proportional to the magnitude of

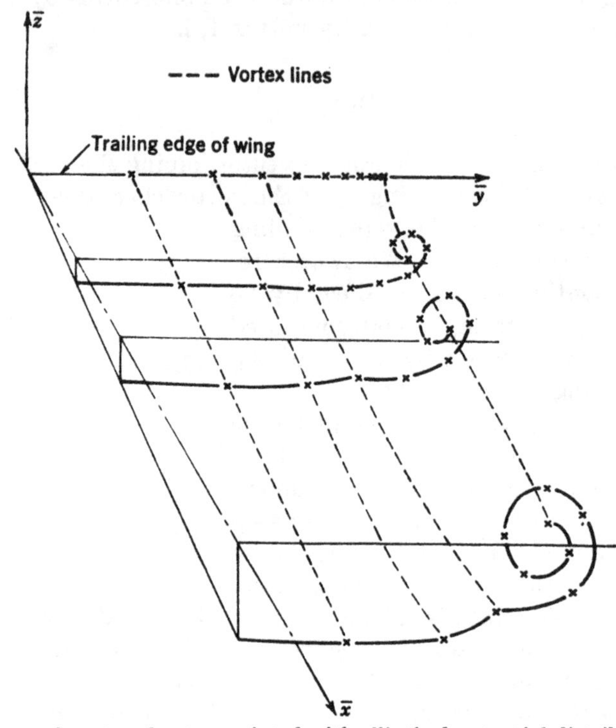

FIG. 6-6. Shape of vortex sheet associated with elliptical potential distribution according to Westwater.

the velocities induced by the vortex system.

$$e \propto \frac{bV_0}{v_i} \tag{6-15}$$

The induced velocities vary directly as the vortex strength and inversely as the vortex span b_v.

$$e \propto \frac{bb_vV_0}{\Gamma} \tag{6-16}$$

The vortex strength is related to the lift by

$$L = \rho_0 V_0 \Gamma b_v \tag{6-17}$$

Thus

$$\frac{e}{b} \propto b_v{}^2 \frac{\rho_0 V_0{}^2}{L} \propto \frac{b_v{}^2}{S_W} \frac{\rho_0 V_0{}^2 S_W}{L} \tag{6-18}$$

Since b_v is a constant fraction of b for a given shape of potential difference curve, we can write

$$\frac{e}{b} = k\frac{A}{C_L} \tag{6-19}$$

where k depends on the shape of the curve. According to Spreiter and Sacks,[6] the distance e given by Kaden's formula seems to be low.

For purposes of computing the downwash and sidewash velocities behind a lifting configuration, it is sometimes not critical or even important whether the vortex sheet is flat or rolled up, as we will see in the next

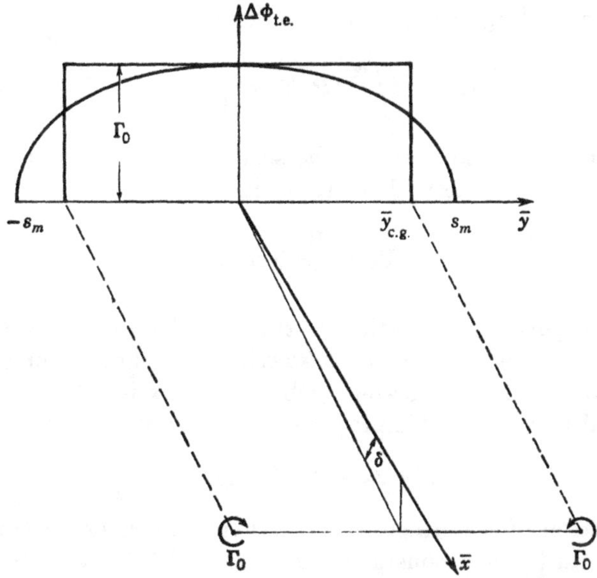

Fig. 6-7. Horseshoe vortex representing elliptical potential distribution.

chapter. Under such circumstances a precise knowledge of how far behind the trailing edge the vortex sheet is rolled up is not required.

From a qualitative picture of the vortex wake behind the wing, let us proceed to the calculation of the strength and lateral position of the rolled-up vortex pair associated with an elliptical potential-difference distribution. Consider first the strength of the vortices Γ_0. With reference to Fig. 6-7, the potential difference at the trailing edge is

$$(\Delta\phi)_{te} = 2\alpha V_0(s_m{}^2 - \bar{y}^2)^{1/2} \tag{6-20}$$

Since the spanwise rate of change of bound vortex strength is the same as the rate for $(\Delta\phi)_{te}$, the total strength of all trailing vortices across the semispan is equal to $(\Delta\phi)_{te}$ at the root chord. Thus

$$\Gamma_0 = (\Delta\phi)_{te} \quad \text{at } \bar{y} = 0 \tag{6-21}$$
$$\Gamma_0 = 2\alpha V_0 s_m$$

It is also possible to relate Γ_0 to the span-load distribution in those cases where the span loading and potential-difference distributions are similar. A *sufficient* condition for this to be true is that the pressure coefficient be given by

$$P^+ = \frac{-2\phi_{\bar{x}}^+}{V_0} \qquad P^- = \frac{-2\phi_{\bar{x}}^-}{V_0} \tag{6-22}$$

The loading is

$$P^+ - P^- = \Delta P = \frac{-4\phi_{\bar{x}}^+}{V_0} \tag{6-23}$$

and the span loading at any spanwise distance is

$$(cc_l) = \int_{le}^{te} \Delta P \, d\bar{x} = \frac{2}{V_0} (\Delta\phi)_{te} \tag{6-24}$$

where c is the local chord and c_l the section lift coefficient. From Eqs. (6-21) and (6-24) the desired relationship is obtained.

$$\Gamma_0 = \frac{V_0}{2} (cc_l)_{\bar{y}=0} \tag{6-25}$$

Let us now turn our attention to the lateral positions of the vortices. To this end we use the Kutta-Joukowski law. For a horseshoe vortex of strength Γ_i the lift associated with the vortex is $\rho_0 V_0 \Gamma_i$ per unit span of the bound portion. This lift for one horseshoe vortex is

$$L_i = 2\rho_0 V_0 \Gamma_i \bar{y}_i \tag{6-26}$$

For a collection of n horseshoe vortices that represent a trailing-vortex sheet the total lift is a constant. The sum of $\Gamma_i \bar{y}_i$ over all the vortices must be a constant independent of distance behind the trailing edge.

$$\sum_{i=1}^{n} \Gamma_i \bar{y}_i = \text{constant} \tag{6-27}$$

It is thus clear that the "lateral center of gravity" of the vortex sheet on each side of the streamwise axis does not change because of the rolling up of the sheet, nor does it depend on how many vortices the sheet forms. For our model of one vortex for each half of the sheet, we obtain the lateral center of gravity of the fully rolled-up vortex.

$$\bar{y}_{cg} = \frac{\sum_{i=1}^{n} \Gamma_i \bar{y}_i}{\Gamma_0} \tag{6-28}$$

Since the strength of the trailing vortices is $d\Gamma/d\bar{y}$ per unit span, we get, with the aid of Eq. (6-12), letting $n \rightarrow \infty$,

$$\bar{y}_{cg} = \frac{\int_0^{s_m} \bar{y}\, \dfrac{d\Gamma}{d\bar{y}}\, d\bar{y}}{\Gamma_0} = \frac{2\alpha V_0}{\Gamma_0} \int_0^{s_m} \frac{\bar{y}^2\, d\bar{y}}{(s_m^2 - \bar{y}^2)^{1/2}} \qquad (6\text{-}29)$$

or

$$\bar{y}_{cg} = \frac{2\alpha V_0 s_m}{\Gamma_0} \frac{\pi s_m}{4} = \frac{\pi}{4} s_m \qquad (6\text{-}30)$$

With the vortex strength and position determined by Eqs. (6-21) and (6-30), we can calculate the angle at which the vortices move downward because of their mutual induction. The downward velocity on the center line due to one vortex is $\Gamma_0/2\pi\bar{y}_{cg}$, so that the angle δ (Fig. 6-7) is

$$\delta = \frac{\Gamma_0}{4\pi V_0 \bar{y}_{cg}} = \frac{8\alpha}{\pi^2} \qquad (6\text{-}31)$$

6-3. Calculation of Induced Velocities of Trailing-vortex System

From the trailing-vortex system the induced velocities in crossflow planes behind the wing can be calculated by several methods, including

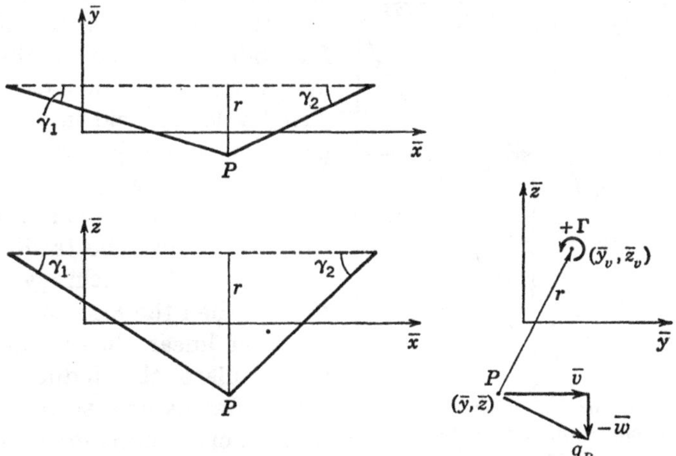

FIG. 6-8. Vortex line segment of Biot-Savart law.

those of two-dimensional incompressible vortices and supersonic horse-shoe vortices. It is of interest to compare these two methods. The induced velocities due to two-dimensional incompressible vortices, which we will generally use, are given by the *Biot-Savart law*. With reference to Fig. 6-8, the induced velocity q_P at point P due to a vortex line of finite length is

$$q_P = \frac{\Gamma}{4\pi r} (\cos \gamma_1 + \cos \gamma_2) \qquad (6\text{-}32)$$

For an infinite line vortex, the induced velocity patterns are similar in all crossflow planes and may be calculated by

$$q_P = \frac{\Gamma}{2\pi r} \tag{6-33}$$

The downwash $-\bar{w}$ and the sidewash $\bar{v}$ are easily obtained by resolving the velocity q_P perpendicular to the radius vector r into components downward and to the right. Thus

$$-\bar{w} = \frac{\Gamma}{2\pi r}\frac{\bar{y}_v - \bar{y}}{r}$$
$$= \frac{\Gamma(\bar{y}_v - \bar{y})}{2\pi[(\bar{y}_v - \bar{y})^2 + (\bar{z}_v - \bar{z})^2]}$$
$$\bar{v} = \frac{\Gamma}{2\pi r}\frac{\bar{z}_v - z}{r}$$
$$= \frac{\Gamma(\bar{z}_v - z)}{2\pi[(\bar{y}_v - \bar{y})^2 + (\bar{z}_v - z)^2]} \tag{6-34}$$

The contours of constant downwash and sidewash for an incompressible infinite line vortex are shown in Fig. 6-9. The use of the infinite line vortex for calculating the induced velocity field in the crossflow planes is compatible with the use of slender-body theory.

If the vortex system representing the flow behind the trailing edge is known to the accuracy of linear theory, then the *supersonic horseshoe vortex* of linear theory can be used to calculate the induced velocity field. Let us now turn to this subject. For a horseshoe vortex the downwash at a point depends on the region of influence in which it lies.[12] With reference to Fig. 6-10, in the Mach forecone from point A, an observer sees the bound vortex as if it were of infinite aspect or two-dimensional. The downwash in the region occupied by A accordingly is zero, as in two-dimensional supersonic flow. Point B sees one trailing vortex and has downwash

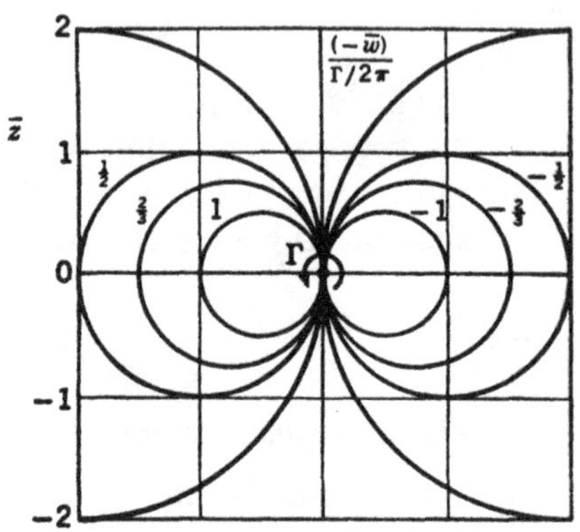

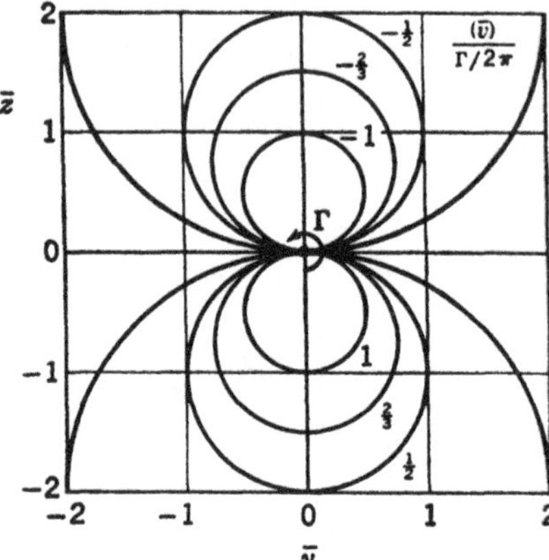

FIG. 6-9. Contours of constant downwash and sidewash associated with infinite line vortex in streamwise direction.

$$\bar{w}_B = \frac{-\Gamma\bar{x}(s_v - \bar{y})[\bar{x}^2 - B^2(\bar{y} - s_v)^2 - 2B^2\bar{z}^2]}{2\pi[(s_v - \bar{y})^2 + \bar{z}^2](\bar{x}^2 - B^2\bar{z}^2)[\bar{x}^2 - B^2(\bar{y} - s_v)^2 - B^2\bar{z}^2]^{1/2}} \tag{6-35}$$

For a point such as C which sees both trailing vortices, the downwash is

$$\bar{w}_C = \frac{-\Gamma\bar{x}(s_v - \bar{y})[\bar{x}^2 - B^2(\bar{y} - s_v)^2 - 2B^2\bar{z}^2]}{2\pi(\bar{x}^2 - B^2\bar{z}^2)[(s_v - \bar{y})^2 + \bar{z}^2][\bar{x}^2 - B^2(\bar{y} - s_v)^2 - B^2\bar{z}^2]^{1/2}}$$

$$- \frac{\Gamma\bar{x}(s_v + \bar{y})[\bar{x}^2 - B^2(\bar{y} + s_v)^2 - 2B^2\bar{z}^2]}{2\pi(\bar{x}^2 - B^2\bar{z}^2)[(s_v + \bar{y})^2 + \bar{z}^2][\bar{x}^2 - B^2(\bar{y} + s_v)^2 - B^2\bar{z}^2]^{1/2}} \quad (6\text{-}36)$$

For large values of x we obtain

$$\bar{w}_C = \frac{-\Gamma(s_v - \bar{y})}{2\pi[(s_v - \bar{y})^2 + \bar{z}^2]} - \frac{\Gamma(s_v + \bar{y})}{2\pi[(s_v + \bar{y})^2 + \bar{z}^2]} \quad (6\text{-}37)$$

This crossflow plane at infinity, the so-called *Trefftz plane*, has a downstream pattern identical with that given by Eq. (6-34) for vortices located

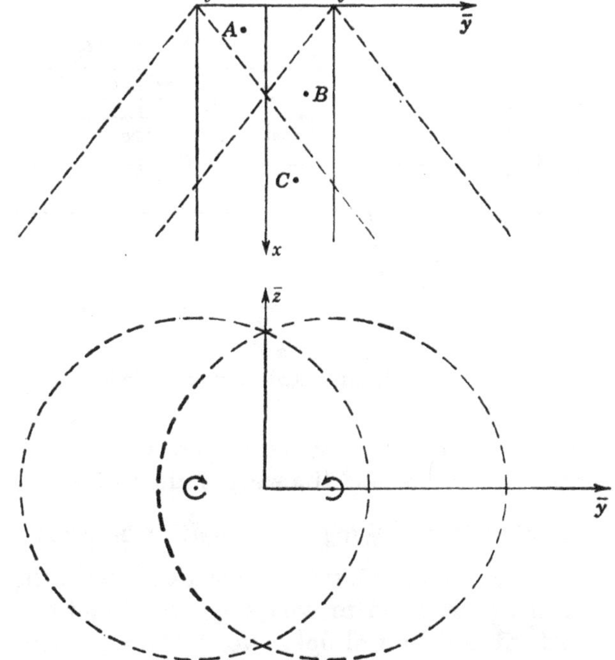

FIG. 6-10. Regions of influence of supersonic horseshoe vortex.

at $\bar{y}_v = \pm s_v$, $\bar{z}_v = 0$. Thus, at distances far downstream, the supersonic horseshoe vortex gives a downwash field identical with that obtained from two-dimensional incompressible vortices.

The foregoing behavior suggests that, at some definite distance behind the trailing edge, the downwash, as calculated by supersonic horseshoe vortices and by the incompressible two-dimensional vortices, should be practically identical. Figure 6-11 compares the downwash on the $\bar{x}$ axis behind a lifting line on the basis of the two methods of

calculation just described.   At a distance $\bar{x}/Bs_v$ of about 2.5 behind the
bound vortex (lifting line) the difference between the downwash calculated by the two methods is about 8 per cent.   Let us interpret this
distance in terms of chord lengths behind the trailing edge for a rectangular wing with the lifting line located at the midchord.   For an effective
aspect ratio BA of 2, this downstream distance for $\bar{x}/Bs_v$ of 2.5 would be
about two chord lengths, and, for an effective aspect ratio of unity, the
distance would be about three-fourths of a chord length.   It is clear that,
for low effective aspect ratios that characterize slender configurations, the

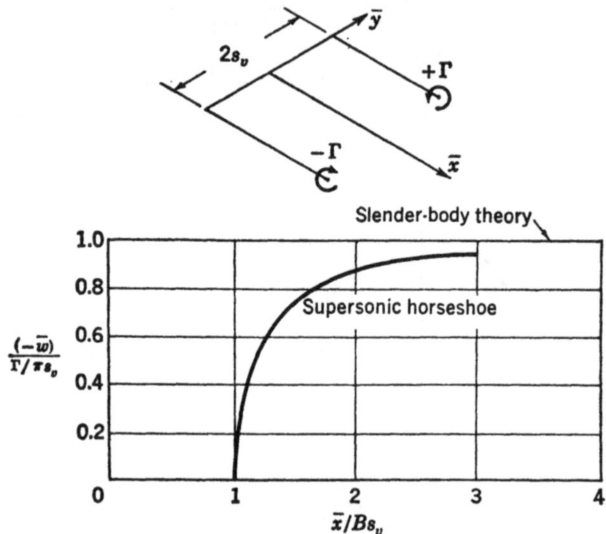

Fig. 6-11. Comparison of downwash calculated by supersonic horseshoe-vortex theory
and slender-body theory.

difference between downwash calculated by the two methods is small for
reasonably large distances behind the wing trailing edge.

6-4. Vortex Model of Planar Wing and Body Combination

The same principles used to construct a vortex model of the flow behind
a wing alone can be extended to wing-body combinations.   The only
additional ingredient is the set of image vortices occasioned by the presence of the body.   Let us first construct the trailing-vortex system associated with the wing panels.   For this purpose let us use Eq. (5-3), and
consider the wing panel for which $\mathfrak{z} = \bar{y}$.   For the potential at the panel
trailing edge, we obtain

$$\phi_{W(B)} = V_0 \left\{ b_0(\bar{x}) + a\frac{da}{d\bar{x}} \log \bar{y} \pm \alpha \left[\left(s_0 + \frac{a^2}{s_0} \right)^2 - \left(\bar{y} + \frac{a^2}{\bar{y}} \right)^2 \right]^{\frac{1}{2}} \right\}$$

$$(6\text{-}38)$$

The plus sign refers to the upper surface, and the minus sign to the lower.

The potential difference at the panel trailing edge is thus

$$(\Delta\phi)_{te} = \phi^- - \phi^+ = 2\alpha V_0 \left[\left(s_m + \frac{a^2}{s_m}\right)^2 - \left(\bar{y} + \frac{a^2}{\bar{y}}\right)^2\right]^{1/2}$$

$$= \frac{2\alpha V_0}{s_m y}\left[(s_m^2\bar{y}^2 - a^4)(s_m^2 - \bar{y}^2)\right]^{1/2} \tag{6-39}$$

The trailing vortices are of strength $d\Gamma/d\bar{y}$ per unit span as given by Eq. (6-6)

$$\frac{d\Gamma}{d\bar{y}} = -\frac{d(\Delta\phi)_{te}}{d\bar{y}} \tag{6-40}$$

Several points of interest arise in connection with the distribution of the potential difference across the wing panel.   The distribution is given by

$$\frac{\Gamma}{\Gamma_0} = \frac{\Delta\phi}{(\Delta\phi)_{\bar{y}=a}}$$

$$= \frac{[(s_m^2\bar{y}^2 - a^4)(s_m^2 - \bar{y}^2)]^{1/2}}{\bar{y}(s_m^2 - a^2)} \tag{6-41}$$

The distribution depends only on the radius-semispan ratio a/s_m, and the fraction exposed semispan

$$(\bar{y} - a)/(s_m - a)$$

It is in fact insensitive to a/s_m, as shown by Fig. 6-12. For a/s_m of zero, the distribution is precisely elliptical as for the wing-alone case, Eq. (6-11). For a/s_m approaching unity, the wing panel is effectively mounted on a vertical reflection plane, so that in this limit the distribution

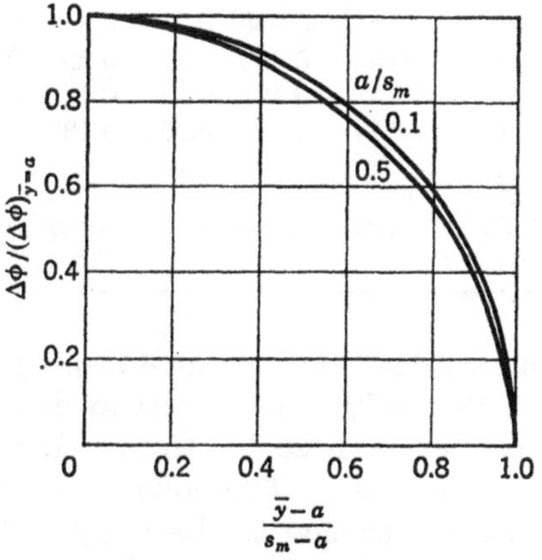

FIG. 6-12. Potential-difference distribution at trailing edge of wing in combination with body according to slender-body theory.

is again elliptical.   The assumption of an elliptical distribution for all values of a/s_m is a good approximation.   The shape of the distribution is tabulated as a function of a/s_m in Table 6-1.

An additional point of interest is that on the basis of slender-body theory the potential difference at the panel trailing edge is independent of the rate of body expansion. This result is a consequence of the fact that the potential due to the body expansion is symmetrical above and below the horizontal plane of symmetry, and thus can add nothing to the potential difference at the trailing edge. The span-load distribution at the wing trailing edge is known to be affected by the rate of body expansion. In this instance, therefore, the potential difference and the span loading at the panel trailing edge are different.

Having now established the strength of the trailing-vortex sheet

TABLE 6-1. NONDIMENSIONAL CIRCULATION DISTRIBUTION OF WING PANEL, Γ/Γ_0

$\dfrac{\bar{y} - a}{s_m - a}$	a/s_m										
	0	0.1	0.2	0.3	0.4	0.5	0.6	0.7	0.8	0.9	1.0
0.0	1.000	1.000	1.000	1.000	1.000	1.000	1.000	1.000	1.000	1.000	1.000
0.2	0.980	0.969	0.966	0.966	0.968	0.970	0.972	0.974	0.976	0.978	0.980
0.4	0.917	0.897	0.887	0.884	0.886	0.889	0.894	0.900	0.906	0.911	0.917
0.5	0.866	0.843	0.832	0.827	0.828	0.832	0.837	0.844	0.851	0.859	0.866
0.6	0.800	0.776	0.762	0.757	0.756	0.760	0.766	0.773	0.782	0.791	0.800
0.7	0.714	0.690	0.676	0.669	0.668	0.671	0.677	0.685	0.694	0.704	0.714
0.8	0.600	0.578	0.565	0.558	0.556	0.558	0.563	0.571	0.580	0.589	0.600
0.85	0.527	0.507	0.494	0.488	0.486	0.488	0.492	0.499	0.507	0.517	0.527
0.90	0.436	0.419	0.408	0.402	0.400	0.402	0.406	0.411	0.419	0.427	0.436
0.92	0.392	0.376	0.366	0.361	0.359	0.360	0.364	0.369	0.376	0.384	0.392
0.94	0.341	0.328	0.319	0.314	0.312	0.313	0.316	0.321	0.327	0.334	0.341
0.96	0.280	0.269	0.261	0.257	0.256	0.257	0.259	0.263	0.268	0.274	0.280
0.98	0.199	0.191	0.186	0.182	0.182	0.182	0.184	0.187	0.190	0.194	0.199
0.99	0.141	0.135	0.131	0.129	0.129	0.129	0.130	0.132	0.135	0.138	0.141
1.00	0	0	0	0	0	0	0	0	0	0	0
$\dfrac{\bar{y}_{cg} - a}{s_m - a}$	0.785	0.769	0.760	0.757	0.757	0.759	0.763	0.768	0.774	0.780	0.785

directly behind the panel trailing edge, we are ready to consider the effect of the body. The vortices due to the body cannot form in the same manner as those due to the wing because of the absence of a well-defined trailing edge. The body imposes the condition that crossflow have zero velocity normal to the body. This condition can be satisfied by introducing an image vortex inside the body for each external vortex. The image vortex is placed on the radius vector to the external vortex a distance a^2/r from the axis. It has the opposite sense of rotation of the external vortex. Let us now prove that the velocity induced normal to the body by the combined actions of the external and image vortices is zero.

The velocities induced at any point on the circle by the external and image vortices as shown in Fig. 6-13 are

$$v_1 = \frac{\Gamma}{2\pi\lambda_1} \qquad v_2 = \frac{\Gamma}{2\pi\lambda_2} \tag{6-42}$$

The outward velocity normal to the body is

$$v_n = \frac{-\Gamma}{2\pi}\left(\frac{-\cos\theta_2}{\lambda_2} + \frac{\cos\theta_1}{\lambda_1}\right) \tag{6-43}$$

The geometric relationships of the figure based on the similarity of triangles OIP and EOP include

$$\frac{\lambda_1}{\lambda_2} = \frac{r_0}{a} \qquad \frac{\cos\theta_1}{\cos\theta_2} = \frac{\lambda_1}{\lambda_2} \tag{6-44}$$

It is thus clear that

$$v_n = 0 \tag{6-45}$$

The application of the boundary condition in this fashion is consistent with the slender-body assumption of the independence of the flow in the various crossflow planes.   Near the wing trailing edge the assumption is only approximate, as discussed in Sec. 6-3.

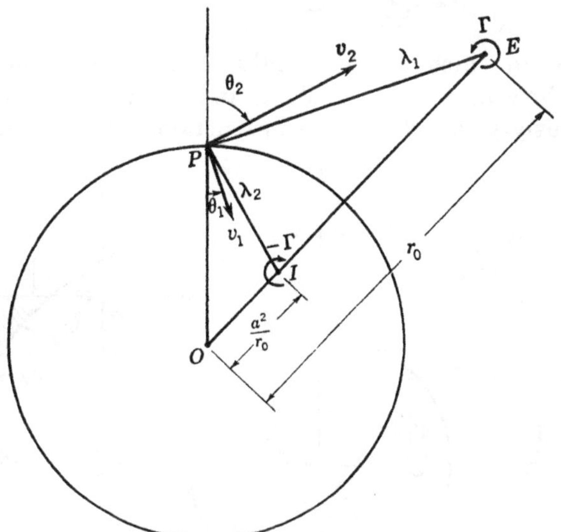

FIG. 6-13. External vortex with internal image vortex.

So far we have established the distribution of the strengths of the panel trailing-vortex sheet and of the image sheet.   Let us replace the continuous sheet by a finite number of trailing vortices, starting with the panel sheet.   We have at our disposal the number, strength, and spacing of the vortices.   The latter two quantities are not independent, but must be chosen so that, for the panel, the sum of the strengths of the vortices equals the circulation at the wing-body juncture,

$$\sum_{i=1}^{n} \Gamma_i = \Gamma_0 \tag{6-46}$$

where n is the number of external vortices per wing panel.   Another condition is that the lateral "center of gravity" of the panel vortex sheet must be constant, as discussed in connection with Eq. (6-27).   For a panel mounted on a body, the lift of the panel is $\rho_0 V_0 \Gamma_0$ per unit of exposed

semispan.   Since the system of a finite set of vortices must represent the
same panel lift as the continuous distribution, we have the lateral center-
of-gravity condition

$$\sum_{i=1}^{n} \Gamma_i(\bar{y}_i - a) = \int_{\Gamma_0}^{0} (\bar{y} - a)\, d\Gamma$$

$$= \text{area under panel circulation curve} \qquad (6\text{-}47)$$

or
$$\bar{y}_{cg} = \frac{\displaystyle\sum_{i=1}^{n} \Gamma_i \bar{y}_i}{\displaystyle\sum_{i=1}^{n} \Gamma_i} = \text{constant} \qquad (6\text{-}48)$$

The manner in which the two conditions Eqs. (6-46) and (6-47) can be
satisfied for the case of $n = 2$ is illustrated in Fig. 6-14.   The first condition is obviously satisfied by the construction.   The second condition is

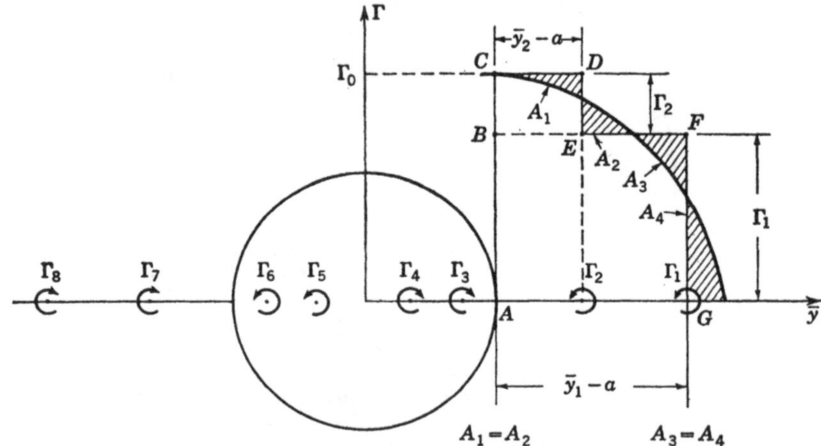

FIG. 6-14 Vortex model utilizing two external vortices per panel.

satisfied by making the crosshatched areas equal as shown.   Then the
areas $BCDE$ and $BFGA$ will add up to the area under the circulation
curve for the panel.   This fitting can be done graphically, or the spacings
can be calculated analytically if the theoretical shape of the circulation
curve is known.   Having established Γ_1, Γ_2, $\bar{y}_1$, and $\bar{y}_2$, we can readily
supply the remaining vortex strengths and positions.   In fact, the
strengths are

$$\Gamma_3 = -\Gamma_2 \qquad \Gamma_4 = -\Gamma_1 \qquad \Gamma_5 = \Gamma_1 \qquad \Gamma_6 = \Gamma_2$$
$$\Gamma_7 = -\Gamma_2 \qquad \Gamma_8 = -\Gamma_1 \qquad (6\text{-}49)$$

The positions are given for the more general case where the vortices may
not lie initially on the horizontal plane of symmetry, as for a high wing

condition or a banked condition.   If the coordinates of the vortices 1
and 2 are given by $\bar{y}_1$, $\bar{z}_1$, and $\bar{y}_2$, $\bar{z}_2$, then the positions of the other vortices
are

$$\bar{y}_3 = \frac{a^2\bar{y}_1}{\bar{y}_1{}^2 + \bar{z}_1{}^2} \qquad \bar{z}_3 = \frac{a^2\bar{z}_1}{\bar{y}_1{}^2 + \bar{z}_1{}^2}$$

$$\bar{y}_4 = \frac{a^2\bar{y}_2}{\bar{y}_2{}^2 + \bar{z}_2{}^2} \qquad \bar{z}_4 = \frac{a^2\bar{z}_2}{\bar{y}_2{}^2 + \bar{z}_2{}^2} \qquad (6\text{-}50)$$

$$\bar{y}_5 = -\bar{y}_4 \qquad \bar{y}_6 = -\bar{y}_3 \qquad \bar{y}_7 = -\bar{y}_2 \qquad \bar{y}_8 = -\bar{y}_1$$

$$\bar{z}_5 = \bar{z}_4 \qquad \bar{z}_6 = \bar{z}_3 \qquad \bar{z}_7 = \bar{z}_2 \qquad \bar{z}_8 = \bar{z}_1$$

For Fig. 6-14 we have $\bar{z}_1 = \bar{z}_2 = 0$.   Having thus constructed a system
of vortices to represent the wing-body combination, we can now calculate
the downstream paths of the vortices.

Before a calculation of the downstream paths, it is necessary to set up a
system of downstream wind and body axes.   Let the origin of the wind

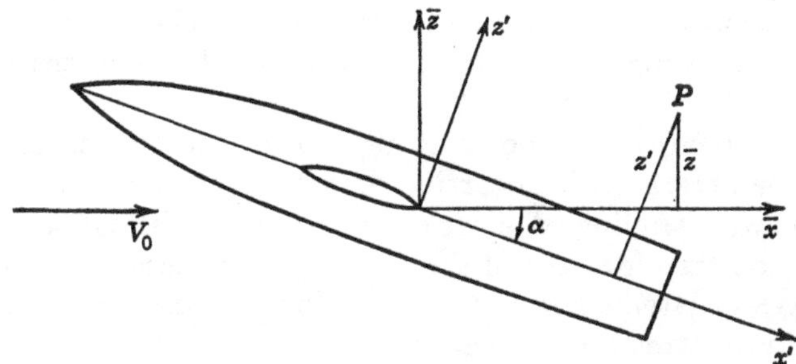

Fig. 6-15. Wind and body axes.

axes and the body axes coincide at the trailing-edge station of the wing,
as shown in Fig. 6-15.   Let $\bar{x}$, $\bar{y}$, and $\bar{z}$ be wind axes and x', y', and z' be
body axes.   Then, for reasonably small values of α, we have

$$x' = \bar{x}$$
$$y' = \bar{y} \qquad (6\text{-}51)$$
$$z' = \bar{z} + \alpha\bar{x}$$

To trace the paths of the vortices downstream, we must calculate
the downwash and sidewash velocities induced on each vortex by the
other external and image vortices, as well as the velocities induced by
the body crossflow.   For the body crossflow the downwash and sidewash
angles at any point $\bar{y}$, $\bar{z}$ are

$$\frac{-\bar{w}_c}{V_0} = -\alpha a^2 \frac{\bar{y}^2 - z'^2}{(\bar{y}^2 + z'^2)^2}$$

$$\frac{\bar{v}_c}{V_0} = -2\alpha a^2 \frac{\bar{y}z'}{(\bar{y}^2 + z'^2)^2} \qquad (6\text{-}52)$$

The downwash angle $-\bar{w}_{j(i)}/V_0$ induced on vortex j by i and the corresponding sidewash angle are, from Eq. (6-34),

$$\frac{-w_{j(i)}}{V_0} = \frac{\Gamma_i}{2\pi V_0} \frac{\bar{y}_j - \bar{y}_i}{(\bar{y}_j - \bar{y}_i)^2 + (\bar{z}_j - \bar{z}_i)^2}$$

$$\frac{v_{j(i)}}{V_0} = \frac{\Gamma_j}{2\pi V_0} \frac{\bar{z}_j - \bar{z}_i}{(\bar{y}_j - \bar{y}_i)^2 + (\bar{z}_j - \bar{z}_i)^2} \tag{6-53}$$

The total downwash and sidewash angles of vortex j are then

$$\epsilon = -\frac{\bar{w}_j}{V_0} = -\frac{\bar{w}_c}{V_0} + \sum_{\substack{i=1 \\ i \neq j}}^{4n} -\frac{\bar{w}_{j(i)}}{V_0}$$

$$\sigma = \frac{\bar{v}_j}{V_0} = \frac{\bar{v}_c}{V_0} + \sum_{\substack{i=1 \\ i \neq j}}^{4n} \frac{\bar{v}_{j(i)}}{V_0} \tag{6-54}$$

The summation is over the $4n$ vortices forming the external and image systems of each wing panel with the exception of the vortex in question, vortex j.

From Eq. (6-54) the velocity at any vortex location in any crossflow plane can be determined. Starting with the vortex strengths and positions at the wing trailing edge, we can calculate the initial angles of downwash and sidewash for each of the n external vortices of one wing panel. The changes in lateral and vertical positions $\Delta\bar{y}$ and $\Delta\bar{z}$ of these vortices in a short downstream distance $\Delta\bar{x}$ are

$$\Delta\bar{y} = \sigma(\Delta\bar{x})$$

$$\Delta\bar{z} = -\epsilon(\Delta\bar{x}) \tag{6-55}$$

for each of the n external vortices corresponding to one wing panel. The new positions of the image vortices are calculated with the help of Eq. (6-50). The process is again repeated for the new crossflow plane a distance $\Delta\bar{x}$ downstream, and the path of the vortex thus constructed in a step-by-step fashion.

Illustrative Example

As an example to fix some of the foregoing ideas, let us calculate the strengths and positions of the vortices representing the configuration of Fig. 6-16, and then make the initial calculation of the directions of the downstream vortex path.

The following data are given:

$$A = \tfrac{2}{3} \qquad \frac{a}{s_m} = 0.6 \qquad M_0 = 2$$

$$\alpha = 0.1 \text{ radian} \qquad n = 1 \qquad a = 1$$

The body radius is taken as unity so that the other dimensions can be considered as multiples of the body radius. We consider the simplest case of one vortex per wing panel.

A. Initial vortex strengths and positions by slender-body theory [from Eqs. (6-39) and (6-46)]:

$$\frac{\Gamma_0}{2\pi V_0 a} = \frac{\alpha(s_m{}^2 - a^2)}{\pi s_m a}$$

$$\frac{\Gamma_0}{2\pi V_0 a} = \left(\frac{s_m}{a} - \frac{a}{s_m}\right)\frac{\alpha}{\pi} = \frac{(1.667 - 0.600)\alpha}{\pi}$$

$$= \frac{1.067\alpha}{\pi}$$

$$\frac{\Gamma_1}{2\pi V_0 a} = \frac{\Gamma_0}{2\pi V_0 a} = \frac{1.067\alpha}{\pi}$$

$$\Gamma_2 = -\Gamma_1 \qquad \Gamma_3 = \Gamma_1 \qquad \Gamma_4 = -\Gamma_1$$

Since we have only one vortex per wing panel, it must lie at the lateral center of gravity of the vortex sheet $\bar{y}_{cg}$. Since the circulation distribution is nearly elliptical, this lateral distance is about at $\pi/4$ or 0.785 of the

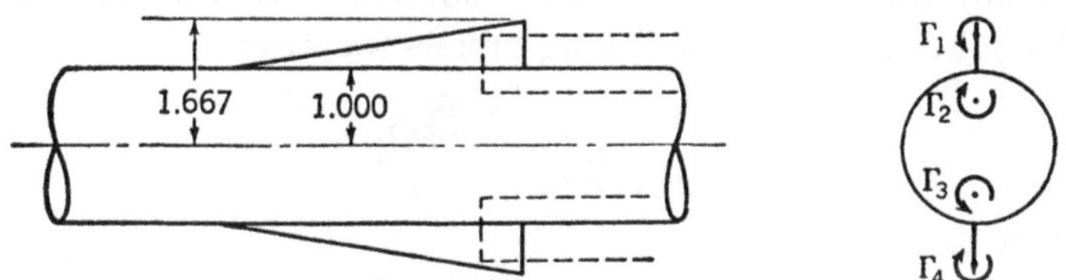

FIG. 6-16. Configuration of example calculation.

exposed semispan. The precise value from Table 6-1 is 0.763, but let us use the more approximate value.

$$\bar{y}_{cg} = a + \pi \frac{s_m - a}{4}$$

$$\bar{y}_{cg} = \bar{y}_1 = 1.000 + 0.785(0.667) = 1.525$$

$$\bar{y}_2 = \frac{a^2}{\bar{y}_1} \qquad \bar{y}_2 = \frac{1}{1.525} = 0.656$$

$$\bar{y}_3 = -\bar{y}_2 \qquad \bar{y}_4 = -\bar{y}_1$$

B. Initial downwash and sidewash angles: The downwash and sidewash angles of vortex 1 are given by Eq. (6-54). At the trailing edge of the wing panel, we have

$$\bar{z}_1 = z_1{}' = 0 \qquad \bar{z}_2 = z_2{}' = 0;$$

$$\bar{z}_3 = z_3{}' = 0 \qquad \bar{z}_4 = z_4{}' = 0$$

Thus for vortex 1

$$\left(\frac{\epsilon}{\alpha}\right)_1 = \frac{-1^2(1.525^2 - 0^2)}{(1.525^2 + 0^2)^2} + \frac{1.067}{\pi}\left[\frac{1}{1.525 - 0.656}\right.$$
$$\left. - \frac{1}{1.525 - (-0.656)} + \frac{1}{1.525 - (-1.525)}\right]$$
$$= -0.430 + 0.347 = -0.083 \text{ radian} = -4.8°$$
$$\sigma_1 = 0$$

The initial downwash angle of the vortex is negative, indicating that it is inclined above the free-stream direction. This is a result of the relatively large body, the upwash of which more than offsets the downwash induced by the other vortices. The initial sidewash is zero since the vortices all lie on a horizontal line. To continue the process, we determine the new values of $\bar{y}_1$ and $\bar{z}_1$ a short distance downstream by Eq. (6-55) for an arbitrarily chosen increment $\Delta\bar{x}$. We relocate the image vortex, and repeat the calculation. The second step will give a nonzero sidewash angle. Whether the size of the chosen downstream increment $\Delta\bar{x}$ is sufficiently small can be determined by inspecting the calculated path.

C. Calculation of initial vortex strength by method of Sec. 5-3: According to Eq. (6-25) the circulation at the root chord of the triangular wing formed by joining the exposed wing panels together is

$$\left(\frac{\Gamma_0}{2\pi V_0 a}\right)_W = \frac{(cc_l)_{\bar{y}=0}}{4\pi a}$$

where from Eq. (2-39),

$$(cc_l)_{\bar{y}=0} = \frac{16(\tan \omega)s\alpha/A}{E(1 - B^2 \tan^2 \omega)^{1/2}}$$

and where ω = semiapex angle
$$B^2 = M_0^2 - 1$$
$\quad\quad s$ = wing semispan
$\quad\quad E$ = complete elliptic integral of the second kind
$\quad\quad A$ = wing aspect ratio

and

$$(cc_l)_{\bar{y}=0} = \frac{16(\frac{1}{6})(0.667)(\frac{3}{2})\alpha}{E(1 - \frac{3}{36})^{1/2}}$$
$$= \frac{2.667\alpha}{1.093} = 2.44\alpha$$
$$\left(\frac{\Gamma}{2\pi V_0 a}\right)_W = \frac{2.44\alpha}{4(1)\pi} = \frac{0.61\alpha}{\pi}$$

In Sec. 5-3 it was shown that the lift on the wing panels is greater by a multiplicative factor K_W than the lift on the wing alone. If we neglect the small difference in the shape of the panel potential distribution and the elliptical distribution of the wing alone, then we can multiply the circulation of the wing alone by K_W to obtain that for the panels of the wing-body combination

$$\left(\frac{\Gamma}{2\pi V_0 a}\right)_{W(B)} = K_W \left(\frac{\Gamma}{2\pi V_0 a}\right)_W$$
$$K_W = 1.555 \qquad \text{(Table 5-1)}$$
$$\left(\frac{\Gamma}{2\pi V_0 a}\right)_{W(B)} = \frac{1.555(0.61)\alpha}{\pi} = \frac{0.95\alpha}{\pi}$$

It is to be noted that this value of $0.95\alpha/\pi$ is slightly lower than the value of $1.067\alpha/\pi$ calculated by slender-body theory. This result might be expected since slender-body theory is known to overpredict the lift of wings alone. See Fig. 5-15. This latter procedure of determining vortex strength is definitely to be preferred to the slender-body method for large aspect ratios. In fact, if a more accurate determination of the potential difference at the panel trailing edge is known than that based on slender-body theory, it should be used in determining the initial vortex strengths.

Slender-body theory, or any linear potential theory for that matter, yields a simple result for the effects of roll angle on the vorticity distribution along the panel trailing edge. Under the combined effects of pitch and roll, the crossflow velocity can be resolved into components $V_0 \alpha_c \cos \varphi$ normal to the plane of the wing and $V_0 \alpha_c \sin \varphi$ parallel to it, as in Fig. 6-17. The velocity component parallel to the plane of the wing produces no potential difference across the wing. Only the normal velocity component produces a potential difference at the panel trailing edge, a difference which is the same at corresponding points on each panel. The vortex pattern is thus symmetrical since the potentials due to the normal and parallel velocity components are additive. This is not to say that the load on each panel is the same; in fact, the downgoing panel carries more load than the upgoing panel. The example is another one where the span loading and potential distributions are not similar, because the loading includes a coupling effect between the two potentials caused by the squared terms of Bernoulli's equation.

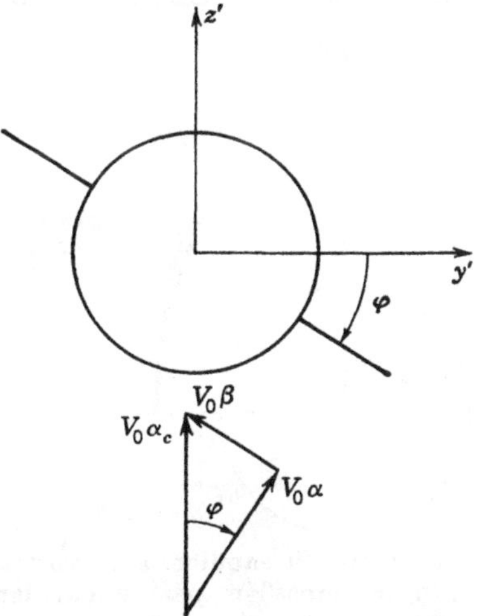

FIG. 6-17. Angle of attack and angle of sideslip components of crossflow velocity.

6-5. Factors Influencing Vortex Paths and Wake Shape behind Panels of Planar Wing and Body Combinations

From the calculation procedure of the preceding section, a number of interesting results have been obtained concerning the characteristics of the vortex paths and wake shape behind the panels of a wing-body

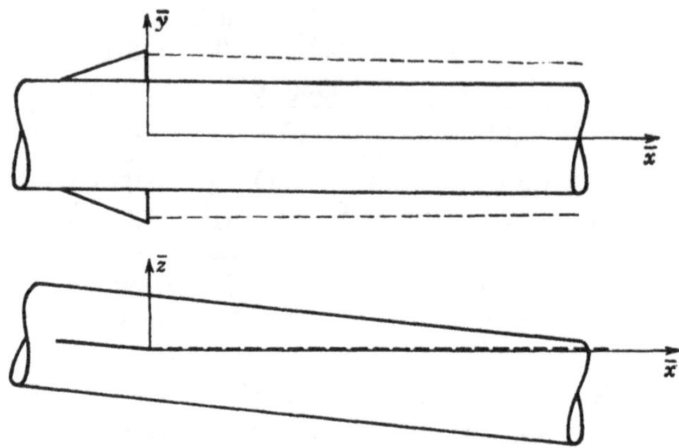

FIG. 6-18. Paths of vortices behind panel of wing-body combination; $a/s_m = 0.6$.

combination. Let us first consider the paths of the vortices used in the preceding example. The actual paths as computed in accordance with the sample calculation are shown in Fig. 6-18. The vortex lines leaving the trailing edge have an upwash component. The associated upward motion carries it out of the high upwash field close to the body to a lower upwash field above. As a consequence the vortex path acquires a component of downwash velocity—but always lies above the extended chord plane. The vortex moves continuously inward toward an asymptotic spacing given by Eq. (4-89). Let us observe the vortex paths in the crossflow plane. It is possible to calculate these paths from Eq. (4-88). It is of interest to note that the paths in the crossflow plane do not depend on angle of attack. The slope of the paths in the crossflow plane is ϵ/σ, which by Eqs. (6-52) to (6-54) is independent of α since Γ increases linearly with α. Thus a calculation for a specific angle of attack can be utilized for all angles of attack and needs to be done only once. A vortex path calculated by the

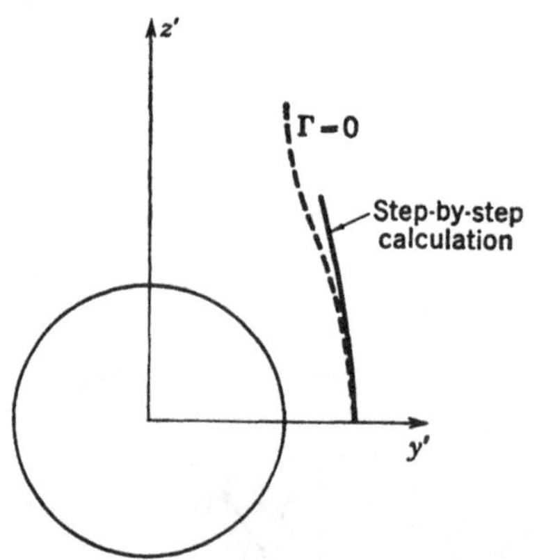

FIG. 6-19. Streamline and vortex path for crossflow past a circular cylinder.

step-by-step method is compared in Fig. 6-19 with the crossflow stream-line for flow past a cylinder.

The factors determining the vortex paths in the crossflow plane are the parameters $\Gamma/2\pi V_n a$ and y_1/a, where V_n is the crossflow velocity, and y_1 is the vortex spanwise location on the horizontal plane of symmetry. The paths for various values of $\Gamma/2\pi V_n a$ are shown in Fig. 6-20. For a

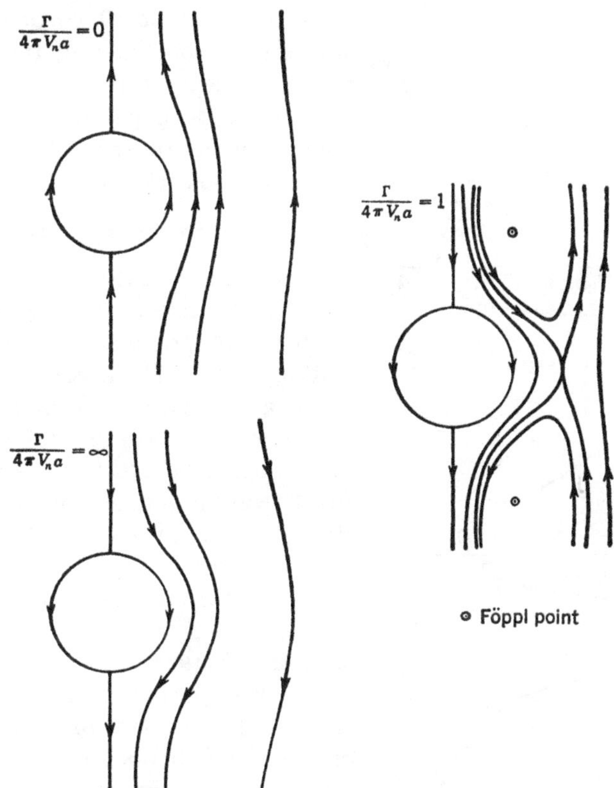

FIG. 6-20. Vortex paths in crossflow past a circular cylinder.

value of the vortex parameter of zero, the paths are simply the stream-lines for potential flow past a circular cylinder. The paths for a vortex parameter of infinity correspond to the motion of a pair of vortices in the presence of a cylinder in still air. They move downward in two straight parallel paths with no body present, but their paths are bulged out by the presence of the cylinder.[5] For finite nonzero values of the vortex parameter there are stationary points on each side of the cylinder. The stationary points not on the axis correspond to "Föppl points" as given by Eqs. (4-82) and (4-83). The stationary points on the axis mark the lateral boundaries, outside of which the vortices move upward as for the zero case, and inside of which the vortices move downward as for the infinite case. Other parameters which can have an effect on the type of vortex motion behind the panels of a wing-body combination include the number of vortices into which the vortex sheet rolls up, the cross-sectional shape of the body, and variations of vortex strength with axial position.

Turning now to the vortex shape, we pass from a model of one vortex per panel to one of many vortices per panel. A step-by-step calculation made with about 10 vortices per panel will give a good idea of the manner in which the wake rolls up. Such calculations have been performed by Rogers,[4] where the wake shape behind a wing-body combination of

$$A = \tfrac{2}{3} \quad \text{and} \quad a/s_m = 0.2$$

is compared with the wake shape calculated by Westwater for an elliptical distribution of potential difference at the trailing edge. The wake shapes are very much alike in the two cases. The criterion for the rate of rolling up of the vortex sheet behind wings given by Eq. (6-14) can be applied with the same degree of accuracy to the wake behind the panels of wing-body combinations, provided the parameter a/s_m is not too large. It should be borne in mind, however, that the shape of the circulation distribution is important in determining wake shape. The manner in which the shape of the curve affects the shape of the vortex wake is shown qualitatively by Fig. 6-21. The elliptical circulation distribution rolls up into a single vortex in the well-known manner. A triangular distribution must roll up in the same manner at both ends, and will eventually form two vortices rotating in the same

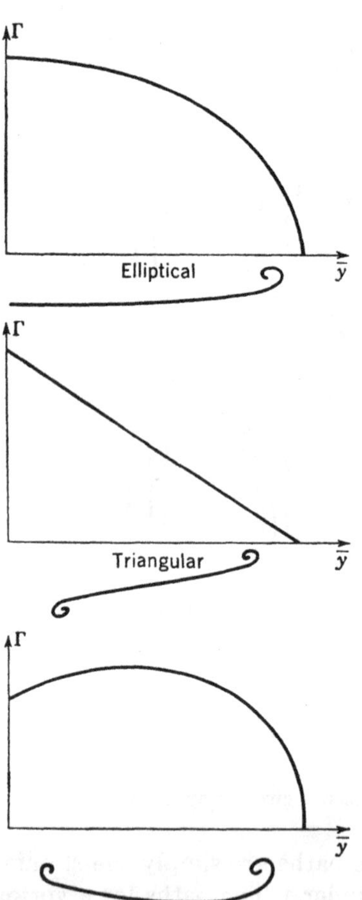

Fig. 6-21. Effect of circulation distribution on wake shape.

direction. A distribution for which the circulation is a maximum somewhere on the span will probably form two vortices of unequal strength, rotating in opposite directions. The shape of the circulation distribution thus has an effect on the number of vortices into which the wake rolls up, and their direction of rotation.

6-6. Factors Influencing Downwash Field behind Panels of Planar Wing and Body Combinations

One purpose of determining the vortex paths and wake shapes behind wing panels is to obtain the downwash and sidewash fields. Consider the downwash distribution across the span of a hypothetical tail surface located on the horizontal plane of symmetry a distance of 10 body radii behind the panel trailing edge. One question of interest is how important the contributions of the body crossflow, the external vortices, and the image vortices are to the downwash at the tail surface. In this connection let us evaluate the contribution of these items to the downwash for the following example.

Illustrative Example

$$A = \tfrac{2}{3} \qquad \frac{a}{s_m} = 0.6 \qquad \alpha = 5°$$

For this case the step-by-step calculation gives the vortex strengths and positions:

$$\frac{\Gamma}{2\pi V_0 a} = \frac{0.95\alpha}{\pi} \qquad \Gamma_2 = -\Gamma_1 \qquad \Gamma_3 = \Gamma_1 \qquad \Gamma_4 = -\Gamma_1$$

$$\bar{y}_1 = 1.39 \qquad \bar{z}_1 = 0.044 \qquad z_1' = 0.919$$

$$\bar{y}_2 = \frac{a^2\bar{y}_1}{\bar{y}_1{}^2 + z_1'{}^2} = 0.501 \qquad z_2' = \frac{a^2 z_1'}{\bar{y}_1{}^2 + z_1'{}^2} = 0.330$$

$$\bar{y}_3 = -.501 \qquad\qquad z_3' = 0.330$$

$$\bar{y}_4 = -1.39 \qquad\qquad z_4' = 0.92$$

The downwash will be calculated at the point $\bar{y} = 2$, $z' = 0$. The downwash at this point calculated from Eq. (6-54) is, for the various components:

Body crossflow:

$$\left(\frac{\epsilon}{\alpha}\right)_B = -1^2 \frac{2^2 - 0^2}{(2^2 + 0^2)^2} = -\frac{1}{4}$$

Vortex 1:

$$\left(\frac{\epsilon}{\alpha}\right)_1 = 0.95 \frac{1.392 - 2.000}{\pi[(1.392 - 2.000)^2 + (0.919)^2]} = -0.152$$

Vortex 2:

$$\left(\frac{\epsilon}{\alpha}\right)_2 = -0.95 \frac{0.500 - 2.000}{\pi[(0.500 - 2.000)^2 + (0.330)^2]} = 0.192$$

Vortex 3:

$$\left(\frac{\epsilon}{\alpha}\right)_3 = 0.95 \frac{-0.500 - 2.000}{\pi[(-0.500 - 2.000)^2 + (0.330)^2]} = -0.119$$

Vortex 4:

$$\left(\frac{\epsilon}{\alpha}\right)_4 = -0.95 \frac{-1.392 - 2.000}{\pi[(-1.392 - 2.000)^2 + (0.919)^2]} = 0.083$$

The downwash due to the external vortices is

$$\epsilon_0 = \left[\left(\frac{\epsilon}{\alpha}\right)_1 + \left(\frac{\epsilon}{\alpha}\right)_4\right]\alpha = -0.35°$$

The downwash due to the image vortices is

$$\epsilon_i = \left[\left(\frac{\epsilon}{\alpha}\right)_2 + \left(\frac{\epsilon}{\alpha}\right)_3\right]\alpha = 0.37°$$

The downwash distributions across the horizontal plane of symmetry are shown in Fig. 6-22. The contributions of the external and image

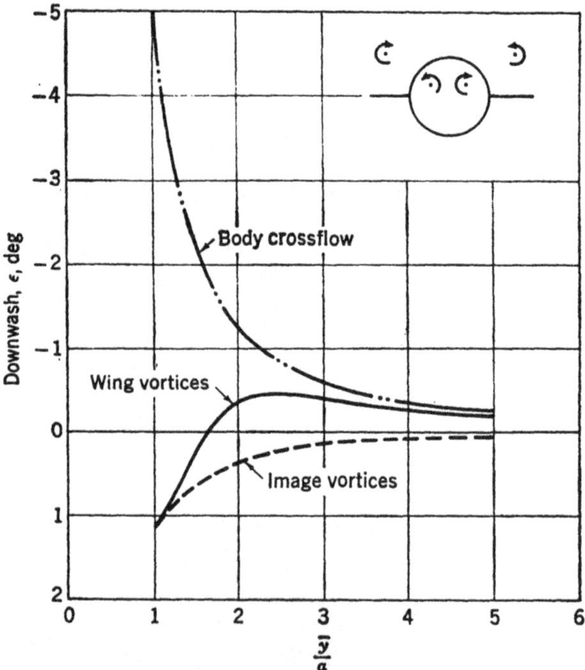

Fig. 6-22. Components of downwash on horizontal plane of symmetry 10 radii behind wing panel; $A = \frac{2}{3}$, $M_0 = 2$, $a/s_m = 0.6$.

vortices in this case are largely compensating, and the total downwash field is dominated by the body upwash. This result is typical of combinations with large ratios of body radius to combination semispan, since the body is then important in comparison to the panels. Figure 6-23 is presented for the same conditions as Fig. 6-22, except that a/s_m is 0.2

rather than 0.6. It is clear that here the wing vortices dominate the downwash field as might be expected.

One question that arises is: How accurate is the downwash field calculated by a model based on one vortex per wing panel, in comparison with one based on many vortices per panel? The answer to this question depends on several factors, one of which is the use to which the downwash is to be put. If it is to be used to calculate the gross tail load, then its average effect on the tail is important, and the precise shape of the downwash variation across the tail is of secondary concern. Here one vortex will usually give sufficiently precise answers in many practical cases.

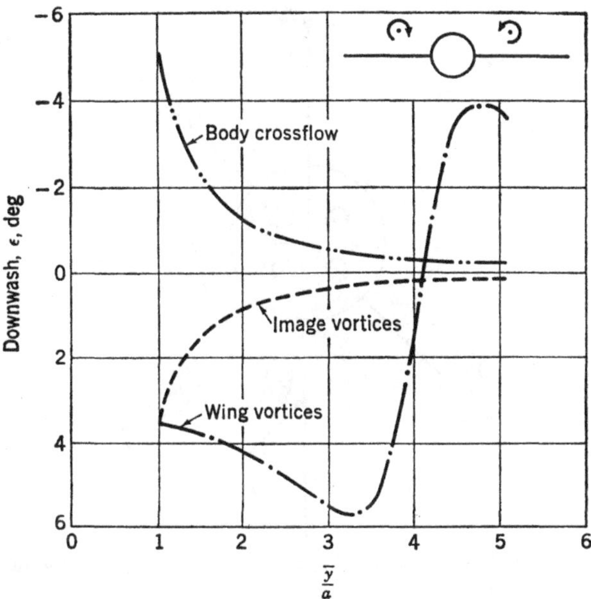

FIG. 6-23. Components of downwash on horizontal plane of symmetry 10 radii behind wing panel; $A = \frac{2}{3}$, $M_0 = 2$, $a/s_m = 0.2$.

Also, if the vortex sheet is essentially rolled up into one vortex—as for an elliptical circulation distribution—then a model using one vortex per wing panel is a good one, being in good accord with the physical facts. On the other hand, if the sheet rolls up into two vortices—as for a triangular distribution—then a model using two vortices per wing panel would give a good representation. In cases where the precise distribution of downwash across the tail panel is important, a scheme using many vortices per panel will be required.

6-7. Cruciform Arrangements

In this section we will discuss the application of the step-by-step procedure to the calculation of vortex paths behind cruciform configurations and in addition will present an analytical solution for a vortex system representing a cruciform wing arrangement. The section discusses wake shapes and "leapfrogging."

The step-by-step calculative process presented in connection with planar wing-body combinations can be readily adapted to cruciform arrangements, provided the initial vortex positions and strengths are specified. The procedure is adaptable to any bank angle, ratio of vertical to horizontal spans, ratio of body radius to configuration semispan, numbers and positions of vortices, subject only to the usual conditions on the sum of the vortex strengths and lateral center of gravity. To calculate the vortex strengths and positions, consider the model shown in Fig.

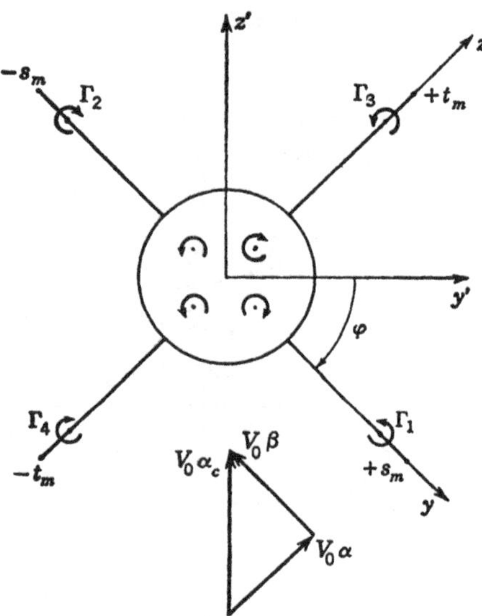

FIG. 6-24. Vortex model of cruciform wing-body arrangement.

6-24. It should be noted that y' and z' are the body axes for zero roll angle. The crossflow velocity is broken into a component $V_0\alpha$ normal to the "horizontal" panels and a component $V_0\beta$ normal to the "vertical" panels. The potential produced by $V_0\alpha$ will be different on the upper and lower sides of the horizontal surfaces, and will thus cause a potential difference between them. The potential produced by $V_0\beta$, on the other hand, is the same on the upper and lower sides of the horizontal panels. Since the potentials due to $V_0\alpha$ and $V_0\beta$ are additive in linear potential theory, it is clear that only the velocity component $V_0\alpha$ produces potential difference across the horizontal panels, while $V_0\beta$ produces potential difference across the vertical panels. In the computation of the potential difference across the wing panels, the horizontal and vertical panels can thus be treated as planar configurations acting at their own angles of attack. The circulation distributions of Table 6-1

are directly applicable to cruciform arrangements.   If we desire to use a
model based on a single vortex per wing panel, then the vortex strengths
can be calculated from Eq. (6-39).

$$\Gamma_1 = -\Gamma_2 = 2V_0\alpha_c \cos\varphi \frac{s_m{}^2 - a^2}{s_m}$$

$$\Gamma_3 = -\Gamma_4 = 2V_0\alpha_c \sin\varphi \frac{t_m{}^2 - a^2}{t_m}$$

(6-56)

The lateral positions at the trailing edge given with reference to the y',z'
coordinate system (Fig. 6-24) are

$$y_1' = -y_2' = \left[a + \frac{y_{cg} - a}{s_m - a}(s_m - a) \right] \cos\varphi$$

$$y_3' = -y_4' = \left[a + \frac{y_{cg} - a}{t_m - a}(t_m - a) \right] \sin\varphi$$

(6-57)

The parameters involving y_{cg} can be obtained from Table 6-1.   With the
initial vortex strengths and positions now determined for a model of one
vortex per panel, we can carry out the step-by-step calculation.   This

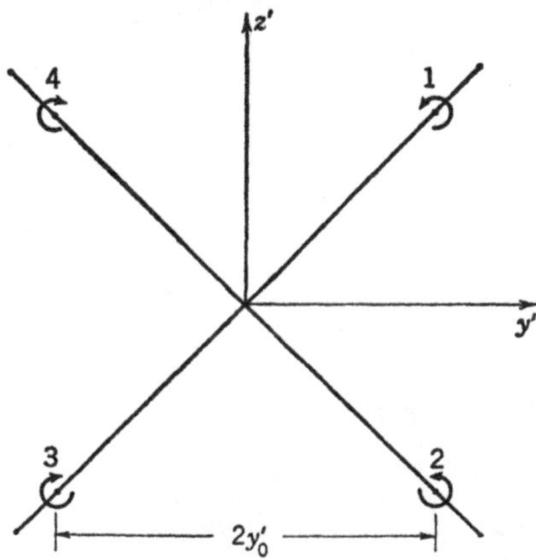

FIG. 6-25. Initial vortex positions for cruciform wing; $\varphi = 45°$.

model is sufficiently accurate for most downwash and sidewash determi-
nations.   For $\varphi = 45°$ and panels of equal span, the vortex strengths will
all be the same.

An analytical solution due to Spreiter and Sacks[6] is known for the
vortex paths for four equal vortices associated with an equal-span cruciform wing arrangement for a roll angle of 45°, as illustrated in Fig. 6-25.
The conditions for this case are

$$a = 0 \qquad s_m = t_m \qquad \varphi = 45° \qquad \frac{y_{cg}}{s_m} = \frac{\pi}{4}$$

If the lateral position of vortex 1 at the trailing edge is denoted by y_0', then

$$\frac{y_0'}{s_m} = \frac{\pi}{4(2)^{1/2}} \tag{6-58}$$

Let the parameter y_1', specifying the lateral position of vortex 1 at any downstream position, be given by

$$\sin^2 \tau = 1 - 3\left(\frac{y_1'}{y_0'} - 1\right)^2 \tag{6-59}$$

where $\tau = \pi/2$ corresponds to the trailing edge, and greater values of τ correspond to positions behind the trailing edge. The downstream distance d corresponding to the lateral position y_1' is

$$\frac{d}{y_0'} = \frac{\pi^3 A}{4C_L'}\left[-1.0834 + {}^{16}\!/_3 E(\tfrac{1}{2}, \tau) - 4F(\tfrac{1}{2}, \tau) - \frac{\sin \tau \cos \tau}{(1 - \tfrac{1}{4}\sin^2 \tau)^{1/2}} \right] \tag{6-60}$$

where F and E are incomplete elliptic integrals of the first and second kinds, and C_L' is the Z' force coefficient based on the area of one set of wing panels, the force being in the combined plane $x'z'$. The corresponding vertical position of vortex 1 is

$$\frac{z_1'}{y_0'} = {}^4\!/_3\,[1.4675 - E(\tfrac{1}{2}, \tau)] + \frac{\sin \tau(\cos \tau + 3^{1/2})}{2(1 - \tfrac{1}{4}\sin^2 \tau)^{1/2}} + \frac{2}{\pi}\frac{d}{y_0'}\frac{C_L'}{A} \tag{6-61}$$

The lateral position of vortex 2 is given by the condition that the lateral center of gravity of vortices 1 and 2 remain unchanged.

$$y_2' = 2y_0' - y_1' \tag{6-62}$$

and the vertical position by the condition

$$\left(\frac{z_1' - z_2'}{2y_0'}\right)^2 = \frac{3y_1'/y_0' - \tfrac{3}{2}(y_1'/y_0')^2 - 1}{\tfrac{1}{2}(y_1'/y_0)^2 - y_1'/y_0' + 1} \tag{6-63}$$

Illustrative Example

To show how the vortex positions can be calculated in a particular instance, consider the problem of determining the downstream distance and vortex positions for which vortex 1 has decreased its y' to 80 per cent of its value at the trailing edge.

$$\frac{y_1'}{y_0} = 0.8 \qquad \frac{y_0'}{s_m} = \frac{\pi}{4(2)^{1/2}} = 0.555 \qquad \frac{y_0'}{b} = 0.2775$$

$$\frac{y_1'}{s_m} = (0.8)(0.555) = 0.444$$

From Eq. (6-59)

$$\sin^2 \tau = 1 - 3(0.8 - 1.0)^2 = 0.880$$
$$\sin \tau = 0.938 \qquad \cos \tau = -0.347$$
$$\tau = 110°20'$$

In preparation for calculating the downstream position, we note that

$$E(\tfrac{1}{2}, \tau) = 2E(\tfrac{1}{2}, \pi/2) - E(\tfrac{1}{2}, \pi - \tau) = (2)(1.467) - 1.158$$
$$= 1.776$$
$$F(\tfrac{1}{2}, \tau) = 2K(\tfrac{1}{2}) - F(\tfrac{1}{2}, \pi - \tau) = 2(1.686) - 1.280$$
$$= 2.090$$

From Eq. (6-60) the downstream position is

$$\frac{d}{y_0'} \frac{C_L'}{A} = \frac{\pi^3}{4}\left[-1.083 + \tfrac{16}{3}(1.776) - 4(2.090) - \frac{0.938(-0.347)}{[1 - \tfrac{1}{4}(0.88)]^{1/2}} \right]$$
$$= 3.04$$
$$\frac{d}{b} \frac{C_L'}{A} = 3.04(0.2775) = 0.84$$

The vertical position of vortex 1 from Eq. (6-61) is

$$\frac{z_1'}{y_0'} = \tfrac{4}{3}(1.467 - 1.776) + \frac{0.938(1.732 - 0.347)}{2(1 - 0.22)^{1/2}} + \frac{2}{\pi}(3.04)$$
$$= 2.256$$
$$\frac{z_1'}{s_m} = 2.256(0.555) = 1.251$$

The vertical position of vortex 2 from Eq. (6-63) is

$$\left(\frac{z_1' - z_2'}{2y_0'}\right)^2 = \frac{3(0.8) - \tfrac{3}{2}(0.8)^2 - 1}{\tfrac{1}{2}(0.8)^2 - 0.8 + 1}$$
$$\frac{z_1' - z_2'}{2y_0'} = 0.92$$
$$\frac{z_2'}{y_0'} = 2.256 - 2(0.92) = 0.416$$
$$\frac{z_2'}{s_m} = 0.416(0.555) = 0.230$$

The lateral position of vortex 2 from the constancy the lateral center of gravity is

$$\frac{y_2'}{s_m} = \frac{2y_0'}{s_m} - \frac{y_1'}{s_m} = 1.110 - 0.444 = 0.666$$

The vortex paths as calculated from the analytical solution are shown in Fig. 6-26 together with the pattern in the crossflow plane. The upper vortices tend to pass downward and inward between the two lower vortices and to "leapfrog." If the leapfrog distance is taken as that when all four vortices are in the same horizontal plane, Eq. (6-60) yields the relationship for d_L.

$$\frac{d_L}{s_m} = \frac{4.66A}{C_L{}'} \tag{6-64}$$

This result is the same form as that of Eq. (6-14) for the distance for an elliptical vortex sheet to roll up, according to Kaden. Such a form would

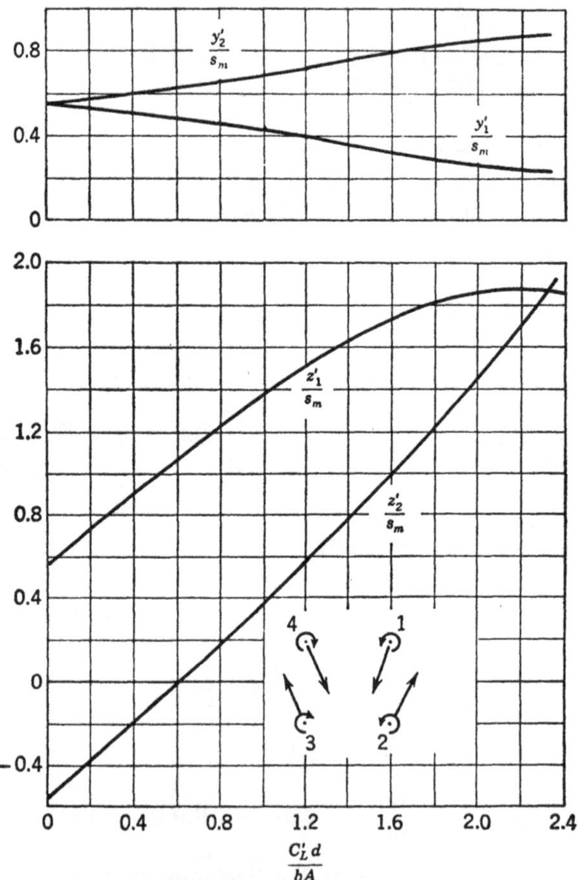

FIG. 6-26. Vortex positions behind panels of cruciform wing; $\varphi = 45°$.

be expected on the basis of the previous dimensional reasoning. For the distance to roll up, the constant is 0.56, so that this distance is about one-eighth the distance to leapfrog.

So far we have concerned ourselves with models based on one vortex per panel. Some calculations by Spreiter and Sacks[6] of 10 vortices per panel of a cruciform wing arrangement give some insight into the shape of the vortex sheet. For panels of equal span and for $\varphi = 45°$, the wake shapes are shown in Fig. 6-27 for several distances behind the trailing edge.   The vortex sheets curl up into two wakes similar to those observed behind wings with elliptical span loading.   The upper vortices move inward and downward with respect to the outer vortices, which move outward and upward with respect to the wing axis.   The upper vortices in approaching one another are accelerated downward between the lower vortices and produce the so-called "leapfrogging."

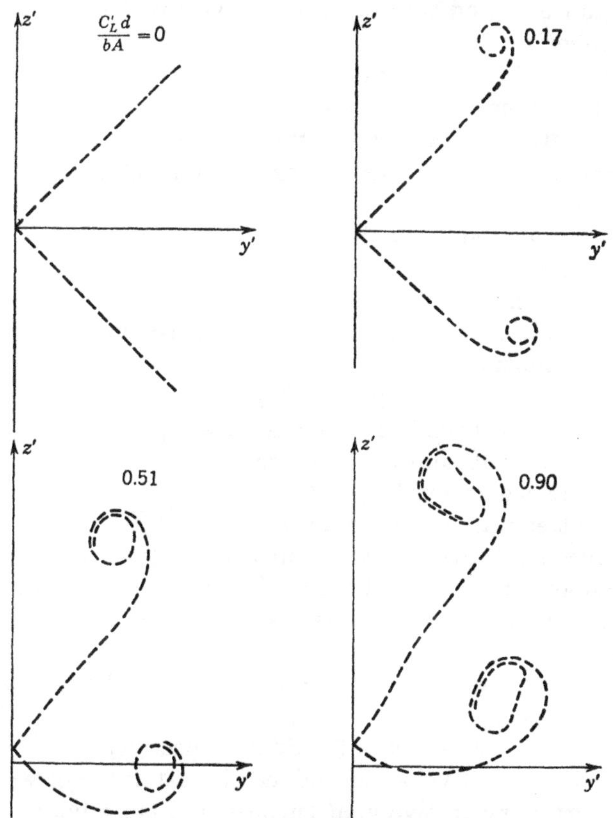

FIG. 6-27. Wake shapes behind a cruciform wing; $\varphi = 45°$.

Series solutions for vortex paths can be developed that provide sufficient accuracy for many missile applications, as for example, those of Alksne.[7]

SYMBOLS

a	radius of body
A	aspect ratio of wing alone, or wing alone formed by two opposing wing panels
b	span of wing alone
b_0	function of x occurring in $W(\mathfrak{z})$

b_v	span of horseshoe vortex
B	$(M_0{}^2 - 1)^{1/2}$
c	wing chord at arbitrary spanwise station
c_l	section lift coefficient at arbitrary spanwise station
C_L	lift coefficient
$C_L{}'$	coefficient for force along z' axis
d	distance behind trailing edge of cruciform wing
d_L	value of d for "leapfrogging"
e	distance behind trailing edge that vortex sheet is "completely" rolled up
E	elliptic integral of second kind
F	elliptic integral of first kind
i, j	summation indices for several vortices
k	numerical constant depending on shape of span-load distribution
K_W	lift ratio given in Table 5-1
L	lift force
M_0	free-stream Mach number
n	number of external vortices per wing panel replacing span-load distribution
P	pressure coefficient $(p - p_0)/q_0$
P^+	pressure coefficient on lower surface
P^-	pressure coefficient on upper surface
ΔP	loading coefficient $P^+ - P^-$
q_0	free-stream dynamic pressure
q_p	induced velocity due to segment of vortex line
q_t	component of flow velocity tangent to circulation contour
r	distance between two vortices
s_0	local semispan of wing alone
s_m	maximum semispan of wing alone or wing-body combination
S_W	area of wing alone
$\bar{u}, \bar{v}, \bar{w}$	flow velocities along $\bar{x}$, $\bar{y}$, and $\bar{z}$, respectively
v_1, v_2	magnitudes of velocities induced by external vortex and image vortex, respectively, at surface of circular body
$v_{2(1)}$	velocity induced at vortex 2 by vortex 1
v_i	velocity induced by vortex system
v_n	velocity induced normal to circular body by external vortex and its image vortex
$\bar{v}_c, \bar{w}_c$	values of $\bar{v}$ and $\bar{w}$ due to potential crossflow around a circular cylinder
V	local flow velocity along streamline
V_0	free-stream velocity
V_n	$V_0 \sin \alpha_c$, crossflow velocity
$\bar{w}_B, \bar{w}_C$	values of $\bar{w}$ at points B and C

$W(\mathfrak{z})$ complex potential, $\phi + i\psi$

x, y, z missile axes of symmetry for $\alpha_c \neq 0$ and $\varphi \neq 0$

x', y', z' missile axes of symmetry for $\alpha_c \neq 0$ and $\varphi = 0$

$\bar{x}, \bar{y}, \bar{z}$ missile axes of symmetry for $\alpha = 0$ and $\varphi = 0$

y_0' lateral position of vortex 1 at trailing edge in x', y', z' coordinates (cruciform wing)

y_1' lateral position of vortex 1 at stations behind trailing edge (cruciform wing)

y_{cg} lateral center of gravity of right half of trailing-vortex sheet

y_i lateral position of ith vortex

$\bar{y}_v, \bar{z}_v$ values of $\bar{y}$ and $\bar{z}$ for infinite line vortex in streamwise direction

$\mathfrak{z}$ $\bar{y} + i\bar{z}$

α angle of attack

α_c included angle

β angle of sideslip

γ_1, γ_2 see Fig. 6-8

Γ vortex strength of circulation, positive counterclockwise

Γ_0 magnitude of circulation at wing-body juncture or root chord of wing alone

Γ_i vortex strength of ith vortex

$d\Gamma/d\bar{y}$ strength of trailing vorticity per unit span

δ angle from $\bar{x}$ axis to plane of vortex sheet, positive downward

ϵ downwash angle

θ_1, θ_2 see Fig. 6-13

λ_1, λ_2 see Fig. 6-13

ρ_0 free-stream density

σ sidewash angle

τ parameter specifying distance of vortex behind trailing edge of cruciform wing

φ angle of bank

ϕ_α potential due to $V_0 \cos \beta$, Fig. 6-4

ϕ_β potential due to $V_0 \sin \beta$, Fig. 6-4

ϕ^+ potential on lower surface

ϕ^- potential on upper surface

$\phi_{W(B)}$ potential of wing panel in presence of body

$\Delta\phi_{te}$ $(\phi^- - \phi^+)$ at trailing edge of wing

ω semiapex angle of triangular wing

REFERENCES

1. Kaden, H.: Aufwicklung einer unstabilen Unstetigkeitsflache, *Ingr.-Arch.*, vol. 11, pp. 140–168, 1931.

2. Westwater, F. L.: Rolling Up of a Surface of Discontinuity behind an Airfoil of Finite Span, *Brit. ARC R & M* 1692, 1935.

3. Spreiter, John R., and Alvin H. Sacks: The Rolling Up of the Trailing Vortex

Sheet and Its Effect on the Downwash behind Wings, *J. Aeronaut. Sci.*, vol. 18, no. 1, pp. 21–32, 72, 1951.

4. Rogers, Arthur Wm.: Application of Two-dimensional Vortex Theory to the Prediction of Flow Fields behind Wings of Wing-Body Combinations at Subsonic and Supersonic Speeds, *NACA Tech. Notes* 3227, 1954.

5. Milne-Thompson, L. M.: "Theoretical Hydrodynamics," 2d ed., pp. 330–331, The Macmillan Company, New York, 1950.

6. Spreiter, John R., and Alvin H. Sacks: A Theoretical Study of the Aerodynamics of Slender Cruciform-wing Arrangements and Their Wakes, *NACA Tech. Repts.* 1296, 1957.

7. Alksne, Alberta Y.: Determination of Vortex Paths by Series Expansion Technique with Application to Cruciform Wings, *NACA Tech. Repts.* 1311, 1957.

8. Trockenbrodt, E.: Experimentelle Theoretische Untersuchungen an symmetrisch angestromten Pfeil- und Deltaflugeln, *Z. Flugwiss.*, August, 1954.

9. DeYoung, John: Theoretical Additional Span Loading Characteristics of Wings with Arbitrary Sweep, Aspect Ratio, and Taper Ratio, *NACA Tech. Notes* 1491, 1947.

10. Lagerstrom, P. A., and M. E. Graham: Aerodynamic Interference in Supersonic Missiles, *Douglas Aircraft Co. Rept.* SM-13743, 1950.

11. Silverstein, Abe, and S. Katzoff: Design Charts for Predicting Downwash Angles and Wake Characteristics behind Plain and Flapped Wings, *NACA Tech. Repts.* 648, 1939.

12. Mirels, Harold, and Rudolph C. Haefeli: The Calculation of Supersonic Downwash Using Line Vortex Theory, *J. Aeronaut. Sci.*, vol. 17, no. 1, 1950.

13. Lagerstrom, P. A., and Martha E. Graham: Methods for Calculating the Flow in the Trefftz Plane behind Supersonic Wings, *Douglas Aircraft Co. Rept.* SM-13288, July, 1948.

14. Jones, Robert T.: Properties of Low-aspect-ratio Wings at Speeds Below and Above the Speed of Sound, *NACA Tech. Repts.* 835, 1946.

15. Ward, G. N.: "Linearized Theory of Steady High-speed Flow," Cambridge University Press, New York, 1955.

CHAPTER 7

WING-TAIL INTERFERENCE

While the present chapter is entitled wing-tail interference, it could equally well have been entitled lifting-surface, vortex interference. Vortices passing close to a lifting surface can cause significant changes in the aerodynamic characteristics of the surface. An important example is the loss of tail effectiveness, which results from wing vortices which pass in close proximity to the tail. Figure 7-1 pictures the physical situation

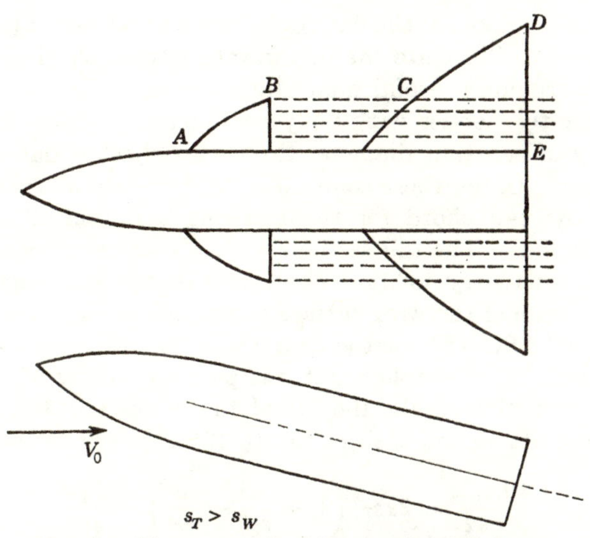

FIG. 7-1. Flat vortex sheet intersecting tailplane.

that gives rise to such wing-tail interference. Consider first the complete configuration minus the wing panels. The tail panels will then develop lift that can be calculated by the wing-body interference methods of Chap. 5. Now add the wing panels to the configuration. The addition of these surfaces causes a general downwash field in the region of the tail panels, and thereby reduces their lifting effectiveness. The loss of tail lift can be directly ascribed to the modification of the flow field produced by the vortices shed by the wing panels together with their images inside the body. It is clear that any vortices, regardless of their origin, will in passing close to the tail produce interference effects similar to those produced by wing vortices. In particular, the vortices originating near the body nose at high angles of attack will produce such effects.

Methods for calculating the nonuniform flow field behind the panels of a wing-body combination have been described in Chap. 6. The next problem is to determine the reaction of the tail section to the nonuniform flow field. The simple case wherein the wing wake is considered to be a flat vortex sheet is treated in Sec. 7-1, and the reaction of the tail section is calculated. In Secs. 7-2 and 7-3 we pass to the case of a completely rolled-up vortex sheet and determine the loading and tail effectiveness on the basis of slender-body theory. The idea of a tail interference factor is developed in Sec. 7-4, and its application to engineering calculation of tail loads is considered in Sec. 7-5. In Sec. 7-6, we consider some useful results based on reverse-flow theorems for determining tail loads in a non-uniform stream and the division of load between tail panels and body. In Sec. 7-7, the subject of shock-expansion interference is considered.

7-1. Wing-Tail Interference; Flat Vortex Sheet

The simplified model of the flat vortex sheet to represent the wing wake can be utilized to illustrate the important features of wing-tail interference and to provide a useful quantitative measure of tail effectiveness. For a sufficiently slender wing-body combination, Ward[1] finds that for an afterbody of constant diameter the vortex sheet is flat and coplanar with the wing. Under these conditions the flow behind the wings will be parallel to the tail chord for the midwing and midtail configuration shown in Fig. 7-1. The tail sections within a semispan equal to that of the wing will thus carry no lift. The vortex sheet can, in fact, be thought of as an extension of the wing surface to the tail to form a single panel of wing and tail ($ABCDE$) for the case when the tail span is greater than the wing span. This simple result was pointed out by Morikawa.[2] In accordance with this result, the lift of the wing-body-tail combination including nose lift on the basis of Eq. (5-13) is (a = constant)

$$\frac{L_{\mathrm{BWT}}}{q_0\alpha} = 2\pi s_T{}^2\left(1 - \frac{a^2}{s_T{}^2} + \frac{a^4}{s_T{}^4}\right) = \frac{L_{\mathrm{BT}}}{q_0\alpha} \tag{7-1}$$

when $s_T > s_W$. The lift for the wing-body combination is

$$\frac{L_{\mathrm{BW}}}{q_0\alpha} = 2\pi s_W{}^2\left(1 - \frac{a^2}{s_W{}^2} + \frac{a^4}{s_W{}^4}\right) \tag{7-2}$$

and the lift of the body alone is

$$\frac{L_B}{q_0\alpha} = 2\pi a^2 \tag{7-3}$$

A convenient measure of the degree of severity of the wing-tail interference is the *tail effectiveness*. The tail effectiveness is defined as the ratio of the lift developed by adding the tail to the wing-body combination to that developed by adding the tail to the body alone.

$$\eta_T = \frac{L_{BWT} - L_{BW}}{L_{BT} - L_B} \tag{7-4}$$

The tail effectiveness is thus a measure of how much the tail lift has been reduced by interference from the wing panels.   If wing-tail interference does not reduce the tail lift, η_T is unity.   If, however, the tail lift is entirely canceled by interference, η_T is zero.   While the tail effectiveness

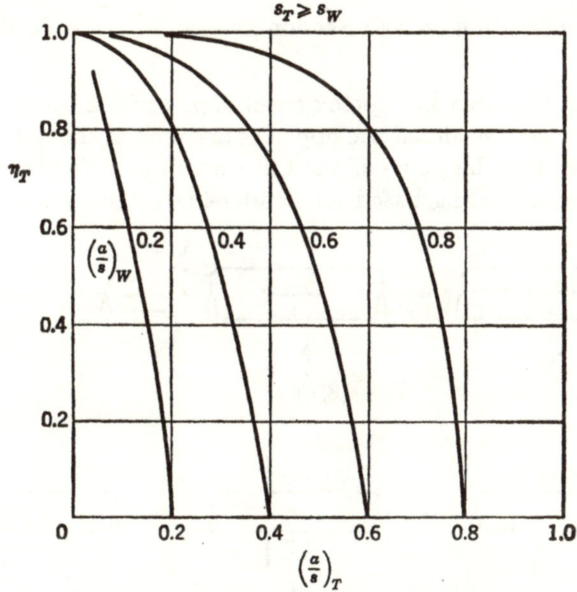

FIG. 7-2. Tail effectiveness for flat vortex sheet.

has been defined on the basis of lift, it is clear that a similar effectiveness can be defined for pitching moment.   The moment effectiveness will differ slightly from the lift effectiveness, since the tail center of pressure will in general not be the same with wing-tail interference as without it.

For the case $s_T > s_W$ the tail effectiveness is

$$\eta_T = \frac{(1 - a^2/s_T^2)^2 s_T^2/a^2 - (1 - a^2/s_W^2)^2 s_W^2/a^2}{(1 - a^2/s_T^2)^2 (s_T^2/a^2)} \tag{7-5}$$

For the case $s_W > s_T$ the vortex sheet from the wing passes through the plane of the tail for the present model.   Since vortex lines follow streamlines, the tail is at zero angle of attack locally and generates no lift.   For this case, then,

$$\eta_T = 0 \tag{7-6}$$

When the tail span is less than the wing span, the tail is thus totally ineffective. How the tail efficiency for the present model actually depends on the parameters a/s_T and a/s_W is shown in Fig. 7-2.

The foregoing calculations of tail efficiency are based on a model of the vortex wake which is not fully representative of a missile on several grounds. In the first place, the vortex sheet is not flat but has rolled up at least in part by the time it has reached the tail. Second, the vortex generally lies closer to the free-stream direction than the extended chord plane, as shown by Fig. 6-18 and by many schlieren photographs. Thus for positive angles of attack the wing vortices will generally lie above an inline tail and thereby produce less adverse interference.

7-2. Pressure Loading on Tail Section Due to Discrete Vortices in Plane of Tail

The fully rolled-up wing vortex sheet represents a model of the wing wake that can be considered the opposite extreme of the flat vortex wake. For the vortices in the plane of the tail we will now derive a solution for the tail pressure loading based on slender-body theory. The model for

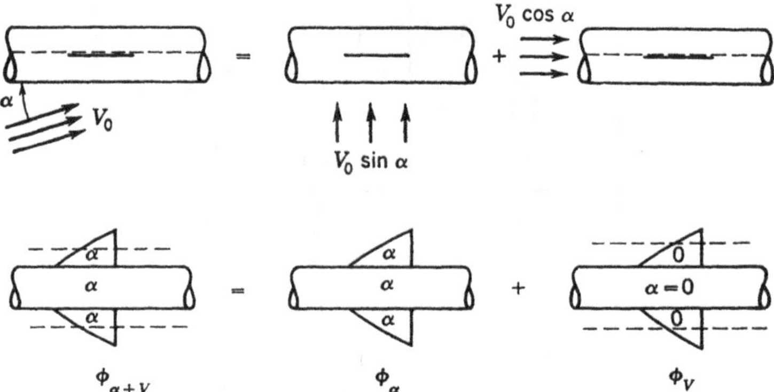

Fig. 7-3. Discrete line vortices intersecting tailplane.

which the solution is obtained is shown in Fig. 7-3. This model is decomposed into a tail-body combination acting at angle of attack α without vortices, and one acting at $\alpha = 0$ with vortices. In this decomposition we note that the angles of attack of tail and body add up to α and the free-stream velocities add up to V_0 acting at angle of attack α. With regard to the vortex itself, it can be replaced by a small solid cylindrical boundary. The circulation around this small boundary is zero for ϕ_α and produces no effect on the flow; but for ϕ_V the circulation is taken as Γ. Since the vortex is specified to lie in the plane of the tail, it must be held in position by a force. Because the potentials ϕ_α and ϕ_V are applied to the same physical boundaries and obey the same linear differential equation, they can be added to obtain the potential for the complete combination at angle of attack α including the vortices. But it is not obvious that the loading coefficients are additive, even if the potentials are, because of the usual squared terms in Bernoulli's equation. The velocity components in the x, y, and z directions have simple symmetry for the vortices lying in the plane of the wing, and it is shown in Appendix A at the end of the chapter that the square terms do not contribute to the loading under these circumstances. Thus the loading coefficients associated with ϕ_α and ϕ_V are additive. If P^+ and P^- denote the pressure coefficients on the lower and upper surfaces, respectively, then

$$\Delta P_\alpha = P_\alpha{}^+ - P_\alpha{}^- = - \frac{4}{V_0} \frac{\partial}{\partial x} (\phi_\alpha{}^+)$$

$$\Delta P_V = P_V{}^+ - P_V{}^- = - \frac{4}{V_0} \frac{\partial}{\partial x} (\phi_V{}^+)$$

(7-7)

$$\Delta P_{\alpha+V} = \Delta P_\alpha + \Delta P_V \tag{7-8}$$

It is appropriate at this time to specify more precisely the limitations of the solution arising from the fact that the vortices are assumed to lie in the plane of the tail. If the vortices attempt to move vertically out of the plane of the tail, it is necessary to apply lateral forces parallel to the y axis to keep them in the plane. Thus, if the vortices are free to move laterally in the plane of the tail, there will be no change in the Z force or loading due to constraining them to lie in the plane.

Let us now turn to the problem of determining ϕ_V and ΔP_V. Consider the cross section of the actual tail in the $\mathfrak{z}$ plane with a pair of symmetrically disposed vortices of equal but opposite strength, as shown in Fig. 7-4. Because the external vortices produce velocity normal to both body and panels, a fairly complicated image system must be put inside the cross section to cancel the normal velocity. Image vortices at the inverse points inside the body will satisfy this requirement for the body but not for the panels. Images will thus be required which satisfy the panel normal velocity condition without at the same time violating the normal velocity condition for the body. A simple means of determining this image system is to transform the tail cross section into the unit circle for which the image system is known. The required transformation is

$$\mathfrak{z} + \frac{a^2}{\mathfrak{z}} = \frac{1}{2}\left(s + \frac{a^2}{s}\right)\left(\sigma + \frac{1}{\sigma}\right) \tag{7-9}$$

Solving the transformation for σ yields

$$\sigma = \frac{1}{A}\left(\mathfrak{z} + \frac{a^2}{\mathfrak{z}}\right) \pm \left[\frac{(\mathfrak{z} + a^2/\mathfrak{z})^2}{A^2} - 1\right]^{\frac{1}{2}} \tag{7-10}$$

where
$$A = s + \frac{a^2}{s} \tag{7-11}$$

The *plus sign* is to be taken for the top surface and the *minus sign* for the lower surface. The points 1, 2, 3, and 4 are shown in the two planes.

The image system in the σ plane is obtained in the usual manner by introducing images inside the circle. The complex potential due to the complete vortex system is with reference to Eq. (4-77)

$$
\begin{aligned}
W_V = \phi_V + i\psi_V &= -\frac{i\Gamma}{2\pi} \log \frac{(\sigma - \sigma_V)(\sigma + 1/\sigma_V)}{(\sigma + \bar{\sigma}_V)(\sigma - 1/\bar{\sigma}_V)} \\
&= -\frac{i\Gamma}{2\pi} \log \frac{(\sigma - 1/\sigma) - (\sigma_V - 1/\sigma_V)}{(\sigma - 1/\sigma) + (\bar{\sigma}_V - 1/\bar{\sigma}_V)} \quad (7\text{-}12)
\end{aligned}
$$

In the transformation back to the $\mathfrak{z}$ plane, symmetrical external vortices appear, together with the necessary internal images. This transforma-

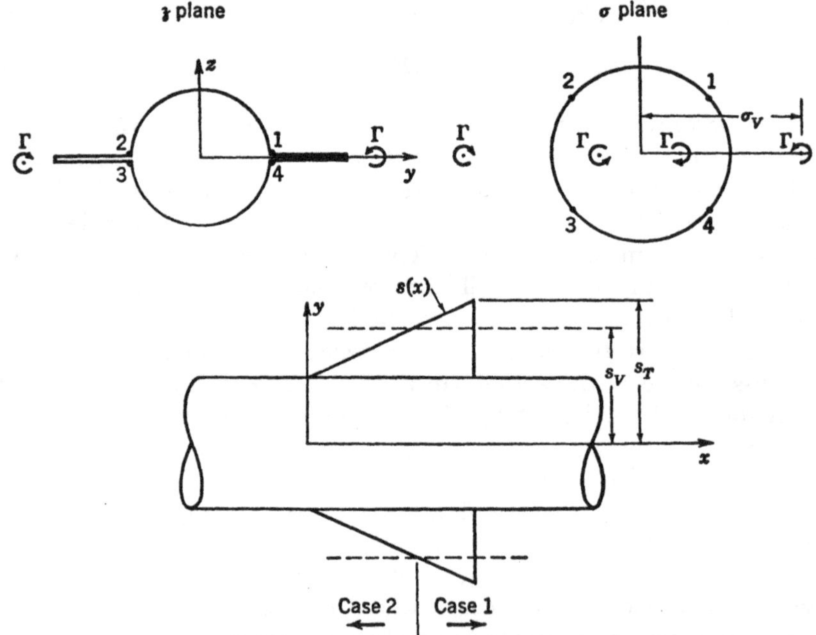

Fig. 7-4. Transformation of tail cross section into unit circle.

tion is accomplished with the aid of the following equation which refers to the upper surface

$$
\sigma - \frac{1}{\sigma} = 2\left[\frac{(\mathfrak{z} + a^2/\mathfrak{z})^2}{A^2} - 1\right]^{1/2} \quad (7\text{-}13)
$$

The potential in the $\mathfrak{z}$ plane depends on whether the vortex span is less than or greater than the local tail span.

Case 1: $s_V < s$

The two cases arising in the present solution are illustrated in Fig. 7-17. In case 1, the vortex is inboard of the leading edge of the tail panel, so that $s_V < s$. In the σ plane the external vortices lie outside the unit circle just opposite their images. If the external vortices actually move onto the circle, they are identically canceled by their images. The complex potential thus becomes identically zero. To show this mathematically, let σ_V lie on the unit circle. Then it is easy to see that

$$\sigma_V - \frac{1}{\sigma_V} = -\left(\bar{\sigma}_V - \frac{1}{\bar{\sigma}_V}\right) \tag{7-14}$$

As a result Eq. (7-12) yields

$$W_V = \phi_V + i\psi_V = 0 \tag{7-15}$$

Case 2: $s_V > s$

A. Tail panels: Instead of Eq. (7-13) for $\sigma_V - 1/\sigma_V$ we now have

$$\sigma_V - \frac{1}{\sigma_V} = 2\left[\frac{(s_V + a^2/s_V)^2}{A^2} - 1\right]^{\frac{1}{2}} \tag{7-16}$$

The logarithm is now a complex quantity, and the complex potential possesses a real part,

$$\phi_V^- = \frac{\Gamma}{\pi}\tan^{-1}\frac{m}{n}$$
$$m = \left[\frac{(s_V + a^2/s_V)^2}{(s + a^2/s)^2} - 1\right]^{\frac{1}{2}} \tag{7-17}$$
$$n = \left[1 - \frac{(y + a^2/y)^2}{(s + a^2/s)^2}\right]^{\frac{1}{2}}$$

wherein positive roots are to be taken and the value of $\tan^{-1}(m/n)$ ranges between 0 and $\pi/2$.

B. Body: On the body we have

$$\sigma - \frac{1}{\sigma} = 2i\left(1 - \frac{4y^2}{A^2}\right)^{\frac{1}{2}} \tag{7-18}$$

and the potential is

$$\phi_V^- = \frac{\Gamma}{\pi}\tan^{-1}\left[\frac{(s_V + a^2/s_V)^2 - (s + a^2/s)^2}{(s + a^2/s)^2 - 4y^2}\right]^{\frac{1}{2}} \tag{7-19}$$

We now pass to the determination of the loading coefficient ΔP_V.

$$\Delta P_V = \frac{4}{V_0}\frac{\partial}{\partial x}(\phi_V^-) = \frac{4}{V_0}\left[\frac{\partial}{\partial s}(\phi_V^-)\frac{ds}{dx} + \frac{\partial}{\partial s_V}(\phi_V^-)\frac{ds_V}{dx} \right.$$
$$\left. + \frac{\partial}{\partial a}(\phi_V^-)\frac{da}{dx}\right] \tag{7-20}$$

Loading coefficients can thus be associated with the rate of change of panel semispan, with lateral movement of the vortices, and with changes in body radius. We consider only the first two effects. For the panels the loading coefficient is

$$\Delta P_V = \frac{-4\Gamma}{\pi V_0 a} \frac{(s/a + a/s)(1 - a^2/s^2)\, ds/dx}{[(s_V/a + a/s_V)^2 - (s/a + a/s)^2]^{1/2}}$$

$$\frac{1}{[(s/a + a/s)^2 - (y/a + a/y)^2]^{1/2}}$$

$$+ \frac{4\Gamma}{\pi V_0 a} \frac{(s_V/a + a/s_V)(1 - a^2/s_V^2)\, ds_V/dx}{[(s_V/a + a/s_V)^2 - (y/a + a/y)^2]}$$

$$\left[\frac{(s/a + a/s)^2 - (y/a + a/y)^2}{(s_V/a + a/s_V)^2 - (s/a + a/s)^2} \right]^{1/2} \quad (7\text{-}21)$$

and for the body

$$\Delta P_V = \frac{-4\Gamma}{\pi V_0 a} \frac{(s/a + a/s)(1 - a^2/s^2)\, ds/dx}{[(s/a + a/s)^2 - 4y^2/a^2]^{1/2}}$$

$$\frac{1}{[(s_V/a + a/s_V)^2 - (s/a + a/s)^2]^{1/2}}$$

$$+ \frac{4\Gamma}{\pi V_0 a} \frac{(s_V/a + a/s_V)(1 - a^2/s_V^2)\, ds_V/dx}{[(s_V/a + a/s_V)^2 - 4y^2/a^2]}$$

$$\left[\frac{(s/a + a/s)^2 - 4y^2/a^2}{(s_V/a + a/s_V)^2 - (s/a + a/s)^2} \right]^{1/2} \quad (7\text{-}22)$$

The loading given by these expressions is illustrated in Fig. 7-5 for triangular panels.　It is noteworthy that the loading associated with posi-

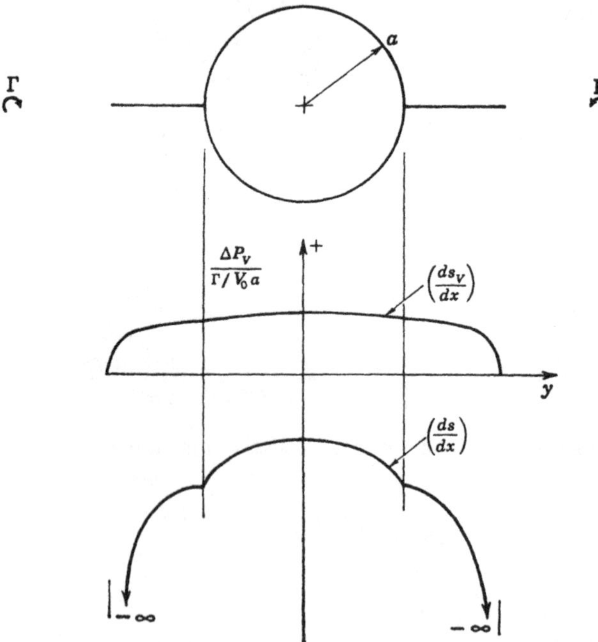

Fig. 7-5. Types of loadings associated with discrete vortices in plane of tail.

tive ds/dx is negative, and that associated with positive ds_V/dx is positive. Significant differences exist in the shape of the two loadings.

7-3. Lift on Tail Section and Tail Efficiency for Discrete Vortices in Plane of Tail

Having determined the potential and loading on tail and body for discrete vortices lying in the plane of the tail, we are now in a position to determine the tail lift and effectiveness. It will also be interesting to compare the tail effectiveness for a fully rolled-up sheet with that for a flat sheet. The tail lifts for the panels and the body can conveniently be set up in terms of the potentials given by Eqs. (7-15), (7-17), and (7-19). With reference to Fig. 7-4, the lift on the tail panels due to the vortex is

$$\frac{L_{T(B)V}}{q_0} = 2 \int_a^{s_T} \int_{le}^{te} \Delta P_V \, dx \, dy$$

$$= -\frac{8}{V_0} \int_a^{s_T} [(\phi_V^+)_{te} - (\phi_V^+)_{le}] \, dy$$

Correspondingly the lift on the body due to the vortex is

$$\frac{L_{B(T)V}}{q_0} = 2 \int_0^a \int_{le}^{te} \Delta P_V \, dx \, dy$$

$$= -\frac{8}{V_0} \int_0^a [(\phi_V^+)_{te} - (\phi_V^+)_{le}] \, dy \tag{7-23}$$

In the foregoing formulas one integration has essentially been performed by passing to the potential, and it remains for us to perform another integration. We will confine our attentions to the case where the vortex intersects the tail, $s_V < s_T$.

Consider the potential field acting on the tail panels. With reference to Fig. 7-4 we have the potential at the leading edge

$$(\phi_V^+)_{le} = -\frac{\Gamma}{2} \qquad s < s_V$$

$$(\phi_V^+)_{le} = 0 \qquad s > s_V \tag{7-24}$$

The lift on the panels is thus

$$\frac{L_{T(B)V}}{q_0} = -\frac{8}{V_0} \int_a^{s_V} \left[0 - \left(-\frac{\Gamma}{2}\right)\right] dy = -\frac{4\Gamma a}{V_0}\left(\frac{s_V}{a} - 1\right) \tag{7-25}$$

For the body, the "leading edge" is taken as the diameter joining the leading edges of the tail-body junctures. From Eqs. (7-15) and (7-19) we have

$$(\phi_V^+)_{le} = -\frac{\Gamma}{\pi} \tan^{-1} \frac{s_V/a_T - a_T/s_V}{2(1 - y^2/a_T^2)^{1/2}} \qquad s_V < s \tag{7-26}$$

$$(\phi_V^+)_{te} = 0$$

The lift on the body is thus

$$\frac{L_{B(T)_V}}{q_0} = -\frac{8\Gamma}{\pi V_0} \int_0^a \tan^{-1} \frac{s_V/a_T - a_T/s_V}{2(1 - y^2/a_T{}^2)^{1/2}} \, dy$$

$$= -\frac{4\Gamma a_T}{V_0}\left(1 - \frac{a_T}{s_V}\right) \qquad (7\text{-}27)$$

The shapes of the span loadings given by $(\phi_V{}^+)_{te} - (\phi_V{}^+)_{le}$ are illustrated

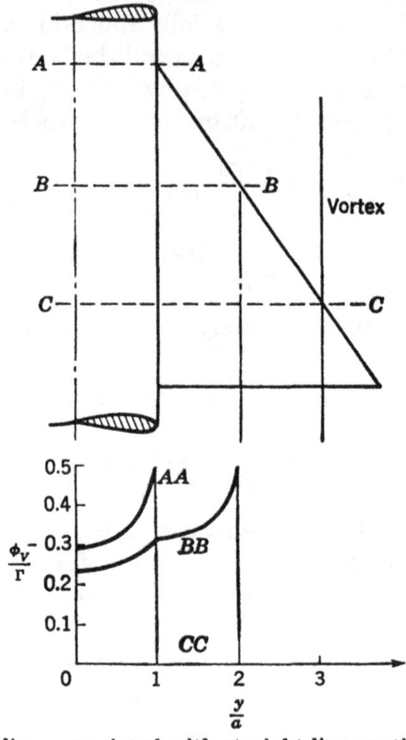

Fig. 7-6. Span loadings associated with straight-line vortices in plane of tail.

in Fig. 7-6 for the trailing edge taken at various streamwise positions. The total loss of lift due to the vortex is

$$\frac{L_V}{q_0} = \frac{L_{T(B)_V} + L_{B(T)_V}}{q_0} = -\frac{4\Gamma a}{V_0}\left(\frac{s_V}{a_T} - \frac{a_T}{s_V}\right) \qquad (7\text{-}28)$$

The tail effectiveness can now be determined with the aid of the following equation:

$$1 - \eta_T = \frac{-L_V}{L_{BT} - L_B} \qquad (7\text{-}29)$$

By apparent-mass methods (Sec. 10-8), it is easy to show that

$$L_{BT} - L_B = 2\pi s_T^2 q_0 \alpha \left(1 - \frac{a_T^2}{s_T^2}\right)^2 \tag{7-30}$$

so that the tail effectiveness is

$$\eta_T = \frac{(s_T/a_T - a_T/s_T)^2 - (2\Gamma/\pi a V_0 \alpha)(s_V/a_T - a_T/s_V)}{(s_T/a_T - a_T/s_T)^2} \tag{7-31}$$

The foregoing result for the tail efficiency is in terms of vortex strength and position as parameters, and it is of no particular importance how the vortices arise. However, when the vortices arise by virtue of the wing panels, the tail effectiveness can be expressed solely in terms of the dimensions of the wing-body-tail combination. Based on the vortex model of

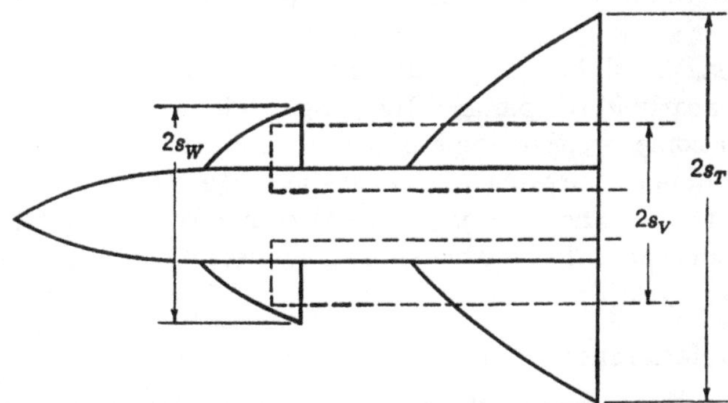

Fig. 7-7. Model of wing vortices impinging on tail.

Fig. 7-7, the vortex strength is equal to the circulation at the wing-body juncture given by Eq. (6-39).

$$\frac{\Gamma}{2V_0 a_W \alpha} = \frac{s_W}{a_W} - \frac{a_W}{s_W} \tag{7-32}$$

Another relationship between the lifts and vortex strength based on slender-body theory and lifting-line theory is

$$\frac{L_{WB} - L_B}{q_0 \alpha} = \frac{2\rho_0 V_0 \Gamma(s_V - a_W^2/s_V)}{q_0 \alpha} = 2\pi s_W^2 \left(1 - \frac{a_W^2}{s_W^2}\right)^2 \tag{7-33}$$

The foregoing equation yields the relationship between wing span and vortex span

$$\frac{s_V}{a_W} - \frac{a_W}{s_V} = \frac{\pi}{4}\left(\frac{s_W}{a_W} - \frac{a_W}{s_W}\right) \tag{7-34}$$

The tail effectiveness can now be expressed in two alternate forms

$$\begin{aligned} \eta_T &= \frac{(s_T/a_T - a_T/s_T)^2 - (s_W/a_W - a_W/s_W)^2}{(s_T/a - a/s_T)^2} \qquad s_V < s_T \\[2ex] \eta_T &= \frac{(s_T/a_T - a_T/s_T)^2 - (16/\pi^2)(s_V/a_W - a_W/s_V)^2}{(s_T/a_T - a_T/s_T)^2} \qquad s_V < s_T \end{aligned} \tag{7-35}$$

It is interesting to compare the tail effectiveness for the present case, Eq. (7-35), with that calculated for a flat vortex sheet, Eq. (7-5). Algebraically, the expressions are identical, but the flat-sheet case was derived on the basis that $s_W < s_T$, whereas the present case holds for $s_V < s_T$. The minimum tail effectiveness is 0 for the flat-sheet case and -0.62 for the present case. As the vortex sheet rolls up, more vorticity is concentrated inboard where it can intersect the tail panel. Thus, the lower minimum tail effectiveness results for the rolled-up vortex case.

Wing-tail interference is usually most adverse for a vortex of *fixed strength* when the vortex lies in the plane of the tail panels. For a combination with the wing and tail mounted centrally on the fuselage, wing deflection can produce a vortex in the plane of the tail. If, however, the tail is above the wing, this condition will prevail at some positive angle of attack. For a midwing-midtail combination with no wing incidence, the vortex strength will be very small for the small values of α for which the vortex lies nearly in the plane of the wing. The interference will be most adverse for some positive angle of attack (for which the vortex does not lie in the plane of the wing), because the vortex strength increases with angle of attack. The case of the vortices not lying in the plane of the wing will subsequently be discussed in connection with the tail interference factor.

7-4. Tail Interference Factor

The interference produced by a vortex on a lifting surface depends on the strength of the vortex and its position relative to the lifting surface. It seems desirable to set up a nondimensional measure of this interference which depends on vortex position but which is independent of vortex strength. Tail effectiveness depends on vortex strength and is thus not such a measure. Therefore, let us consider a quantity called the *tail interference factor* i_T. Now for a fixed vortex position the local induced velocities at the lifting surface will be proportional to the vortex strength. They will produce effective twist and camber of the surface proportional to vortex strength. Thus, for a vortex of fixed position the lift on the lifting surface is proportional to the product of the vortex strength and lift-curve slope of the lifting surface. Let the strength be expressed in the nondimensional form $\Gamma/V_0 l_r$ where l_r is a reference length. Let the lift per degree of the tail alone be written L_T/α. The lift on the tail section due to the fixed vortex is then

$$L_{BT_V} \propto \frac{\Gamma}{V_0 l_r} \frac{L_T}{\alpha} \tag{7-36}$$

It is convenient for our purposes to use a reference length based on tail dimensions, namely, $2\pi(s_T - a_T)$. The constant of proportionality in Eq. (7-36) is then defined as the tail interference factor i_T.

$$L_{\mathrm{BT}_V} = i_T \frac{\Gamma}{2\pi V_0(s_T - a_T)} \frac{L_T}{\alpha} \qquad (7\text{-}37)$$

or

$$i_T = \frac{L_{\mathrm{BT}_V}/L_T}{\Gamma/2\pi V_0 \alpha (s_T - a_T)} \qquad (7\text{-}38)$$

It should be mentioned that L_{BT_V} is usually negative so that i_T is negative.

The tail interference factor can be interpreted as the ratio of two nondimensional quantities, the first of which is a lift ratio and the second a

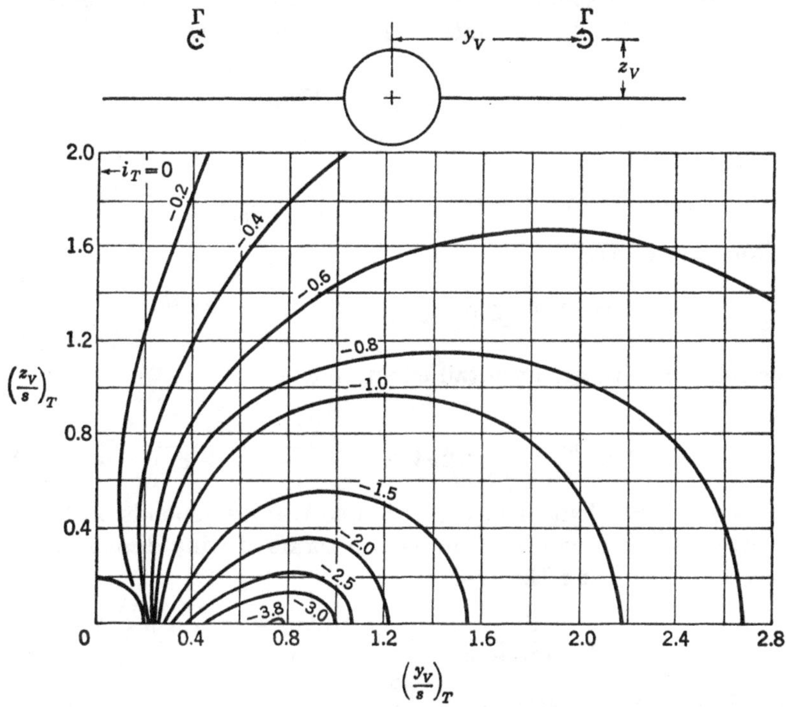

$\left(\dfrac{z_V}{s}\right)_T$

$\left(\dfrac{y_V}{s}\right)_T$

FIG. 7-8. Chart of tail interference factor based on strip theory; $\lambda_T = 0$, $(a/s)_T = 0.2$.

nondimensional vortex strength. The appearance of a lift ratio has the effect that, even if the particular theory used to obtain i_T is known to predict the *lift* incorrectly, still it may be suitable for evaluating i_T since it may predict the *lift ratio* correctly. It will be recalled that the use of this stratagem in predicting K_W and K_B by slender-body theory led to practical interference calculations for nonslender configurations.

The tail interference factor can be calculated by such methods as slender-body theory, strip theory, and reverse-flow methods. A particularly simple method for calculating i_T is through the use of strip theory, and it has the flexibility of not being dependent on Mach number. A set of i_T charts based on strip theory[3] has been drawn up for various values of the parameters $(a/s)_T$ and tail taper ratio λ_T. A typical chart from this group is shown in Fig. 7-8. It is to be noted that the chart is for a pair of equal and opposite external vortices symmetrically disposed with respect to the vertical plane of symmetry, and that the chart is similar in all four quadrants. The magnitude of i_T is greatest near the tip of the tail panels. Thus, care should be taken in design to keep strong vortices from entering the region of the tip of the tail panels.

Illustrative Example

Calculate on the basis of slender-body theory the relationship between the tail effectiveness and the tail interference factor for the vortex model shown in Fig. 7-7. The definition of tail effectiveness gives

$$1 - \eta_T = -\frac{L_{BT_V}}{L_{BT} - L_B}$$

From Eq. (7-30)

$$L_{BT} - L_B = 2\pi s_T^2 q_0 \alpha \left(1 - \frac{a_T^2}{s_T^2}\right)^2$$

and from Eq. (7-37)

$$L_{BT_V} = i_T \frac{\Gamma}{2\pi V_0 (s_T - a_T)} [2\pi (s_T - a_T)^2 q_0]$$

These relationships give the tail effectiveness

$$1 - \eta_T = -i_T \frac{\Gamma s_T^2}{2\pi V_0 \alpha (s_T - a_T)(s_T + a_T)^2} \tag{7-39}$$

Since i_T is independent of Γ, the tail effectiveness depends on Γ. When the vortex strength can be expressed in terms of wing dimensions as for the present case using Eq. (7-32)

$$\frac{\Gamma}{2\pi V_0 \alpha (s_T - a_T)} = \frac{a_W (s_W/a_W - a_W/s_W)}{\pi (s_T - a_T)} \tag{7-40}$$

The tail interference factor is

$$i_T = -\pi (1 - \eta_T) \frac{(s_T - a_T)(s_T + a_T)^2}{a_W s_T^2 (s_W/a_W - a_W/s_W)} \tag{7-41}$$

7-5. Calculation of Tail Lift Due to Wing Vortices

The tail interference factor discussed in the preceding section simplifies the calculation of tail lifts. Let us set up the necessary equations for calculating the tail lift for a planar missile (or a cruciform missile at zero bank angle) due to wing vortices. Consider one vortex per wing panel. This vortex strength is obtained by equating the lift on the wing panels calculated by use of K_W to the lift as calculated by lifting-line theory. On the basis of the wing-body interference methods of Chap. 5, using K_W, we obtain

$$L_{W(B)} = K_W (C_{L\alpha})_W S_W q_0 \alpha \tag{7-42}$$

Likewise the lift of the panels based on the lifting-line model of Fig. 5-17 is

$$L_{W(B)} = 2\rho V_0 \Gamma_0 (y_v - a_W) \tag{7-43}$$

where y_v is the lateral position of the wing vortex, and Γ_0 is the circulation strength at the wing-body juncture.   Thus we have the vortex strength

$$\Gamma_0 = \frac{V_0(\alpha K_W)(C_{L\alpha})_W S_W}{4(y_v - a_W)} \tag{7-44}$$

When this vortex strength is introduced into Eq. (7-37), the lift on the tail section due to the wing vortices takes the form

$$L_{BT_V} = i_T \frac{\alpha K_W (C_{L\alpha})_W (C_{L\alpha})_T S_W S_T q_0}{8\pi (s_T - a_T)(y_v - a_W)} \tag{7-45}$$

We convert to lift coefficient form on the basis of a reference area S_R.

$$(C_{L_{BT}})_V = \frac{L_{BT_V}}{q_0 S_R} = i_T \frac{K_W (C_{L\alpha})_W (C_{L\alpha})_T (s_T - a_T) S_W / S_R}{2\pi A_T (y_v - a_W)} \tag{7-46}$$

The angle of attack is in radians, and the lift-curve slopes are per radian.

   To make engineering calculations using Eq. (7-46) requires a knowledge of the vortex position at the tail section.   The factor i_T depends on such information.   Let $(y_v/s)_T$ and $(z_v/s)_T$ be the lateral and vertical positions of the vortex associated with the right tail panel in the crossflow plane through the centroid of area of the tail panels.   These quantities can be determined in several ways.   First, the positions can be determined by the step-by-step procedure described in the preceding chapter.   Such a procedure is lengthy, and some simplifying assumption is usually warranted.   Thus, we come to the second method based on the assumption that the vortices trail back in the streamwise direction from the wing trailing edge.   To demonstrate the use of Eq. (7-46) and the simplifying assumption, let us perform a sample calculation.

Illustrative Example

   With reference to the wing-body-tail combination of Fig. 7-9, calculate the lift on the tail section due to the wing vortices and the corresponding tail effectiveness for the conditions $M_0 = 2$, $\alpha = 5°$, and $\beta = 0$.

   As a first step let us ascertain the vortex lateral and vertical locations in the crossflow plane at the centroid of the tail panels.   The lateral position at the wing trailing edge for a single vortex per panel is obtained with the help of Table 6-1.

$$\left(\frac{a}{s}\right)_W = \frac{0.5625}{2.812} = 0.2$$

$$\left(\frac{y_v - a}{s_m - a}\right)_W = 0.76 \qquad \text{Table 6-1}$$

$$(y_v)_W = 0.5625 + 0.76(2.25) = 2.27$$

We neglect any lateral motion of the vortex which is assumed to trail back in a streamwise direction. Thus

$$\left(\frac{y_V}{s}\right)_T = \frac{2.27}{1.812} = 1.25$$

The vertical height at the wing trailing edge is 0, and the tail centroid is

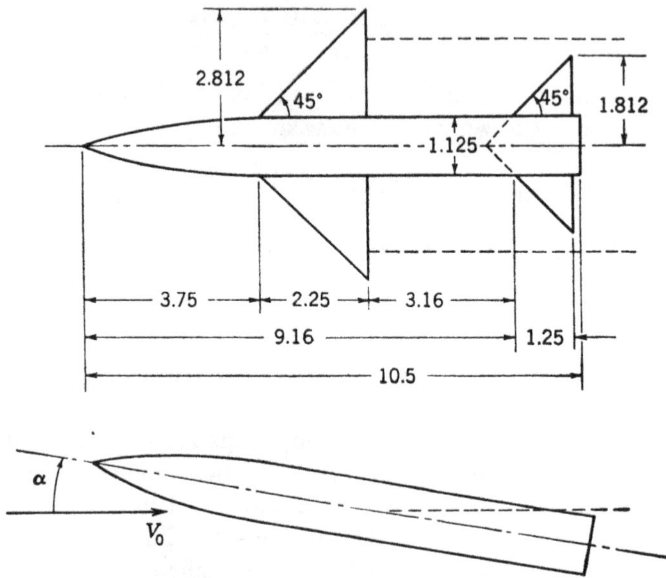

Fig. 7-9. Wing-body-tail combination of illustrative example.

3.99 units behind the trailing edge of the wing measured parallel to the body axis. Thus

$$(z_V)_T = 3.99 \tan 5° = 0.348$$
$$\left(\frac{z_V}{s}\right)_T = \frac{0.348}{1.812} = 0.19$$

The value of λ_T is 0, and of $(a/s)_T$ is 0.31. Actually, i_T is a slowly varying function of λ_T and of $(a/s)_T$ for small values of $(a/s_m)_T$. For instance, the value of i_T for $\lambda_T = 0$, $(a/s)_T = 0.2$, from Fig. 7-8, is -1.85, and for $(a/s)_T = 0.4$ from Pitts et al.[3] is -1.75. Let us use a value of $i_T = -1.8$.

Turning now to the other quantities in Eq. (7-46) we obtain

$$K_W = 1.16 \qquad \text{Table 5-1}$$

Since the leading edges of the wing and tail panels are supersonic for $M_0 = 2$, the lift-curve slopes of the wing alone and tail alone are equal to the two-dimensional lift-curve slope

$$(C_{L\alpha})_W = (C_{L\alpha})_T = \frac{4}{(M_0{}^2 - 1)^{1/2}} = \frac{4}{3^{1/2}} = 2.31 \text{ per radian}$$

Taking the reference area as S_W, we now have all the quantities necessary for evaluating the lift increment on the tail section due to the vortices.

$$(C_{L_{\mathrm{BT}}})_V = -1.8\,\frac{(5/57.3)1.16(2.31)^2(1.812 - 0.561)(1)}{2\pi(4)(2.27 - 0.56)}$$

$$= -0.029$$

We now consider the tail effectiveness. Based on the wing-alone area as a reference area, the tail-body combination and the body alone have lift-curve slopes of 0.0282 and 0.0069 per degree, respectively, calculated by the method of Chap. 5. The tail effectiveness is

$$1 - \eta_T = \frac{-(C_{L_{\mathrm{BT}}})_V}{(C_L)_{\mathrm{BT}} - (C_L)_B} = \frac{0.029}{5(0.0282 - 0.0069)} = 0.27$$

$$\eta_T = 0.73$$

The tail effectiveness has been reduced 27 per cent as a result of the adverse effect of wing-tail interference.

Let us examine the lift and moment curves for the example configuration as presented in Fig. 7-10. The effect of wing-tail interference on the lift is not large, which is not surprising in view of the fact the wing is much larger than the tail. The effect on pitching moment, however, is considerable, because of the large "lever arm" of the tail. The moment curve is now nonlinear, and the combination becomes more stable as the angle of attack increases, although wing-tail interference decreases the static margin by about 3 per cent of the combination length at $\alpha = 0°$.

Let us consider the nonlinearity exhibited by the moment curve in greater detail. As the angle of attack increases, the vortex strength increases linearly. If the vortex position remained fixed with respect to the tail, the adverse effect of interference would also be proportional to α, and the moment-curve slope would be constant. However, as the tail

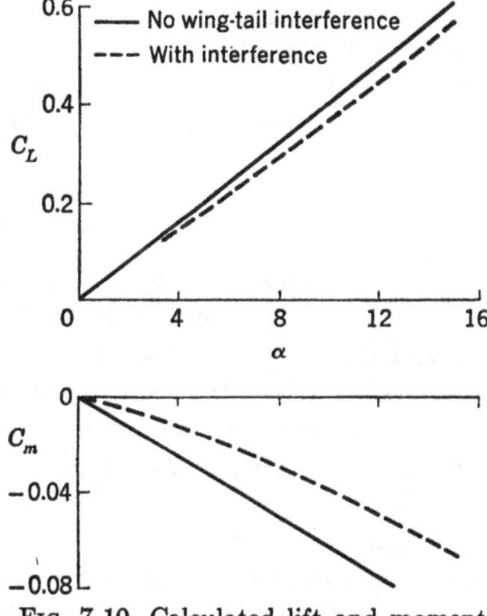

FIG. 7-10. Calculated lift and moment curves of example configuration.

moves down with increasing angle of attack, it moves away from the vortex, which trails back streamwise from the wing trailing edge. The lift on the tail due to the vortex is thus less than proportional to α, as exhibited by the difference between the two moment curves in Fig. 7-10. The net effect is an increase in stability $-dC_m/d\alpha$ as α increases.

It would be inferred from the foregoing argument that, if the tail approaches the vortex as α increased, there would result a decrease in stability. Such an effect would in fact occur for a high tail position as illustrated in Fig. 7-11. After the tail passes through the vortex and starts moving away from it, the stability would again increase as illustrated. It should be noted that this effect can also occur for a cruciform

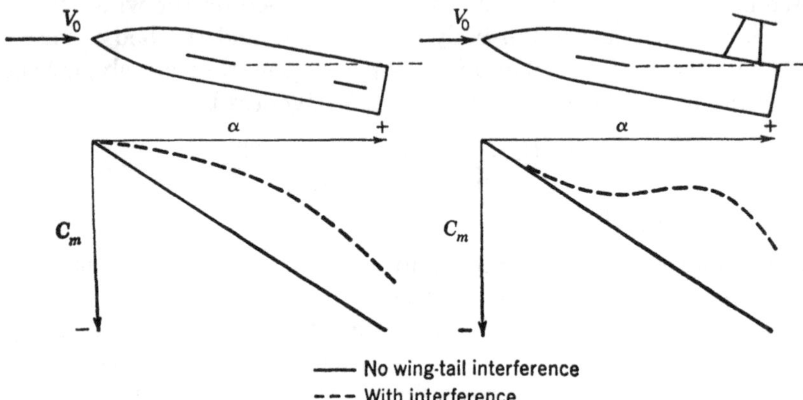

FIG. 7-11. Effect of tail height on moment curve of complete configuration.

tail interdigitated with respect to the wing panels. The upper panels of the tail will move toward the wing wake as α increases and exhibit the typical high-tail nonlinearity in the moment curve.

7-6. Use of Reverse-flow Method for Calculating Aerodynamic Forces on Tail Section in Nonuniform Flow

We have considered methods for calculating the aerodynamic forces on tail sections based on slender-body theory and on strip theory. Another powerful method for this purpose is to be found in the reverse-flow theorems of linearized theory. These theorems are based on an application of Green's theorem to second-order partial differential equations with certain mathematical symmetries. We forego the pleasure of reproducing the elegant derivation of the reverse-flow theorem we will use, but refer the reader to Heaslet and Lomax[6] instead. We will be particularly interested in the reversibility theorem involving pressure coefficients and angle of attack. Consider a tail-body combination consisting of circular cylinder and flat panels inclined at such small angles to the flow direction that the boundary condition of no flow normal to the solid boundaries can be applied on surfaces parallel to the flow direction. Let the combination in direct flow (Fig. 7-12) have local pressure coefficient P_1 and local angle of attack α_1. Let the corresponding quantities at the same point in reverse flow be P_2 and α_2. Then the particular reverse-flow theorem germane to our purpose is

$$\iint_{S_B+S_T} P_2\alpha_1\, dS = \iint_{S_T+S_B} P_1\alpha_2\, dS \tag{7-47}$$

Here the quantities P_2, α_2 are measured at the same point as the quantities P_1, α_1. Because all surfaces are taken in the streamwise direction, Eq. (7-47) applies equally to flows obeying Laplace's equation at subsonic speeds and to flows obeying the wave equation at supersonic speeds.

Let us now consider how the theorem is used to determine the aerodynamic forces on the tail in the nonuniform field produced by a vortex. Consider the tail panels to be at zero incidence, and let body and tail be

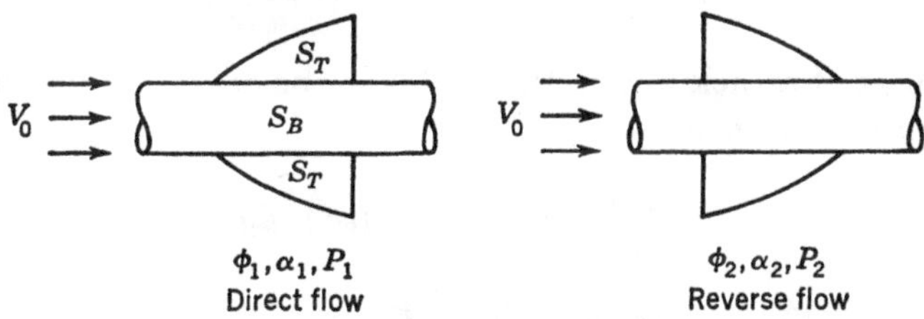

FIG. 7-12. Configuration in direct and reverse flow.

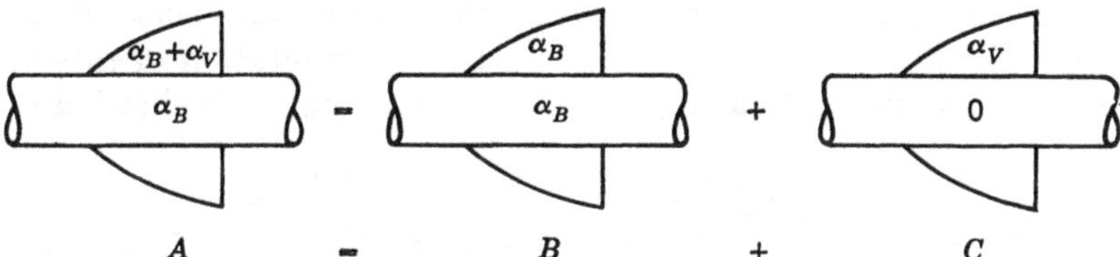

FIG. 7-13. Decomposition of twisted tail configuration into components.

at angle of attack α_B. The vortex system external to the body will produce velocities normal to the wing and body. The velocities normal to the body are canceled by introducing image vortices in the usual manner. The vortex system and its images then produce angle of attack α_V on the tail panels. The tail panels can be considered to be cambered and twisted to conform to α_V. If α_V depends only on lateral position and does not vary chordwise, then the tail panels are only twisted but not cambered. The final configuration thus has the body at angle of attack α_B and the tail panels effectively at angle of attack $\alpha_B + \alpha_V$ as shown in Fig. 7-13. The configuration is decomposed into two configurations B and C. The lift of configuration C is that due to the vortex, and it will now be calculated using reverse-flow methods.

Let us take for the conditions of direct flow the boundary conditions shown for configuration C of Fig. 7-13.

$$\begin{aligned} \alpha_1 &= 0 &&\text{on } S_B \\ \alpha_1 &= \alpha_V &&\text{on } S_T \end{aligned} \tag{7-48}$$

Take as the boundary conditions for reverse flow unit incidence of the tail panels as shown in Fig. 7-14.

$$\alpha_2 = 0 \qquad \text{on } S_B$$
$$\alpha_2 = 1 \qquad \text{on } S_T \tag{7-49}$$

A direct application of the reverse-flow theorem, Eq. (7-47), yields

$$\iint_{S_T} P_1 \, dS_T = \iint_{S_T} P_2 \alpha_V \, dS_T \tag{7-50}$$

Since the first integral of the preceding equation represents the force on one surface of the tail panels due to the vortex, we have

$$\frac{L_{T(B)V}}{q_0} = 2 \iint_{S_T} P_2 \alpha_V \, dS_T \tag{7-51}$$

Let the angle of attack due to the vortex α_V be one of pure twist. Then integration with respect to x yields

$$\frac{L_{T(B)V}}{q_0} = 2 \int_a^{s_T} \alpha_V (cc_l)_2 \, dy \tag{7-52}$$

where
$$(cc_l)_2 = 2 \int_{le}^{te} P_2 \, dx \tag{7-53}$$

The quantity $(cc_l)_2$ is the span loading due to unit incidence of the tail panels. Equation (7-52) can be given the interpretation of integration across the span of the local angle of attack with $(cc_l)_2$ as weighting factor.

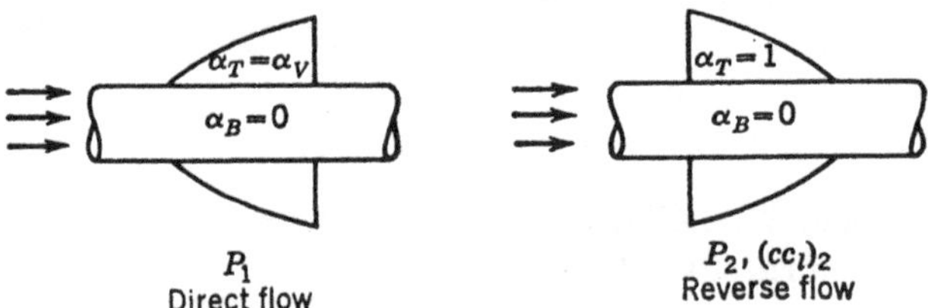

Fig. 7-14. Special configuration in direct and reverse flow.

Thus, if the span loading is known in reverse flow for a tail-body combination for unit tail incidence, then it is possible to determine the gross lift on the tail panels due to the vortex. It should be noted that the simplification of calculation achieved by the use of the reverse-flow theorem is made possible because we desire to obtain a gross force. No details concerning the span loading on the tail panels due to the vortex are given.

It must not be forgotten that the body will carry lift which is transferred to it by the tail panels, and we now proceed to a consideration of this problem. For the reverse-flow condition, let us take

$$\alpha_3 = 1 \qquad \text{on } S_B$$
$$\alpha_3 = 0 \qquad \text{on } S_T \tag{7-54}$$

and retain the same direct-flow conditions. A direct application of Eq. (7-47) yields

$$\iint_{S_B} P_1 \, dS_B = \iint_{S_T} P_3 \alpha_V \, dS_T \qquad (7\text{-}55)$$

where P_3 is the pressure coefficient due to unit angle of attack of the body with $\alpha_T = 0$. Performing an integration with respect to x yields

$$\frac{L_{B(T)_V}}{q_0} = 2 \iint_{S_T} P_3 \alpha_V \, dS_T = 2 \int_a^{s_T} (cc_l)_3 \alpha_V \, dy \qquad (7\text{-}56)$$

The lift on the body due to the vortex can thus be calculated using strip integration with $(cc_l)_3$ as weighting factor. We can obtain the total lift in one integration if we assume that both body and tail are at unit incidence in reverse flow

$$\begin{aligned}
\alpha_4 &= 1 \qquad \text{on } S_B \\
\alpha_4 &= 1 \qquad \text{on } S_T
\end{aligned} \qquad (7\text{-}57)$$

and let $(cc_l)_4$ be the corresponding span loading

$$\frac{L_{T(B)_V} + L_{B(T)_V}}{q_0} = \frac{L_{BT_V}}{q_0} = 2 \int_a^{s_T} (cc_l)_4 \alpha_V \, dy \qquad (7\text{-}58)$$

It is possible to evaluate the moment by considering the reverse-flow condition

$$\alpha_5 = x \qquad \text{on } S_B \text{ and } S_T \qquad (7\text{-}59)$$

A direct application of Eq. (7-50) yields

$$\frac{M_{BT_V}}{q_0} = \iint_{S_B + S_T} P_1 x \, dS = 2 \int_a^{s_T} (cc_l)_5 \alpha_V \, dy \qquad (7\text{-}60)$$

Likewise, if the reverse-flow conditions are taken to be

$$\alpha_6 = y \qquad \text{on } S_T \qquad (7\text{-}61)$$

then the rolling moment due to the vortex is

$$\frac{L'_{BT_V}}{q_0} = \iint_{S_T} P_1 y \, dS = 2 \int_a^{s_T} (cc_l)_6 \alpha_V \, dy \qquad (7\text{-}62)$$

7-7. Shock-expansion Interference

If the emphasis on wing-tail interference due to vortices has created the impression that other types of wing-tail interference do not exist, such an impression is unintentional. For a missile employing a horizontal tail above the wing, another type of interference is quite possible, namely, interference due to the action of the wing shock-expansion field on the tail. If the Mach number is sufficiently high, a horizontal tail in line with the wing can also fall within the wing shock-expansion field at angle of attack. An approximate analysis of shock-expansion interference is to be found in Nielsen and Kaattari.[7]

Let us now briefly consider the qualitative aspects of shock-expansion interference of the wing on the tail.   Consider the high-tail missile shown in Fig. 7-15 for $\alpha = 5°$ and $\alpha = 20°$.   The shock-expansion field shown is that of the wing on the assumption of a two-dimensional flow field.   At $\alpha = 5°$ the tail is shown partially between the expansion fan from the leading edge and the shock wave from the trailing edge.   Note the path of the streamline going through the expansion fan.   It is approximately parallel to the wing chord.   If the tail is set parallel to the wing chord, the tail will be essentially at zero angle of attack with respect to the local flow direction.   As a result it will develop very little nose-down moment. The tail effectiveness is thus nearly zero.

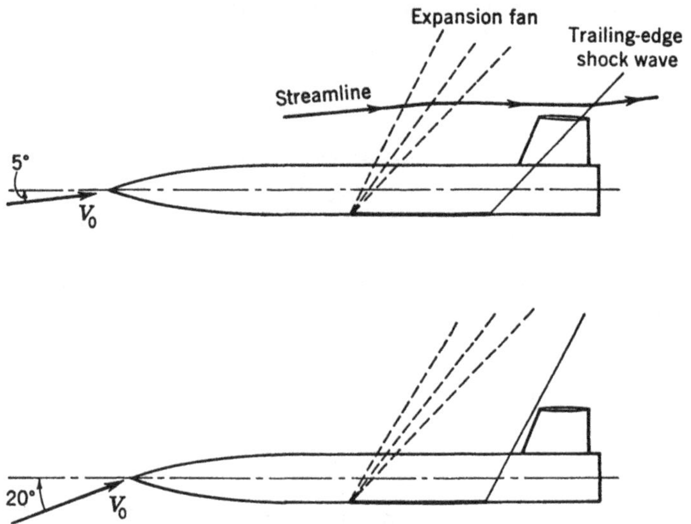

FIG. 7-15. Interference of wing shock-expansion field on tail.

If the angle of attack is now increased to 20°, the situation is altered greatly.   As the angle of attack changes, it is a property of the shock-expansion field that the trailing-edge shock wave remains nearly fixed in direction with respect to the free-stream velocity.   The tail therefore moves beneath the shock wave.   Here the flow direction is very closely in the free-stream direction so that the tail has recovered almost all its effectiveness.   There is some slight loss due to changes in Mach number and dynamic pressure resulting from the entropy increase through the shock wave.

From the preceding description of the flow changes at the tail due to the wing shock-expansion field, it is possible to see qualitatively the influence of the interference on the contribution of the tail to the pitching moment $(\Delta C_m)_T$.   With reference to Fig. 7-16, the values of $(\Delta C_m)_T$ for $\eta_T = 0$ and $\eta_T = 1$ are shown as a function of α.   For $\alpha = 5°$ the tail

effectiveness is near zero, and for $\alpha = 20°$ near unity.   The corresponding curve is sketched qualitatively on the figure.   If the missile were in
trim at $\alpha = 5°$, the interference in this case would have a stabilizing effect
for increases in angle of attack.

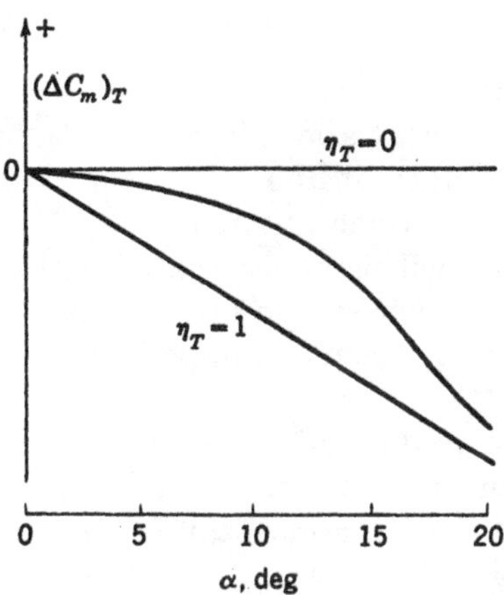

FIG. 7-16. Tail pitching-moment contribution with shock-expansion interference.

One fact that is clear from the discussion is that shock-expansion interference can be sensitive to Mach number.   This characteristic differenti-
ates it from wing-tail interference due to wing vortices.

SYMBOLS

a	radius of circular cylinder
a_T	radius of circular body at tail
a_W	radius of circular body at wing
A	$s + a^2/s$
A_T	aspect ratio of surface formed by joining tail panels together
c	local chord of wing or tail panel
c_l	section lift coefficient
(cc_l)	span loading
$(cc_l)_2$	span loading due to unit incidence of tail panels in reverse flow
$(cc_l)_3$	span loading due to unit incidence of body with tail panels at zero angle of attack in reverse flow
$(cc_l)_4$	span loading due to unit incidence of tail-body combination in reverse flow
$(cc_l)_5$	span loading for tail-body combination in reverse flow cambered so that the angle of attack equals x
$(cc_l)_6$	span loading for tail-body combination in reverse flow twisted so that the angle of attack equals y

C_L	lift coefficient
$C_{L\alpha}$	lift-curve slope
C_m	pitching moment about y axis
i_T	tail interference factor
K_W	wing-body interference lift ratio
l_r	reference length
le	leading edge
L	lift force
L'	rolling moment about body longitudinal axis
M	pitching moment about y axis
M_0	free-stream Mach number
$P_2, P_3,$ etc.	pressure coefficients associated with $(cc_l)_2$, $(cc_l)_3$, etc.
$P+$	pressure coefficient on lower surface
$P-$	pressure coefficient on upper surface
ΔP	$P+ - P-$
q_0	free-stream dynamic pressure
s	local semispan of wing-body or tail-body combination
s_T	semispan of horizontal tail in combination with body
s_V	semispan of vortex
s_W	semispan of wing in combination with body
S_B	area of body planform in empennage
S_R	reference area (arbitrary)
S_T	area of horizontal tail panels
S_W	area of wing panels
te	trailing edge
u, v, w	components of flow velocity along x, y, and z axes
V	velocity in crossflow plane at vortex location
V_0	free-stream velocity
W_V	complex potential due to vortices
x, y, z	system of axes lying in planes of symmetry of the missile with the origin on the body axis at the location of the leading edge of the tail-body juncture, Fig. 7-4
y_V, z_V	coordinates of vortex associated with right tail panel
$\mathfrak{z}$	$y + iz$
α	angle of attack
$\alpha_2, \alpha_3, \ldots$	angles of attack associated with $(cc_l)_2$, $(cc_l)_3$, $\ldots$
α_V	angle of attack produced at given spanwise position by vortex system
β	angle of sideslip
Γ	vortex strength, circulation
Γ_0	strength of bound vorticity at wing-body juncture
η_T	tail effectiveness
λ_T	tail taper ratio
ρ_0	free-stream density

	complex variable of transformed plane in which empennage cross section is a unit circle, Fig. 7-4
σ_V	position of right external vortex in σ plane
ϕ_α	potential of tail-body combination in absence of vortices
$\phi_V + i\psi_V$	W_V, complex potential due to vortex system

Subscripts:

B	body alone
$B(T)$	body section influenced by presence of horizontal tail
BT	body-tail combination; empennage
BW	body-wing combination
BWT	body-wing-tail combination
T	surface formed by joining horizontal tail panels together
$T(B)$	tail panels in presence of body
V	due to vortices
W	wing alone
α	due to angle of attack

REFERENCES

1. Ward, G. N.: Supersonic Flow past Slender Pointed Bodies, *Quart. J. Mech. and Appl. Math.*, vol. 2, part 1, p. 94, 1949.

2. Morikawa, George: Supersonic Wing-Body-Tail Interference, *J. Aeronaut. Sci.*, vol. 19, no. 5, 1952.

3. Pitts, W. C., Jack N. Nielsen, and George E. Kaattari: Lift and Center of Pressure of Wing-Body-Tail Combinations at Subsonic, Transonic, and Supersonic Speeds, *NACA Tech. Repts.* 1307, 1957.

4. Bryson, Arthur E.: Stability Derivatives for a Slender Missile with Application to a Wing-Body-Vertical Tail Configuration, *J. Aeronaut. Sci.*, vol. 20, no. 5, pp. 297–308, 1953.

5. Alden, Henry L., and Leon H. Schindel: The Calculation of Wing Lift and Moments in Nonuniform Supersonic Flows, *MIT Meteor Rept.* 53.

6. Heaslet, Max A., and Harvard Lomax: Supersonic and Transonic Small Perturbation Theory, sec. D, chap. 7, in "General Theory of High-speed Aerodynamics," vol. VI of "High-speed Aerodynamics and Jet Propulsion," Princeton University Press, Princeton, 1954.

7. Nielsen, Jack N., and George E. Kaattari: The Effects of Vortex and Shock-expansion Fields on Pitch and Yaw Instabilities of Supersonic Airplanes, *Inst. Aeronaut. Sci. Preprint* 743, 1957.

APPENDIX 7A. PRESSURE COEFFICIENT FOR COMBINED INFLUENCES OF ANGLE OF ATTACK AND VORTICES

For the body coordinates used in the present analysis, the pressure coefficient is given by Eq. (3-52) with $\alpha = \alpha_c$ and $\varphi = 0$,

$$P = -2(u + \alpha w) - (v^2 + w^2) \tag{7A-1}$$

where u, v, and w are the perturbation velocities along x, y, and z for unit free-stream velocity. The vortex patterns in the σ plane for case 1, $s_V < s$, and for case 2, $s_V > s$, are shown in Fig. 7-17. For case 1, we can immediately conclude that there is no loading due to the vortices. This result follows from the fact that the image vortex is as far inside the circle as the external vortex is outside, so that, when the external vortex approaches the circle in the limit, the two vortices annihilate each other

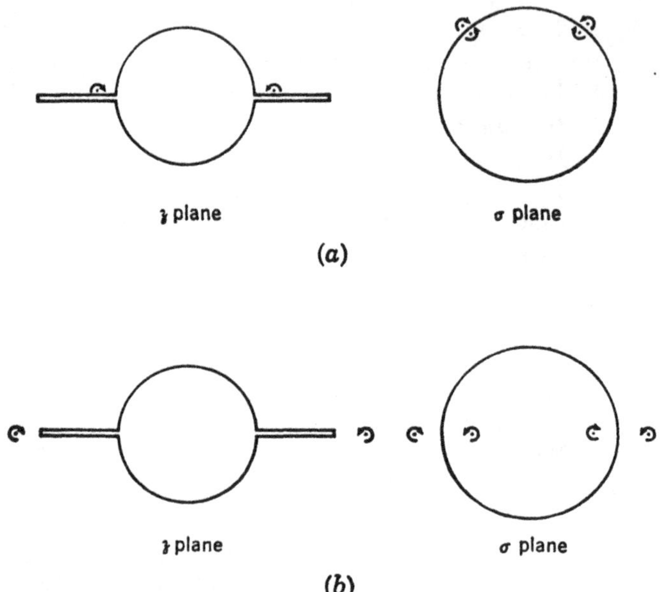

FIG. 7-17. Vortices in plane of tail panels. (a) Case 1, $s_v < s$; (b) case 2, $s_v > s$.

in pairs. This result, of course, also follows from the fact that ϕ_V is zero if $s_V < s$ as given by Eq. (7-15). We need now be concerned only with case 2.

For the tail panel in case 2 the velocity components have the following properties for the vortex in the $z = 0$ plane:

$$
\begin{aligned}
u_\alpha^+ &= -u_\alpha^- & u_V^+ &= -u_V^- \\
v_\alpha^+ &= -v_\alpha^- & v_V^+ &= -v_V^- \\
w_\alpha^+ &= w_\alpha^- = -\alpha & w_V^+ &= w_V^- = 0
\end{aligned}
\tag{7A-2}
$$

The pressure coefficient on the impact surface $P_{\alpha+V}^+$ for the combined effects of angle of attack and vortices is

$$
P_{\alpha+V}^+ = -2(u_\alpha^+ + u_V^+ - \alpha^2) - (v_\alpha^+ + v_V^+)^2 - (-\alpha)^2 \tag{7A-3}
$$

and for the suction surface

$$
P_{\alpha+V}^- = -2(-u_\alpha^+ - u_V^+ - \alpha^2) - (-v_\alpha^+ - v_V^+)^2 - (-\alpha)^2 \tag{7A-4}
$$

The loading coefficient is therefore

$$\Delta P = P_{\alpha+V}^{+} - P_{\alpha+V}^{-} = -4u_{\alpha}^{+} - 4u_{V}^{+} \qquad (7\text{A-}5)$$

It follows that the panel loadings due to angle of attack and due to the vortices are additive and do not involve the squared terms in Bernoulli's equation.

On the body the velocity components have the following symmetries and properties:

$$
\begin{aligned}
u_{\alpha}^{+} &= -u_{\alpha}^{-} & u_{V}^{+} &= -u_{V}^{-} \\
v_{\alpha}^{+} &= -v_{\alpha}^{-} & v_{V}^{+} &= -v_{V}^{-} \\
w_{\alpha}^{+} &= w_{\alpha}^{-} & w_{V}^{+} &= w_{V}^{-}
\end{aligned}
\qquad (7\text{A-}6)
$$

$$(v_{\alpha}^{+} + iw_{\alpha}^{+} + i\alpha)\|(v_{V}^{+} + iw_{V}^{+})$$

On the body the pressure on the impact surface $P_{\alpha+V}^{+}$ for the combined effects of angle of attack and vortices is

$$
\begin{aligned}
P_{\alpha+V}^{+} = -2[u_{\alpha}^{+} + u_{V}^{+} + \alpha(w_{\alpha}^{+} + w_{V}^{+})] \\
- (v_{\alpha}^{+} + v_{V}^{+})^{2} - (w_{\alpha}^{+} + w_{V}^{+})^{2} \quad (7\text{A-}7)
\end{aligned}
$$

Similarly

$$
\begin{aligned}
P_{\alpha+V}^{-} = -2[-u_{\alpha}^{+} - u_{V}^{+} + \alpha(w_{\alpha}^{+} + w_{V}^{+})] \\
- (-v_{\alpha}^{+} - v_{V}^{+})^{2} - (w_{\alpha}^{+} + w_{V}^{+})^{2} \quad (7\text{A-}8)
\end{aligned}
$$

The loading is thus

$$\Delta P = P_{\alpha+V}^{+} - P_{\alpha+V}^{-} = -4(u_{\alpha}^{+} + u_{V}^{+}) \qquad (7\text{A-}9)$$

Again, for the body, the loadings due to angle of attack and to the vortices are additive and do not involve the square terms in Bernoulli's equation.

CHAPTER 8

AERODYNAMIC CONTROLS

The choice of controls to effect changes in the angles of attack, sideslip, and bank of a missile is a problem of great importance to the missile designer.   This choice must take into account a large number of considerations such as the altitude, attitude, and speed of the missile; available positions and space for controls and control actuators; and type of guidance system.   A wide variety of missile controls exists, and others are being invented all the time.   A complete discussion of all control types is thus not possible and probably would not be desirable.   However, we can consider a few common types that make up a large fraction of the controls that occur in practice.   It is the primary purpose of this chapter to show how various theoretical methods can be used to predict the aerodynamic characteristics of some of these common types.   The title of the chapter indicates that we will confine our attention to controls that depend primarily on the surrounding atmosphere for their operation, in contradistinction to reaction controls needed for flight outside the earth's atmosphere.

It is of interest to note in a general way the role that theory plays in the prediction of control characteristics.   The theory to be used depends on the type of control, the quantity to be calculated, and the ranges of angles of attack and control deflection, as well as the Mach number. For controls such as all-movable ones, which can produce significant interference fields on the body, slender-body theory offers a powerful means of analysis, particularly when coupled with reverse-flow theorems. For types of controls where interference effects are not usually important, such as many trailing-edge controls, the extensive results of supersonic wing theory are available.   Our general attack on problems of control-characteristic prediction is first to calculate the linear characteristics on the basis of linear theory.   However, the large control deflections called for in many maneuvering missiles introduce a number of nonlinearities. The next step in our general attack is to consider the modification of the linear characteristics in the light of the nonlinearities.   Some of the non-linearities can be calculated, but for others all we can hope to do is to determine their qualitative effects.

Our first consideration will be to classify the types of missile controls (not completely), and then to specify certain conventions regarding control deflections, control effectiveness, etc. We then discuss the characteristics of all-movable controls for planar and cruciform configurations in Secs. 8-2 and 8-3, respectively, illustrating therein several methods based on slender-body theory. In Sec. 8-4 various types of couplings are considered that can occur between control functions, such as roll induced by pitch control. The general subject of trailing-edge controls is discussed in Sec. 8-5. Trailing-edge controls cover such a wide range, for which such extensive results are available, that the treatment of the section is to classify the results in a general way, and to refer to original sources for full details.

The results through Sec. 8-5 are based on linear theory; therefore, in Sec. 8-6, we consider a number of important nonlinearities. One particular control characteristic which can be handled only to a limited degree by theory is hinge moment. Some discussion of this general problem is contained in Sec. 8-7. An important constant of a missile is the time it takes to respond to a sudden change in control setting. A simplified analysis of the missile response to a step input in pitch control is presented in Sec. 8-8. On the basis of this analysis, the effect of altitude on missile response is indicated.

8-1. Types of Controls; Conventions

Many types of controls are available to the missile designer. The following list is by no means exhaustive and includes many types with which we will not be directly concerned. (See Fig. 8-1.)

All-movable panels	Nose controls
All-movable tip controls	Shock-interference controls
Trailing-edge controls	Jet controls
Canard controls	Air-jet spoilers

No unanimity of opinion prevails with regard to the definitions of control types; in fact, the technical literature contains many inconsistencies. Furthermore, the control types are not mutually exclusive. One of the principal controls with which we will be concerned is the *all-movable panel* or *all-movable control*. By this we mean an entire wing or tail panel free to rotate about a lateral axis (which may be swept). By an *all-movable tip control* is meant an outboard section of a wing or tail panel free to rotate about a lateral axis. A *trailing-edge control* is a rearward section of a wing or tail panel free to rotate about a lateral axis, with the control trailing edge forming all or part of the panel trailing edge. It is clear from the foregoing definitions that a control could be an all-movable tip control and a trailing-edge control at the same time.

A possible basis of control classification is the control location with reference to the missile center of gravity. If the controls are located well behind the center of gravity, as for conventional aircraft, then the term *tail control* applies.   If, however, the controls are placed forward of the center of gravity, the term *canard control* applies.   When the control is mounted on the main lifting surface near the center of gravity, the term *wing control applies*.

A number of control types with which we are not particularly concerned are, nevertheless, of interest.   A *nose control* is one mounted on the nose of the missile and may comprise all or part of the nose.   A *shock interference control* is a type designed for using interference pressure fields to produce control.   It is so located that it throws a pressure field

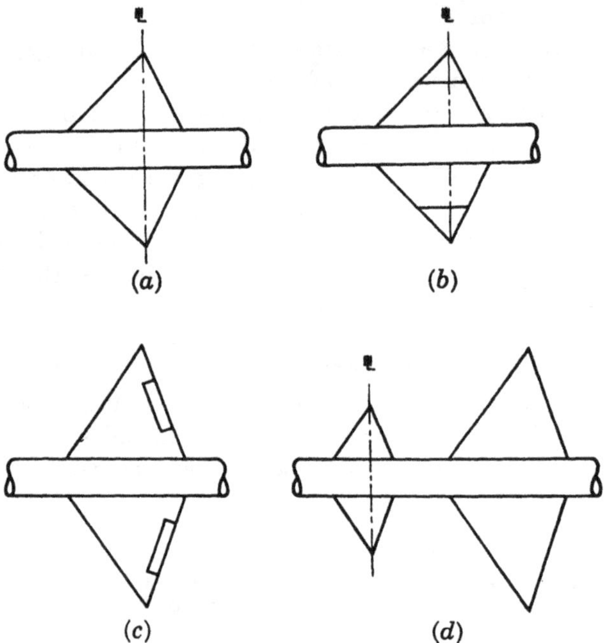

FIG. 8-1. Common types of missile controls.   (*a*) All-movable; (*b*) all-movable tip; (*c*) trailing edge; (*d*) canard.

onto some adjacent surface.   A type of control particularly useful at extreme altitudes is the *jet control*.   Actually, this type includes the *reaction jet*, which depends on the reaction of the jet for its effectiveness, and the *jet vane*, which depends on deflecting a propulsive jet for its effectiveness.   Another interesting type of control is the *air-jet spoiler*. With this, jets of air are ejected more or less normal to a surface to cause changes in the external air flow which augment the reaction of the jets.

It is desirable for the purposes of this book to standardize notation and sign convention for control deflection angles and control effectiveness. Let us consider a *horizontal reference plane*, which is the horizontal plane through the missile axis for zero bank angle, and corresponds to the horizontal plane of symmetry when one exists.   The *vertical reference plane* is a plane through the missile axis normal to the horizontal reference plane, and corresponds to the vertical plane of symmetry. Now, with reference to Fig. 8-2, let the control deflection angles for the right and left horizontal controls looking forward be δ_1 and δ_2, respectively. These angles are measured between the horizontal reference plane and the chord plane of the controls (assuming no camber) in a plane parallel to the vertical reference plane. Trailing edge down is taken to be positive so that negative pitching moment is produced for tail control. Let the control angles for the upper and lower vertical controls be δ_3 and δ_4, respectively. The angles are measured between the vertical reference plane and the chord plane of the controls in a plane parallel to the horizontal

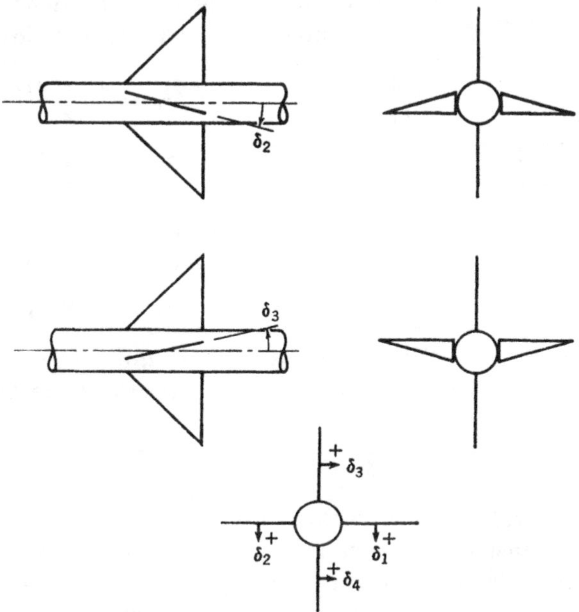

FIG. 8-2. Positive deflection angles. Top left, side view; center left, top view; bottom, end view.

reference plane. Positive values of δ_3 and δ_4 correspond to a movement of the trailing edges of the controls to the right so that a positive yawing moment is produced for tail control. These conventions with regard to control deflection angle hold equally for all-movable controls, all-movable tip controls, and trailing-edge controls.

Let us specify precisely what we mean by pitch control, yaw control, and roll control. Let the control deflections δ_1 and δ_2, not necessarily equal, of the horizontal panels be resolved into pitch deflection δ_e and roll deflection δ_a defined as follows:

$$\delta_e \equiv \frac{\delta_1 + \delta_2}{2}$$
$$\delta_a \equiv \frac{\delta_1 - \delta_2}{2}$$

$$(8\text{-}1)$$

If the deflections of the controls are equal in magnitude and sign so that

$$\delta_e = \delta_1 = \delta_2$$

we have *pitch control* as shown in Fig. 8-3. If, on the other hand, the deflections are of equal magnitude but opposite sign so that

$$\delta_a = \delta_1 = -\delta_2$$

we have *roll control* with the horizontal controls. Now let the control deflections δ_3 and δ_4 of the vertical controls be decomposed into yawing deflections δ_r and rolling deflections $\delta_{a'}$ as follows:

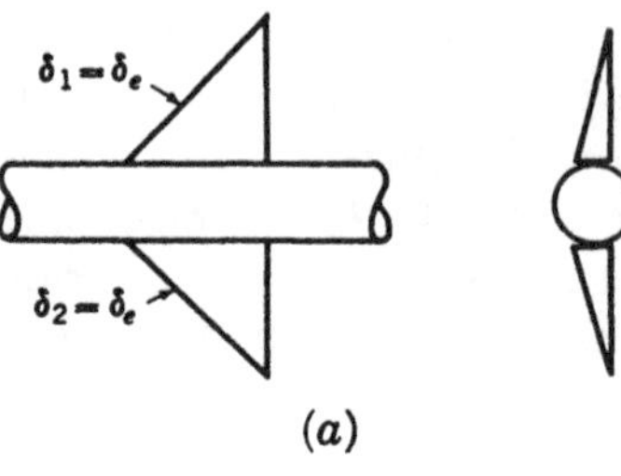

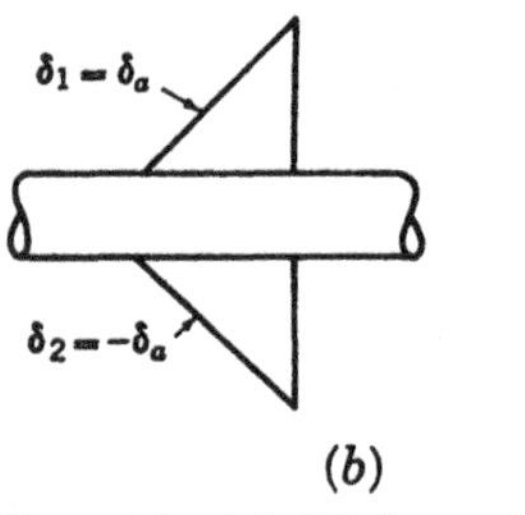

FIG. 8-3. (a) Pitch control and (b) roll control with horizontal controls.

$$\delta_r = \frac{\delta_3 + \delta_4}{2}$$

$$\delta_{a'} = \frac{\delta_3 - \delta_4}{2} \qquad (8\text{-}2)$$

If the deflections of the vertical controls are of equal magnitude and sign so that

$$\delta_r = \delta_3 = \delta_4$$

we have *yaw control* as shown in Fig. 8-4.

But, if the deflections are of equal magnitude but opposite sign so that

$$\delta_{a'} = \delta_3 = -\delta_4$$

we have *roll control* with the vertical controls.

Let us now consider what we mean by the pitching and rolling effectiveness of the horizontal controls. The *pitching effectiveness* is measured by the rate of change of pitching-moment coefficient C_m with pitch control.

$$\text{Pitching effectiveness} = \frac{\partial C_m}{\partial \delta_e} \qquad (8\text{-}3)$$

The *rolling effectiveness* of the horizontal controls is measured similarly on the basis of rolling-moment coefficient C_l.

$$\text{Rolling effectiveness} = \frac{\partial C_l}{\partial \delta_a} \qquad (8\text{-}4)$$

The parameter $\partial C_m/\partial \delta_e$ is normally negative with tail control and positive with canard control. The parameter $\partial C_l/\partial \delta_a$ is usually negative. A change in sign of the control effectiveness is known as *control reversal*.

Consider now the yawing and rolling effectiveness of the vertical panels. The *yawing effectiveness* is measured by the rate of change of yawing moment with yaw control.

$$\text{Yawing effectiveness} = \frac{\partial C_n}{\partial \delta_r} \tag{8-5}$$

The rolling effectiveness is measured in the same manner for the horizontal controls.

$$\text{Rolling effectiveness} = \frac{\partial C_l}{\partial \delta_{a'}} \tag{8-6}$$

The sign of the yawing effectiveness is usually opposite to that of the pitching effectiveness. The sign of the rolling effectiveness for both horizontal and vertical panels should be negative. The use of canard

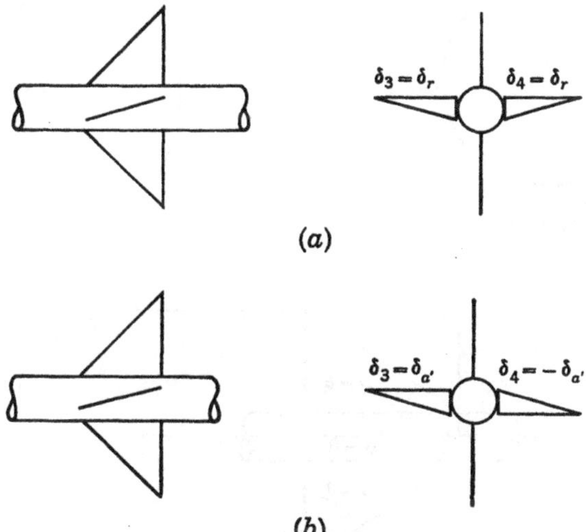

FIG. 8-4. (a) Yaw control and (b) roll control with vertical controls.

controls for roll control tends to give a positive value of $\partial C_l/\partial \delta_a$ because of interference effects of the main lifting surface. It is clear that effectiveness derivatives can be defined on the basis of forces as well as moments, but we will not be concerned with such derivatives. When there is a coupling or "cross talk" between two controls, then certain *cross-coupling derivatives* can be defined. While we will not make precise definitions of cross-coupling derivatives, we will consider their qualitative behavior in some detail.

8-2. All-movable Controls for Planar Configurations

The all-movable control used in canard, wing, or tail control applications is an important type of missile control. One reason for its importance is the simple method it provides for obtaining a control of large area for fast response at high altitudes. In the ensuing analysis of the properties of the all-movable control, we will approach the problem of the pitching effectiveness by constructing the crossflow potential and applying slender-body theory to obtain detailed loadings. This approach demonstrates certain tricks in constructing the potential. The approach to the problem of rolling effectiveness will be along the lines of reverse-flow theorems and slender-body theory, to show the great simplification occurring in the analysis when only gross quantities are determined in contrast to detailed loadings.

Let us calculate the pitching effectiveness of all-movable controls mounted on a body to produce a planar configuration as shown in Fig.

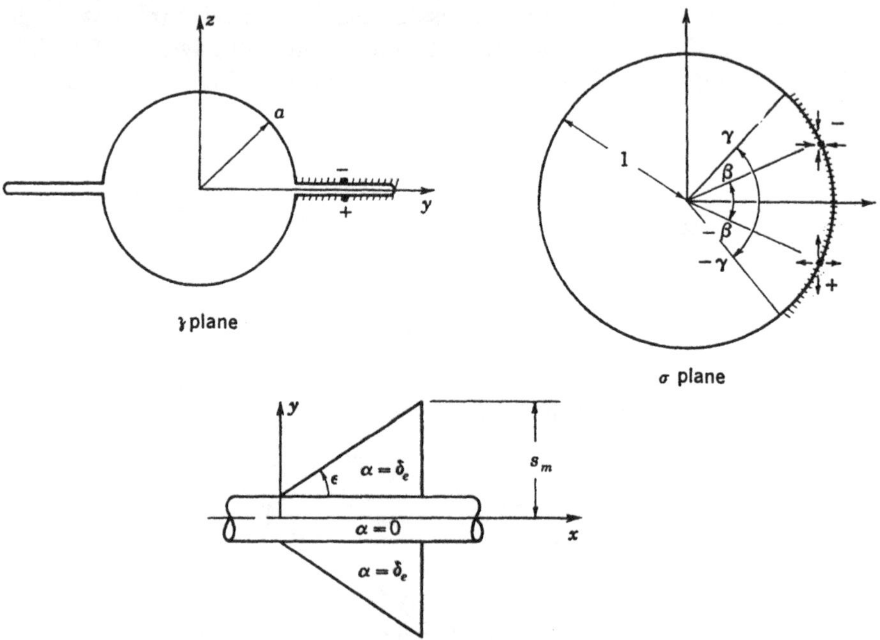

Fig. 8-5. Crossflow plane and transformed plane for pitch control of planar configuration.

8-5. Although this calculation will be made on the basis of slender-body theory, it will be extended to nonslender configurations using the same general methods of wing-body interference described in Chap. 5. Let us focus attention on the crossflow plane, the ʓ plane, and construct the potential for the flow. Since the control panels are deflected to angles δ_e, and the body is at zero angle of attack, the boundary conditions are

$$\frac{\partial \phi}{\partial z} = -\delta_e V_0 \qquad -s \leq y \leq a \qquad \text{and} \qquad a \leq y \leq s \qquad \text{for } z = 0$$

$$\frac{\partial \phi}{\partial r} = 0 \qquad r = a \tag{8-7}$$

$$\phi = 0 \qquad |ʓ| \to \infty$$

The potential must be continuous throughout the flow field except for possible singular points on the boundary where ϕ_z or ϕ_y may be singular. If a solution could be found producing a local velocity normal to the panel, and zero velocity normal to the panel and body everywhere else, the solution could be used to construct the potential for any arbitrary variation of normal velocity across the span of the control. The usual doublet does not satisfy these conditions, but a doubletlike solution having such properties can be constructed. Let us first transform the cross section of the missile in the ζ plane into the unit circle by an application of Eq. (7-13) as shown in Fig. 8-5. Now introduce a source and sink on the surface of the unit circle into which the panel is transformed as shown in the figure. The family of circles passing through the source and sink form the streamlines of their combined flow. In particular, the unit circle is a streamline so that no flow is induced normal to it. Let us now transform the flow in the σ plane back to the ζ plane. In the transformation the source and sink are brought into close proximity, forming a doubletlike solution. In the transformation, the property of no flow normal to the solid boundaries is preserved, with the exception of the point where the source and sink come into confluence. At this point the doubletlike solution, henceforth called a doublet, produces a velocity normal to the panel surface and continuous through it. Our next step is to determine this local normal velocity in terms of the doublet strength.

Let the strength of the source and sink in the σ plane be of magnitude $2\delta_e V_0\,dy$, where dy is the element of control span at the doublet location y. The complex potential in the σ plane is then

$$W_d(\sigma) = -\frac{\delta_e V_0\,dy}{\pi} \log \frac{\sigma - e^{i\beta}}{\sigma - e^{-i\beta}} \tag{8-8}$$

where $e^{i\beta}$ and $e^{-i\beta}$ are points in the σ plane where the sink and source are placed. For a point $\sigma = e^{i\Theta}$ on the unit circle the complex potential can be written

$$W_d = \phi_d(\Theta) + i\psi_d(\Theta) = -\frac{\delta_e V_0\,dy}{\pi}\left[\log \frac{\cos\beta - \cos\Theta}{1 - \cos(\Theta + \beta)} + i\beta\right] \tag{8-9}$$

Since ψ_d is constant on the unit circle, it is a streamline, as formerly stated. The potential on the unit circle is

$$\phi_d(\Theta) = -\frac{\delta_e V_0\,dy}{\pi} \log \frac{\cos\beta - \cos\Theta}{1 - \cos(\Theta + \beta)} \tag{8-10}$$

Now in the transformation the source strength has remained unaltered. Half the fluid due to the source flows upward through the span element of width dy at velocity $w(y)$ into the sink. Applying the equation of continuity locally, we thus obtain

$$w(y)\,dy = -\delta_e V_0\,dy$$

or
$$w(y) = -\delta_e V_0 \tag{8-11}$$

Now our boundary conditions call for $w(y)$ to be $-\delta_e V_0$, so that the potential for a velocity $-\delta_e V_0$ at points y and $-y$ on the control panels is

$$\phi_d(\Theta) + \phi_d(\pi - \Theta) = \frac{-\delta_e V_0}{\pi}\left[\log \frac{\cos \beta - \cos \Theta}{1 - \cos (\Theta + \beta)} + \log \frac{\cos \beta + \cos \Theta}{1 + \cos (\Theta - \beta)}\right] dy$$

Since the panels are at a uniform deflection angle, we can carry out the integration across the panel span to obtain the potential for the entire flow as follows:

$$\phi(\Theta) = \frac{-\delta_e V_0}{\pi}\int_0^{\beta=\gamma}\left[\log \frac{\cos \beta - \cos \Theta}{1 - \cos (\Theta + \beta)} + \log \frac{\cos \beta + \cos \Theta}{1 + \cos (\Theta + \beta)}\right] dy \quad (8\text{-}12)$$

where
$$\frac{y}{a} = \frac{1}{\cos \gamma}[\cos \beta + (\cos^2 \beta - \cos^2 \gamma)^{1/2}] \quad (8\text{-}13)$$

and
$$\cos \gamma = \frac{2a}{s + a^2/s} = a \quad (8\text{-}14)$$

The integration is tedious, as the final answer for the potential shows. For the top surface of the wing the potential according to a solution of Gaynor J. Adams is

$$\phi_{W(B)} = \frac{V_0 \delta}{4\pi}\left\{-4r_1 \tanh^{-1}\left(\frac{s_1{}^2 - y_1{}^2}{s_1{}^2 - r_1{}^2}\right)^{1/2}\right.$$
$$+ 4y_1 \tanh^{-1}\left[\frac{r_1}{y_1}\left(\frac{s_1{}^2 - y_1{}^2}{s_1{}^2 - r_1{}^2}\right)^{1/2}\right]$$
$$\left. + 2\pi(s_1{}^2 - y_1{}^2)^{1/2}\left(1 + \frac{2}{\pi}\cos^{-1}\frac{r_1}{s_1}\right)\right\} \quad (8\text{-}15)$$

wherein
$$s_1 = s + \frac{a^2}{s}$$
$$y_1 = y + \frac{a^2}{y} \quad (8\text{-}16)$$
$$r_1 = 2a$$

Equation (8-15) is also given by Dugan and Hikido.[2]

Having determined the potential, we can obtain the forces and loadings on the panels due to deflecting the panels. It is convenient first to consider the lifts on the panels and the body, before considering the loading distributions. To specify these lifts, we introduce two new lift ratios k_W and k_B analogous to the two lift ratios K_W and K_B defined in Chap. 5. If $L_{W(B)}$ is the lift on the panels due to its own deflection and L_W is the lift on the wing alone formed by joining the two panels together ($\alpha_W = \delta_e$), then the ratio k_W is defined as

$$k_W = \frac{L_{W(B)}}{L_W} \qquad \alpha_B = 0 \qquad \alpha_W = \delta_e \qquad (8\text{-}17)$$

The subscript W applies equally to all-movable canard or tail panels as to wing panels. An analogous ratio k_B is defined for the lift $L_{B(W)}$ carried over onto the body as a result of the panel deflection:

$$k_B = \frac{L_{B(W)}}{L_W} \qquad \alpha_B = 0 \qquad \alpha_W = \delta_e \qquad (8\text{-}18)$$

To evaluate k_W we must find the loading on the panel. The loading is given by Eq. (8-30). This loading is integrated across one panel and doubled to obtain $L_{W(B)}$. With reference to Fig. 8-5, the lift is given by

$$\frac{L_{W(B)}}{q_0 \delta_e} = 2 \int_0^{\frac{s_m - a}{\tan \epsilon}} dx \int_a^s \left(\frac{\Delta P}{\delta_e} \right)_{W(B)} dy \qquad (8\text{-}19)$$

With the aid of the value for

$$L_W = 2q_0 \delta_e \pi (s_m - a)^2 \qquad (8\text{-}20)$$

the value of k_W obtained by integrating Eq. (8-19) is

$$\begin{aligned}
k_W = \frac{1}{\pi^2} \Bigg\{ &\frac{\pi^2}{4} \left(\frac{\lambda + 1}{\lambda} \right)^2 + \pi \left[\frac{\lambda^2 + 1}{\lambda(\lambda - 1)} \right]^2 \sin^{-1} \frac{\lambda^2 - 1}{\lambda^2 + 1} \\
&- \frac{2\pi(\lambda + 1)}{\lambda(\lambda - 1)} + \frac{(\lambda^2 + 1)^2}{\lambda^2(\lambda - 1)^2} \left(\sin^{-1} \frac{\lambda^2 - 1}{\lambda^2 + 1} \right)^2 \\
&- \frac{4}{\lambda} \frac{\lambda + 1}{\lambda - 1} \sin^{-1} \frac{\lambda^2 - 1}{\lambda^2 + 1} + \frac{8}{(\lambda - 1)^2} \log \frac{\lambda + 1/\lambda}{2} \Bigg\}
\end{aligned}$$

$$\lambda = \frac{s_m}{a} \qquad (8\text{-}21)$$

If the loading is integrated over the body and the lift ratio k_B is formed, there is obtained

$$\begin{aligned}
k_B = \frac{2}{\pi(1 - \lambda)^2} \Bigg[&\left(\frac{1 - \lambda^2}{\lambda} \right)^2 \frac{\pi}{4} + \frac{1 - \lambda^2}{\lambda} + \frac{(\lambda^2 + 1)^2}{2\lambda^2} \\
&\sin^{-1} \frac{\lambda^2 - 1}{1 + \lambda^2} \Bigg] - k_W \qquad (8\text{-}22)
\end{aligned}$$

As a matter of interest, this equation coupled with Eq. (5-17) yields a simple relationship among three important lift ratios:

$$K_W = k_W + k_B \qquad (8\text{-}23)$$

Let us examine these lift ratios to obtain an over-all idea of the gross lift forces due to panel deflection. The values of k_W and k_B are shown as a function of a/s_m in Fig. 8-6. It is of interest to note that k_W is not much less than unity for all values of a/s_m. What this means is that all-movable panels in the presence of a body for all practical purposes develop

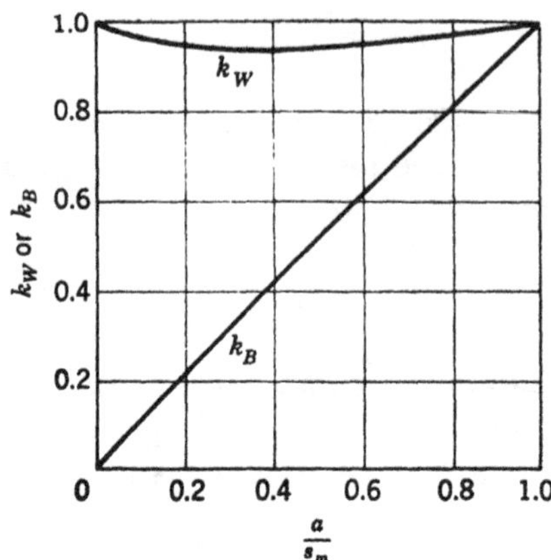

FIG. 8-6. Lift ratios for symmetrical deflection of all-movable panels on circular cylindrical body.

almost as much lift as the wing formed by joining them together. (Some discussion of the effect of gaps at the body-control junctures will subsequently be given.) The lift ratio k_B for the body shows an almost linear variation with a/s_m. In fact, a simple rule of thumb is that the fractional part of the panel lift carried over onto the body is equal to a/s_m for all-movable controls. The values of k_W and k_B are listed in Table 8-1 for general use. It is again noted in passing that these values can be applied to nonslender configurations on the same basis as K_W and K_B.

The lift coefficient for the complete configuration due to control deflection $(L_C)_\delta$ can be expressed simply in terms of k_W and k_B for tail control.

$$(L_C)_\delta = L_{W(B)} + L_{B(W)} = q_0 \delta_e S_W (C_{L\alpha})_W (k_W + k_B) \qquad (8\text{-}24)$$

However, for wing control or canard control there will usually be wing or tail surfaces in the wake of the control. In such cases a loss of lift effectiveness will occur that can be calculated using the wing-tail interference methods of Chap. 7. Let us now consider the pitching effectiveness.

TABLE 8-1. NONDIMENSIONAL RATIOS FOR SYMMETRICAL DEFLECTION OF ALL-MOVABLE WINGS MOUNTED ON CIRCULAR BODY

$\dfrac{a}{s_m}$	K_W	k_W	k_B	$\dfrac{k_B}{k_W}$	$\dfrac{K_B}{K_W}$	$\left(\dfrac{\bar{x}}{c_r}\right)_{W(B)}$ *
0	1.000	1.000	0	0	0	0.667
0.1	1.077	0.963	0.114	0.118	0.123	0.669
0.2	1.162	0.944	0.218	0.231	0.239	0.668
0.3	1.253	0.936	0.317	0.338	0.349	0.666
0.4	1.349	0.935	0.414	0.442	0.454	0.665
0.5	1.450	0.940	0.510	0.542	0.551	0.664
0.6	1.555	0.948	0.607	0.641	0.646	0.663
0.7	1.663	0.958	0.705	0.736	0.737	0.664
0.8	1.774	0.971	0.803	0.827	0.827	0.666
0.9	1.887	0.985	0.902	0.916	0.915	0.667
1.0	2.000	1.000	1.000	1.000	1.000	0.667

* Triangular panel.

To obtain the pitching effectiveness of an all-movable control (or its hinge moment) requires a knowledge of the center of pressure of the lift due to the panel deflection as well as a knowledge of the lift itself. We are free to calculate the center of pressure by integration of the loading distribution given by Eqs. (8-30) and (8-31). Unfortunately, such an integration will yield different results for each planform of the panel. The panel center of pressure has been calculated for a triangular panel, but only the final results are reproduced herein. The center-of-pressure position of the control panel is given in fractional parts of the chord at the panel-body juncture c_r measured behind the leading edge of the juncture. These values of $(\bar{x}/c_r)_{W(B)}$ are listed in Table 8-1. The interesting fact is noted from these results that, because of panel-body interference, the center of pressure of the panel in the presence of the body has not been changed by more than $0.005c_r$ from its wing-alone value of $0.667c_r$. On the basis of this result, we might surmise that the wing-alone center of pressure is a good approximation to the center of pressure of the panel in the presence of the body. This is so, and we shall assume below that

$$\left(\frac{\bar{x}}{c_r}\right)_{W(B)_\delta} = \left(\frac{\bar{x}}{c_r}\right)_W \tag{8-25}$$

(It should be remembered here that the wing alone as used here refers to the wing formed by joining the two all-movable panels together.) For the reasons discussed in Sec. 5-6, the center of pressure of the lift on the body due to the control panels cannot be accurately calculated using slender-body theory. Actually, the center of pressure of the body lift resulting from the panel is not sensitive to the precise shape of the span-load distribution of the panel, and will be nearly the same whether the lift is developed by angle of attack or by panel deflection. It is, however, sensitive to afterbody length and Mach number. On the basis of these facts we may write approximately

$$\left(\frac{\bar{x}}{c_r}\right)_{B(W)_\delta} = \left(\frac{\bar{x}}{c_r}\right)_{B(W)_\alpha} \tag{8-26}$$

The pitching effectiveness for tail control can now be written

$$-\frac{\partial C_m}{\partial \delta_e} = \frac{k_W[(\bar{x}/c_r)_W - (x/c_r)_{cg}] + k_B[(\bar{x}/c_r)_{B(W)_\alpha} - (x/c_r)_{cg}]}{l_r/c_r}(C_{L_\alpha})_W \tag{8-27}$$

where l_r is the reference length for pitching moment, and $(x/c_r)_{cg}$ gives the position of the missile center of gravity. For canard or wing control it is necessary to consider also the increment in pitching moment due to interference associated with the control wake.

We have purposely avoided any discussion of the loading distribution due to panel deflection until now, to avoid breaking up the foregoing discussions of the lift and pitching effectivenesses. However, we now apply Bernoulli's equation including quadratic terms to the calculation of the loading coefficients. Let u^+, v^+, and w^+ be the perturbation velocities on the lower surface of the control panels due to deflection δ_e, and let u^-, v^-,

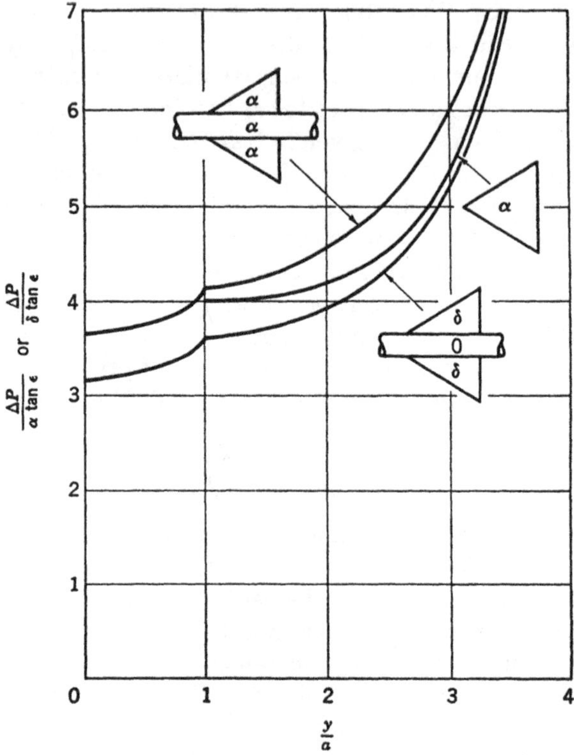

Fig. 8-7. Loading coefficient at trailing edge of various configurations employing triangular panels.

and w^- be the corresponding values for the upper surface. Then, neglecting coupling effects due to any other perturbation velocities (which we shall discuss later), we have

$$\frac{\Delta p}{q_0} = \frac{+2(u^- - u^+)}{V_0} + \frac{(v^-)^2 - (v^+)^2 + (w^-)^2 - (w^+)^2}{V_0^2} \qquad (8\text{-}28)$$

The symmetry properties of the velocity components yield the loading coefficient

$$\Delta P = \frac{+4u^-}{V_0} \qquad (8\text{-}29)$$

From the potential for the panel, Eq. (8-15), the panel loading coefficient is found to be

$$\frac{(\Delta P)_{W(B)}}{\delta_e} = \frac{2}{\pi} \tan \epsilon \, \frac{s^4 - a^4}{s^3} \frac{\pi + 2 \cos^{-1}[2a/(s^2 + a^2/s)]}{[(s + a^2/s)^2 - (y + a^2/y)^2]^{\frac{1}{2}}} \qquad (8\text{-}30)$$

Similarly for the body loading coefficient, it can be shown that

$$\frac{(\Delta P)_{B(W)}}{\delta_e} = \frac{2}{\pi} \tan \epsilon \, \frac{s^4 - a^4}{s^3} \frac{\pi + 2 \cos^{-1}[2a/(s + a^2/s)]}{[(s + a^2/s)^2 - 4y^2]^{\frac{1}{2}}} \qquad (8\text{-}31)$$

It is interesting to compare the loading coefficients at the trailing edge of the control panel due to unit angle of attack and unit deflection angle. This is done in Fig. 8-7 for a body with triangular panels. Also included for comparison is the loading coefficient distribution for the wing alone. As might be surmised, the loading due to angle of attack is greater than that due to the wing alone, while that due to panel deflection is less. Qualitatively the loading distributions are similar, aside from the obvious fact that the wing alone has no body loading.

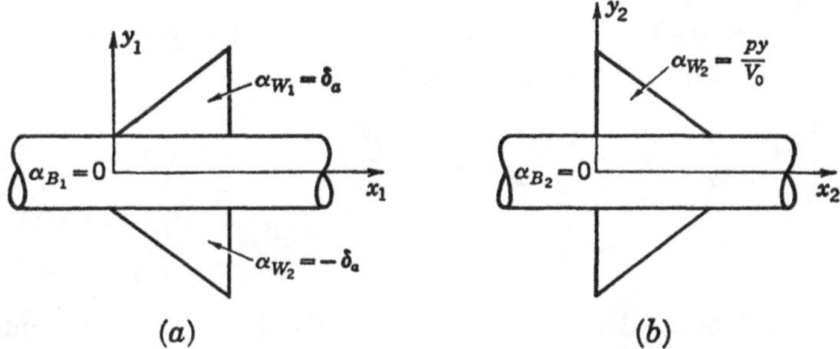

FIG. 8-8. Direct and reverse flows for calculating rolling effectiveness of planar configurations. (a) Case 1; (b) case 2.

In the treatment of the pitching effectiveness of all-movable controls, we have used a straightforward application of slender-body theory. The amount of work involved is, however, considerable. To calculate the rolling effectiveness by the same method would lead to complicated elliptic integrals as seen in Adams and Dugan.[1] We shall forego such a calculation and confine ourselves to reverse-flow methods. We shall apply reverse-flow methods to the calculation of the rolling effectiveness of panels with straight trailing edges. The configurations in direct and reverse flow germane to such a calculation are shown in Fig. 8-8. The reverse-flow theorem, Eq. (7-47), gives

$$\iint_{S_W} P_1 \alpha_{W_2} \, dS_W = \iint_{S_W} P_2 \alpha_{W_1} \, dS_W \qquad (8\text{-}32)$$

where P_1 and P_2 are the panel loading coefficients for cases 1 and 2, respectively. For case 1, the rolling moment L' is

$$-\frac{L'}{q_0} = \iint_{S_W} P_1 y \, dS_W = 2 \iint_{S_P} P_1 y \, dS_W \qquad (8\text{-}33)$$

We have chosen α_{W_2} such that

$$\alpha_{W_2} = y \frac{p}{V_0} \tag{8-34}$$

so that

$$-\frac{L'}{q_0} = \frac{2}{p/V_0} \iint_{S_P} P_1 \alpha_{W_2}\, dy = 2\delta_a \iint_{S_P} \frac{P_2}{p/V_0}\, dS_{W_2} \tag{8-35}$$

If we denote the span-load distribution for case 2 by $(cc_l)_2$, we obtain for the rolling moment

$$-\frac{L'}{q_0} = 2\delta_a \int \frac{(cc_l)_2}{p/V_0}\, dy \tag{8-36}$$

This result can be interpreted as a relationship between the rolling effectiveness of an all-movable wing and the span loading associated with the damping in roll of a planar wing and body combination in reverse flow. In examining this reverse-flow case, we note that the maximum span is at the leading edge, so that in accordance with slender-body theory all the load is concentrated along the leading edge. The span loading for the rolling combination from Heaslet and Spreiter[3] is

$$\frac{(cc_l)_2}{p/V_0} = \left\{ \left(1 + \frac{2}{\pi}\cos^{-1}\frac{2as}{s^2 + a^2}\right)\left(y + \frac{a^2}{y}\right)\left[(s^2 - y^2)\left(1 - \frac{a^4}{y^2 s^2}\right)\right]^{\frac{1}{2}} \right.$$
$$\left. + \frac{2}{\pi}\left(y - \frac{a^2}{y}\right)^2 \cosh^{-1}\frac{(y^2 + a^2)(s^2 - a^2)}{(y^2 - a^2)(s^2 + a^2)} \right\} \tag{8-37}$$

This result introduced into Eq. (8-36) yields the rolling moment due to differential deflection δ_a. If the rolling effectiveness parameter is taken to be

$$(C_l)_{\delta_a} = \frac{L'}{q_0 \delta_a S_W (2s_m)}$$

then

$$-(C_l)_{\delta_a} = \frac{1}{s_m S_W} \int_a^{s_m} \frac{(cc_l)_2}{p/V_0}\, dy \tag{8-38}$$

The result for the rolling effectiveness based on the exposed panel areas as reference area and the total span of the combination as reference length is

$$-\frac{(C_l)_{\delta_a}}{A} = \frac{1}{2(1-\lambda)^2}\left\{\left[\left(1 + \frac{2}{\pi}\cos^{-1}\frac{2\lambda}{\lambda^2 + 1}\right)\left(1 - \frac{2\lambda^2}{3} + \lambda^4\right)\frac{\lambda^2}{2}\right.\right.$$
$$\left. - \frac{2\lambda^3}{3\pi}(1 - \lambda^2)\right] F(\phi,k) + \frac{1}{2}\left(\frac{1}{3} - 2\lambda^2 + \frac{\lambda^4}{3}\right)$$
$$\left(1 + \frac{2}{\pi}\cos^{-1}\frac{2\lambda}{1 + \lambda^2}\right) E(\phi,k) + \frac{3\lambda}{\pi}(1 - \lambda^2)$$
$$\left. + \frac{8\lambda^3}{3\pi}\log\frac{2\lambda}{\lambda^2 + 1} + \frac{\lambda^2}{3}(1 - \lambda^2)\left(1 + \frac{2}{\pi}\cos^{-1}\frac{2\lambda}{1 + \lambda^2}\right)\right\} \tag{8-39}$$

wherein

$$\lambda = \frac{a}{s_m} \qquad\qquad A = \frac{4(s_m - a)^2}{S_W}$$

$$\phi = \sin^{-1}\frac{1}{(\lambda^2 + 1)^{1/2}} \qquad k = (1 - \lambda^4)^{1/2} \tag{8-40}$$

For the extreme values of $(C_l)_{\delta_a}$ we have

$$-(C_l)_{\delta_a} = \frac{1}{6A} \qquad\qquad \lambda = 0$$

$$(C_l)_{\delta_a} = \left(\frac{\pi}{4}\right) A \qquad \lambda = 1 \tag{8-41}$$

To illustrate the rolling-effectiveness properties of all-movable controls the values of $(C_l)_{\delta_a}$ as determined from Eq. (8-39) are plotted in Fig. 8-9

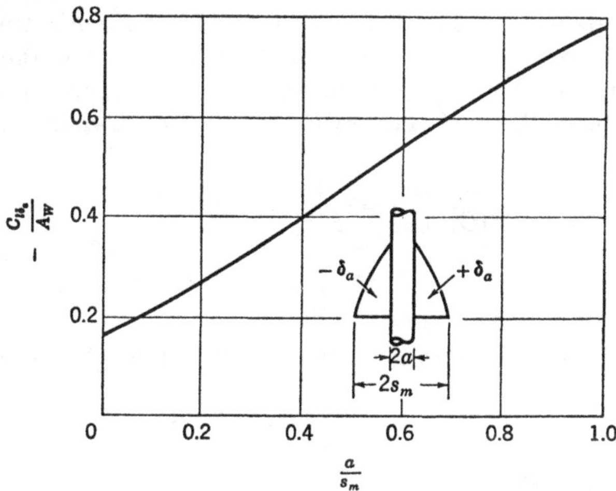

FIG. 8-9. Rolling effectiveness of planar all-movable controls.

against a/s_m. It is seen that the same control panels mounted on a body of large diameter produce nearly four times the rolling effectiveness as the same panels acting on a very small body. There are two reasons for this behavior: panel-panel interference and outboard movement of the lateral center of pressure. Consider the influence of panel-panel interference for $a/s_m = 0$. If a vertical reflection plane were placed between the two panels they would act independently as half of a wing. The rolling effectiveness due to forces on the two panels under these conditions would correspond to a value of $\frac{1}{3}$ for $(C_l)_{\delta_a}/A$. Removal of the reflection plane reduces this value to $\frac{1}{6}$. (Testing a semispan model on a reflection plate would give a rolling effectiveness too great by a factor of 2.) As the panels are spread apart, the adverse effect of the panel-panel interference largely disappears. The second effect is the obvious one that the panel lifts are concentrated at a greater per cent semispan as a/s_m increases.

Several points need mentioning before ending this discussion of rolling effectiveness.   Figure 8-9 as it stands applies directly to tail control.   It also applies to wing control and canard control, with the important proviso that the interference effects due to the control wake be also considered.   These effects are usually such that the surfaces behind the controls tend to produce roll in opposition to that developed by the controls themselves.   For canard control the reverse roll can be large enough to produce control reversal.   As a result, canard controls are not well-suited to roll control.   It should be noted that, on the basis of slender-body theory, the derivation of the rolling effectiveness applies equally to a full-span trailing-edge control as to an all-movable control.   This is a direct consequence of the fact that the loading of the control in reverse flow is all concentrated at the leading edge.   This result applies, of course, only to very slender configurations.   In the use of Fig. 8-9 a correction should be applied to these values, to account for the fact that slender-body theory overpredicts force coefficients for nonslender configurations.   For this reason the results of Fig. 8-9 should be scaled down in the ratio of the lift-curve slopes of the wing alone, as calculated by supersonic wing theory and slender-body theory.   This ratio for triangular wings is

$$
\begin{aligned}
\frac{(C_{L_\alpha})_{\text{LT}}}{(C_{L_\alpha})_{\text{SBT}}} &= \frac{1}{E(k)} \qquad BA \leq 4 \\
&= \frac{8}{\pi BA} \qquad BA \geq 4
\end{aligned}
\tag{8-42}
$$

wherein $E(k)$ is the complete elliptic integral of the second kind of modulus

$$
k = \left[1 - \left(\frac{BA}{4} \right)^2 \right]^{1/2}
\tag{8-43}
$$

Illustrative Example

Let us calculate the pitching effectiveness and the rolling effectiveness of the wing-body-tail combination treated in the illustrative example of Sec. 7-5 if the tail panels are used as all-movable controls at $M_0 = 2$. Let the center of gravity be a distance 3.95 length units in front of the leading edge of the tail-body juncture from which all x distances are measured.   Let us first calculate the pitching effectiveness using Eq. (8-27). Since

$$
\left(\frac{a}{s_m} \right)_T = \frac{0.5625}{1.812} = 0.31
$$

Table 8-1 yields

$$
k_T = 0.936 \qquad k_B = 0.327
$$

The subscript T is used instead of W in this example since we are considering tail control, not wing control.   The tail center of pressure is, for all practical purposes, at the two-thirds root chord at the juncture so that

$$\left(\frac{\bar{x}}{c_r}\right)_T = 0.67$$

The center of pressure of body lift can, on the basis of Eq. (8-26), be taken from Table 5-1. This slender-body value will be sufficiently accurate since there is no afterbody

$$\left(\frac{\bar{x}}{c_r}\right)_{B(T)\alpha} = 0.56$$

If we take the reference length l_r to be the mean aerodynamic chord of the wing panels and take $(C_{L_\alpha})_T$ from the former illustrative example, Eq. (8-27) yields

$$-\frac{\partial C_m}{\partial \delta_e} = \frac{[0.936(0.67 + 3.95/1.25) + 0.327(0.56 + 3.95/1.25)]2.31}{1.5/1.25}$$
$$= 9.24 \text{ per radian}$$

For the rolling effectiveness, slender-body theory (Fig. 8-9) yields

$$\frac{(C_l)_{\delta_a}}{A_T} = -0.34$$

With $A_T = 4$ and $B = 1.732$, Eq. (8-42) gives a factor

$$\frac{8}{\pi B A} = \frac{8}{\pi(4)(1.732)} = 0.367$$

to be applied to the rolling effectiveness. Therefore, we have

$$(C_l)_{\delta_a} = 4(-0.34)0.367 = -0.50 \text{ per radian}$$

8-3. All-movable Controls for Cruciform Configurations

The results for pitching effectiveness of planar configurations can be applied unchanged to cruciform configurations if we neglect the panel-panel interference terms that arise because of the square terms in Bernoulli's equation. However, we cannot apply the rolling-effectiveness results for planar configurations directly to cruciforms since the panel-panel interference in this case is associated with the linear terms of Bernoulli's equation. In the next section the qualitative effects of coupling due to the quadratic terms will be discussed, but in this section, as in the previous one, we neglect such effects. Let us start with a discussion of pitching effectiveness.

Consider the cruciform configuration with all-movable controls shown in Fig. 8-10 under angles of pitch and sideslip. If α_c is the included angle in the plane of the body axis and free-stream direction, then the angles of attack and sideslip are obtained by the decomposition

$$\alpha = \alpha_c \cos \varphi$$
$$\beta = \alpha_c \sin \varphi \tag{8-44}$$

The effectiveness in pitch in the α plane can then be calculated directly from Eq. (8-27) without any regard for the sideslip velocity. Again a calculation of the yaw effectiveness can be made based on β without any regard for α.

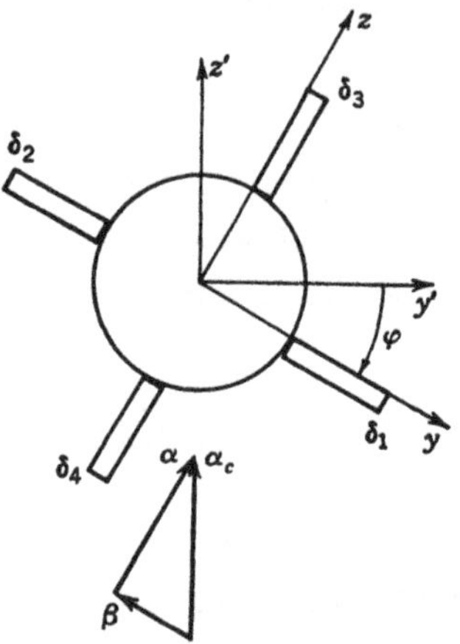

Fig. 8-10. Combined pitch and yaw control of cruciform configuration.

It is interesting to note the resultant forces due to certain combinations of pitch control and yaw control for cruciform configurations. Consider the pitch control of a cruciform configuration at zero bank angle. If both horizontal panels are deflected to δ, a force F results in the vertical plane. If now the configuration is rolled to 45° and all four panels are deflected to δ, a force $(2)^{\frac{1}{2}}F$ will be developed in the vertical plane due to the controls. As a result, the pitch effectiveness in the vertical plane has been significantly increased. Therefore, to obtain the largest force in response to a command for acceleration in a given plane, a missile must roll to a bank angle of 45° with respect to the plane, and then deflect all four panels. Since the missile has such a low inertia in roll compared to that in pitch, such a maneuver can result in fast pitch control. However, we should not lose sight of the fact that one of the characteristic features of a cruciform arrangement is its ability to perform a maneuver in any plane without banking. Let us now turn to the subject of roll control.

Panel-panel interference produces such sizable modifications to the rolling effectiveness of cruciform arrangements that the rolling-effectiveness results for planar configurations are inapplicable. The nature of this interference is made clear by an examination of the general features of the flow in the crossflow plane of the panels as shown in Fig. 8-11. For pitch control of the horizontal panels, the flow symmetry about the vertical panels is such as to produce no sideforce. However, when aileron deflections are applied to obtain roll control, the figure shows how positive pressure is created on one side of the vertical panels and negative pressure on the other. It is to be noted that the resulting rolling moment always opposes the rolling moment called for by the control deflection. Therefore, this panel-panel interference phenomenon is termed *reverse roll*. It is possible to calculate the magnitude of the reverse roll by applying the first method of Sec. 8-2 and superimposing solutions of the type given by Eq. (8-10). In fact, this is precisely the method used by

Adams and Dugan[1] to solve this problem. The actual mathematics leads to elliptic integrals of a complex nature so that only the final results will be considered here. The rolling-effectiveness parameters $(C_l)_{\delta_a}$ based on the area of the two deflected panels as reference area and the total span $2s_m$ as reference length are shown in Fig. 8-12 as a function of a/s_m. The contribution of the horizontal panels to direct roll and the contribution of the vertical panels to reverse roll are both shown. At a value of zero for a/s_m the horizontal panels produce a value $-(C_l)_{\delta_a}/A$ of 0.28, while the vertical panels produce a value of -0.15. As a result, a value of about 0.13 is obtained for a cruciform as compared to a value of $\frac{1}{6}$ for a planar arrangement. However, the use of the vertical panels for roll

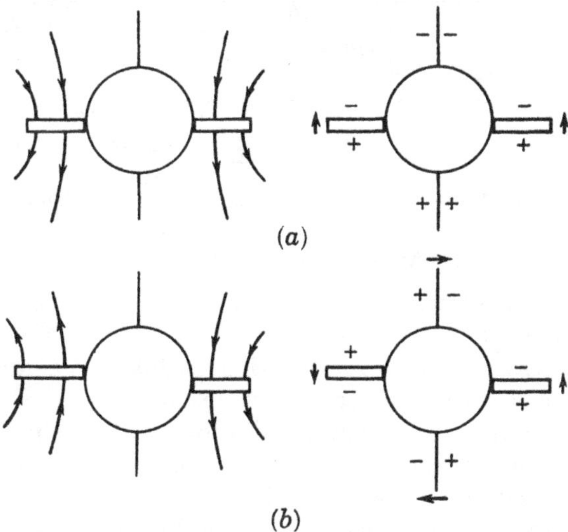

FIG. 8-11. Induced flow and direction of panel forces due to (a) pitch control and (b) roll control of horizontal control panels.

control, as well as the horizontal panels, increases the potential rolling effectiveness of a cruciform configuration compared to that of a planar one. If the rolling-effectiveness parameters of Figs. 8-9 and 8-12 are compared, it is clear that, for large values of a/s_m, the adverse effects of panel-panel interference in producing reverse roll are small, so that planar and cruciform arrangements have essentially the same rolling effectiveness for the horizontal panels.

Again it should be noted that for panels of large aspect ratio the results of Fig. 8-12, which are based on slender-body theory, should be scaled down by the factors given by Eq. (8-42). Bleviss[6] has calculated the rolling effectiveness of all-movable triangular panels in planar and cruciform arrangement on the basis of supersonic wing theory for the case of supersonic leading edges. For arrangements having small bodies and large panels these results can be used.

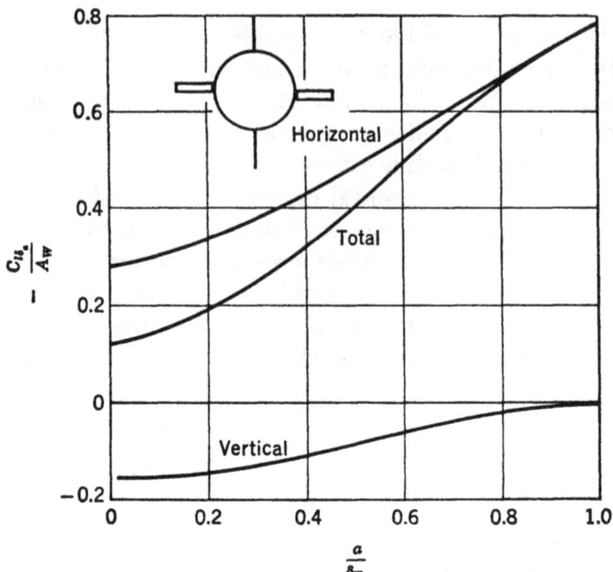

FIG. 8-12. Rolling effectiveness of cruciform all-movable controls.

8-4. Coupling Effects in All-movable Controls

In the preceding sections, the discussion of the control effectiveness of planar and cruciform missiles equipped with all-movable controls was based on the linearized Bernoulli equation.   Since the quadratic terms of Bernoulli's equation are significant in slender-body theory, we can, using this theory, deduce information in addition to that already presented. Coupling effects fall into the category of such information.   An example of a coupling effect would be the rolling moment developed by a missile with planar all-movable wings due to sideslip of the missile at a fixed angle of attack.   Such a rolling moment is produced by an interaction or coupling between angles of attack and sideslip, and is proportional to the product $\alpha\beta$.   The coupling effects between α and β in wing-body interference were discussed in Sec. 5-5.   Coupling effects associated with controls are studied, using methods analogous to those of that section.   In this section we will consider the types of coupling that can occur among the effects of thickness, angle of attack, angle of sideslip, symmetrical deflection of the horizontal panels, and differential deflection of the horizontal panels.   It will be possible to classify completely the types of coupling that occur, and to derive formulas for evaluating the couplings.

With reference to Fig. 8-13, consider the free-stream velocity V_0 inclined at angle α_c to the body axis.   Let the component of V_0 parallel to the body axis produce perturbation potential ϕ_t.   Let the velocity $V_0\alpha_c$ normal to the missile longitudinal axis be resolved into components $V_0\alpha$ and $V_0\beta$ as shown, and let ϕ_α and ϕ_β be the perturbation potentials for unit velocities in the two directions.   Furthermore, let ϕ_{δ_s} and ϕ_{δ_a} be the potentials for unit symmetrical and unit asymmetrical deflections of the horizontal panels. The total perturbation velocity for unit V_0 can then be written

$$\phi = \phi_t + \alpha\phi_\alpha + \beta\phi_\beta + \delta_e\phi_{\delta_e} + \delta_a\phi_{\delta_a} \tag{8-45}$$

In terms of perturbation velocities along the y' and z' axes the pressure coefficient is

$$\frac{p - p_0}{q_0} = -2\left(\frac{\partial\phi}{\partial x} + \alpha_c\frac{\partial\phi}{\partial z'}\right) - \left[\left(\frac{\partial\phi}{\partial y'}\right)^2 + \left(\frac{\partial\phi}{\partial z'}\right)^2\right] \tag{8-46}$$

With respect to the body axis system x, y, z the pressure coefficient is given by

$$\frac{p - p_0}{q_0} = -2\left(\frac{\partial\phi}{\partial x} + \alpha_c\cos\phi\frac{\partial\phi}{\partial z} - \alpha_c\sin\phi\frac{\partial\phi}{\partial y}\right)$$
$$- \left[\left(\frac{\partial\phi}{\partial y}\right)^2 + \left(\frac{\partial\phi}{\partial z}\right)^2\right] \tag{8-47}$$

It is on the basis of this equation that we evaluate the coupling effects.

To study the coupling effects we will put the velocity components into Eq. (8-47) and form the local pressure difference across the horizontal and

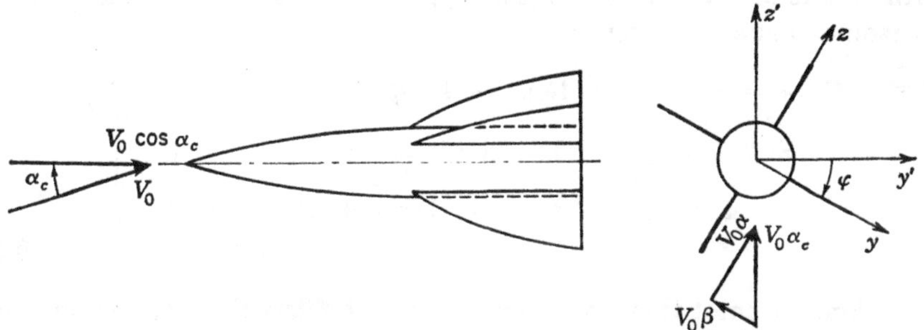

FIG. 8-13. Cruciform missile under combined pitch and bank.

vertical panels. The symmetry properties of the velocity components and the panel boundary conditions simplify the resultant loading considerably. Let us designate the velocity components and pressure on the lower side of the horizontal panels by a plus superscript, and the same quantities on the upper surface by a minus superscript. The panel boundary conditions and the symmetry properties of the velocity components then yield

$$
\begin{aligned}
u_t^+ &= u_t^- & u_\alpha^+ &= -u_\alpha^- & u_\beta^+ &= u_\beta^- & u_{\delta_e}^+ &= -u_{\delta_e}^- & u_{\delta_a}^+ &= -u_{\delta_a}^- \\
v_t^+ &= v_t^- & v_\alpha^+ &= -v_\alpha^- & v_\beta^+ &= v_\beta^- & v_{\delta_e}^+ &= -v_{\delta_e}^- & v_{\delta_a}^+ &= -v_{\delta_a}^- \\
w_t^+ &= -w_t^- & w_\alpha^+ &= w_\alpha^- & w_\beta^+ &= w_\beta^- & w_{\delta_e}^+ &= w_{\delta_e}^- & w_{\delta_a}^+ &= w_{\delta_a}^- \\
&= \left(\frac{dz}{dx}\right)^+ & &= -1 & &= 0 & &= -1 & &= \mp 1
\end{aligned}
$$

$$\tag{8-48}$$

The upper sign of w_{δ_a} refers to the right panel, and the lower sign to the left panel. The panel section has been assumed symmetrical. For the lower surfaces of the horizontal panels, the velocity components are

$$
\begin{aligned}
u^+ &= u_t{}^+ + \alpha u_\alpha{}^+ + \beta u_\beta{}^+ + \delta_e u_{\delta_e}^+ + \delta_a u_{\delta_a}^+ \\
v^+ &= v_t{}^+ + \alpha v_\alpha{}^+ + \beta v_\beta{}^+ + \delta_e v_{\delta_e}^+ + \delta_a v_{\delta_a}^+ \\
w^+ &= \left(\frac{dz}{dx}\right)^+ - \alpha - \delta_e \mp \delta_a
\end{aligned}
\tag{8-49}
$$

and, for the upper surfaces of the horizontal panels, the components are

$$
\begin{aligned}
u^- &= u_t{}^+ - \alpha u_\alpha{}^+ + \beta u_\beta{}^+ - \delta_e u_{\delta_e}^+ - \delta_a u_{\delta_a}^+ \\
v^- &= v_t{}^+ - \alpha v_\alpha{}^+ + \beta v_\beta{}^+ - \delta_e v_{\delta_e}^+ - \delta_a v_{\delta_a}^+ \\
w^- &= -\left(\frac{dz}{dx}\right)^+ - \alpha - \delta_e \mp \delta_a
\end{aligned}
\tag{8-50}
$$

The pressure coefficient for the lower surfaces is

$$
P^+ = \frac{p^+ - p_0}{q_0} = -2(u^+ + \alpha_c w^+ \cos\varphi - \alpha_c v^+ \sin\varphi)
$$
$$
- [(v^+)^2 + (w^+)^2] \tag{8-51}
$$

with a similar expression for the upper surfaces. The loading on the horizontal panels is given by

$$
\begin{aligned}
(P^+ - P^-)_H ={}& -4\alpha u_\alpha{}^+ - 4\delta_e u_{\delta_e}^+ - 4\delta_a u_{\delta_a}^+ \\
&+ 4\delta_e\left[\left(\frac{dz}{dx}\right)^+ - v_{\delta_e}^+ v_t{}^+\right] + 4\delta_a\left[\pm\left(\frac{dz}{dx}\right)^+ - v_{\delta_a}^+ v_t{}^+\right] \\
&- 4\alpha v_\alpha{}^+ v_t{}^+ + 4\beta(1 - v_\beta{}^+)\alpha v_\alpha{}^+ + 4\beta(1 - v_\beta{}^+)\delta_e v_{\delta_e}^+ \\
&+ 4\beta(1 - v_\beta{}^+)\,\delta_a v_{\delta_a}^+ \tag{8-52}
\end{aligned}
$$

An examination of this result reveals that the first three terms are linear terms representing the direct effects of angle of attack, pitch control, and roll control. However, the last six terms are coupling terms. Before we explore the nature of these coupling terms, let us find the loading for the vertical panels.

For the vertical panels denote the right side as the plus side and the left side as the minus side. The velocity components possess the following properties:

$$
\begin{array}{ccccc}
u_t{}^+ = u_t{}^- & u_\alpha{}^+ = u_\alpha{}^- & u_\beta{}^+ = -u_\beta{}^- & u_{\delta_e}^+ = u_{\delta_e}^- & u_{\delta_a}^+ = -u_{\delta_a}^- \\
v_t{}^+ = -v_t{}^- & v_\alpha{}^+ = v_\alpha{}^- & v_\beta{}^+ = v_\beta{}^- & v_{\delta_e}^+ = v_{\delta_e}^- & v_{\delta_a}^+ = v_{\delta_a}^- \\
\quad = \left(\dfrac{dy}{dx}\right)^+ & \quad = 0 & \quad = 1 & \quad = 0 & \quad = 0 \\
w_t{}^+ = w_t{}^- & w_\alpha{}^+ = w_\alpha{}^- & w_\beta{}^+ = -w_\beta{}^- & w_{\delta_e}^+ = w_{\delta_e}^- & w_{\delta_a}^+ = -w_{\delta_a}^-
\end{array}
$$
$$
\tag{8-53}
$$

By methods similar to those for the horizontal panels, it can be shown that the loading on the vertical panels is

$$(P^+ - P^-) = -4\beta u_\beta{}^+ - 4\delta_a u_{\delta_a}{}^+ - 4\beta w_\beta{}^+ w_t{}^+ - 4\delta_a w_{\delta_a}{}^+ w_t{}^+$$
$$- 4\alpha\beta w_\beta{}^+(1 + w_\alpha{}^+) - 4\alpha\delta_a(w_{\delta_a}{}^+)(1 + w_\alpha{}^+)$$
$$- 4\delta_e w_{\delta_e}{}^+ \beta w_\beta{}^+ - 4\delta_e w_{\delta_e}{}^+ \delta_a w_{\delta_a}{}^+ \qquad (8\text{-}54)$$

The first two terms represent the direct effects of sideslip and of roll control using the horizontal panels. The δ_a term represents, in fact, the reverse roll of the vertical panels due to the panel-panel interference illustrated in Fig. 8-11. Again we have six coupling terms. The couplings for the horizontal and vertical panels are summarized in the boxes of Fig. 8-14.

From the foregoing coupling terms for the horizontal and vertical panels, we can determine the qualitative nature of all the cross-coupling terms. Let us consider these under the categories of no control, pitch control, and roll control. Under the category of no control we have $\alpha\beta$ coupling and a pair of couplings due to αt and βt. It will be remembered that the subject of $\alpha\beta$ coupling was treated

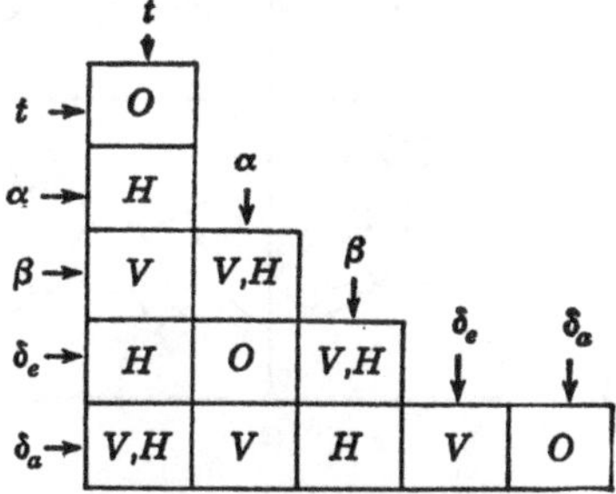

FIG. 8-14. Types of coupling between horizontal and vertical control panels.

both qualitatively and quantitatively for planar and cruciform configurations in Sec. 5-5. The αt coupling for the horizontal panels produces a force along the axis of z as follows:

$$Z_{\alpha t} \propto -4\alpha v_\alpha{}^+ v_t{}^+ \qquad (8\text{-}55)$$

Actually, an integration over the panel must be performed to evaluate $Z_{\alpha t}$. For the right panel $v_\alpha{}^+$ is positive, and for the left panel negative. For the right panel $v_t{}^+$ can be positive or negative, but for the left panel it has the opposite sign. Thus $Z_{\alpha t}$ is symmetrical about the xz plane. It is shown as positive for both planar and cruciform configurations in Fig. 8-15, which summarizes, in simple form, the types of forces developed by the panels as a result of the various couplings. The argument for $Y_{\beta t}$ coupling is analogous to that for $Z_{\alpha t}$ coupling.

The use of pitch control with the horizontal panels induces $t\delta_e$ and $\beta\delta_e$ couplings which are also illustrated in Fig. 8-15. Consider first $t\delta_e$ coupling for the horizontal panels. The Z force corresponds to a coupling as follows:

$$Z_{t\delta_e} \propto 4\delta_e \left[\left(\frac{dz}{dx}\right)^+ - v_{\delta_e}{}^+ v_t{}^+ \right] \qquad (8\text{-}56)$$

The $Z_{t\delta_e}$ force can be positive or negative, but it is symmetrical left to right since $(dz/dx)^+$ is the same at corresponding points on each panel and $v_{\delta_e}^+$ and v_t^+ both change signs together. The force is shown as positive in the figure. There is no $t\delta_e$ coupling for the vertical panels of the cruciform configuration. The $\beta\delta_e$ coupling produces a Z force on the horizontal panels.

$$Z_{\beta\delta_e} \propto 4\beta(1 - v_\beta^+)\delta_e v_{\delta_e}^+ \qquad (8\text{-}57)$$

Since v_β^+ is negative for both panels, whereas $v_{\delta_e}^+$ is positive for the right panel and negative for the left, $Z_{\beta\delta_e}$ changes sign from the right to the left panel and produces a rolling moment. The above discussion is valid for both planar and cruciform configurations, but the magnitude of $Z_{\beta\delta_e}$ will be dif-

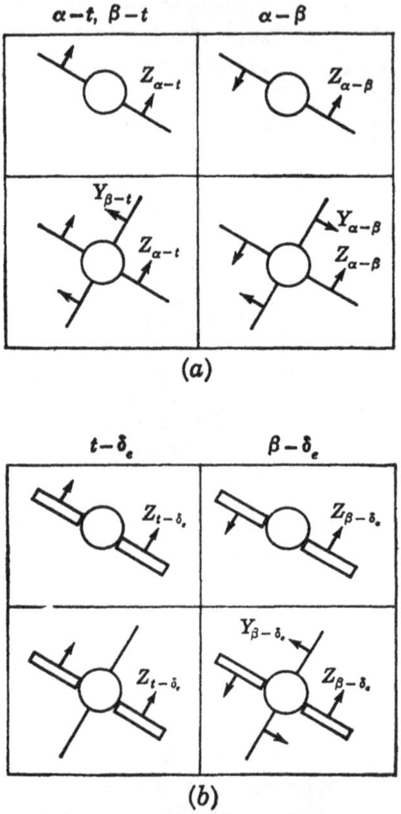

FIG. 8-15. Qualitative effects of coupling on panel forces for (a) no control and (b) pitch control.

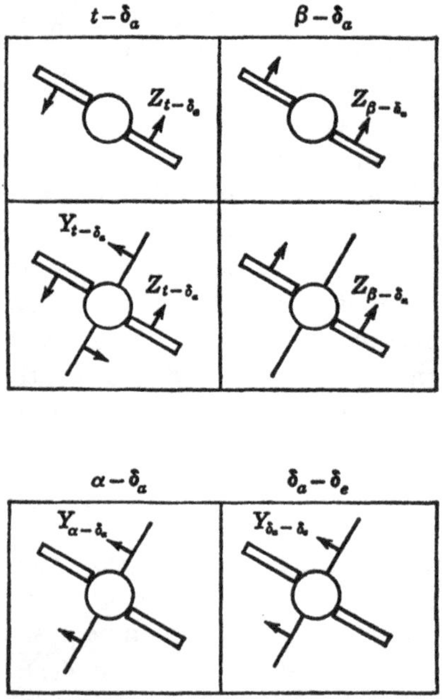

FIG. 8-16. Qualitative effects of coupling on panel forces for roll control.

ferent for each. The vertical panels of the cruciform configuration produce a Y force

$$Y_{\beta\delta_e} \propto 4\beta\delta_e w_{\delta_e}^+ w_\beta^+ \qquad (8\text{-}58)$$

In this relationship account has been taken of the fact that a positive loading produces a negative Y force. Now, as shown in Fig. 8-11, $w_{\delta_e}^+$ is negative for both vertical panels, whereas w_β^+ is positive for the upper panel but negative for the lower one. The result is that a negative rolling moment is developed by the vertical panels. It can thus be said that the use of pitch control under conditions of sideslip produces a negative rolling moment for both planar and cruciform configurations.

We now consider the coupling effects that can develop when the horizontal panels are used as ailerons. Couplings involving $t\delta_a$, $\alpha\delta_a$, $\beta\delta_a$, and $\delta_e\delta_a$ can occur. The last coupling can, of course, be considered under pitch control. The qualitative natures of these coupling terms are illustrated in Fig. 8-16. The $t\delta_a$ coupling involves a term as follows:

$$Z_{t\delta_a} \propto 4\delta_a \left[\pm \left(\frac{dz}{dx}\right)^+ - v_t v_{\delta_a}^+ \right] \tag{8-59}$$

The upper sign of $(dz/dx)^+$ refers to the right panel, and the lower sign to the left panel. The symmetry of the velocity products is such that the force is asymmetrical, producing a rolling moment. For the vertical panels the $t\delta_a$ coupling term is

$$Y_{t\delta_a} \propto 4\delta_a w_{\delta_a}^+ w_t^+ \tag{8-60}$$

The asymmetry of w_t^+ between the bottom and top panel has the effect of producing a rolling moment. The net effect of $t\delta_a$ coupling is thus to modify the rolling effectiveness.

With regard to the $\beta\delta_a$ coupling, only the deflected panels of the planar of cruciform configurations are involved. The coupling term is

$$Z_{\beta\delta_a} \propto 4\beta(1 - v_\beta^+)\delta_a v_{\delta_a}^+ \tag{8-61}$$

For both panels v_β^+ is negative and $v_{\delta_a}^+$ is positive, so that upward forces are developed on both. The net effect of $\beta\delta_a$ coupling is to produce pitch control with the application of roll control for planar or cruciform configurations.

While $\beta\delta_a$ coupling affects only the horizontal panels, $\alpha\delta_a$ coupling affects only the vertical panels as follows:

$$Y_{\alpha\delta_a} \propto 4\alpha\delta_a w_{\delta_a}^+(1 + w_\alpha^+) \tag{8-62}$$

Since $w_{\delta_a}^+$ is negative for both panels while w_α^+ is positive, the result is a negative Y force for both panels. The application of roll control thus results in yaw control for a cruciform configuration through $\beta\delta_a$ coupling. The coupling introduced through the simultaneous application of pitch and roll control produces sideforce on the vertical panels of a cruciform configuration in a similar fashion as $\alpha\delta_a$ coupling. The coupling term

$$Y_{\delta_e\delta_a} \propto 4\delta_e w_{\delta_e}^+ \delta_a w_{\delta_a}^+ \tag{8-63}$$

has the same symmetry properties as $Y_{\alpha\delta_a}$.

In summary, roll control in planar configurations is influenced through coupling terms by a modification of the rolling effectiveness and the appearance of pitch control. For cruciform configurations the same effects occur, but yaw control is also introduced. While the foregoing results have been derived from a consideration of panel forces alone, they are qualitatively true when a body is present. For instance, if the panel forces are symmetrical left to right or top to bottom, the lift carried over onto the body is such that the body forces possess the same direction and symmetry as the panel forces. When the panel forces are asymmetrical, the lift carried over is such that the body develops no resultant force. No quantitative change in the rolling moment can result from body forces. It should be noted that we here consider deflection of the horizontal panels only, and that the use of vertical panels for yaw or roll control introduces new coupling effects. These can be analyzed in the same manner as those for the horizontal panels. For panels of large aspect ratio to which slender-body theory does not apply directly, it is to be anticipated that the coupling effects may be significantly different from those just discussed.

8-5. Trailing-edge Controls

We have considered at some length the characteristics of all-movable controls, and now we take up *trailing-edge controls:* that is, controls free to rotate about a lateral axis, and forming all or part of the panel trailing edge. Various types of trailing-edge controls are illustrated in Fig. 8-17. An examination of these types shows that the all-movable control can be considered a trailing-edge control under our definition. However, our concern in this section is primarily for those controls which form only a fractional part of the panel surface. A number of theoretical approaches have been used to estimate the aerodynamic characteristics of trailing-edge controls. If the control characteristics are not substantially affected by wing-body interference, then the extensive results of supersonic wing theory are available. For those controls where wing-body interference has an important influence on the aerodynamic characteristics, reverse-flow theorems, combined with slender-body theory, provide a powerful theoretical tool, as we have seen for all-movable controls. For controls of high aspect ratio, simple sweep theory provides a useful theoretical approach. Because the geometric parameters characterizing trailing-edge controls are numerous, large numbers of specialized results and design charts are to be found in the literature. It is clearly impractical to reproduce these results, but it will be our objective to classify the types of results available and to rely on the original references for details.

We now consider that class of trailing-edge controls to which the extensive results of supersonic wing theory can be applied. Among the early papers devoted to supersonic controls are those of Frick[7] and Lagerstrom and Graham.[11] To illustrate how supersonic wing theory can be applied to controls, let us consider the approach of Frick, whose work is based on a combination of the line-source solutions of R. T.

Jones,[8] and the lift-cancellation technique of Lagerstrom.[9] The line source (Sec. 2-5) is a solution for linearized supersonic flow which produces a change in flow direction across any line along which it is placed. A line source will produce a wedge, the leading edge of which is coincident with the line source (which may be swept). A line sink will cause the diverging flow of a wedge to converge if placed, for instance, at the ridge line of a double-wedge wing. The plane containing the line source and lying in the free-stream direction is a plane of symmetry of the flow.

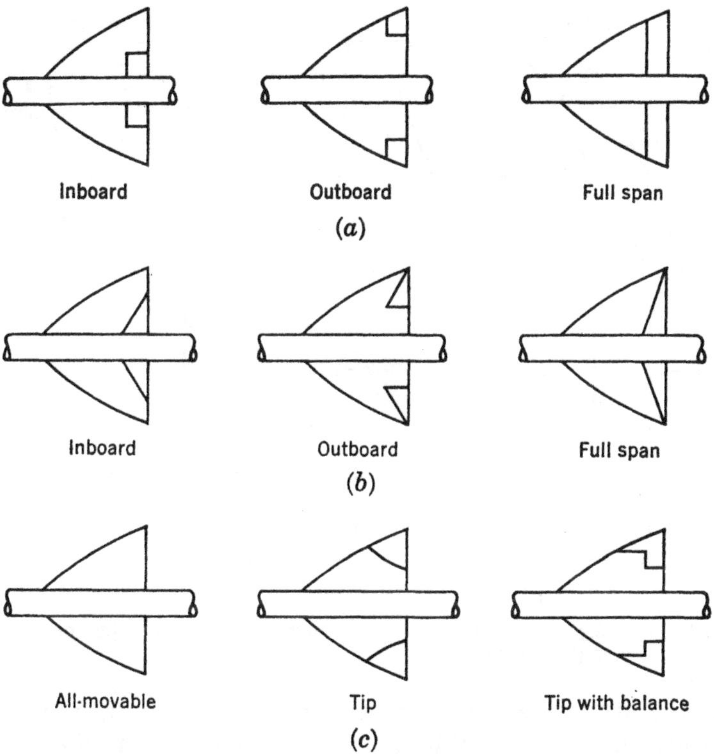

FIG. 8-17. Types of trailing-edge controls. (a) Constant chord; (b) constant taper; (c) others.

Now the flow produced by deflecting a control is not symmetrical about the plane of the control. However, to the extent that one surface of the control does not communicate pressure pulses to the other surface of the control, we can use the symmetrical line-source and sink solutions of Jones to represent the flow due to the control. We simply take the solution for the line source or sink and give the pressure field a positive or negative sign, in accordance with the deflection of the control and the side under consideration. For those areas of the control surface affected by pressure communication between the top and bottom of the control, the pressure field so constructed will be incorrect. The corrections to the pressure fields acting on such areas are obtained by the lift-cancellation technique.

To fix ideas, consider the case of a control with a supersonic hinge line as shown in Fig. 8-18.   A line sink placed along the hinge line will produce a deflection δ of the flow crossing the hinge line if the strength of the sink is suitably chosen.   The resultant pressure field will be conical from point A; that is, the pressure will be uniform along each ray emanating from A.   The pressure remains constant between the hinge line and the Mach line at a value corresponding to simple sweep theory (Sec. 2-7).

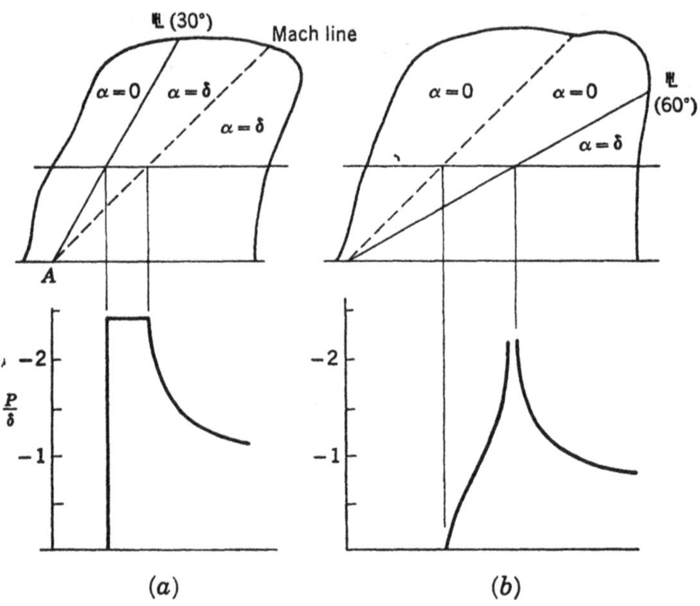

FIG. 8-18. Typical pressure distributions associated with flap-type controls utilizing (a) supersonic and (b) subsonic hinge lines for $M_0{}^2 = 2$.

Behind the Mach line, the pressure starts an asymptotic approach back to free-stream pressure.   With a subsonic hinge line, a different type of pressure distribution prevails, as shown in the figure.   The infinite pressure at the hinge line corresponds to a pressure which in reality has a large but finite magnitude.   It represents an integrable singularity which contributes a finite amount to the normal force acting on the surface.   The pressure fields calculated from line sources and sinks apply to the entire surface of controls similar to that pictured in Fig. 8-19a.   For this control there is no pressure communication between its upper and lower surfaces around the wing tip or wing trailing edge, and there is no pressure field from the control on the opposite wing panel.   No corrections by the lift-cancellation technique are thus required.   It should be noted that part of the pressure field, due to control deflection, is "caught" by the wing. It has been assumed that the effects of pressure communication through control-surface gaps are negligible.

Let us now consider controls affected, at least in part, by pressure communication between upper and lower surfaces. Such a control is the one with a subsonic hinge line shown in Fig. 8-19b. The pressure field due to a line sink along the hinge line OC extending indefinitely to the right will act as indicated in the previous figure. Since the pressure field due to the line sink has been taken as positive on one side of the control and negative on the other, a pressure difference will act in the area outboard of the tip. Such a pressure difference cannot be supported without a solid surface. The corresponding lift will alter the pressure field on the wing and control behind the Mach line AB. Lagerstrom[9] shows how to construct the necessary pressure fields to cancel the lift of an outboard tip sector of the present kind.

Another case requiring use of the lift-cancellation technique is shown in Fig. 8-19c. Here both the hinge line and trailing edge are subsonic. A line sink is introduced at the hinge line to deflect the flow downward through the angle δ, and a line source is placed along the trailing edge to straighten the flow out in the free-stream direction. Both the line

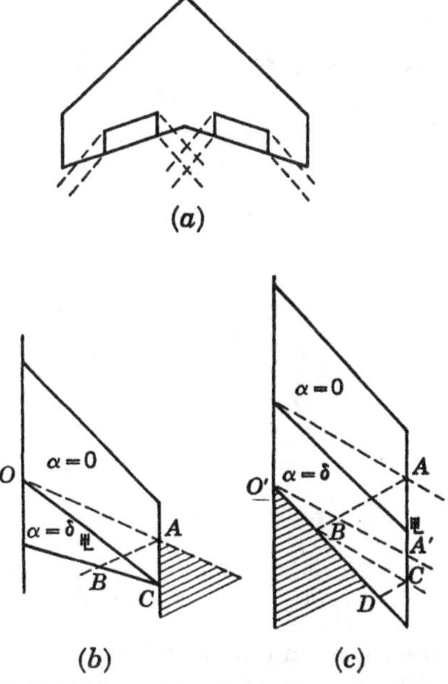

FIG. 8-19. Some cases encountered in using the lift-cancellation technique. (a) No lift cancellation; (b) tip sector; (c) trailing-edge sector.

source and sink produce lift in the trailing-edge sector. The cancellation of this lift will influence the pressure distribution behind the line $O'A'$. Cohen[10] has studied the application of lift cancellation to such sectors. Multiple reflections AB, BC, CD, etc., make application of the lift-cancellation technique to the tip of the control impractical. Reverse-flow techniques offer a means of overcoming this difficulty.

Some of the sources of control-surface formulas and design charts based on supersonic wing theory are now considered. For triangular tip controls, the analytical results of Lagerstrom and Graham[11,12] are available. Various combinations of supersonic and subsonic leading edges and hinge lines are considered. Goin[13] has studied a wide class of trailing-edge controls, the characteristics of which depend on control planform and Mach number independent of wing planform. A sufficient set of assumptions for this to be the case is

(1) Supersonic control leading and trailing edges.

(2) Streamwise tips.

(3) The control extending to the wing tip *or* located sufficiently far inboard so that the outmost Mach cone of the control does not intersect the wing tip.

(4) The innermost control Mach cone does not intersect the wing root chord.

Extensive charts and tables for such controls have been presented by Goin.[13]   The characteristics of trailing-edge controls on triangular wings have been extensively studied by Tucker.[14–16]

In all the above references, few, if any, analytical results accurate to the order of linear theory are available for control surfaces with subsonic trailing edges.   The difficulty associated with obtaining such solutions is

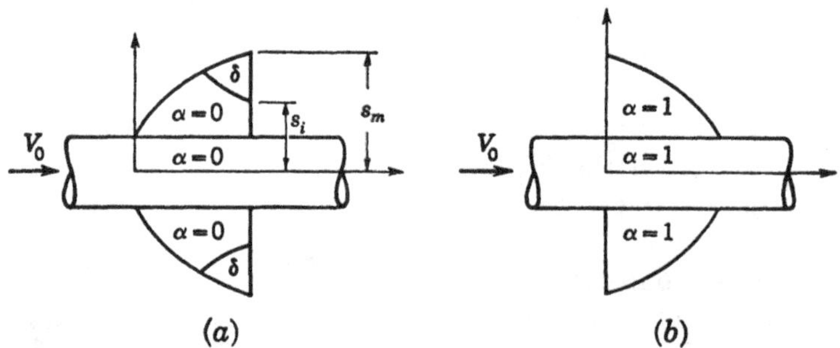

Fig. 8-20. Direct and reverse flows for calculating lift effectiveness of all-movable tip controls.   (a) Case 1; (b) case 2.

due to the multiple-reflection phenomena shown in Fig. 8-19c.   However, with the use of reverse-flow methods, closed analytical results for gross control forces and moments can be obtained.   Frost[17] has used such methods to obtain the lift effectiveness of trailing-edge controls mounted on swept pointed wings and swept tapered wings for both subsonic and supersonic trailing edges.   The methods are also applicable to other control surfaces and to pitching and rolling effectivenesses.

While supersonic wing theory is a valuable tool for many trailing-edge controls, it is of limited usefulness when appreciable interference exists between control and body.   For this class of controls the combined use of slender-body theory and reverse-flow theorems presents a more useful tool, particularly for a trailing edge of no sweep.   Let us consider the cases shown in Fig. 8-20.   The trailing-edge control is supposed to occupy the trailing edge of the wing between s_i and s_m, its precise planform being otherwise unimportant.   With reference to Eq. (7-47) the reverse-flow theorem for the particular circumstance here is

$$\iint_{S_F} \left(\frac{\Delta p}{q_0}\right)_2 \delta \, dS_F = \iint_{S_B + S_W} \left(\frac{\Delta p}{q_0}\right)_1 dS \qquad (8\text{-}64)$$

where S_F is the total control-surface area, S_W the wing area, and S_B the body area. Since the lift is concentrated at the leading edge in the reverse flow, we can write for the total lift on the missile due to the control

$$L_F = q_0 \iint_{S_B+S_W} \left(\frac{\Delta p}{q_0}\right)_1 dS = q_0 \iint_{S_F} \left(\frac{\Delta p}{q_0}\right)_2 \delta dS_F$$

$$= 2q_0\delta \int_{s_i}^{s_m} (cc_l)_2\, dy \tag{8-65}$$

The span loading $(cc_l)_2$ for the wing-body combination in reverse flow is the same as that for a rectangular wing of span s_m mounted on a body of radius a for unit angle of attack. This span loading from Eq. (6-39) is

$$(cc_l)_2 = \frac{2(\Delta\phi)_{te}}{V_0} = \frac{4(s_m^2 y^2 - a^4)^{1/2}(s_m^2 - y^2)^{1/2}}{s_m y} \tag{8-66}$$

We can express the lift due to the control as

$$\frac{L_F}{q_0\delta} = \frac{8}{s_m} \int_{s_i}^{s_m} \frac{(s_m^2 y^2 - a^4)^{1/2}(s_m^2 - y^2)^{1/2}}{y}\, dy \tag{8-67}$$

The integration yields the desired result.

$$\frac{L_F}{q_0\delta} = 4s_m^2 \left[\frac{\pi}{4}\left(1 - \frac{a^2}{s_m^2}\right)^2 - \left(\frac{s_i^2}{s_m^2} - \frac{a^4}{s_m^4}\right)^{1/2}\left(1 - \frac{s_i^2}{s_m^2}\right)^{1/2} \right.$$

$$+ \frac{1}{2}\left(1 + \frac{a^4}{s_m^4}\right)\sin^{-1}\frac{1 - 2s_i^2/s_m^2 + a^4/s_m^4}{1 - a^4/s_m^4}$$

$$\left. + \frac{a^2}{s_m^2}\sin^{-1}\frac{(1 + a^4/s_m^4)(s_i^2/s_m^2) - 2a^4/s_m^4}{(1 - a^4/s_m^4)(s_i^2/s_m^2)} \right] \tag{8-68}$$

The foregoing result can be used to illustrate how the lift effectiveness of the control depends on the ratio of the body radius and wing semispan and on the lateral position of the control on the wing. To illustrate these interesting effects, let us consider the ratio of the lift due to the control to the lift of a wing alone formed by joining the two controls together, assuming that the controls have streamwise edges as shown in Fig. 8-21. It is interesting to note that the controls can develop several times the lift of the isolated wing. For a very large body-radius–wing-semispan ratio, the ratio L_F/L_W approaches 2. For this case the control has a lift L_W acting on it and induces another L_W on the body. This result indicates the importance the body can play in increasing control lift effectiveness by acting as a "lift catcher." For the condition $s_i = a$ we have an all-movable control for which the ratio L_F/L_W is k_W plus k_B. For $s_i > a$ we have tip controls. As the value of s_i/a is increased for a constant value of a/s_m, the lift effectiveness increases. This behavior illustrates the inherent effectiveness of tip controls. Their good effectiveness is associated with the large wing area that exists to "catch" the lift developed by a tip control.

With regard to inboard trailing-edge flaps, their effectiveness can be calculated from the results of Fig. 8-21, by considering the control to be the difference between two outboard controls extending to the wing tip. Though the rolling effectiveness of trailing-edge controls can be evaluated by using the slender-body theory and reverse-flow theorems, we will not carry out the general calculation here. We observe only that the case $s_i = a$ for planar configurations is treated in Sec. 8-2. The method can be applied to partial-span trailing-edge controls.

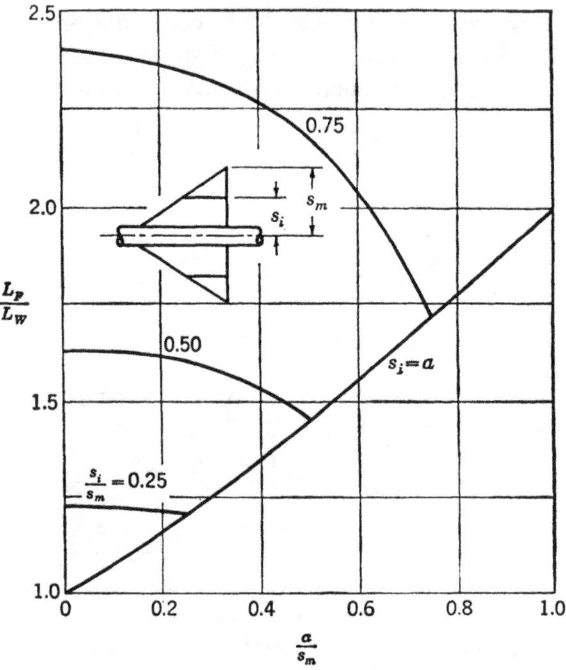

FIG. 8-21. Lift effectiveness of all-movable tip controls with unswept trailing edges.

As a final subject in trailing-edge controls, let us consider the simple effects of sweep on control effectiveness, using two-dimensional theory. Such an analysis applies to trailing-edge controls of large aspect ratio. Now, with reference to Fig. 8-22 let the sweep of the control hinge line be variable, but let the deflection of the control δ_0 in the streamwise direction be constant as the sweep angle varies. Let the subscript 0 refer to the condition of no sweep, and let the subscript n refer to conditions taken normal to the control hinge line when swept but supersonic. The simple sweep theory (Sec. 2-7) yields the result

$$\frac{L_n}{L_0} = \frac{q_n}{q_0}\left(\frac{\delta_n}{\delta_0}\right)\frac{(C_{L\alpha})_n}{(C_{L\alpha})_0} \tag{8-69}$$

Now the following relationships are valid

$$q_n = q_0 \cos^2 \Lambda$$
$$\delta_n = \frac{\delta_0}{\cos \Lambda} \tag{8-70}$$

where Λ is the sweep angle of the hinge line. The two-dimensional lift curve slopes are

$$(C_{L\alpha})_0 = \frac{4}{(M_0{}^2 - 1)^{1/2}}$$
$$(C_{L\alpha})_n = \frac{4}{(M_n{}^2 - 1)^{1/2}} \tag{8-71}$$

The result of introducing the foregoing relationships into Eq. (8-69) is

$$L_n = \frac{L_0(M_0{}^2 - 1)^{1/2}}{(M_0{}^2 - 1/\cos^2 \Lambda)^{1/2}} \tag{8-72}$$

Thus, if the control hinge line is swept while a constant control deflection is maintained in the streamwise direction, the lift effectiveness will

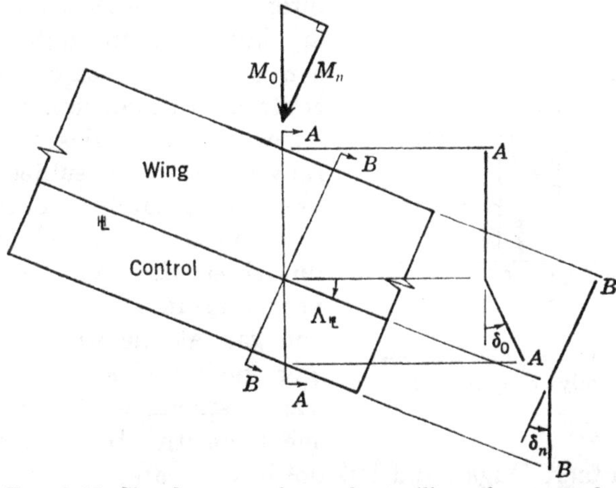

FIG. 8-22. Simple sweep theory for trailing-edge controls.

increase. It is, however, necessary to avoid boundary-layer separation to realize this effectiveness. Equation (8-72) is singular when the hinge line is sonic; that is, when $M_0 \cos \Lambda$ is unity. Physically, this corresponds to a detachment of the shock from the control hinge line. Actually, because of wing and control thickness, the hinge line must be somewhat supersonic to avoid shock detachment. The precise limits can be calculated by the shock-expansion theory given in the Ames supersonic handbook.[20]

8-6. Some Nonlinear Effects in Aerodynamic Control

A number of nonlinear phenomena appear in the characteristics of aerodynamic controls, and theory is only partially successful in accounting for these effects. A knowledge of the nonlinearities is a useful guide in the judicious use of the theoretical results presented in the preceding sections of this chapter. Most controls possess gaps at their inboard side edges, their hinge lines, or elsewhere. Under certain conditions such gaps can produce nonlinear behavior of the control. There is a tendency to use large control deflections for missiles required to maneuver at high altitudes. This tendency accounts for the importance of a number of nonlinearities. First, there is a tendency for the control characteristics to depart from linearity if the control is at a large angle of attack. The effects are termed *higher-order effects* of angle of attack and control deflection. The extreme angles also act to produce an interaction between the control boundary layer and the outer flow, which can cause separation of the flow on the control. In addition, the use of extreme control angles of attack naturally brings up the subject of the maximum lift capabilities of controls.

Let us start our discussion of nonlinearities with gap effects. One gap occurring in all-movable controls is the gap at the wing-body juncture. For small angles the effect of such a gap is amenable to theoretical treatment on the basis of slender-body theory. In fact, Dugan and Hikido[2] have treated this problem, as has Mirels[19] also. Although these treatments neglect the effects of viscosity, which is probably of overriding importance for small gaps, they are, nevertheless, of considerable interest as standards by which the importance of viscosity is to be judged. The qualitative effects of a gap at the wing-body juncture are shown in Fig. 8-23. For the slightest gap, inviscid fluid theory requires that the span loading in the juncture fall to zero as shown. As a result, the smallest gap will produce a substantial loss of lift effectiveness on the basis of inviscid fluid theory. However, it is known that, with such gaps, large losses of lift effectiveness

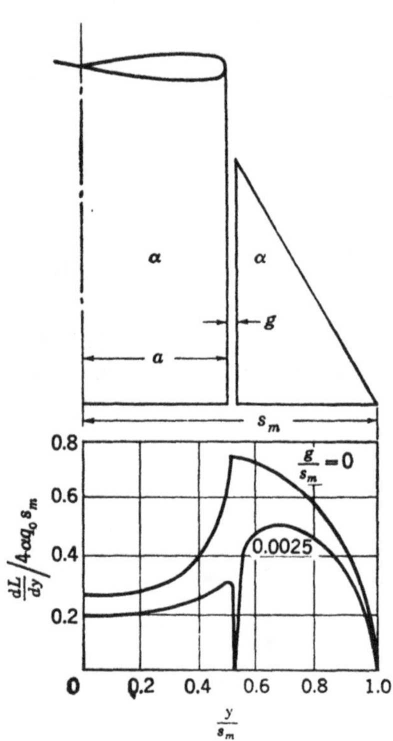

FIG. 8-23. Effect of gaps on span loading of wing-body combination; no viscosity.

do not occur in real fluids because of the effects of viscosity. Only when the gap width is large will the results of inviscid theory be valid.

Yet another type of gap occurs in the use of all-movable controls. For extreme deflections the forward or aft part of the control may pass above or below the body in side view, as shown in Fig. 8-24. For such gaps the results of the previous investigators are clearly not applicable. The positive pressures existing beneath the control leading edge can produce a download on the body, and the negative pressures above the trailing edge can produce an upload. The net result will be a large couple.

The so-called higher-order effects of angles of attack and deflection or of thickness can produce departures of the control characteristics from linear theory at moderate angles. A general theory of higher-order effects for wings of low to moderate aspect ratios has not been developed. However, for controls of sufficiently large aspect ratio to be considered two-dimensional, the effects of higher order can be calculated by Busemann's

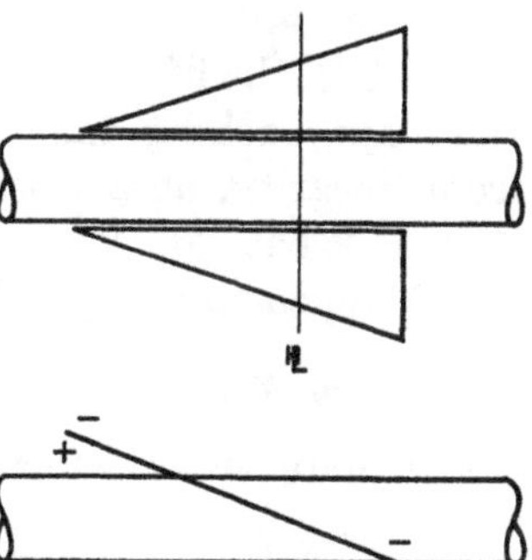

FIG. 8-24. Gaps associated with large deflections of all-movable controls.

second-order theory and by shock-expansion theory. In fact, Goin,[13] and the Ames staff[20] have considered such application of Busemann's theory. As an example of the use of this theory, let us consider the modification as the result of section thickness to the lift effectiveness of a trailing-edge control of symmetrical section such as that shown in Fig. 8-25. The Busemann second-order theory gives for the pressure coefficients of the upper and lower surfaces

$$P^- = -C_1(\delta - \theta) + C_2(\delta - \theta)^2$$
$$P^+ = C_1(\delta + \theta) + C_2(\delta + \theta)^2 \tag{8-73}$$

where

$$\theta = \frac{dz_u}{dx}$$

$$C_1 = \frac{2}{(M_0{}^2 - 1)^{1/2}} \tag{8-74}$$

$$C_2 = \frac{(\gamma + 1)M_0{}^4 - 4(M_0{}^2 - 1)}{2(M_0{}^2 - 1)^2}$$

The control lift coefficient based on the flap chord is

$$c_l = \frac{1}{c - x_H} \int_{x_H}^{c} (P^+ - P^-) \, dx$$
$$= 2C_1\delta - \frac{2C_2\delta(t_H - h)}{c - x_H} \tag{8-75}$$

Despite the fact that the pressure coefficients are nonlinear in the angle δ,

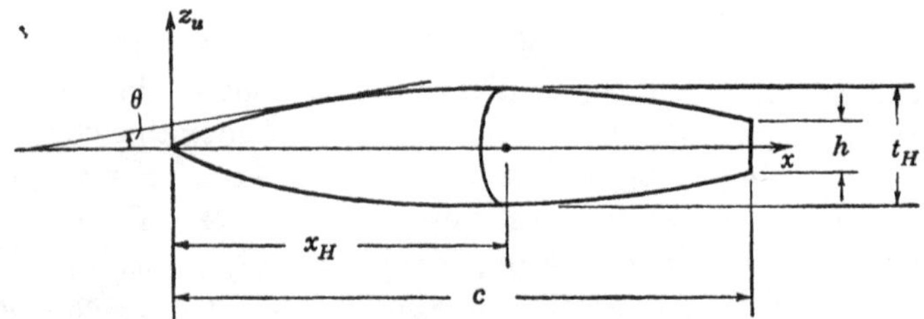

FIG. 8-25. Blunt trailing-edge control with symmetrical section.

Eq. (8-73), the lift due to the control is linear in δ.　For very large deflections approaching the shock detachment angle, the lift would depart from linearity, as a calculation by shock-expansion theory will readily show.　What Eq. (8-75) does show is that the lift developed by the control is dependent on its thickness distribution.

Illustrative Example

Calculate the lift effectiveness for the following example:

$$M_0 = 1.54 \qquad \frac{x_H}{c} = 0.8 \qquad \frac{t_m}{c} = 0.05$$

Biconvex airfoil section:

$$C_1 = \frac{2}{(1.54^2 - 1)^{1/2}} = 1.708$$

$$C_2 = \frac{2.4(1.54)^4 - 4(1.54^2 - 1)}{2(1.54^2 - 1)^2} = 2.129$$

The thickness distribution is given by

$$\frac{t}{c} = 4 \frac{t_m}{c} \frac{x}{c}\left(1 - \frac{x}{c}\right)$$

so that

$$\left(\frac{t}{c}\right)_H = 0.64 \frac{t_m}{c} = 0.032$$

Now from Eq. (8-75) the ratio of the lift of the control to the lift with zero thickness is

$$\frac{c_l}{(c_l)_{t=0}} = 1 - \frac{C_2}{C_1} \frac{(t/c)_H}{1 - (x/c)_H}$$

$$= 1 - \frac{2.129(0.32)}{1.708(0.20)} = 0.801$$

The moderate thickness of the present control thus causes a loss of lift effectiveness of 20 per cent at all angles of deflection.　Results for biconvex sections with various hinge-line positions have been presented by Goin,[13] and results for general airfoil sections are presented in the Ames supersonic handbook.[20]

An important viscous phenomenon occurring with all-movable and trailing-edge controls is the separation of the flow over the control that can result from so-called boundary-layer–shock-wave interaction. The interaction involved is, in reality, one between the boundary layer and the outer flow. Some fundamental work of Chapman, Kuehn, and Larson,[21] among others, provides quantitative information for estimating when boundary-layer separation will occur. One of the significant conditions influencing the type of boundary-layer separation is the location of the transition point relative to the points of separation and reattachment. For "purely laminar" separation the transition point is downstream of the reattachment point, and for "purely turbulent" separation the transition point is upstream of the separation point. An intermediate type of

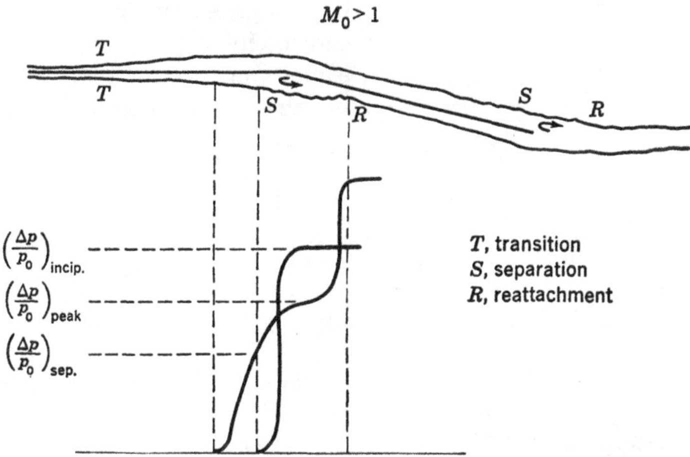

FIG. 8-26. Separation of supersonic turbulent boundary layer on trailing-edge control.

separation occurs when the transition point is between the separation and reattachment points. We will concern ourselves only with the purely turbulent type shown qualitatively for the control in Fig. 8-26. Separation has taken place on both surfaces of the control as a result of the pressure rise occurring downstream of the separation point. The pressure distributions just before separation and some time after are both sketched in Fig. 8-26. Just before separation the relatively sharp step in the pressure distribution predicted for a wedge by supersonic shock theory is manifest. If the control is now deflected to a slightly greater angle, the sharp step changes into a gradual rise across the region of separated flow. In front of the separation point, the pressure rises to its first plateau value of $(\Delta p)_{\text{peak}}$ above p_0, and then rises sharply to its final value at the reattachment point. Chapman et al.[21] have presented data for the pressure rise $(\Delta p)_{\text{incip}}$ necessary to bring about a condition of incipient separation, and the corresponding flap deflection angle can readily be calculated. The pressure rises to bring about incipient separation together with the corresponding flap deflection angles are given in Fig. 8-27. In applying these data one should keep in mind that they refer to a sharp change in slope as for a wedge. If the deflection of the flow is achieved by means of a fairing with a gradual curvature, high pressure rises may be obtained before separation.

As a final topic in nonlinearities let us consider the maximum lifting capabilities of controls, particularly all-movable controls. Some indication of the maximum lift capabilities of all-movable controls can be obtained by examining data on the maximum lift coefficients of wings alone at supersonic speeds as presented by Gallagher and Mueller.[23] The typical lift and drag curves for wings at supersonic speeds are shown in Fig. 8-28. The supersonic wing does not develop a stall in the usual subsonic sense but

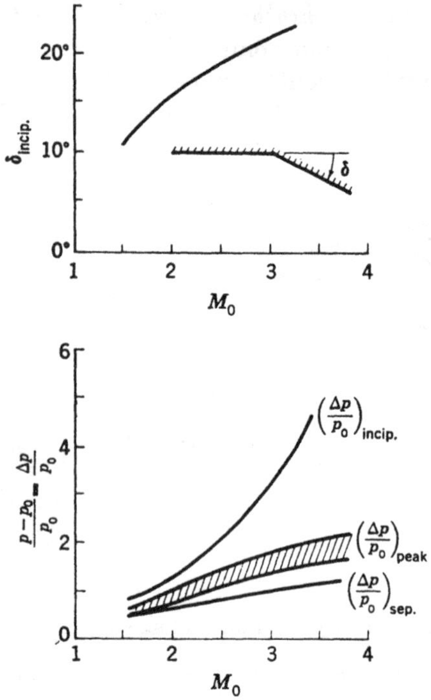

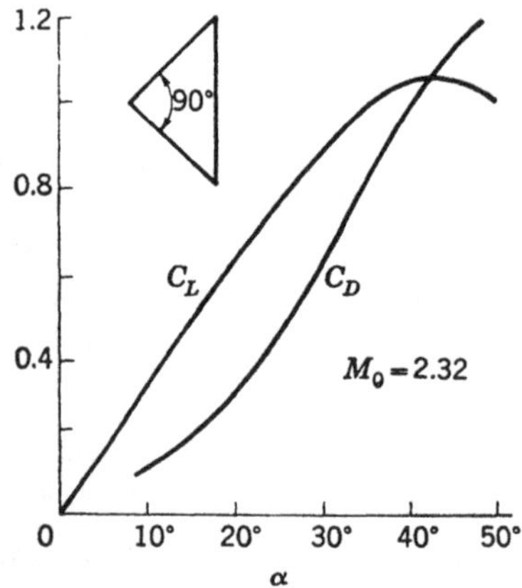

FIG. 8-27. Control deflection and pressure ratio for incipient separation of supersonic turbulent boundary layer.

FIG. 8-28. Maximum lift characteristics of triangular wing at supersonic speed.

continues to develop lift up to angles of attack of about 40 or 45°. The falling off of the lift thereafter is not abrupt. Now, if the wing of Fig. 8-28 were an all-movable control, it would probably develop its maximum lift at a body angle of attack plus angle of deflection of somewhat less than 40° since body upwash would tend to increase the aerodynamic angle of attack above the geometric angle of attack. Also, gap effects of the type illustrated in Fig. 8-24 may well influence the geometric angle of attack of the control at which maximum lift is developed when the lift acting on the body is considered. One of the interesting findings of Gallagher and Mueller is that the triangular, rectangular, sweptback, and trapezoidal wings tested by them at Mach numbers between 1.55 and

2.32 all had maximum lift coefficients in the range 1.05 ± 0.05 at an angle of attack close to 40°.   The wings had aspect ratios ranging upward from 1.37.

8-7. Notes on Estimating Hinge Moments

We have deferred consideration of hinge-moment coefficients to a separate section because of the special nature of these coefficients.   It is often contended that calculations of hinge moments are not reliable because of the frequent nonlinear variation of hinge-moment coefficient with control deflection and angle of attack.   Much can, however, be done to estimate or explain hinge moments.   Two characteristics usually sought are linear dependence of hinge moment on such parameters as control deflection and angle of attack, and low values of the hinge-moment coefficient.   These two requirements can be mutually contradictory.   Consider a hinge line located a large distance from the center of pressure of a control.   The nonlinearities due to movement of the center of pressure will be masked by the large moment arm, but the hinge-moment coefficients will be large.   Now locate the hinge line through the center of pressure.   The small migrations of the center of pressure will cause large nonlinearities in the hinge-moment coefficient which now is small.   Thus, for a closely balanced control, it will be difficult to predict accurately the nonlinear hinge moments of the control, but this difficulty is alleviated by the small magnitudes of the hinge moments.

Let us consider estimating the hinge moments of an all-movable triangular control.   The important quantity to determine in this respect is the center-of-pressure position of the control panel.   Our general approach is to assume as a first approximation that the center of pressure acts at the same position as for a lifting surface with the wing-alone planform. Then we apply corrections to this position to account for control-body interference and for control-section effects.   The corrections due to control-body interference effects associated with changes in α can be assessed from the values in Table 5-1.   This table shows that the shift is a maximum of about 2 per cent of the root chord.   The corrections in center-of-pressure position due to the interference between control and body accompanying control deflection are given in Table 8-1, where a maximum correction of less than about 1 per cent of the root chord is indicated. The change in the control section center-of-pressure position due to thickness can be readily estimated by the Busemann second-order theory described in the preceding section.   The thickness correction can amount to 3 or 4 per cent of the root chord, and it is applied to the control by strip theory.   On the basis of these considerations, we then have the following procedure for estimating the hinge moment.   Calculate the lifts due to angle of attack and control deflection by the methods of Secs. 5-6 and 8-2.   Assume that the lift due to angle of attack acts at the center of pressure of the wing alone, corrected for thickness effects and for interference effects by Table 5-1. Assume that the lift due to control deflection acts at the center of pressure of the wing alone (no thickness), corrected for thickness effects and for interference effects by Table 8-1. The hinge moment is then the combined moment due to the lifts for angle of attack and deflection angle. After discussing the hinge moments of all-movable rectangular controls, we will consider a calculative example for a triangular control.

It is clear that the general approach just discussed is applicable in principle to all-movable controls of many planforms. In practice, the applicability of the method depends on the availability of the necessary theoretical data. For rectangular all-movable controls, slender-body theory gives the obviously inaccurate result that the lift of the control is all concentrated at its leading edge. Thus, slender-body theory gives no basis for estimating the shifts in panel center of pressure due to interference. For rectangular panels, results based on linear theory[5] are available for the effect of control deflection on lift and center of pressure. For low aspect ratios they show as much as 4 per cent shift in center of pressure as against 2 per cent for triangular controls. Rectangular all-movable controls will thus show larger effects of interference on center-of-pressure position than triangular controls, and we are in a position to calculate this shift for control deflection (but not for angle of attack).

Illustrative Example

As an illustrative example, let us estimate the hinge-moment coefficient for the all-movable triangular control shown in Fig. 8-29. Assume a biconvex section 5 per cent thick in the streamwise direction. The hinge-moment coefficient based on the control area and its mean aerodynamic chord $\bar{c}$ is

$$C_h = \frac{c_r}{\bar{c}} \left\{ (C_L)_\alpha \left[\left(\frac{x}{c_r}\right)_H - \left(\frac{\bar{x}}{c_r}\right)_\alpha \right] + (C_L)_\delta \left[\left(\frac{x}{c_r}\right)_H - \left(\frac{\bar{x}}{c_r}\right)_\delta \right] \right\} \quad (8\text{-}76)$$

All quantities refer to the panel in the presence of the body, the subscript α denoting quantities associated with body angle of attack, and δ quantities associated with control deflection. Let M_0 be 2, α be 0.1 radian, and δ be 0.2 radian. The lift coefficients associated with α and δ are

$$(C_L)_\alpha = K_W \alpha (C_{L\alpha})_W$$
$$(C_L)_\delta = k_W \delta (C_{L\delta})_W$$

Since the triangular wing formed by the two panels has a supersonic edge,

$$(C_{L\alpha})_W = \frac{4}{(M_0{}^2 - 1)^{1/2}} = 2.31$$

From Tables 5-1 and 8-1 we have

$$K_W = 1.21 \qquad k_W = 0.94$$

Turning now to the centers of pressure for α and δ, we note that the wing alone has its center of pressure at the two-thirds root chord for no thickness.   Let us now evaluate the shift in center of pressure due to

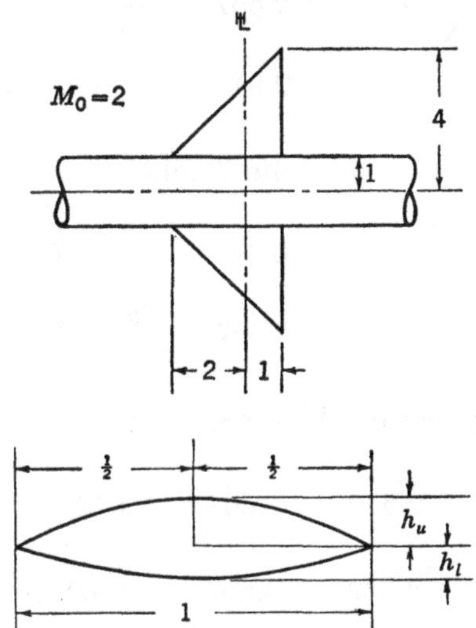

Fig. 8-29. Calculative example for hinge-moment coefficient.

thickness, using the Busemann second-order theory of the previous section.   On the basis of the Ames supersonic handbook,[20] the section lift and moment coefficients are

$$c_l = 2C_1\alpha + {}^{16}\!/_3 C_2(h_l{}^2 - h_u{}^2) \tag{8-77}$$

$$(c_m)_{\frac{1}{2}} = {}^2\!/_3 C_1(h_l - h_u) + {}^4\!/_3 C_2\alpha(h_l + h_u) \tag{8-78}$$

where c_m is taken about the midchord, and h_l and h_u are the distances shown in Fig. 8-29.   The center-of-pressure position for the symmetrical biconvex section is thus

$$\frac{1}{2} - \frac{\bar{x}}{c} = \frac{(c_m)_{\frac{1}{2}}}{cc_l} = \frac{2C_2}{3C_1}\frac{t}{c}$$

$$\frac{1}{2} - \frac{\bar{x}}{c} = \frac{2}{3}\left(\frac{1.467}{1.155}\right)0.05 = 0.042 \tag{8-79}$$

As a result of thickness, the center of pressure of each streamwise section of the control has been shifted forward by 0.042 of the local chord on a strip-theory basis.   For the wing alone as a whole the thickness has moved the center of pressure forward an amount $0.042\bar{c}$.   Thus, the center-of-pressure position corrected for thickness is

$$\frac{\bar{x}}{c_r} = \frac{2}{3} - 0.042\,\frac{\bar{c}}{c_r} = 0.667 - 0.028 = 0.639$$

Let us now apply the corrections due to control-body interference. For angle of attack we have from Table 5-1

$$\left(\frac{\bar{x}}{c_r}\right)_{W(B)} = 0.667 \qquad \frac{a}{s_m} = 0$$
$$= 0.648 \qquad \frac{a}{s_m} = 0.25$$

For control deflection we have from Table 8-1

$$\left(\frac{\bar{x}}{c_r}\right)_{W(B)} = 0.667 \qquad \frac{a}{s_m} = 0$$
$$= 0.667 \qquad \frac{a}{s_m} = 0.25$$

Applying the shifts given by these results, we have

$$\left(\frac{\bar{x}}{c_r}\right)_\alpha = 0.639 - (0.667 - 0.648) = 0.620$$
$$\left(\frac{\bar{x}}{c_r}\right)_\delta = 0.639 - (0.667 - 0.667) = 0.639$$

We now have all the quantities necessary to estimate C_h for the hinge line through the centroid.

$$C_h = \tfrac{3}{2}(2.31)[0.1(1.21)(0.667 - 0.620) + 0.2(0.94)(0.667 - 0.639)]$$
$$= 0.038$$

8-8. Change in Missile Attitude Due to Impulsive Pitch Control; Altitude Effects

An important quantity in missile control is the rapidity with which a missile changes attitude in response to an impulsive application of control deflection. From the change in attitude the necessary normal force is derived to change the missile flight path direction. Let us consider a missile flying along approximately level in equilibrium, and let the deflection of the pitch control be impulsively changed. We will determine the change in angle of attack of the missile as a function of time due to the control change on the basis of a simplified analysis. The essential features of the simplified analysis are that the missile is assumed to respond in pitch like a two-degree-of-freedom harmonic oscillator with damping and an impulsive forcing function. It is physically tenable that the pitching behavior of the missile can be closely approximated by such a system. It is possible for any particular missile to evaluate the stability derivatives, and to see if these equations can be simplified to an approximate one-degree-of-freedom second-order equation for the angle of attack. One of the pertinent assumptions is that the changes in flight speed are negligible. Since significant changes in flight speed occur only in a time of the same order of the phughoid period, and since we are concerned with times of the order of the short period (which for a missile is very much less than the phughoid period), this assumption is almost always warranted. We also assume that the missile is stiff, and that the control forces are developed in times small comparable to the short period. In writing the equation of motion, we consider the missile inertia, the damping, the spring constant, and the forcing function.

In Appendix A at the end of the chapter the equation of motion governing the angle of attack is derived:

$$mK_y^2\ddot{\alpha} - \left(\frac{K_y^2 Z_\alpha}{V_0} + M_{\dot{\alpha}} + M_q\right)\dot{\alpha} - \left(M_\alpha - \frac{M_q Z_\alpha}{mV_0}\right)\alpha = M(\delta) \quad (8\text{-}80)$$

In this equation m is the mass of the missile, and K_y is the radius of gyration about the y axis through the center of gravity. The various derivatives such as M_q are simply partial derivatives, i.e., $\partial M/\partial q$. The term $M(\delta)$ represents the moment contributed by the pitch control and is a function of time. In particular we will take $M(\delta)$ equal to zero for t less than zero, and constant for t greater than zero. Ignore $M_q Z_\alpha/mV_0$ in comparison with M_α for simplicity, even though the assumption is not necessary.

It is now our purpose to put Eq. (8-80) into coefficient form, and then to reduce it to a specialized form in terms of natural frequency and damping parameter. The derivatives with respect to w are simply expressed in terms of C_{L_α} and C_{m_α} for the complete missile as follows:

$$Z_\alpha = -C_{L_\alpha}(q_0 S_R) \quad (8\text{-}81)$$
$$M_\alpha = +C_{m_\alpha}(q_0 S_R l_r) \quad (8\text{-}82)$$

(Note that the Z force is downward in accordance with the usual practice in dynamic stability.)

The derivatives C_{m_q} and $C_{m_{\dot{\alpha}}}$ are defined in Chap. 10 as follows:

$$C_{m_q} = \frac{\partial C_m}{\partial(q l_r/2V_0)} \qquad C_{m_{\dot{\alpha}}} = \frac{\partial C_m}{\partial(\dot{\alpha} l_r/2V_0)}$$

so that for the present case

$$(M_{\dot{\alpha}} + M_q) = (C_{m_q} + C_{m_{\dot{\alpha}}})\frac{q_0 S l_r^2}{2V_0} \quad (8\text{-}83)$$

When the foregoing three equations are used, the equation of motion becomes

$$\ddot{\alpha} + \left[C_{L_\alpha}\left(\frac{q_0 S_R}{V_0 m}\right) - (C_{m_q} + C_{m_{\dot\alpha}})\left(\frac{q_0 S_R l_r^2}{2V_0 m K_y^2}\right) \right] \alpha$$
$$- C_{m_\alpha}\left(\frac{q_0 S_R l_r}{m K_y^2}\right)\alpha = C_{m_\delta}\left(\frac{q_0 S_R l_r}{m K_y^2}\right)\delta(t) \qquad (8\text{-}84)$$

Introducing the natural frequency ω_n and damping parameter ζ as follows,

$$\omega_n^2 = -C_{m_\alpha}\left(\frac{q_0 S_R l_r}{m K_y^2}\right) \qquad (8\text{-}85)$$

$$\zeta = \frac{[C_{L_\alpha}/V_0 - (C_{m_q} + C_{m_{\dot\alpha}})l_r^2/2V_0 K_y^2]q_0 S_R/m}{2[-C_{m_\alpha}(q_0 S_R l_r/m K_y^2)]^{1/2}} \qquad (8\text{-}86)$$

and the final missile angle of attack

$$\alpha^* = \frac{C_{m_\delta}}{-C_{m_\alpha}} \qquad (8\text{-}87)$$

we can write Eq. (8-84) in the common form for dynamical analysis

$$\ddot{\alpha} + 2\zeta\omega_n\dot{\alpha} + \omega_n^2\alpha = \alpha^*\omega_n^2 H(t) \qquad (8\text{-}88)$$

where $H(t)$, the variation of δ with time, is a unit step function in the present case.

The solution of Eq. (8-88) will be given subject to the initial conditions

$$\alpha(0) = 0 \qquad \dot{\alpha}(0) = 0 \qquad (8\text{-}89)$$

The form of the solution depends on whether $\zeta < 1$ or $\zeta > 1$. For $\zeta < 1$, less than critical damping, there is obtained

$$\frac{\alpha}{\alpha^*} = 1 - \frac{e^{-\omega_n\zeta t}}{(1-\zeta^2)^{1/2}}\cos[\omega_n(1-\zeta^2)^{1/2}t - \gamma] \qquad (8\text{-}90)$$

with $$\gamma = \sin^{-1}\zeta$$

For $\zeta > 1$ the solution is

$$\frac{\alpha}{\alpha^*} = 1 - \frac{e^{-\omega_n\zeta t}}{(\zeta^2-1)^{1/2}}\sinh[\omega_n(\zeta^2-1)^{1/2}t + \gamma'] \qquad (8\text{-}91)$$

with $$\gamma' = \cosh^{-1}\zeta$$

Let us examine the missile to see how it attains its final pitch attitude for subcritical and supercritical damping. The solutions can conveniently be plotted in the form shown in Fig. 8-30. For no damping, the missile overshoots its equilibrium value of α^* and performs a steady periodic oscillation of amplitude α^* about a mean value of α^*. As the damping is increased, the missile takes somewhat longer to reach its equilibrium value, but the overshoot is less. As ζ becomes greater than unity, the approach to α^* is asymptotic from below with no overshoot.

In the foregoing analysis we have considered only the missile angle of attack. It is possible to determine the variation of w and θ with time from Eqs. (8A-1) and (8A-3).

One of the consequences that can be derived from the solutions of Eqs. (8-90) and (8-91) is the deterioration of the missile response rate as the altitude increases. Let us consider the effect of altitude on missile response rate for unit control deflection for a constant Mach number. We first observe that the natural frequency of the missile varies as the square root of the dynamic pressure. Also, ζ will vary in the same manner if we neglect the change in V_0 with altitude for constant Mach number, an approximation sufficiently accurate for our present purpose.

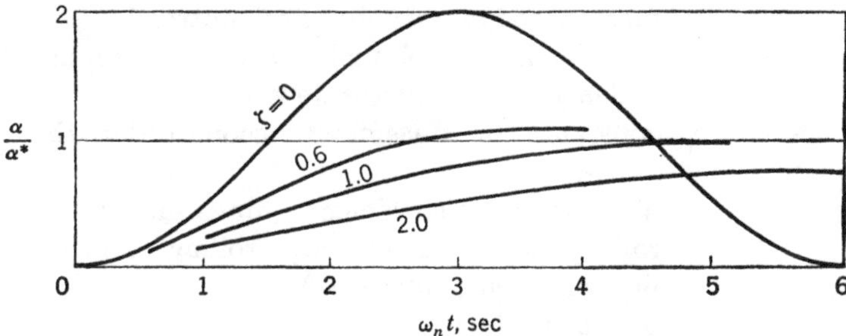

Fig. 8-30. Change in attitude of missile due to sudden application of pitch control.

We will now proceed to calculate the time to reach α^* as a function of altitude for the following numerical values at sea level.

$$(\omega_n)_0 = 2 \text{ cycles per second}$$
$$\zeta_0 = 0.6$$

Subscript 0 refers to sea level, and no subscript indicates any altitude. We have

$$\frac{\zeta_0}{\zeta} = \frac{(\omega_n)_0}{\omega_n} = \left(\frac{q_0}{q}\right)^{1/2} = \left(\frac{M_0{}^2 p_0}{M^2 p}\right)^{1/2} = \left(\frac{p_0}{p}\right)^{1/2} \tag{8-92}$$

If τ^* is the time for the missile attitude to attain α^*, then from Eq. (8-90) there is obtained

$$\tau^* = \frac{\pi/2 + \sin^{-1}\zeta}{\omega_n(1 - \zeta^2)^{1/2}} \tag{8-93}$$

The following tabulation indicates the effect of altitude on τ^* for the values of $(\omega_n)_0$ and ζ_0 above.

h, ft	$(p/p_0)^{1/2}$	ω_n	ζ	τ^*
0	1.000	2.000	0.600	1.37
30,000	0.545	1.090	0.327	1.85
50,000	0.338	0.676	0.203	2.67
100,000	0.103	0.206	0.062	7.94

The reduction of missile response rate at high altitudes can be overcome in part by the use of big controls and large control deflections.

SYMBOLS

a	radius of circular body
A_P	aspect ratio of two control panels joined together
A_T	aspect ratio of horizontal tail panels joined together
B	$(M_0{}^2 - 1)^{1/2}$
c	local chord of flap
c_l	section lift coefficient
c_r	root chord, chord at juncture of control and body
$\bar{c}$	mean aerodynamic chord of control
$(cc_l)_2$	span loading of rolling body with control panel at zero deflection in reverse flow
C_1, C_2	constants in Busemann second-order theory, Eq. (8-74)
C_h	hinge-moment coefficient, (hinge moment$/q_0 S_R l_r$)
C_l	rolling-moment coefficient, (rolling moment$/q_0 S_R l_r$)
C_L	lift coefficient, (lift$/q_0 S_R$)
C_{L_α}	lift-curve slope
$(C_L)_\alpha, (C_L)_\delta$	lift coefficients associated with angle of attack and control deflection, respectively
$(C_{L_\alpha})_n, (C_{L_\alpha})_0$	two-dimensional lift-curve slopes based on normal Mach number and free-stream Mach number, respectively
C_m	pitching-moment coefficient, $$\dfrac{\text{pitching moment}}{q_0 S_R l_r}$$
C_{m_q}	$\dfrac{\partial C_m}{\partial(q l_r/2 V_0)}$
$C_{m_{\dot\alpha}}$	$\dfrac{\partial C_m}{\partial(\dot\alpha l_r/2 V_0)}$
C_n	yawing-moment coefficient, $$\dfrac{\text{yawing moment}}{q_0 S_R l_r}$$
$(dz/dx)^+$	slope of upper surface of control with respect to chord line
$E(\phi,k)$	incomplete elliptic integral of second kind of amplitude ϕ and modulus k
F	force developed in vertical plane by cruciform missile with horizontal panels deflected an amount δ
$F(\phi,k)$	incomplete elliptic integral of first kind of amplitude ϕ and modulus k

h	trailing-edge thickness of control
h_l	maximum thickness of lower control surface measured from chord line
h_u	maximum thickness of upper control surface measured from chord line
$H(t)$	Heaviside unit step function
k	$(1 - \lambda^4)^{1/2}$; also modulus of elliptic integral
k_B	lift ratio; ratio of lift on body due to control deflection to lift of control alone
k_T	lift ratio; ratio of lift on tail control in presence of body to lift of control alone
k_W	lift ratio; ratio of lift on wing control in presence of body to lift of control alone
K_B	body lift interference ratio, Table 5-1
K_W	wing panel lift interference ratio, Table 5-1
K_y	radius of gyration of missile about lateral y axis through center of gravity
l_r	reference length
L	lift force
L_0	lift due to control for two-dimensional flow based on conditions in streamwise direction
$L_{B(W)}$	lift on body in presence of wing-control panels
L_C	lift of complete missile
L_F	total lift of missile due to control deflection
L_n	lift due to control for two-dimensional flow based on conditions normal to hinge line
$L_{W(B)}$	lift on control panel in presence of body
L'	rolling moment about missile longitudinal axis, positive right wing downward
m	mass of missile
M	pitching moment
M_0	free-stream Mach number
M_n	Mach number based on flow normal to hinge line
M_q	$\partial M/\partial q$
M_w	$\partial M/\partial w$
M_α	$\partial M/\partial \alpha$
p	rolling velocity about missile longitudinal axis, positive right wing down; also local static pressure
p_0	free-stream static pressure; also pressure at sea level
Δp	$p - p_0$
P	pressure coefficient, $\Delta p/q_0$
P_H	pressure coefficient of horizontal control panel
ΔP	difference in pressure coefficients
$(\Delta P)_{\text{incip}}$	see Fig. 8-26

$(\Delta P)_{\text{peak}}$	see Fig. 8-26
q	missile angular velocity about y axis
q_0	free-stream dynamic pressure
q_n	dynamic pressure based on flow velocity normal to hinge line
r	radial distance from x axis
s	local semispan of control
s_i	semispan of inner edge of tip control
s_m	maximum semispan of control
S_B	planform area of body
S_F	planform area of controls
S_P	planform area of one panel
S_R	reference area
S_W	planform area of entire wing panels including controls
t	local thickness of airfoil section
t_H	airfoil thickness at hinge line
t_m	maximum thickness of airfoil section
u, v, w	perturbation velocity components along x, y, and z axes
V_0	free-stream velocity
w	see u, v, w
W_d	complex potential of doublet
x_H	value of x for hinge axis, Fig. 8-25
x, y, z	principal axes of symmetry of missile, Fig. 8-5
x', y', z'	principal axes of symmetry for α_c with $\varphi = 0$, Fig. 8-13
$\bar{x}$	center-of-pressure location
Y, Z	forces along y and z axes
z_u	z coordinate of upper surface of control
Z_w	$\partial Z/\partial w$
$\mathfrak{z}$	$y + iz$
α	angle of attack
α_B	angle of attack of body
α_c	included angle between V_0 and missile longitudinal axis
α_W	angle of attack of all-movable control
α^*	final missile angle of attack after impulsive pitch control
β	angle of sideslip
δ	general symbol for control deflection
δ_0	deflection of control measured in free-stream direction
$\delta_1, \delta_2, \delta_3, \delta_4$	control deflections of horizontal and vertical all-movable controls, Fig. 8-2
δ_a	$(\delta_1 - \delta_2)/2$
δ_e	$(\delta_1 + \delta_2)/2$

$\delta_{a'}$	$(\delta_3 - \delta_4)/2$
δ_r	$(\delta_3 + \delta_4)/2$
δ_n	deflection of control measured normal to control hinge line
ϵ	semiapex angle of triangular wing formed from two triangular controls
ζ	damping parameter, Eq. (8-86)
ζ_0	value of ζ at sea level
θ	pitch angle of missile ($\dot{\theta} = q$)
λ	a/s_m
Λ	sweep angle of hinge line
σ	plane in which missile cross section transforms into unit circle
τ	a/s_m; also dummy variable for time
τ^*	time for missile to attain α^* with impulsive control action
ϕ	velocity potential
φ	angle of bank
ϕ_d	velocity potential for doublet
$\phi_\alpha, \phi_\beta, \phi_{\delta_e}, \phi_{\delta_a}, \phi_t$	velocity potentials associated with α, β, δ_e, δ_a, and t, respectively
ψ_d	stream function for doublet
ω_n	natural frequency of missile, Eq. (8-85)
ω_{n_0}	natural frequency at sea level

Subscripts:

B	body
$B(W)$	body in presence of wing
cg	center of gravity
C	complete combination
H	hinge line
t	associated with airfoil thickness
W	wing alone or wing panels
$W(B)$	wing panels in presence of body
α	associated with angle of attack
β	associated with angle of sideslip
δ	associated with control deflection
$+$	impact surface
$-$	suction surface

REFERENCES

1. Adams, Gaynor J., and Duane W. Dugan: Theoretical Damping in Roll and Rolling Moment Due to Differential Wing Incidence for Slender Cruciform Wings and Wing-Body Combinations, *NACA Tech. Repts.* 1088, 1952.

2. Dugan, Duane W., and Katsumi Hikido: Theoretical Investigation of the Effects upon Lift of a Gap between Wing and Body of a Slender Wing-Body Combination, *NACA Tech. Notes* 3224, August, 1954.

3. Heaslet, Max A., and John R. Spreiter: Reciprocity Relations in Aerodynamics, *NACA Tech. Repts.* 1119, 1953.

4. Lomax, Harvard, and Max A. Heaslet: Damping-in-roll Calculations for Slender Swept-back Wings and Slender Wing-Body Combinations, *NACA Tech. Notes* 1950, 1949.

5. Nielsen, Jack N.: Quasi-cylindrical Theory of Interference at Supersonic Speeds and Comparison with Experiment, *NACA Tech. Repts.* 1252, 1955.

6. Bleviss, Zegmund O.: Some Roll Characteristics of Plane and Cruciform Delta Ailerons and Wings in Supersonic Flow, *Douglas Aircraft Co. Rept.* SM-13431, June, 1949.

7. Frick, Charles W., Jr.: Application of the Linearized Theory of Supersonic Flow to the Estimation of Control-surface Characteristics, *NACA Tech. Notes* 1554, March, 1948.

8. Jones, Robert T.: Thin Oblique Airfoils at Supersonic Speeds, *NACA Tech. Notes* 1107, 1946.

9. Lagerstrom, P. A.: Linearized Supersonic Theory of Conical Wings, *NACA Tech. Notes* 1685, 1948.

10. Cohen, Doris: Formulas for the Supersonic Loading, Lift, and Drag of Flat Swept-back Wings with Leading Edges behind the Mach Lines, *NACA Tech. Repts.* 1050, 1951.

11. Lagerstrom, P. A., and Martha E. Graham: Linearized Theory of Supersonic Control Surfaces, *J. Aeronaut. Sci.*, vol. 16, no. 1, 1949.

12. Lagerstrom, P. A., and Martha E. Graham: Linearized Theory of Supersonic Control Surfaces, *Douglas Aircraft Co. Rept.* SM-13060, July, 1947.

13. Goin, Kennith L.: Equations and Charts for the Rapid Estimation of Hinge-moment and Effectiveness Parameters for Trailing-edge Controls Having Leading and Trailing Edges Swept Ahead of the Mach Lines, *NACA Tech. Notes* 2221, November, 1950.

14. Tucker, Warren A.: Characteristics of Thin Triangular Wings with Triangular-tip Control Surfaces at Supersonic Speeds with Mach Lines behind the Leading Edge, *NACA Tech. Notes* 1600, 1948.

15. Tucker, Warren A.: Characteristics of Thin Triangular Wings with Constant-chord Full-span Control Surfaces at Supersonic Speeds, *NACA Tech. Notes* 1601, 1948.

16. Tucker, Warren A.: Characteristics of Thin Triangular Wings with Constant-chord Partial-span Control Surfaces at Supersonic Speeds, *NACA Tech. Notes* 1660, July, 1948.

17. Frost, Richard C.: Supersonic Flap Lift Effectiveness for Some General Plan Forms, *J. Aeronaut. Sci.*, vol. 21, no. 9, 1954.

18. Kainer, Julian H., and Jack E. Marte: Theoretical Supersonic Characteristics of Inboard Trailing-edge Flaps Having Arbitrary Sweep and Taper Mach Lines behind Flap Leading and Trailing Edges, *NACA Tech. Notes* 2205, 1950.

19. Mirels, H.: Gap Effect on Slender Wing-Body Interference, *J. Aeronaut. Sci.*, vol. 20, no. 8, Readers' Forum, 1953.

20. Staff of Ames 1- by 3-foot Supersonic Wind Tunnel: Notes and Tables for Use in the Analysis of Supersonic Flow, *NACA Tech. Notes* 1428, December, 1947.

21. Chapman, D. R., Donald M. Kuehn, and Howard K. Larson: Investigation of Separated Flows in Supersonic and Subsonic Streams with Emphasis on Effect of Transition, *NACA Tech. Notes* 3869, 1957.

22. Holder, D. W., and G. E. Gadd: The Interaction between Shock Waves and

Boundary Layers and Its Relation to Base Pressure in Supersonic Flow, *Natl. Phys. Lab. Symposium on Boundary Layer Effects in Aerodynamics*, Paper 8, Teddington, England, Mar. 31 to Apr. 2, 1955.

23. Gallagher, James J., and James N. Mueller: Preliminary Tests to Determine the Maximum Lift of Wings at Supersonic Speeds, *NACA Research Mem.* L7510, December, 1947.   (Declassified)

24. Dugan, Duane W.: Gap Effect on Slender Wing-Body Interference, *J. Aeronaut. Sci.*, vol. 21, no. 1, Readers' Forum, 1954.

25. Tobak, Murray, and H. Julian Allen: Dynamic Stability of Vehicles Traversing, Ascending or Descending Paths through the Atmosphere, *NACA Tech. Notes* 4275, July, 1958.

APPENDIX 8A. EQUATIONS OF MOTION FOR MISSILE WITH PITCH CONTROL

Consider a missile flying straight and level essentially at zero angle of attack as shown in Fig. 8-31.   Apply pitch control to the missile so that it acquires angular velocity about the lateral axis through its center of

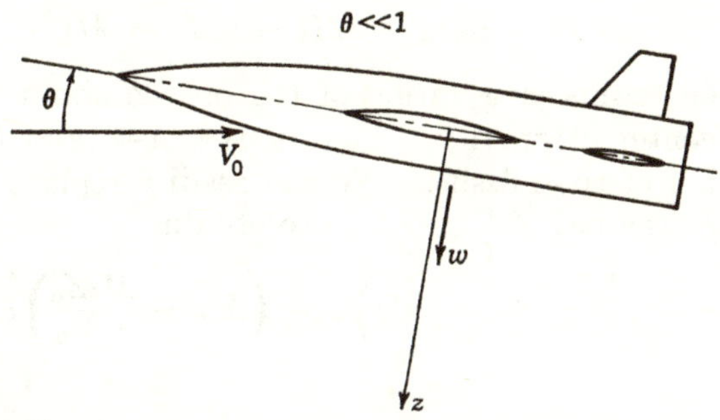

FIG. 8-31. Missile undergoing impulsive pitch control.

gravity and develops a velocity w of the center of gravity along the z axis positive in the downward direction.   Let the inclination of the longitudinal axis to the horizontal be θ.   As a result of θ and of w, the missile undergoes a change in angle of attack α given by

$$\alpha = \theta + \frac{w}{V_0} \tag{8A-1}$$

and

$$\dot{\alpha} = \dot{\theta} + \frac{\dot{w}}{V_0} \tag{8A-2}$$

Assume for simplicity that the lift on the missile depends principally on α and is independent of θ and $\dot{\alpha}$.   Then

$$\dot{w} = \frac{(\partial Z/\partial \alpha)\alpha}{m} \tag{8A-3}$$

where Z is the negative of the lift force, and m is the mass of the missile.

The relationship between $\dot{\alpha}$ and $\dot{\theta}$ is the equation for translation of the center of gravity in the vertical direction.

$$\dot{\alpha} = \dot{\theta} + \frac{Z_\alpha \alpha}{mV_0} \tag{8A-4}$$

Let us now write the equation of motion for rotation about the center of gravity.

The moment acting on the missile due to the control action will result from changes in α, $\dot{\theta}$, $\dot{\alpha}$, and control deflection δ.

$$M = M_\alpha \alpha + M_{\dot{\alpha}}\dot{\alpha} + M_{\dot{\theta}}\dot{\theta} + M(\delta) \tag{8A-5}$$

The quantity $M(\delta)$ is the time-dependent moment due to variable control deflection δ. The term $M_\alpha \alpha$ is the moment associated with static stability. The moments $M_{\dot{\alpha}}\dot{\alpha}$ and $M_{\dot{\theta}}\dot{\theta}$ are damping moments and are due principally to the tail. These moments are precisely those due to $M_{\dot{\alpha}}$ and M_q $(q = \dot{\theta})$ discussed in Sec. 10-11. The equation for the rotation about the center of gravity is now

$$mK_y{}^2\ddot{\theta} - M = 0$$
$$mK_y{}^2\ddot{\theta} - \dot{\alpha}M_{\dot{\alpha}} - \dot{\theta}M_{\dot{\theta}} - \alpha M_\alpha = M(\delta) \tag{8A-6}$$

Here K_y is the radius of gyration of the missile about the lateral axis through the center of gravity. Equations (8A-4) and (8A-6) together give the motion of the missile. We can readily replace $\dot{\theta}$ and $\ddot{\theta}$ in Eq. (8A-6) through the use of Eq. (8A-4) to obtain

$$mK_y{}^2\ddot{\alpha} + \left(-\frac{K_y{}^2 Z_\alpha}{V_0} - M_{\dot{\alpha}} - M_{\dot{\theta}} \right) \dot{\alpha} - \left(M_\alpha - \frac{M_{\dot{\theta}} Z_\alpha}{mV_0} \right) \alpha = M(\delta) \tag{8A-7}$$

The term $M_{\dot{\theta}} Z_\alpha$ will be ignored in comparison with the M_α term. Had we ignored the vertical motion of the center of gravity, we would have α equal to θ, and the Z_α terms would disappear from Eq. (8A-7).

A more sophisticated derivation of Eq. (8A-7) is to be found in Tobak and Allen[25] considering also changes in forward speed.

CHAPTER 9

DRAG

The supersonic drag of projectiles has occupied the attention of ballisticians for many years and achieved importance even before the airplane. In recent years the supersonic airplane and missile have brought about widespread interest in and extensive enlargement of our knowledge of aerodynamic drag at supersonic speeds.   Though great progress has been made, it can safely be said that succeeding years will see further extensive additions to our knowledge of aerodynamic drag at high speeds.   In this chapter we will present some of the important results that have been obtained, with a particular view to their usefulness to missile engineers and scientists.

Of the forces and moments acting on a missile, the drag force is most influenced by the viscosity of the medium in which the missile is traveling.   It is therefore not surprising that the drag force is also the most difficult to predict or to measure accurately.   The theoretical tools used to predict drag must take into account viscosity, and as such they are quite apart from the methods of potential flow usually used to predict the other forces and the moments.   It is therefore fitting that we should devote a special chapter to the study of drag.

In Sec. 9-1 a number of ways are discussed for subdividing the total drag of a missile into components.   One scheme gives as the components of the total drag the *pressure drag exclusive of base drag*, the *base drag*, and the *viscous drag* or the *skin friction*.   The chapter is broken down into these three main sections.   In Sec. 9-2 we consider the analytical properties of drag curves, and describe the basic aerodynamic parameters specifying the drag curves.

The subject of the pressure drag exclusive of base drag, or *pressure foredrag*, is started with a discussion of Ward's drag formula for slender bodies in Sec. 9-3.   The pressure foredrag of bodies of given shape, not necessarily slender, is considered in Sec. 9-4, followed in Sec. 9-5 by a treatment of methods for shaping bodies to achieve least pressure foredrag.   The pressure foredrag of wings alone is the subject of Sec. 9-6, and that of wing-body combinations of given shape is the subject of Sec. 9-7. Methods for minimizing the pressure drag of wing-body combinations at zero angle of attack are considered in Sec. 9-8, particularly area rule

261

methods. In Sec. 9-9 we take up methods for the minimization of the pressure drag due to lift of wings and wing-body combinations.

The second important component of the missile drag, namely, *base drag*, is considered in Sec. 9-10, where the general physical features of flow at a blunt base are described. The physical basis for the correlation of base-pressure measurements is laid in Sec. 9-11, preparatory to a presentation of base-pressure correlations in Sec. 9-12. A number of variables also influencing base pressure are discussed in Sec. 9-13.

The third and final component of the missile drag, namely, *skin friction*, is described in its general aspects in Sec. 9-14. Engineering methods for calculating purely laminar skin friction and purely turbulent skin friction for flat plates are presented in Secs. 9-15 and 9-16, respectively. The chapter concludes with some comments on factors influencing skin friction such as transition and the departure from a flat plate.

9-1. General Nature of Drag Forces; Components of Drag

Of the several significant methods for separating the drag into component parts, the simplest is probably that arising naturally from a consideration of whether the drag is caused by forces acting normal to the

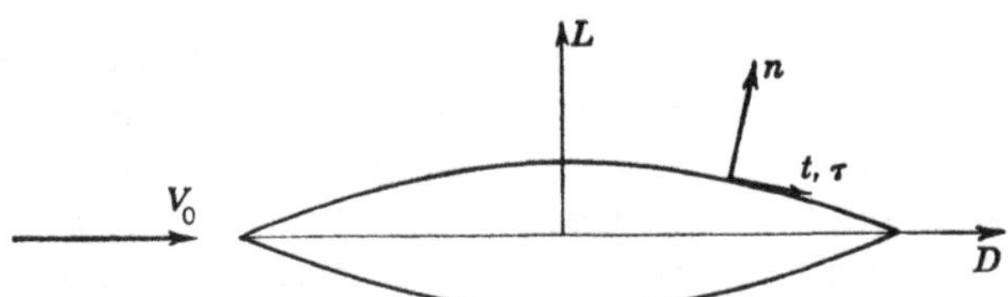

FIG. 9-1. Aerodynamic body subject to normal and tangential stresses.

missile surface or forces acting tangential to it. The drag arising from the pressure forces acting normal to the missile surface is known as *pressure drag*, and that arising from the tangential forces of skin friction acting on the surface by virtue of viscosity is called *viscous drag* or *skin friction*. With reference to Fig. 9-1, the drag due to pressure p at the missile surface is

$$D_P = - \iint_{S_m} p \cos (n, V_0) \, dS_m \tag{9-1}$$

where $\cos (n, V_0)$ is the cosine of the angle between V_0 and the outward normal to the missile surface. The surface S_m comprises the total area of the missile including the base area. If the base contains a jet, the surface is taken straight across the jet exit from the missile. The integral of the pressure over the internal surfaces containing the jet is taken as the propulsive force.

If τ is the local skin friction per unit area due to viscosity, then the viscous drag is

$$D_V = \iint_{S_m} \tau \cos (t, V_0) \, dS_m \tag{9-2}$$

where cos (t, V_0) is the cosine of the angle between V_0 and the tangent to the missile surface in the τ direction. Note that t and τ are in the same direction.

The drag can also be separated into the components of *foredrag* and *base drag*. The foredrag is that part of the total drag acting on the missile surface exclusive of the base area. It contains significant amounts of pressure drag and viscous drag. The base drag, on the other hand, is almost wholly pressure drag. As a consequence the total missile drag can now be subdivided into *pressure foredrag*, *base drag*, and *viscous drag*. It is convenient to consider these quantities as distinct quantities which can be added together to obtain the total missile drag. Though these quantities are distinct one from another that is not to say that they are independent of one another. For instance, the condition of the boundary layer, laminar or turbulent, which specifies the viscous drag also significantly influences the base drag.

The first component of missile drag, pressure foredrag, is amenable to analysis by potential theory in those cases wherein the boundary layer does not separate and cause large alterations in the pressure distribution. (Even with boundary-layer separation, potential flow frequently plays a role in determining the pressure distribution.) The slender-body theory of drag has been well developed for complete configurations, and linear theory has been extensively applied to supersonic wings. For bodies alone, theories of greater accuracy than linear theory are available in the form of the second-order theory of Van Dyke,[15] and the method of characteristics. It is not surprising, in view of the fact that pressure foredrag is amenable to analysis by the highly developed methods of potential theory, that much work has been successfully directed toward minimizing pressure foredrag.

The second component of total missile drag, base drag, is determined by considerations of potential flow and of viscosity. The so-called *dead water region* behind the base of a missile has a static pressure, which depends on how the outer flow closes in behind the missile, and how the boundary layer from the base mixes with the dead water and the outer flow. Although much theoretical work has been done on the problem of base pressure, its engineering determination is still dependent principally on correlations of systematic experimental data. The base pressure is also influenced by any *boattailing* in front of the missile base, by the proximity of tail fins to the base, etc.

The final component of total missile drag, the *viscous drag* or *skin friction*, is difficult to predict or measure accurately. This difficulty stems, in part, from the incomplete understanding of where the boundary layer turns from laminar to turbulent in flight. Even if the transition point in flight were known, it would be hard to measure the skin friction in the wind tunnel for this known transition location, because of unknown amounts of wave drag caused by the mechanism for fixing transition. The transition location depends on Reynolds number, Mach number, pressure distribution, turbulence level, heat-transfer rate, surface roughness, sound level, and other factors which, to understate the case, are imperfectly understood. The point of view we adopt is that, given the transition point, the skin friction can be calculated by methods to be described.

So far we have considered two distinct schemes for subdividing the total missile drag and the relationship between the schemes. Yet another method arises naturally in the application of "control-surface" methods for evaluating drag as illustrated in Fig. 9-2. The decomposition results in the components of *wave drag* and *wake drag*. The drag associated with the momentum transfer through the control surface S_2,

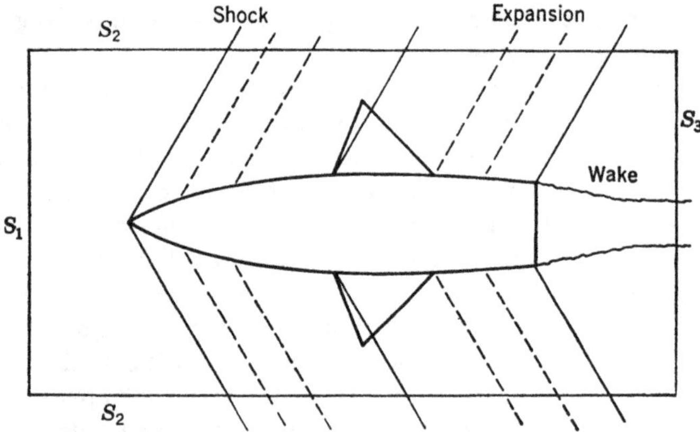

Fig. 9-2. Missile at supersonic speeds enclosed by cylindrical control surface.

as S_1, S_2, and S_3 all move infinitely far away from the missile, is called the *wave drag*. The drag associated with the net momentum transfer through surfaces S_1 and S_3 is termed the *wake drag*. The wake drag in the general case of a viscous fluid will represent in part the skin-friction drag because of mixing in the wake between the boundary layer and the inviscid flow. However, in certain theories such as slender-body theory and linear theory, there is no viscous wake, and the wake drag is due entirely to creation of vortices with kinetic energy. For this case of no skin friction the wake drag is all *vortex drag*. Thus, on the basis of inviscid fluid theory the entire drag is pressure drag composed of wave drag and vortex drag. This particular decomposition is of great importance when we come to the problem of minimizing the drag due to lift of wings and wing-body combinations. Let us now examine the nature of the wave drag and wake drag more closely.

Figure 9-2 shows waves from the body passing through control surface S_2, which is parallel to the free-stream direction, and carrying momentum outward through the surface. The momentum transport per unit time is sometimes called the wave drag, although other definitions of wave drag will shortly be mentioned. The particular usefulness of this definition depends on the possibility of evaluating the momentum by some theoretical means. If the evaluation is made on the basis of slender-body theory, the control-surface radius must be kept small, since slender-body theory is valid only in the immediate neighborhood of the body. If the control-surface radius were permitted to approach infinity in slender-body theory, the wave drag would become infinite. For this reason the radius in the derivation of Ward's drag formula (Sec. 3-9) was kept small although arbitrary. If the wave drag is evaluated on the basis of linear theory, the radius of the control surface can approach infinity, and the wave drag will remain finite. From a broader point of view than the foregoing theoretical one, the wave drag is associated with the energy necessary continuously to form the wave system of the missile as it moves at supersonic speeds. In this context the wave drag is really *wave-making drag* similar to that of a ship. From yet another point of view, wave drag represents the entropy increase of the fluid passing through the shock waves of the missile. It can be calculated in principle if the shapes and strengths of all the shock waves are known, by integrating along all the waves to obtain the total increase in entropy.

The net momentum change per unit time for control areas S_1 and S_3 represents viscous drag of the boundary layer, kinetic energy of vortices generated by lift, and possibly base drag, although some of this appears in the wave drag, too. For blunt-base bodies or wings, these three components are inextricably combined within the limitations of our present knowledge of the flow fields behind such bodies or wings. The wake drag in such cases has no particular significance. However, for missiles with sharp bases and trailing edges, the wake drag is meaningful under certain circumstances. Assume that for such a missile, symmetrical about a horizontal plane, the boundary layer remains attached and does not produce any appreciable alteration in the wave system from that for an inviscid fluid. At zero angle of attack the total drag then consists of the so-called *zero wave drag* and wake drag which is purely viscous drag. However, consider the drag due to lift occurring as a result of an increase in angle of attack. This will consist first of additional wave drag due to an alteration in the strengths and shape of the wave system. It will also consist of an additional drag in the wake associated with vortices appearing there because of the lift.

9-2. Analytical Properties of Drag Curves

For the purposes of predicting drag and of analyzing experimental drag curves, it is desirable to have a standard set of parameters and symbols which define a drag curve. A drag curve, or *drag polar* as it is sometimes called, is a plot of drag coefficient versus lift coefficient. On the basis of linear theory, the drag curve is a parabola. A parabolic drag curve together with certain standard symbols appertaining thereto is shown in Fig. 9-3. The minimum ordinate C_{D_0} is called the *minimum drag coefficient* and the corresponding lift coefficient C_{L_0} is called the *lift coefficient for minimum drag*. The tangent to the parabola from the origin (of which there are two) specifies the *optimum lift coefficient* $C_{L_{opt}}$. At the

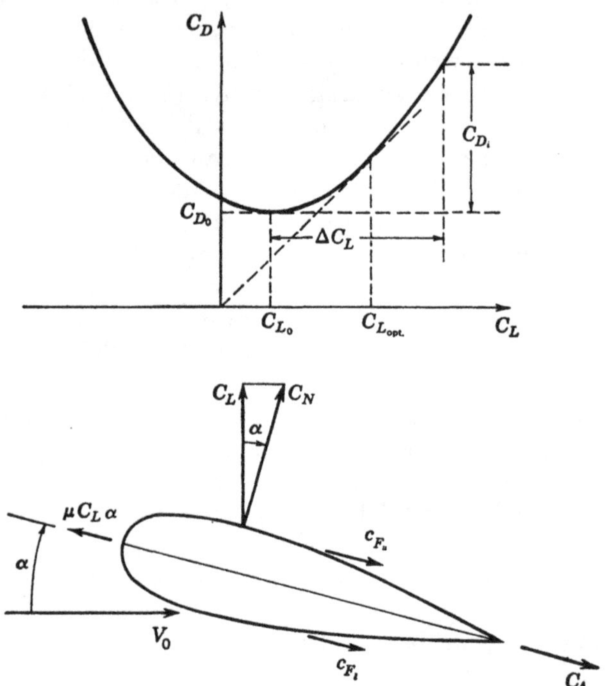

FIG. 9-3. Drag polar and forces acting on aerodynamic body.

optimum lift coefficient the value of C_L/C_D is termed the *maximum lift-drag ratio* and is frequently written $(L/D)_{max}$.

Experimental drag curves are frequently well approximated by parabolas. The drag curve can then be represented by the equation

$$C_D - C_{D_0} = k(C_L - C_{L_0})^2 \qquad (9\text{-}3)$$

The factor k is called the *drag-rise factor*, and its value can be obtained experimentally by plotting $C_D - C_{D_0}$ versus $(C_L - C_{L_0})^2$. If the drag data plotted in this manner fall on a straight line, the drag curve is parabolic, and the slope of the line is the drag-rise factor. Although many experimental drag curves are closely parabolic, the parabolicity should be tested in each instance. The term *drag-rise factor* for k follows the usage of Vincenti[59] and others. It should not be confused with the frequently used term *drag rise*, which refers to the increase in the minimum drag coefficient in the transonic region above its value for incompressible flow (at the same Reynolds number). The drag coefficient increment above that for minimum drag C_{D_0} is written C_{D_i} as given by

$$C_{D_i} = C_D - C_{D_0} \tag{9-4}$$

and is mathematically equivalent to induced drag at subsonic speeds. The lift coefficient above that for minimum drag C_{L_0} is written ΔC_L as given by

$$\Delta C_L = C_L - C_{L_0} \tag{9-5}$$

The drag-rise factor from Eq. (9-3) then has the form

$$k = \frac{C_{D_i}}{\Delta C_L{}^2} \tag{9-6}$$

and will be henceforth written in this fashion. For a parabolic drag curve, the value of C_{L_opt} is found to be

$$C_{L_\text{opt}} = \left(C_{L_0}{}^2 + \frac{C_{D_0}}{C_{D_i}/\Delta C_L{}^2} \right)^{\frac{1}{2}} \tag{9-7}$$

The corresponding maximum lift-drag ratio is

$$\left(\frac{L}{D}\right)_\text{max} = \frac{1}{2(C_{L_\text{opt}} - C_{L_0})(C_{D_i}/\Delta C_L{}^2)}$$
$$= \frac{C_{L_0} + [C_{L_0}{}^2 + C_{D_0}/(C_{D_i}/\Delta C_L{}^2)]^{\frac{1}{2}}}{2C_{D_0}} \tag{9-8}$$

It is clear that the drag curve, the optimum lift coefficient, and the maximum lift-drag ratio are completely determined by C_{D_0}, C_{L_0}, and $C_{D_i}/\Delta C_L{}^2$. For a missile with a horizontal plane of symmetry, the values of C_{L_opt} and $(L/D)_\text{max}$ are simply

$$C_{L_\text{opt}} = \left(\frac{C_{D_0}}{C_{D_i}/\Delta C_L{}^2} \right)^{\frac{1}{2}} \tag{9-9}$$

$$\left(\frac{L}{D}\right)_\text{max} = \frac{1}{2} \left[\frac{1}{C_{D_0}(C_{D_i}/\Delta C_L{}^2)} \right]^{\frac{1}{2}} \tag{9-10}$$

Let us examine the quantities which determine the drag-rise factor, namely, pressure drag due to lift, leading-edge thrust, and skin friction variations due to angle of attack. For this purpose consider the force acting on the symmetrical wing shown in Fig. 9-3. First, the chord-force coefficient in the absence of leading-edge thrust and skin friction is denoted by C_A. The leading-edge thrust[27] is due to suction pressures arising from the high flow velocities around the leading edge in certain cases to be discussed later (Sec. 9-6). It is convenient to specify this thrust as a fraction μ of the drag of a flat plate at angle of attack α and lift coefficient C_L; that is, in coefficient form the leading-edge thrust is

$\mu C_L \alpha$. If the average skin-friction coefficient, based on the same reference quantities as the other coefficients, is c_{F_u} for the upper surface and c_{F_l} for the lower surface, the total chord-force coefficient is

$$C_C = C_A - \mu C_L \alpha + c_{F_u} + c_{F_l} \tag{9-11}$$

Let the subscript zero stand for zero angle of attack for the symmetric wing shown here. Then

$$C_{C_0} = C_{A_0} + (c_{F_u})_0 + (c_{F_l})_0 \tag{9-12}$$

We will now form the drag-rise factor. The drag coefficient is exactly

$$C_D = C_N \sin \alpha + C_C \cos \alpha \tag{9-13}$$

We may substitute the lift coefficient for the normal force coefficient, and the error will be only of the order α^3. Thus

$$C_D = C_L \sin \alpha + C_C \cos \alpha + O(\alpha^3) \tag{9-14}$$

Forming the drag-rise factor from Eqs. (9-11), (9-12), and (9-14), we obtain

$$\frac{C_{D_i}}{\Delta C_L{}^2} = \frac{C_D - C_{D_0}}{C_L{}^2} = \frac{1 - \mu}{C_{L_\alpha}} + \frac{C_A - C_{A_0}}{C_L{}^2}$$
$$+ \frac{(c_{F_u} + c_{F_l}) - (c_{F_u} + c_{F_l})_0}{C_L{}^2} \tag{9-15}$$

An examination of Eq. (9-15) for the drag-rise factor reveals three terms, each representing a distinct physical phenomenon. The first term is the dominant term, and the latter two terms are usually neglected. The first term is essentially the pressure drag due to lift, which appears partly in the wave system of the wing, and partly in the vortex wake as described previously. It is inversely proportional to the lift-curve slope, and increases directly as the leading-edge thrust decreases. For a wing with supersonic leading edges, μ is theoretically zero; but, for a triangular wing of very low aspect ratio or a slender body of revolution, μ is theoretically 0.5. The second term is a change in chord pressure force exclusive of leading-edge thrust. It can arise, for instance, by second-order pressures proportional to the product of thickness and angle of attack. Alternatively, it might arise as a result of boundary-layer separation induced by angle of attack. The third factor represents the change in skin friction with angle of attack. It can arise from changes in the transition points on the lower and upper wing surfaces as the angle of attack changes. It further depends on changes in density and velocity at the outer edge of the wing boundary layer, changes that can become significant at high Mach numbers corresponding to hypersonic flight. Also, any difference in temperature between the upper and lower surfaces as a result of aerodynamic heating, radiation, etc., can enter the third factor in a manner discussed in Secs. 9-15 and 9-16.

PRESSURE FOREDRAG

9-3. Pressure Foredrag of Slender Bodies of Given Shape; Drag Due to Lift

The great analytical simplification of aerodynamic problems brought about by slender-body theory applies to drag problems equally as well as to those of lift and sideforce. In fact, the drag formula of Ward derived

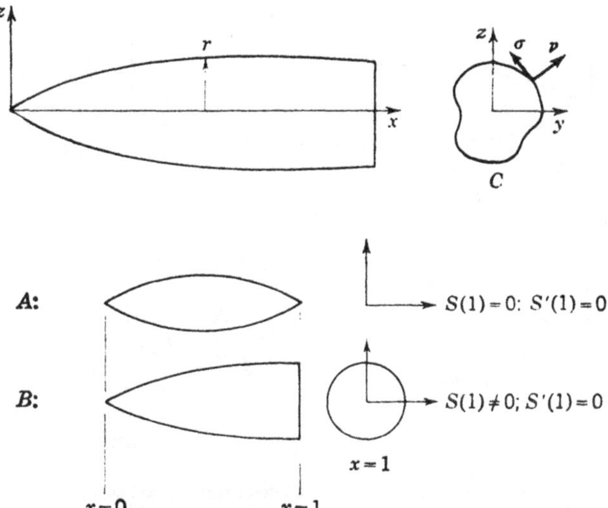

FIG. 9-4. Notation for use in drag formula of Ward.

in Sec. 3-9 allows considerable insight into the drag of slender bodies, including an understanding of the nature of the various components which go to make up the total pressure foredrag. Accordingly, we will apply the drag formula of Ward to a series of bodies of increasing complexity to show how various components of the drag of a slender body arise. With reference to Fig. 9-4 the complete drag formula of Ward can be written for a pointed body as

$$
\frac{D}{q_0} = \frac{1}{2\pi} \int_0^1 \int_0^1 S''(x)S''(\xi) \log \frac{1}{|x - \xi|} \, d\xi \, dx
$$
$$
- \frac{S'(1)}{2\pi} \int_0^1 S''(\xi) \log \frac{1}{|1 - \xi|} \, d\xi - \oint_C \phi \frac{\partial \phi}{\partial \nu} \, d\sigma
$$
$$
- P_B S(1) + O(t^6 \log^2 t) \quad (9\text{-}16)
$$

where $\quad a_0 = \dfrac{S'(x)}{2\pi}$

$$b_0 = \frac{1}{2\pi}\left[S'(x) \log \frac{\beta}{2} - \int_0^x S''(\xi) \log (x - \xi)\, d\xi \right]$$

(9-17)

and $\quad \phi = a_0 \log r + b_0 + \displaystyle\sum_{n=1}^{\infty} \frac{A_n \cos n\theta + B_n \sin n\theta}{r^n}$

(9-18)

The coefficients a_0, b_0, A_n, and B_n are real functions of x, and t is the reciprocal of the body fineness ratio. The slender-body potential ϕ for

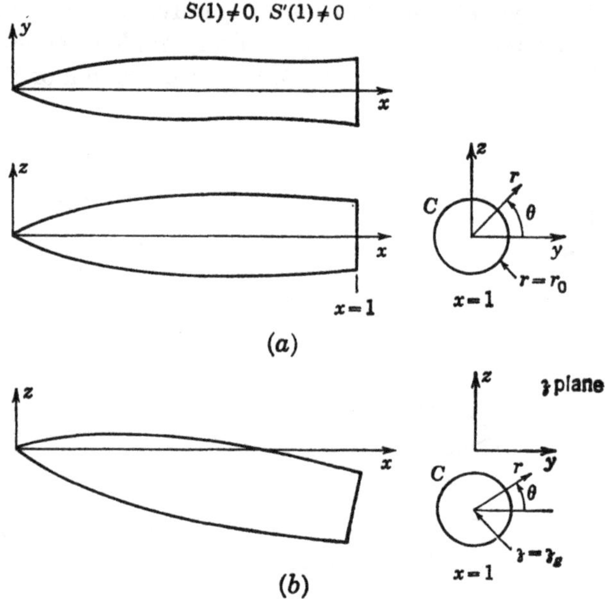

Fig. 9-5. Nonaxisymmetric slender body at zero angle of attack and at angle of attack; circular base. (a) $\alpha = 0$; (b) $\alpha > 0$.

unit free-stream velocity can always be put into the form of Eq. (9-18), and we will examine specific examples in the following four cases. Of the several cases of a body of revolution at zero angle of attack, perhaps the body pointed at both ends is most simple.

Case 1: Body Pointed at Both Ends

For case 1 shown in Fig. 9-4 the body is pointed at both ends. This condition is sufficient to make $S(0) = S'(0) = S(1) = S'(1) = 0$. As a result the entire drag is given by

$$\frac{D}{q_0} = \frac{1}{2\pi} \int_0^1 \int_0^1 S''(x) S''(\xi) \log \frac{1}{|x - \xi|}\, dx\, d\xi$$

(9-19)

This result will subsequently be used to derive the Sear-Haack body of least wave drag subject to certain conditions. Since the body has a sharp base, it will have no wake (in an inviscid fluid), and all the drag will arise as a result of wave formation by the body. For this reason the drag represented by Eq. (9-19) is wave **drag**. It is clearly independent of Mach number.

Case 2: Body with Cylindrical Base at Zero Angle of Attack

By a body with a cylindrical base we mean one which would have no discontinuities in streamwise slope if the base were prolonged by a cylindrical extension. See Fig. 9-4. For this case we have $S'(1) = 0$. Also, since $\partial\phi/\partial\nu$ is the velocity component normal to the body in a plane normal to the body axis, this quantity will be zero at the base. Equation (9-16) therefore reduces to

$$\frac{D}{q_0} = \frac{1}{2\pi}\int_0^1\int_0^1 S''(x)S''(\xi)\log\frac{1}{|x-\xi|}\,d\xi\,dx - P_B S(1) \qquad (9\text{-}20)$$

The drag in this case consists of wave drag, as in case 1, plus a base **drag**. The prediction of base drag is beyond the realm of slender-body theory. The discussion of base drag in the second main part of this chapter shows it to be Mach-number-dependent. Thus, while the drag of a pointed body is independent of Mach number, that of a body with a blunt base varies with Mach number. We will subsequently derive the shape of the Kármán ogive, which is the body with a cylindrical base possessing the least pressure foredrag at zero angle of attack on the basis of slender-body theory.

Case 3: General Body with Circular Base at Zero Angle of Attack

For a general body with a circular base we have that neither $S(1)$ nor $S'(1)$ is zero. Let the body furthermore possess horizontal and vertical planes of symmetry. (The restrictions are imposed merely so that we may obtain a simple analytical answer, and they are easily relaxed by transforming the base section into a circle.) The streamwise slope of the body surface at the base dr/dx will not be zero at the body base as in the previous case, but will vary with angular position around the body as shown in Fig. 9-5a. It is interesting to see how the slanting sides contribute to the drag. For this purpose, let us expand dr/dx at the body base in a Fourier series. By virtue of the symmetry properties of the base, we have with reference to Fig. 9-5

$$\left.\frac{dr}{dx}\right|_{x=1} = f_0 + f_2\cos 2\theta + f_4\cos 4\theta + \cdots \qquad (9\text{-}21)$$

where f_0, f_2, etc., are dimensionless. Since ϕ in Eq. (9-18) is for unit free-stream velocity, the radial velocity at $x = 1$ is

$$\frac{\partial\phi}{\partial r} = \frac{dr}{dx} = \frac{a_0}{r_0} - \sum_{n=1}^{\infty}\frac{nA_n\cos n\theta + nB_n\sin n\theta}{r_0^{n+1}} \qquad (9\text{-}22)$$

with the result that

$$a_0 = f_0 r_0 = \frac{S'(1)}{2\pi}$$

$$A_n = 0 \qquad n \text{ odd}$$

$$A_n = -\frac{f_n r_0^{n+1}}{n} \qquad n \text{ even} \tag{9-23}$$

$$b_0(1) = \frac{1}{2\pi}\left[S'(1) \log \frac{\beta}{2} - \int_0^1 \log (1 - \xi) S''(\xi)\, d\xi \right]$$

$$B_n = 0$$

The first two integrals of Eq. (9-16) both contribute to the drag, as well as the contour integral, which becomes

$$\oint_C \phi \frac{\partial \phi}{\partial \nu}\, d\sigma = \int_0^{2\pi} \left(a_0 \log r_0 + b_0 - \sum_{n=1}^{\infty} \frac{f_{2n} r_0^{2n+1} \cos 2n\theta}{2n r_0^{2n}} \right)$$

$$\left(\frac{a_0}{r_0} + \sum_{m=1}^{\infty} f_{2m} \cos 2m\theta \right) r_0\, d\theta \tag{9-24}$$

The result of evaluating the integral is

$$\oint_C \phi \frac{\partial \phi}{\partial \nu}\, d\sigma = 2\pi a_0 (a_0 \log r_0 + b_0) - \pi \sum_{n=1}^{\infty} \frac{f_{2n}^2 r_0^2}{2n} \tag{9-25}$$

The total drag is

$$\frac{D}{q_0} = \frac{1}{2\pi} \int_0^1 \int_0^1 S''(x) S''(\xi) \log \frac{1}{|x - \xi|}\, dx\, d\xi$$

$$+ \frac{S'(1)}{\pi} \int_0^1 \log (1 - \xi) S''(\xi)\, d\xi - \frac{1}{2\pi} [S'(1)]^2 \log \frac{\beta r_0}{2}$$

$$+ \pi \sum_{n=1}^{\infty} \frac{f_{2n}^2 r_0^2}{2n} - P_B S(1) \tag{9-26}$$

Examination of this result is instructive. Let us interpret each term of the drag physically. For this purpose assume that (1) the body has a tangent-cylindrical base, (2) it has atmospheric base pressure, and (3) it is axially symmetric, and then relax the assumptions one by one. With all three assumptions the only term in the drag is the first term, which represents principally the wave-making drag of the head wave. (Since the base pressure is atmospheric and the body pressure is also closely atmospheric at the base, there is no trailing wave within the scope of inviscid fluid theory.) Relax assumption (1) by letting the body have boattail. The second and third terms are not now zero. Because of the boattail, the flow toward the base has an inward radial velocity. This flow must be straightened out approximately into the free-stream direction for atmospheric base pressure by a conical shock wave. The second and third terms represent the drag associated with this trailing shock wave or the boattail drag. Let us now relax assumptions (1) and (2). The flow behind the base will no longer be approximately in the free-stream direction, but will converge toward a point behind the base. The location of this point is determined by a complicated mixing process between the outer potential flow, the discharged boundary-layer air, and the air in the dead water region. The analysis of this problem is the subject of that part of the chapter entitled "Base Drag." In any event, the trailing shock wave is now not dependent on inviscid considerations alone, but has an intensity determined also by the viscous process behind the base. The trailing shock-wave system therefore represents fractions of boattail drag and base drag. Finally, let us relax the assumption of axial symmetry so that the drag represented by the summation is not zero. Actually, this term can be interpreted as a drag due to kinetic energy of the wake being laid down by the body. The flow leaving the base has local inward and outward radial velocity due to the $\cos 2n\theta$ terms for $n = 1$ or greater, which average out to zero around the body. Nevertheless, the kinetic energy being discharged into the wake by these radial velocities is not zero on the average and represents a positive drag.

Case 4: Drag Due to Lift of General Body with Circular Base

By a general body with a circular base we mean one which is also pointed but which otherwise is general within the scope of slender-body theory. The complex potential for such a body with reference to Fig. 9-5b is

$$W(\mathfrak{z}) = a_0 \log (\mathfrak{z} - \mathfrak{z}_0) + b_0 + \sum_{n=1}^{\infty} \frac{a_n}{(\mathfrak{z} - \mathfrak{z}_0)^n} \qquad (9\text{-}27)$$

If the coordinate system is changed from y,z to r,θ with the new origin, the potential will still have the same form since

$$\mathfrak{z} - \mathfrak{z}_0 = re^{i\theta} \qquad (9\text{-}28)$$

Now an inspection of the terms of Eq. (9-16) shows that drag due to lift must appear either in the contour integral around C or in the base drag term. Therefore, ignoring any change in the base pressure, the pressure foredrag due to lift can be evaluated by that part of the contour integral about C due to the angle of attack α at the base. (If the angle of attack at the base is zero, the lift is also zero, independent of the slope of the body in front of the base) Let the potential at the base crossflow plane be comprised of a part for zero angle of attack plus a part due to angle of attack

$$\phi = \phi_0 + \phi_\alpha \tag{9-29}$$

The boundary condition for the potential due to angle of attack is

$$\left.\frac{\partial \phi_\alpha}{\partial r}\right|_{r=r_0} = -\alpha \sin \theta \tag{9-30}$$

so that ϕ_α is of the form

$$\phi_\alpha = \frac{B_1}{r} \sin \alpha$$

The constant B_1 is readily evaluated with the result

$$\phi_\alpha = \frac{\alpha r_0{}^2}{r} \sin \theta \tag{9-31}$$

The total drag due to the contour integral is

$$\frac{D}{q_0} = -\oint_0^{2\pi} \phi \frac{\partial \phi}{\partial r} r_0 \, d\theta = -r_0 \int_0^{2\pi} (\phi_0 + \phi_\alpha) \frac{\partial}{\partial r} (\phi_0 + \phi_\alpha) \, d\theta \tag{9-32}$$

and that part due to angle of attack (or lift) is

$$\left(\frac{D}{q_0}\right)_\alpha = -r_0 \int_0^{2\pi} \left(\phi_0 \frac{\partial \phi_\alpha}{\partial r} + \phi_\alpha \frac{\partial \phi_0}{\partial r} + \phi_\alpha \frac{\partial \phi_\alpha}{\partial r} \right) d\theta$$
$$= \pi r_0{}^2 \alpha^2$$

Since the lift is

$$\frac{L}{q_0} = 2\pi r_0{}^2 \alpha$$

we have

$$\left(\frac{D}{q_0}\right)_\alpha = \frac{1}{2} \frac{L}{q_0} \alpha \tag{9-33}$$

This result, derived in detail here for a body with cylindrical base, was derived in Sec. 3-10 for a slender body with a base of arbitrary shape.

The interesting fact shown by Eq. (9-33) is that the drag due to lift of a slender body is just one-half that for a flat plate. Since the drag due to lift is proportional to the rearward inclination of the resultant force vector from the normal to the stream direction, the resultant force is inclined rearward at angle $\alpha/2$ for a slender body. In this respect of theory, a body of revolution is equivalent to a very low-aspect-ratio triangular wing with full leading-edge suction.

Viscous Crossflow

The pressure foredrag for a slender body at zero angle of attack or at incidence has been calculated on the basis of slender-body theory. Additional drag due to lift can be incurred by a slender body or a nonslender body because of viscous crossflow of the type discussed in Sec. 4-6. This drag acts through pressure forces accompanying boundary-layer separation and vortex formation and is not due to skin-friction forces. Actually, the pressure forces arising as a result of crossflow are the basis of the definition of the crossflow drag coefficient c_{d_c}, as discussed in Sec. 4-6. The drag due to viscous crossflow is taken as the force normal to the body axis due to crossflow times the angle of attack. For a cylinder this relationship is exact. Thus, if S_c is the planform area of the body subject to viscous crossflow and c_{d_c} is the crossflow drag coefficient, then the drag due to viscous crossflow D_c is

$$D_c = c_{d_c} q_0 \alpha^3 S_c \tag{9-34}$$

Viscous crossflow introduces a cubic dependence of the drag on angle attack. Therefore, the drag curve for a body will not be parabolic above the angle of attack for the onset of viscous crossflow.

Another factor acts to change the parabolic shape of the drag curve of a body, namely, changes in transition point with angle of attack. Suppose the transition point is near the body base at zero angle of attack. Increase in angle of attack will cause the transition point to move forward and may induce separation and vortex formation. The change in skin friction with angle of attack will depend on how fast the transition point moves forward, and how vortex formation influences the skin friction in separated flow. Formally, these influences can be considered as changes in the drag-rise factor through the third term in Eq. (9-15).

9-4. Pressure Foredrag of Nonslender Missile Noses at Zero Angle of Attack

In the previous section the emphasis was on missile bodies of high fineness ratio. For zero angle of attack, missile noses of low fineness ratio can be handled with relative ease because of the simple nonlinear theories that have been developed. We will discuss these nonlinear theories in their general aspects since a detailed consideration of the half dozen or so methods available would be unduly lengthy.

One of the early studies of the drag of missile noses at supersonic speeds is that of Taylor and Maccoll,[5] who calculated the pressure coefficients of cones using the full nonlinear potential equation. Extensive tables of flow around cones are available in an MIT report,[4] and convenient charts for cones are to be found in an Ames report.[6] While cone results are of intrinsic value in themselves, perhaps they have still greater value as a standard against which the accuracy of many approximate theories for conical and nonconical noses may be gauged. The pressure field of a cone depends on two independent parameters: the cone semiapex angle and the Mach number. It would be useful if the pressure field depended strongly on some combination of these two parameters as independent variables, and weakly on any other independent variable.   Such a combination of parameters K, called the *hypersonic similarity parameter*, has been advanced by Tsien[9] for slender pointed bodies in high-speed flow. The hypersonic similarity parameter is the ratio of the free-stream Mach number to the body fineness ratio.

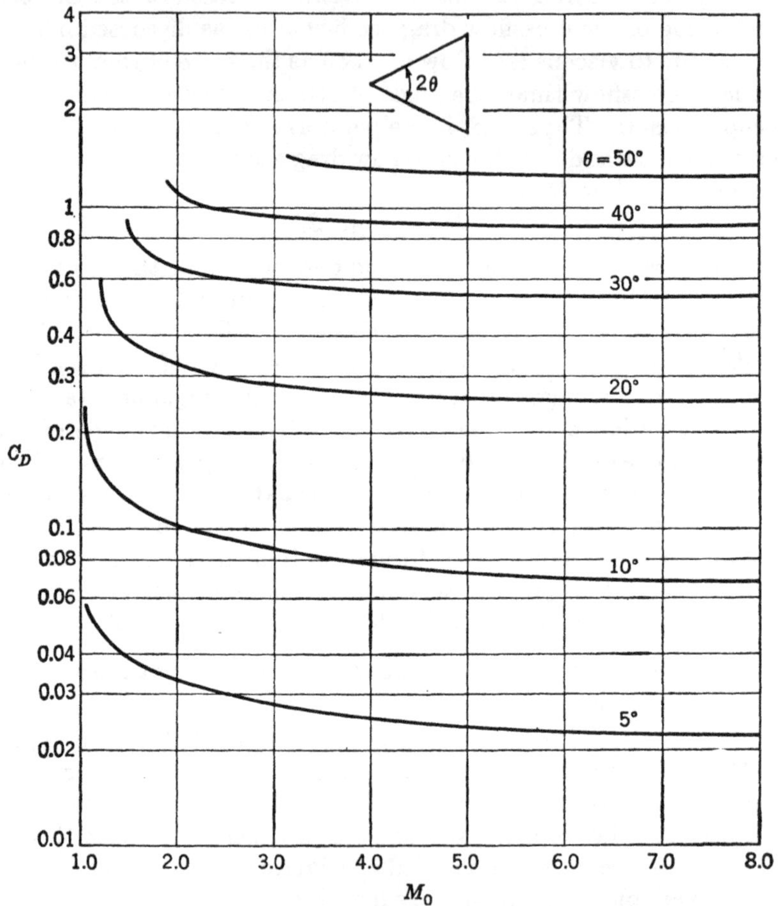

FIG. 9-6. Drag coefficients of cones at supersonic speeds.

The *hypersonic similarity law* applies to a family of bodies which are related one to the other by uniform expansion or contraction of the axial or radial dimensions.   The bodies need not be axially symmetric.   Corresponding points for two such bodies are points which go one into the other when the bodies are brought into coincidence by expansion or contraction.   Let the pressures at two such points be measured by the following ratio $(p - p_0)/p_0$, involving the local static pressure p and the free-stream static pressure p_0.   The hypersonic similarity law then states that the pressure ratios $(p - p_0)/p_0$ at corresponding points for two bodies of the same family differing in fineness ratio will be equal if the Mach numbers for the two bodies are so chosen that the similarity parameter K remains unchanged. Since we are interested in drag, let us see what the implication of the hypersonic similarity law is for drag. For a cone the drag coefficient based on the base area is equal to the usual pressure coefficient

$$C_D = \frac{p - p_0}{p_0}\frac{p_0}{q_0} = \frac{p - p_0}{p_0}\left(\frac{\gamma}{2}M_0{}^2\right)^{-1} \tag{9-35}$$

As a result the similarity law predicts that the parameter $M_0{}^2 C_D$ is a function only of K as the Mach number and fineness ratio are independently

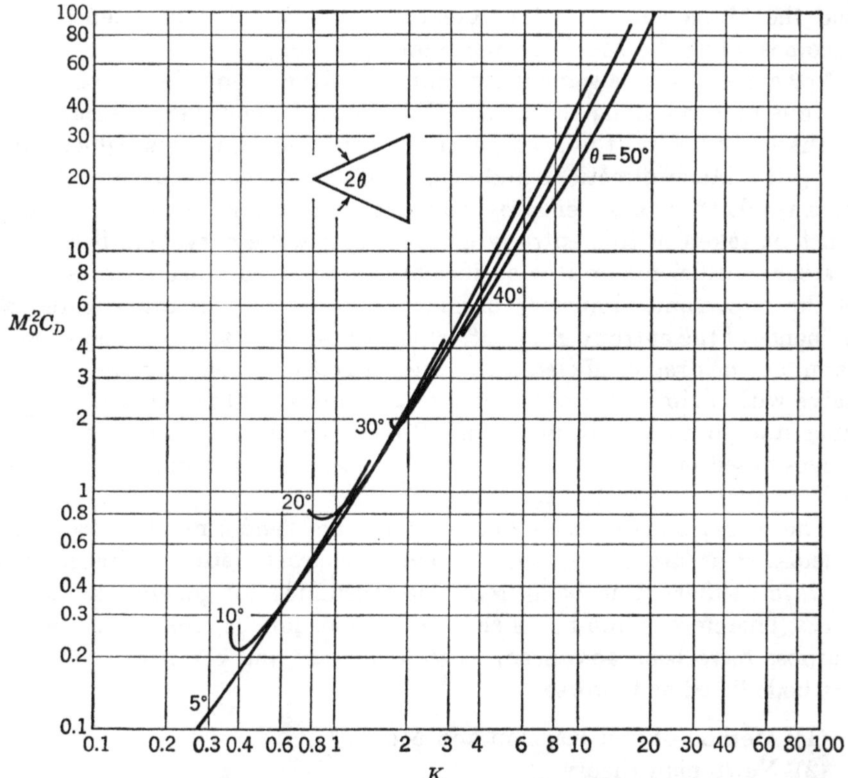

FIG. 9-7. Correlation of drag coefficients of cones at supersonic speeds by hypersonic similarity parameter.

varied. We have a convenient set of data for cones to substantiate this hypothesis.

In Fig. 9-6, the drag coefficients of cones are shown as a function of Mach number M_0 and cone semiapex angle θ. These drag coefficients are correlated on the basis of the hypersonic similarity parameter in Fig. 9-7. For cone angles up to about $\theta = 30°$, the correlation is good for

Mach numbers to 8 or 10 except at the lower ends of the individual curves. As the Mach number approaches the Mach number of shock detachment from above, the individual curve for values of $\theta \leq 30$ turn upward away from the mean correlation curve. For large cone angles, $\theta > 30°$, the hypersonic similarity parameter does not correlate the drag coefficients well. This departure from correlation is associated with approach of the shock wave to the cone surface itself. Ehret, Rossow, and Stevens[7] have studied the problem of correlating ogive as well as cone pressure coefficients on the basis of the hypersonic similarity parameter, and have determined the ranges of Mach number and fineness ratio over which the vertex pressure coefficients can be correlated within ± 5 per cent. As a rough rule of thumb, it can be said that the fineness ratio must be 2 or greater, and the Mach number 1.5 or greater. However, as the fineness ratio becomes large, the Mach number may approach unity.

For cones the nose wave is straight, and the entropy change across the wave is uniform along the wave. For an ogive, however, the curvature of the body behind the apex generates expansion wavelets, which move along the Mach directions and cause the nose wave to curve backward. As a result there is an entropy gradient along the nose wave. Account must be taken of this entropy gradient and wave curvature if accurate pressure coefficients or drag coefficients are to be obtained at large values of the hypersonic similarity parameter. Rossow[8] has investigated the influence of the entropy gradient, which gives rise to the so-called rotation term, on the drag coefficients and pressure coefficients of ogives. For an ogive with a similarity parameter of 2 he finds a 30 per cent increase in drag due to the rotation term. Rossow's drag correlation curve for ogives based on the hypersonic similarity rule is given in Fig. 9-8, where it is compared with that for cones.

The pressure distribution and drag of a nonslender missile nose can be calculated accurately by the method of characteristics. This method, however, suffers from being too time-consuming for general engineering use. Therefore, a number of shorter methods for accomplishing the same purpose have been advanced. Let us discuss and compare the shorter methods listed as follows:

(1) Method of von Kármán and Moore[20]
(2) Newtonian theory
(3) Van Dyke's second-order method[15]
(4) Tangent-cone method 1
(5) Tangent-cone method 2
(6) Conical shock-expansion theory[19]

The method of von Kármán and Moore is one of linearized theory for bodies of revolution. It is based on a step-by-step numerical determination of the source distribution along an axis necessary to shape the body.

The second-order theory of Van Dyke has been developed to the point of a calculative technique using tables and a calculating form. The tangent-cone methods are rules of thumb: Method 1 simply states that the pressure coefficient at any point on a body of revolution corresponds to that of a cone having a semiapex angle equal to the angle between the body axis and the tangent to the body at the point. Method 2 is slightly more sophisticated than method 1; it assumes that the local Mach number is

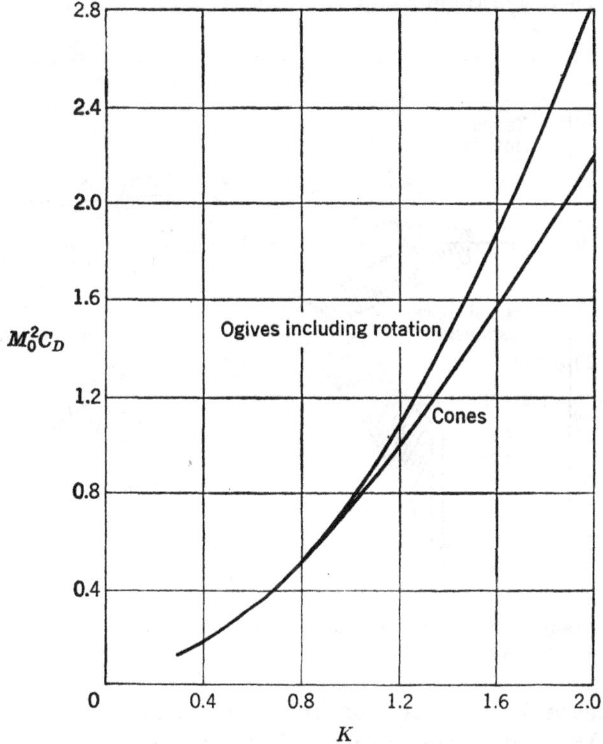

FIG. 9-8. Correlation curves for drag of cones and tangent ogives on basis of hypersonic similarity parameter.

given by the tangent cone. This local Mach number is then used, together with the known loss in stagnation pressure through the shock wave at the apex, to establish the local static pressure. The conical shock-expansion theory is a calculative method developed by Eggers and Savin[19] for large values of the hypersonic similarity parameter. With the exception of Newtonian theory, the foregoing methods apply principally to bodies of revolution, although several of the other methods can formally be applied to other bodies.

A comparative study of the accuracy of the foregoing methods has been made by Ehret.[14] The accuracy of the methods was assessed by comparing the predictions of the approximate theories with the accurate calculations by the method of characteristics for the pressure drag of cones, ogives, and a Sears-Haack body with a pointed nose. The general results of the study are summarized in Fig. 9-9. In the first place it is seen that the Kármán and Moore theory applies at values of the similarity parameters below unity, as would be expected for a linearized theory. The error of Van Dyke's second-order theory also increases as the similarity parameter increases, but the error is generally only about one-third that of the linearized theory. In contrast to these two methods, Newtonian theory, tangent-cone method 1, and conical shock-expansion

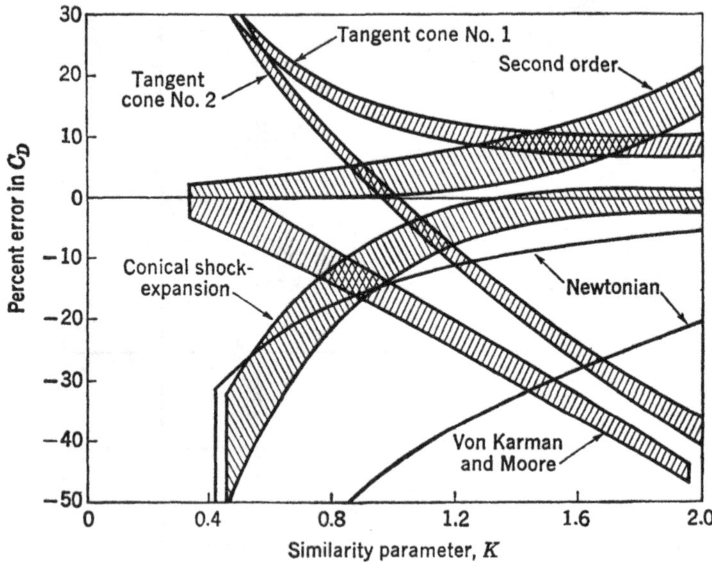

FIG. 9-9. Accuracy of various methods for estimating pressure drag of nonslender missile noses at zero angle of attack.

theory increase in accuracy as the hypersonic similarity parameter increases. It is interesting to note that tangent-cone method 2, which appears more sophisticated than method 1, is more accurate only for similarity parameters less than about 1.2. These results of Ehret serve as a good guide to the choice of a method for the calculation of the drag of a nonslender missile nose in any particular case.

9-5. Shape of Bodies of Revolution for Least Pressure Foredrag at Zero Angle of Attack

We have concerned ourselves at some length with the direct problem of finding the pressure foredrag of a missile nose of prescribed shape. Considerations of aerodynamic efficiency require solutions to the indirect problem of finding the shape of the body for least pressure foredrag for certain prescribed constraints such as fixed length, fixed volume, fixed base area, fixed angle of attack, etc. Bodies of revolution of least pressure foredrag include such bodies as the *Kármán ogive*, the *Sears-Haack body*, *Newtonian bodies*, etc. It is interesting that problems of least pressure foredrag of bodies of revolution are much older than the airplane and were, in fact, studied by Newton himself.[10] Furthermore, such problems have long been popular with mathematicians, like Todhunter, skilled in the calculus of variations.[11]

Such bodies of least pressure foredrag as the above-named bodies are frequently termed *bodies of minimum wave drag*. An understanding of this term is predicated on two considerations. First, the use of the adjective *minimum* in this connection is not to be confused with the use of the adjective *minimum* in reference to C_{D_0}, the minimum drag, as shown in Fig. 9-3. Second, the wave drag is equivalent to the pressure foredrag, which is in actuality minimized, only under special circumstances. These circumstances are that the fluid be inviscid and that the base pressure be free-stream pressure. This equivalence of pressure foredrag and wave drag is discussed in Sec. 9-3 for slender bodies. For these reasons we shall term so-called *bodies of minimum wave drag*, such as the Kármán ogive, *bodies of least pressure foredrag*.

The Sears-Haack body and the Kármán ogive are bodies of least pressure foredrag derivable on the basis of slender-body theory. The method we will use to derive the bodies is one mentioned by von Kármán.[3] It is based on an analogy between the computation of the induced drag of a lifting line of arbitrary span loading and the pressure foredrag of a slender body at zero lift with an arbitrary distribution of area along its length. Consider now slender bodies of the types considered in cases 1 and 2 in Sec. 9-3. The pressure foredrag of such bodies is given by

$$\frac{D}{q_0} = -\frac{1}{2\pi} \int_0^l \int_0^l S''(x)S''(\xi) \log |x - \xi| \, d\xi \, dx \tag{9-36}$$

We are now taking the body to have length l rather than unit length. Since the bodies have pointed noses, and either pointed or cylindrical bases,

$$S'(0) = S(0) = S'(l) = 0 \tag{9-37}$$

The variables x and ξ are changed to θ and ϕ.

$$\frac{x}{l} = \tfrac{1}{2}(1 + \cos \theta)$$

$$\frac{\xi}{l} = \tfrac{1}{2}(1 + \cos \phi) \tag{9-38}$$

The values of θ and ϕ equal to zero refer to the base of the missile as shown in Fig. 9-10, whereas the values of π correspond to the pointed nose.

Our first objective is to obtain an expression for the drag integral of Eq. (9-36) in terms of certain Fourier coefficients which specify the *area distribution* of a slender body. By the area distribution is meant the variation of the body cross-sectional area along the length of its axis. The area distribution, or rather its axial derivative, can be expanded in a sine series convergent in the intervals $0 \leq \theta \leq \pi$ and $0 \leq \phi \leq \pi$

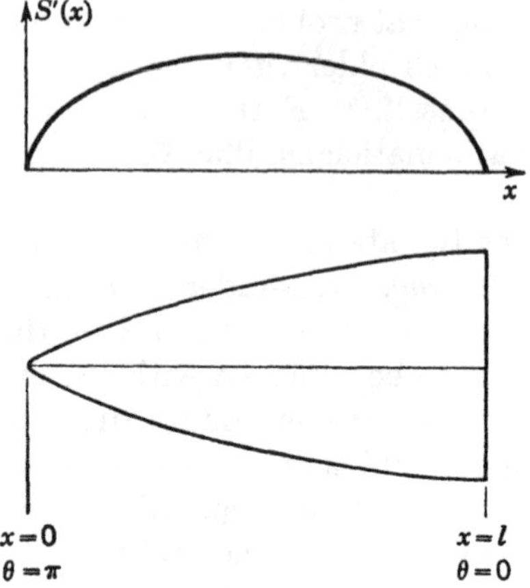

Fig. 9-10. Notation for use in minimizing pressure foredrag of slender bodies at zero angle of attack.

$$S'(x) = \pi l \sum_{n=1}^{\infty} b_n \sin n\theta$$

$$S'(\xi) = \pi l \sum_{n=1}^{\infty} b_n \sin n\phi \tag{9-39}$$

The first integration to obtain the drag is denoted $g(x)$

$$g(x) = \int_0^l S''(\xi) \log |x - \xi| \, d\xi = \pi^2 l \sum_{n=1}^{\infty} b_n \cos n\theta \tag{9-40}$$

The drag can now be evaluated.

$$\frac{D}{q_0} = \frac{\pi^2 l^2}{2} \int_0^\pi \left(\sum_{m=1}^{\infty} m b_m \sin m\theta \right) \left(\sum_{n=1}^{\infty} b_n \cos n\theta \right) d\theta$$

$$= \frac{\pi^3 l^2}{4} \sum_{n=1}^{\infty} n b_n^2 \tag{9-41}$$

Thus, we have a simple result for the pressure foredrag of a slender body subject to the conditions of Eq. (9-37). Its similarity to the formula for induced drag at subsonic speeds is apparent.

The next objective is to determine the body cross-sectional area distribution and the body volume in terms of the values of b_n, preparatory to determining bodies of least pressure foredrag. The area distribution is obtained from an integration of Eq. (9-39) subject to the conditions of Eq. (9-37). The resulting area distribution is

$$S(\theta) = \frac{\pi l^2}{4} \left\{ b_1 \left(\pi - \theta + \frac{\sin 2\theta}{2} \right) + \sum_{n=2}^{\infty} b_n \left[\frac{\sin (n+1)\theta}{n+1} - \frac{\sin (n-1)\theta}{n-1} \right] \right\} \tag{9-42}$$

and the body volume is

$$\text{Vol.} = \frac{\pi^2 l^3}{8}\left(b_1 - \frac{b_2}{2}\right) \tag{9-43}$$

Bodies of minimum wave drag can now be formed for certain restrictions on length, area, and volume by joint consideration of Eqs. (9-41), (9-42), and (9-43). This is the approach used by Sears[18] in his derivation of the Sears-Haack body independently derived by Haack.[17] If we desire to have a body of nonzero volume we must keep either b_1 or b_2 nonzero. Let us explore the case of $b_2 \neq 0$ but all other b_n equal to zero. We have then

$$\begin{aligned}
b_2 &= -\frac{16\ \text{Vol.}}{\pi^2 l^3} \\[2mm]
S(\theta) &= \frac{16\ \text{Vol. } \sin^3 \theta}{3\pi l} \\[2mm]
S(x) &= \frac{16\ \text{Vol.}}{3\pi l}\left[1 - \left(1 - \frac{2x}{l}\right)^2\right]^{3/2} \\[2mm]
S(l) &= \left(\frac{\pi l}{2}\right)^2 b_1 = 0
\end{aligned} \tag{9-44}$$

This body is symmetrical about the midpoint of its axis, being pointed at both ends. This is the *Sears-Haack body*, which is the body of least pressure foredrag (drag) for zero base area and a given length and volume. Since we have specified zero base area, we must have b_1 equal to zero. To have a body with any volume we must have a nonzero value of b_2 given by Eq. (9-44). The values of b_1 and b_2 are thus uniquely fixed by the prescribed conditions. The only other question that now arises is whether inclusion of any other of the b_n terms can reduce the drag. Equation (9-41) answers this question firmly in the negative. As a result the Sears-Haack body is the one and only body for least pressure foredrag under the prescribed conditions. Its drag is simply

$$\frac{D}{q_0} = \frac{128\ \text{Vol.}^2}{\pi l^4} \tag{9-45}$$

and the drag coefficient based on the maximum cross-section area is

$$C_D = \frac{24\ \text{Vol.}}{l^3} \tag{9-46}$$

This simple result is typical of solutions to problems of least pressure foredrag. It indicates the desirability of spreading the volume over as long an axial distance as possible. The area distribution of the Sears-Haack body is given in Table 9-1.

TABLE 9-1. COORDINATES OF BODIES OF LEAST WAVE DRAG

	r/a				
x/l	Sears-Haack*	Kármán ogive†	Three-quarter-power body	Newtonian bodies‡	
				$l/a = 3$	$l/a = 5$
0	0	0	0	0.0073	0.00165
0.02	0.089	0.069	0.053	0.060	0.055
0.04	0.148	0.116	0.089	0.099	0.091
0.06	0.199	0.156	0.121	0.129	0.123
0.08	0.245	0.194	0.150	0.159	0.153
0.10	0.288	0.228	0.178	0.186	0.181
0.20	0.465	0.377	0.299	0.305	0.300
0.30	0.609	0.502	0.405	0.412	0.407
0.40	0.715	0.611	0.503	0.509	0.505
0.50	0.806	0.707	0.595	0.599	0.596
0.60	0.877	0.791	0.682	0.685	0.682
0.70	0.932	0.865	0.765	0.767	0.765
0.80	0.970	0.926	0.846	0.847	0.846
0.90	0.992	0.974	0.924	0.925	0.924
1.00	1.000	1.000	1.000	1.000	1.000

* Given volume and length $2l$; maximum radius a.

† Given base radius a and length l; tangent-cylindrical base.

‡ Given base radius a and length l.

The second body of least pressure foredrag is obtained by letting b_1 be nonzero and all other values of b_n be zero. The various quantities then turn out to be

$$b_1 = \frac{4S(l)}{\pi^2 l^2} = \frac{8\,\text{Vol.}}{\pi^2 l^3}$$

$$S(\theta) = \frac{S(l)}{\pi}\left(\pi - \theta + \tfrac{1}{2}\sin 2\theta\right) \tag{9-47}$$

This body is the *Kármán ogive*, which has the least pressure foredrag for a given length and a cylindrical base of given area. Because the base area and length are prescribed, the value of b_1 is uniquely determined. If b_2 were not zero, the drag would be increased by Eq. (9-41) and the volume changed by Eq. (9-43). Any other nonzero values of b_n would increase the drag without changing the volume. The drag of the body is

$$\frac{D}{q_0} = \frac{16\,\text{Vol.}^2}{\pi l^4} = \frac{4S^2(l)}{\pi l^2} \tag{9-48}$$

and the drag coefficient based on the base area is

$$C_D = \frac{8\,\text{Vol.}}{\pi l^3} = \frac{4S(l)}{\pi l^2} \tag{9-49}$$

A comparison of Eqs. (9-45) and (9-48) reveals that the pressure foredrag of a Kármán ogive is only one-eighth the pressure foredrag of the Sears-Haack body of the same volume and length. This large difference between the two bodies is in part counteracted by the base drag of the Kármán ogive. The area distribution of the Kármán ogive is given in Table 9-1.

Illustrative Example

Compare the pressure foredrag of a Kármán ogive of 5 calibers with the Sears-Haack body of comparable length and volume. If the base-pressure coefficient of the Kármán ogive is -0.2, how do the total pressure drags compare?

For a Kármán ogive of 5 calibers, the base radius is 1 if the length is 10, so that $S(l) = \pi$. By Eq. (9-48) the pressure foredrag is

$$\frac{D}{q_0} = \frac{4\pi^2}{\pi(10)^2} = 0.126$$

and the pressure foredrag coefficient based on the base area by Eq. (9-49) is

$$C_D = \frac{4\pi}{\pi(10)^2} = 0.04$$

The volume from Eq. (9-47) is

$$\text{Vol.} = \frac{10\pi}{2} = 5\pi$$

For a Sears-Haack body of length 10 and volume 5π, Eq. (9-45) gives the pressure foredrag as

$$\frac{D}{q_0} = \frac{128(5\pi)^2}{\pi(10)^4} = 1.005$$

The pressure foredrag of the Kármán ogive is thus one-eighth that of the Sears-Haack body, as previously mentioned. Now for a base drag coefficient of -0.2 the Kármán ogive has five times as much base drag as pressure foredrag. Therefore, the total pressure drag for the Kármán ogive is three-quarters that for the Sears-Haack body.

While bodies of least pressure foredrag are readily derivable on the basis of slender-body theory, the question arises whether similar bodies cannot be derived on the basis of other theories. Actually, such bodies can be found on the basis of Newtonian impact theory, which gives the following simple expression for the local pressure coefficient:

$$P = 2 \sin^2 \delta \tag{9-50}$$

Here δ is the angle between the tangent to the body in the streamwise direction and the streamwise direction itself. Eggers, Dennis, and

Resnikoff[12] have studied Newtonian bodies of least pressure foredrag, and the reader is referred to their paper as well as the book of Todhunter[11] for the mathematical details. The integrals of drag, volume, and surface area of bodies of revolution can easily be evaluated in terms of the equation for the shape of the body. Subject to restraints on base area, length, volume, or surface area, and subject to certain other mathematical conditions, the drag integral can be minimized by the calculus of variations to yield the shape for least pressure foredrag. Five bodies are given by Eggers et al.[12] for different combinations of restraints.

Of the various Newtonian shapes the one of particular interest here is that for prescribed body length and base area, since it is directly comparable to the Kármán ogive. (We could just as well have specified length and base area for the ogive.) Actually, the Newtonian body for given length and base area is flat-nosed. Its shape is given in parametric

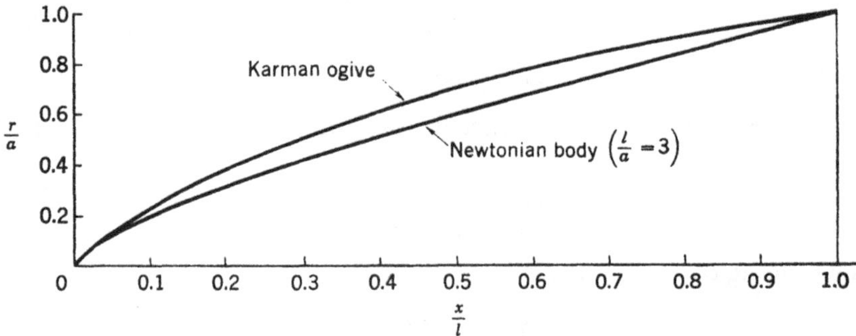

Fig. 9-11. Comparison of shape of Kármán ogive with that of Newtonian body of least pressure foredrag.

form by Eggers et al.,[12] and the shape coordinates are given in Table 9-1 for two fineness ratios. The bluntness for a fineness ratio of 5 is 0.16 per cent of the base diameter, and for a fineness ratio of 3 the bluntness is 0.73 per cent. The bluntness of the nose is judged to have only a small local aerodynamic effect on pressure drag, as long as the fineness ratio does not get much below 3. The actual shape of the Newtonian bodies is approximated closely enough for most purposes by a three-quarter power body.

$$\frac{r}{a} = \left(\frac{x}{l}\right)^{n} \qquad n = \tfrac{3}{4} \tag{9-51}$$

The shape ordinates for the three-quarter power body are also given in Table 9-1, where they can be compared with those of the Newtonian bodies to show the closeness of fit.

A comparison of the Newtonian body of least pressure drag for a given length and base area and the Kármán ogive is made in Fig. 9-11. First, it should be noted that the condition $S'(l) = 0$ used in deriving the Kármán body was not involved for the Newtonian body. As a result the Kármán ogive is somewhat fuller than the Newtonian body. Nevertheless, the two shapes are not greatly different from each other. This result tends to suggest that the shape for least pressure foredrag for prescribed conditions may not be sensitive to the actual physical law used to obtain the pressure coefficient. As a consequence, a shape found to be optimum on the basis of a particular physical law might be expected to be nearly optimum under aerodynamic conditions where the physical law is known to be grossly inadequate.

The question naturally arises whether the pressure foredrag calculated by Newtonian theory is accurate, assuming that the shape is indeed optimum. It has been found[12] that Newtonian impact theory generally gives pressure foredrags which are too low when compared to the experimental foredrags corrected for skin friction. In lieu of accurate absolute drags, it might be asked whether Newtonian impact theory predicts accurately the ratio of the drag of a Newtonian body to that of a cone. This question was investigated by Eggers et al.[12] for bodies of the type given by Eq. (9-51). For $n = 0.75$ it was found that the ratio is indeed accurately predicted. For $n = 0.6$ the error appears to be about 10 per cent in the ratio, and increases rapidly as n decreases further. As a rule of thumb, one would compute the ratio by impact theory and multiply the ratio by the known pressure foredrag coefficient of a cone to get accurate pressure foredrag coefficients for Newtonian bodies of prescribed length and base area. For the low values of n some improvement in prediction of pressure foredrag coefficient can be achieved by the special methods of Eggers et al.[12]

The final question we consider in the comparison of the Newtonian body and the Kármán ogive is: Which has the lower drag? Jorgensen[13] has used Van Dyke's second-order theory to compute the drag of the Newtonian body and the Kármán ogive for a fineness ratio of 3 for $M_0 = 1.5$, 2.0, and 3.0. It might be expected that a particular body would show lower drag in that region where its theoretical basis is known to be superior. Actually, the Newtonian shape exhibited generally lower calculated drags, but the differences appear not to be significant. Jorgensen also proposes some empirical shapes which have slightly lower calculated pressure foredrag than either the Kármán or Newtonian shapes.

9-6. Pressure Drag of Wings Alone

An extensive literature has been built up on the subject of the pressure drag of wings at supersonic speeds, mainly on the basis of supersonic wing theory. The pressure drag of a wing alone at supersonic speeds can be considered to be the result of *thickness drag* and *camber drag* that occur at zero lift and of drag due to lift. To obtain an insight into these various components of the total pressure drag of a wing, it is useful to examine

the two-dimensional pressure drag of airfoils. For this purpose let us
imagine an airfoil with camber and thickness distribution at zero angle of

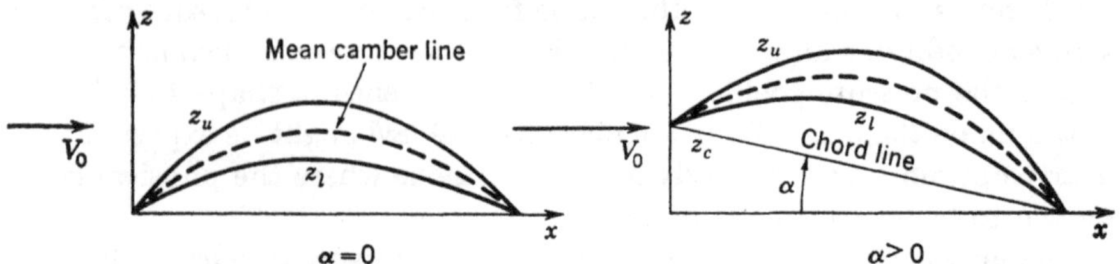

Fig. 9-12. Notation used in specifying thickness and camber distributions.

attack as shown in Fig. 9-12. The chord is the line joining the leading
and trailing edges. The thickness distribution is

$$t(x) = z_u - z_l \tag{9-52}$$

and the camber distribution is

$$\bar{z}(x) = \frac{z_u + z_l}{2} \qquad \alpha = 0 \tag{9-53}$$

At angle of attack the camber distribution is

$$\bar{z}(x) = \frac{z_u + z_l}{2} - z_c \tag{9-54}$$

For combined effects of angle of attack, camber, and thickness, the
coordinates of the upper and lower airfoil surfaces are given by

$$z_u = z_c + \bar{z} + \frac{t}{2}$$
$$z_l = z_c + \bar{z} - \frac{t}{2} \tag{9-55}$$

According to two-dimensional supersonic airfoil theory the pressure coefficients on the upper surface P_u and the lower surface P_l are

$$P_u = \frac{2(dz_u/dx)}{(M_0{}^2 - 1)^{1/2}} = \frac{2}{(M_0{}^2 - 1)^{1/2}}\left(\frac{dz_c}{dx} + \frac{d\bar{z}}{dx} + \frac{1}{2}\frac{dt}{dx}\right)$$
$$P_l = \frac{-2(dz_l/dx)}{(M_0{}^2 - 1)^{1/2}} = \frac{-2}{(M_0{}^2 - 1)^{1/2}}\left(\frac{dz_c}{dx} + \frac{d\bar{z}}{dx} - \frac{1}{2}\frac{dt}{dx}\right) \tag{9-56}$$

and the increments in drag for the top and bottom surfaces are

$$dD_u = P_u\frac{dz_u}{dx}q_0\,dx$$
$$dD_l = -P_l\frac{dz}{dx}q_0\,dx \tag{9-57}$$

The total drag per unit chordwise length is

$$\frac{d(D_u + D_l)}{dx} = \frac{4q_0}{(M_0{}^2 - 1)^{1/2}}\left[\left(\frac{dz_c}{dx} + \frac{d\bar{z}}{dx}\right)^2 + \frac{1}{4}\left(\frac{dt}{dx}\right)^2\right] \qquad (9\text{-}58)$$

At zero angle of attack the drag per unit chord is

$$\frac{d(D_u + D_l)}{dx} = \frac{4q_0}{(M_0{}^2 - 1)^{1/2}}\left[\left(\frac{d\bar{z}}{dx}\right)^2 + \frac{1}{4}\left(\frac{dt}{dx}\right)^2\right] \qquad (9\text{-}59)$$

Illustrative example

Determine the thickness and camber drags of a double-wedge airfoil of maximum thickness t_m at the mid-chord if one side is flat. Compare with the drag of a symmetrical double-wedge airfoil with the same thickness distribution.

With reference to Fig. 9-13 the values of dt/dx and $d\bar{z}/dx$ for the flat-side double-wedge airfoil are

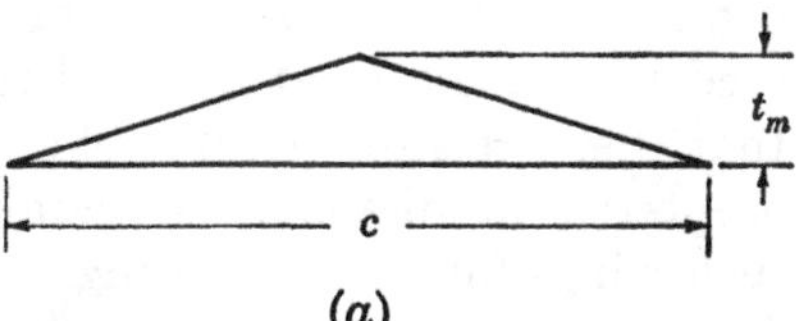

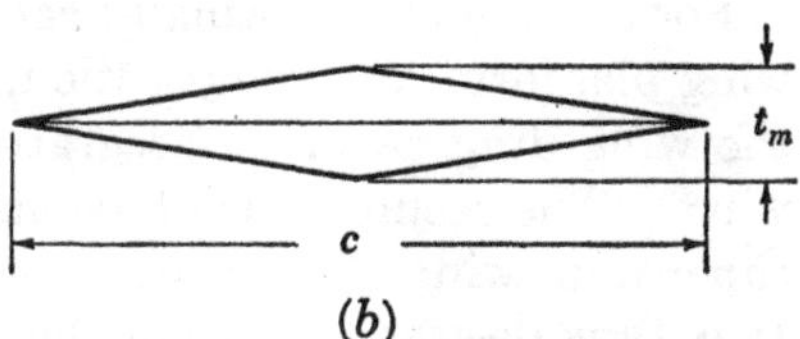

FIG. 9-13. (a) Cambered and (b) un-cambered double-wedge airfoils of identical thickness distribution.

$$\frac{dt}{dx} = \frac{2t_m}{c} \qquad 0 \leq x \leq \frac{c}{2}$$

$$= \frac{-2t_m}{c} \qquad \frac{c}{2} < x \leq c$$

$$\frac{d\bar{z}}{dx} = \frac{1}{2}\frac{dt}{dx} \qquad 0 \leq x \leq c$$

We can thus write from Eq. (9-59)

$$\frac{D_u + D_l}{q_0 c} = c_d = \frac{4}{(M_0{}^2 - 1)^{1/2}}\left[\left(\frac{d\bar{z}}{dx}\right)^2 + \frac{1}{4}\left(\frac{dt}{dx}\right)^2\right]$$

The components of c_d due to camber and thickness are thus

$$c_{d_{\text{cam}}} = \frac{4}{(M_0{}^2 - 1)^{1/2}}\left(\frac{t_m}{c}\right)^2 \qquad \text{due to camber}$$

$$c_{d_t} = \frac{4}{(M_0{}^2 - 1)^{1/2}}\left(\frac{t_m}{c}\right)^2 \qquad \text{due to thickness}$$

Thus, both camber and thickness cause equal increments in drag at zero angle of attack.

For the symmetrical double-wedge airfoil the thickness distribution is the same as for the flat-bottom airfoil, so that its thickness drag coefficient is the same. However, its camber drag is zero. As a result, the use of camber in this particular instance has doubled the drag at zero angle of attack of the airfoil.

The foregoing consideration of the components of the drag of a two-dimensional supersonic airfoil leads to results that are true also for wings. First, the introduction of camber leads to additional drag at zero lift or angle of attack.   In fact, any departure from a horizontal plane of symmetry produces the same result.   However, what is also important is the "coupling" between the drags due to thickness, camber, and angle of attack.   Let us rewrite Eq. (9-58) as

$$\frac{d}{dx}(D_u + D_l) = \frac{4q_0}{(M_0{}^2 - 1)^{1/2}}\left[\alpha^2 - 2\alpha\frac{d\bar{z}}{dx} + \left(\frac{d\bar{z}}{dx}\right)^2 + \frac{1}{4}\left(\frac{dt}{dx}\right)^2\right] \quad (9\text{-}60)$$

It is noted that the thickness distribution produces a drag of its own, independent of the camber and angle of attack.   However, the drag associated with angle of attack and camber are not independent in the sense that they are superposable like the drag due to thickness.   Many wings of interest in missile aerodynamics are symmetrical, and we will consider such wings for the time being.   Later we will return to camber when we discuss cambering and twisting of wings to reduce drag due to lift.

For a symmetrical wing at zero lift the drag coefficient depends on the wing planform, the wing section, and the Mach number.   Calculation of the wing drag requires integration of the pressure distribution over the wing.   The resultant expressions for the drag coefficient on the basis of supersonic wing theory are usually unwieldy, frequently filling a page. It is thus desirable to have the drag coefficients made up in chart form for easy use.   References 22 to 25 are examples of papers containing such charts for a wide range of wing planform, and many others exist.   To reduce the number of charts, it is usual to present the drag results in generalized form.   Consider, for instance, the frequent case of a wing with straight leading and trailing edges, streamwise tips, and a uniform wing section.   The drag results for such a wing can be presented in a form generalized for Mach number, thickness ratio, and planform by giving $\beta C_D/\tau^2$ as a function of βA, λ, and β ctn Λ_{le}.   The symbol τ denotes the thickness ratio of the wing section.   Other sets of geometric parameters besides the above three can be used to specify the planform.   An example of charts of the thickness drag of symmetrical wings is shown in Fig. 9-14 as taken from Puckett and Stewart.[25]   Although drag charts are available for a large number of wings, the range of wing planforms and sections of possible interest is much larger.   Some progress has been made toward developing rapid computing schemes for calculating the thickness drag of wings of arbitrary section.   The work of Grant and Cooper,[21] for instance, permits a rapid determination of thickness drag for arrow wings of arbitrary section for a wide range of leading-edge and trailing-edge sweep angles.   It should be mentioned that the pressure drag of a symmetrical wing at zero angle of attack is all wave drag on the basis of supersonic wing theory. Such pressure drag is therefore frequently termed the *minimum wave drag* of the wing.

Since the pressure drag of wings alone has been widely studied by supersonic wing theory, it is of interest to know how well such drag estimates agree with experiment. A decisive comparison between theory and experiment is not usually possible from force measurements, because the experimental drag coefficients contain an amount of skin friction which must be estimated. If the location of transition is accurately known or if transition is fixed by a device of known drag, then the skin friction can

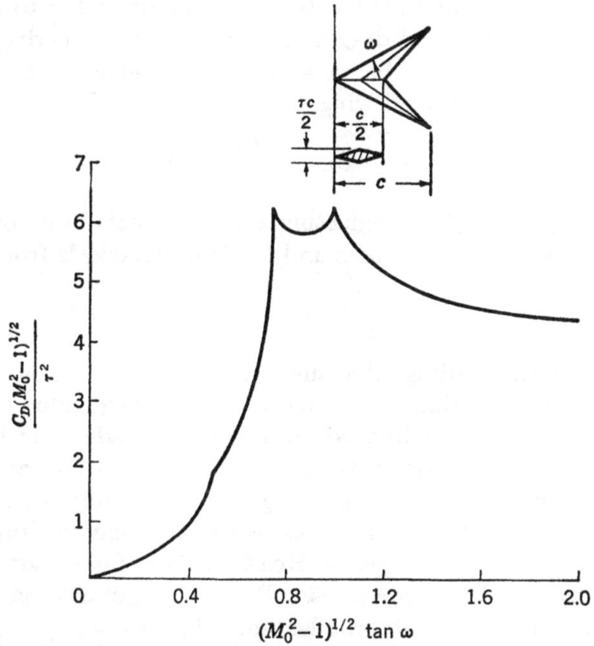

FIG. 9-14. Pressure drag at zero angle of attack for arrow wing on basis of supersonic wing theory.

usually be estimated with fair accuracy. Katzen and Kaattari[26] have made a systematic comparison between the measured and theoretical drag coefficients at zero lift of a series of triangular wings with double-wedge airfoil sections with the maximum thickness of 8 per cent at the midchord. Their calculated drags, obtained by adding the estimated skin friction to the pressure drag from supersonic wing theory, were generally greater than the experimental drags. There is reason to suspect that the drag estimated by supersonic wing theory might be high under certain conditions. An examination of the drag curves of Fig. 9-14 reveals a cusped peak in the drag curve when the leading edge, the line of maximum thickness, or the trailing edge becomes sonic. For the condition of a sonic leading edge, the shock would not be attached as assumed in supersonic wing theory, so that the theory is not valid in the neighborhood of the cusped peaks. It is also physically improbable that the experimental drag data would attempt to form a cusped peak; the data of Katzen and Kaattari exhibit no such tendency. Thus, while the experimental and theoretical drags may be in good accord away from the cusps, the experiment should pass beneath the cusps themselves. The charts of Puckett and Stewart[25] show particularly high and sharp cusps for arrow wings of double-wedge section with the maximum thickness well forward. The theory would be particularly suspect in the region of such cusps.

Let us now turn from pressure drag at zero lift to the drag due to lift for a wing. As we have already discussed, the drag due to lift contains significant amounts of wave drag and vortex drag. The drag due to lift is specified by the drag-rise factor and the lift coefficient as discussed in Sec. 9-2. For a symmetrical wing

$$C_D - C_{D_0} = \frac{C_{D_i}}{\Delta C_L^2} C_L^2 \tag{9-61}$$

where the drag-rise factor, neglecting viscous effects and coupling pressures due to combining thickness and angle of attack, is from Eq. (9-15)

$$\frac{C_{D_i}}{\Delta C_L^2} = \frac{1 - \mu}{C_{L_\alpha}} \tag{9-62}$$

The factor μ is the leading-edge suction factor which measures the per cent reduction in drag due to lift below the flat-plate value of αC_L. For wings with supersonic leading edges, the leading-edge suction factor is zero because of the impossibility of leading-edge suction at least in the mathematical theory. Thus, the drag due to lift follows directly from Eqs. (9-61) and (9-62). For wings with subsonic leading edges the mathematical theory, described by Heaslet and Lomax,[27] gives a leading-edge thrust from suction pressures. For a triangular wing the leading-edge suction factor μ for subsonic leading edges is

$$\mu = \frac{(1 - \beta^2 \tan^2 \omega)^{1/2}}{2E} \tag{9-63}$$

where ω is the semiapex angle, and E is the complete elliptic integral of the second kind with modulus $(1 - \beta^2 \tan^2 \omega)^{1/2}$. The foregoing result can also be used to obtain μ for arrow wings with supersonic trailing edges. In such cases the leading edge of the arrow wing does not "know" what the sweep of the supersonic trailing edge is. As a result we can calculate the leading-edge thrust as though the wing were triangular. The physical force will be unchanged by sweeping the trailing edge as long as it remains supersonic. The only question is one of changes in reference area.

While the leading-edge thrust has a definite value in the mathematical theory, the physical realization of the thrust depends on the shape of the leading edge.   Sharp leading edges which cause flow separation cause a loss of significant fractions of the leading-edge thrust.   If rounding the leading edge will delay separation, some increase in leading-edge thrust can be expected.   If separation is the result of a large upwash angle at the leading edge, such as exists at the tips of sweptback wings at high angles of attack, then the use of camber to turn the nose into the upflow can increase the leading-edge thrust.

Illustrative Example

Determine the values of C_{D_0}, μ, $C_{D_i}/\Delta C_L^2$, $(L/D)_{\max}$, and $C_{L_{opt}}$ for a triangular wing with a double-wedge section having its 8 per cent thickness at the midchord.   Let the wing aspect ratio be 2, the Mach number 1.5, and let the average skin-friction coefficient for the wing be 0.002.

The pressure drag of the wing at $\alpha = 0$ can be obtained directly from the charts of Puckett[22] in the form $\beta C_D/\tau^2 = 4.2$.   Since the skin friction acts on both sides of the wing, the minimum drag coefficient is

$$C_{D_0} = 0.0240 + 2(0.002) = 0.028$$

To obtain the drag-rise factor from Eqs. (9-62) and (9-63), we require the lift-curve slope which for a triangular wing is

$$C_{L_\alpha} = \frac{2\pi \tan \omega}{E(1 - \beta^2 \tan^2 \omega)^{1/2}}$$

where ω = wing semiapex angle
E = elliptic integral of second kind of modulus $(1 - \beta^2 \tan^2 \omega)^{1/2}$
$\beta \tan \omega = 1.119(1/2) = 0.559$
$\sin^{-1} (1 - 0.559^2)^{1/2} = 56°$
$E = 1.249$

Thus

$$C_{L_\alpha} = \frac{2\pi(0.5)}{1.249} = 2.52 \text{ per radian}$$

$$\mu = \frac{(1 - \beta^2 \tan^2 \omega)^{1/2}}{2E} = \frac{(1 - 0.559^2)^{1/2}}{1.249} = 0.332$$

$$\frac{C_{D_i}}{\Delta C_L^2} = \frac{1 - \mu}{C_{L_\alpha}} = \frac{1 - 0.332}{2.52} = 0.265$$

The lift-drag ratio and optimum lift coefficient from Eqs. (9-9) and (9-10) then are

$$(C_L)_{opt} = \left(\frac{C_{D_0}}{C_{D_i}/\Delta C_L^2}\right)^{1/2} = 0.325$$

$$\left(\frac{L}{D}\right)_{\max} = \frac{1}{2}\left(C_{D_0}\frac{C_{D_i}}{\Delta C_L^2}\right)^{-1/2} = 5.8$$

9-7. Pressure Foredrag of Wing-Body Combinations of Given Shape at Zero Angle of Attack

We have essentially a problem of wing-body interference in trying to calculate the pressure foredrag of wing-body combinations of given shape. For slender wing-body combinations there is the drag formula of Ward, which in practice does not differentiate significantly between bodies and wing-body combinations, as we will see in greater detail in the next section. For certain nonslender configurations with the panels mounted on quasi-cylindrical body sections, there are methods exact to the order of linear theory. By a quasicylindrical body section we mean a body section that is closely cylindrical. We will later be concerned with body sections which lie close to circular cylinders.

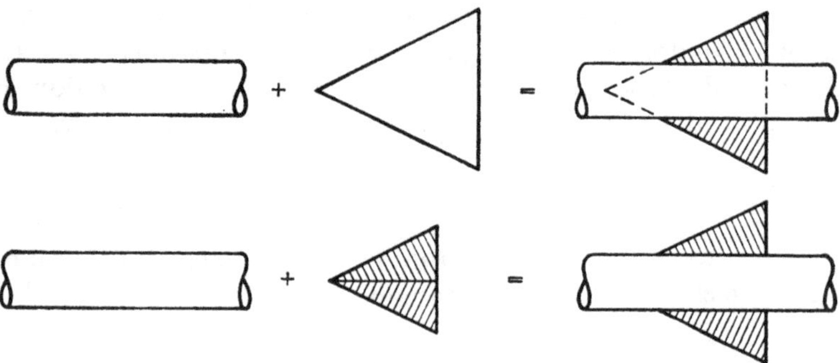

Fig. 9-15. Methods of superposing wing and body to form a wing-body combination.

There are several ways in which a wing-body combination can be formed from a wing alone or a body alone. Two such methods are shown in Fig. 9-15. In the first method the body is added directly to the wing, blanketing it in part. The effect of the body on the wing is thus to reduce the drag by submerging a large part of the wing within the body. Thus, the interference appears favorable. However, if the wing span gets small approaching the body diameter, the exposed panels bear little resemblance to the wing alone. If the wing span is less than the body diameter, the process is meaningless. A second method having closer correspondence to a real wing-body combination, particularly for small wing panels, is also shown in Fig. 9-15. The wing alone is the two panels joined together. In the formation of the wing-body combination, the two panels are drawn apart a distance equal to the body diameter, and the body is inserted between them. We will adopt this second method of forming a wing-body combination and take the wing alone as the two panels joined together. This procedure also has the advantage of a simple rule of thumb as we will see.

By the wing-body interference we mean the difference between the drag of the wing-body combination and the sum of the drags of the wing alone and the body alone. The wing-body interference is composed in part of the change in drag of the panels due to the addition of the body and in part of the change in body drag due to addition of the panels. Thus, the total combination drag has four components:

$$D_C = D_B + D_W + D_{W(B)} + D_{B(W)}$$
$$D_B = \text{drag of body alone}$$
$$D_W = \text{drag of wing alone} \tag{9-64}$$
$$D_{W(B)} = \text{drag of wing due to presence of body}$$
$$D_{B(W)} = \text{drag of body due to presence of wing}$$

The components $D_{W(B)}$ and $D_{B(W)}$ are due to the pressure field of the interference potential ϕ_i as discussed in Sec. 5-1. The component $D_{W(B)}$ is the change in drag of the wing panel by virtue of the difference in its position in the wing alone and its position in the wing-body combination. It is the result of two moves; first, separating the two halves of the wing alone a distance apart equal to the body diameter, and then inserting the body between the two panels. The component $D_{B(W)}$ can be thought of as the change in drag of the body due to bringing up two wing panels from infinity and attaching them to the body.

For a symmetrical wing mounted on a body section of nearly circular cylindrical shape—a so-called circular quasi-cylinder—the drag of the wing-body combination can be accurately calculated within the scope of linear theory by the W-function method described by Nielsen[28] and discussed in Sec. 4-4. This method makes use of a special function $W_m(x,r)$ in a numerical solution of the problem. The method is useful as a standard against which approximate but simple methods can be checked. One such simple rule of thumb is to assume that $D_{W(B)}$ is zero. The basis for this rule is a series of calculations performed by Katzen and Kaattari on the drag of triangular panels of a wide range of sizes and aspect ratio mounted on a circular body. The method used by the investigators is that of Nielsen and Matteson,[29] a forerunner of the more refined W-function method mentioned above. The investigators found that, for panels small compared to the body, the interference drag $D_{W(B)}$ could be a substantial percentage of the panel drag but a negligible percentage of the combination drag. For large panels, the interference drag is a negligible percentage of the panel drag. From the physical point of view this means that the panel mounted on a body of revolution acts as if it were mounted on a vertical reflection plane, or as if it were mounted in the wing alone with the other panel present. It is to be emphasized that the rule of thumb is not applicable to panels mounted on expanding bodies or contracting bodies which develop longitudinal pressure gradients. In such cases a correction should be made for longitudinal pressure gradients by assuming the panels to act in the pressure field of the body alone.

Let us take up the question of estimating $D_{B(W)}$. For the frequent case in which the body is cylindrical, $D_{B(W)}$ is zero. If the body is quasi-cylindrical, the value of $D_{B(W)}$ accurate to the order of linear theory can be calculated by the afore-mentioned W-function method. For a case where the body is not even approximately a circular quasi-cylinder, a first approximation to $D_{B(W)}$ can be obtained by assuming that the body is acting in the pressure field of the wing alone as given by supersonic wing theory.

9-8. Wings and Wing-Body Combinations of Least Pressure Foredrag at Zero Angle of Attack

The problem of shaping a wing or wing-body combination of least pressure foredrag has received much attention. Historically the search for wings of low drag at subsonic speeds has been a long and fascinating story, and long strides have been taken down a similar road at supersonic speeds. It is true that for many missiles the wing pressure drag may not be an important part of the total drag. And for other missiles the drag may be of no importance in the particular tasks for which the missiles were designed. Nevertheless, a large group of missiles exists for which the wing wave drag is important, and the group will become larger as the aerodynamic design of missiles is refined. The growing importance of drag minimization for missiles or for airplanes cannot be doubted.

We will first consider ways of minimizing the pressure foredrag of wings alone and then discuss Jones's criterion. Next we will consider the question of minimizing the drag of slender wing-body combinations by Ward's formula, and show how it prepared the stage for Whitcomb's discovery of the *NACA area rule*. In this connection the early contributions of the following authors to the area rule are recognized: Hayes,[43] Graham,[67] Oswatitsch and Keune,[68] and Legendre.[69] The theoretical extension of the NACA area rule to higher supersonic speeds into the *supersonic area rule* will be discussed. Also, the importance of body cross-sectional shape at high supersonic Mach numbers will be pointed out.

At the onset it must be stated that the minimization of the drag of a wing or wing-body combination can be accomplished under various restrictions, as discussed in connection with bodies. If no restrictions are placed on the wing, for instance, its drag coefficient can be made as small as desired. This can be accomplished by making the wing very thin or by sweeping the wing behind the Mach cone and increasing its aspect ratio. General ways of reducing thickness drag under no restrictions are useful, particularly for suggesting new design trends. If we invoke the restrictions that the wing planform be fixed and that the wing thickness distribution contain a specified volume, then the *Jones criterion for minimum thickness drag* specifies the distribution of thickness over the planform. This criterion[35] says that the thickness distribution will be optimum if the pressure gradient in the *combined flow field* is constant. Let us explain this criterion for the particular wing shown in Fig. 9-16. Let the pressure distribution along the section shown be P_F for forward flight and P_R for reverse flight. The longitudinal perturbation velocity in the combined flow field is the algebraic *sum* of the longitudinal perturbation velocities. However, the pressure P_C in the combined flow field is the *difference* of the pressures $P_F - P_R$. If this difference has a

uniform slope dP_C/dx over the wing planform, then the thickness distribution is that for minimum thickness drag. The Jones criterion can in the direct sense be thought of as a test to see if a proposed thickness distribution gives the least thickness drag. For instance, it is known that a biconvex parabolic-arc airfoil has a linear pressure distribution in two-dimensional supersonic flow. It thus fulfills Jones's criterion and has the least thickness drag for a given volume. In another sense the Jones criterion can be used to determine the optimum thickness distribution. This Jones has done for a wing of elliptic planform.[36,42] The general problem of optimizing planform for minimum thickness drag is difficult to formulate mathematically. How-

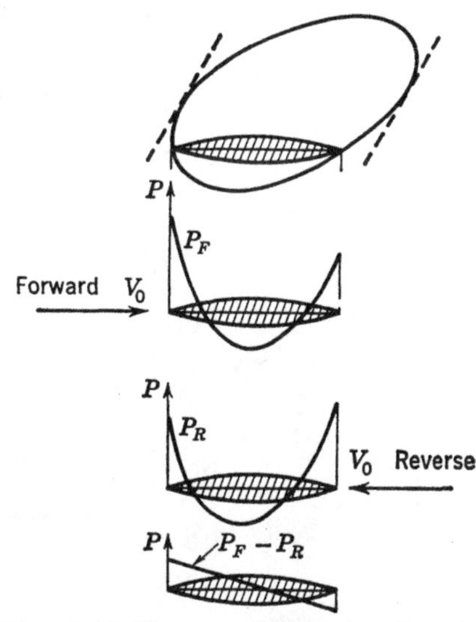

Fig. 9-16. Example illustrating Jones's criterion for minimum thickness drag for a given wing planform and wing volume.

ever, for no restrictions, a planform swept behind the Mach line and of infinite aspect ratio will have zero thickness drag.

For slender wing-body combinations, solutions for the area distribution for least pressure foredrag, or for minimum thickness drag in this case, can be found by using the drag formula of Ward if this drag formula is valid for wing-body combinations. In Ward's original article one of the assumptions was that the curvature must be order $1/d$ at all points where the body cross section is convex outward, and d is the maximum diameter of the section. For wing leading edges the curvature is generally much larger, and it is not clear that the theory applies. However, Ward discusses the reasonableness of relaxing the assumption in the special case of a "flat laminar wing of small aspect ratio with highly sweptback leading edges." Also, he points out that the wing can come from the body at a finite angle without the necessity of introducing further approximations (into slender-body theory). Nevertheless, either it was not realized that slender-body theory could give significant results for the effect of wing thickness on the thickness drag of wing-body combinations, or else the technological implications of such an application were not realized. Otherwise, the theory already discussed in Sec. 9-5 would have been applied to wing-body combinations, and the NACA area rule would have

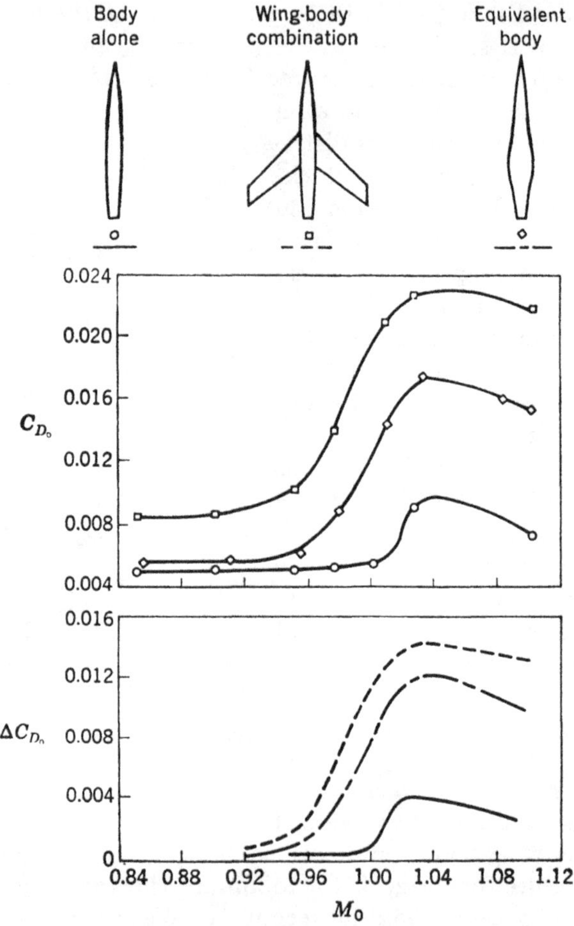

FIG. 9-17. Experimental results illustrating equivalent-body concept of Whitcomb.

had birth in theory rather than experiment. The experimental discovery of the NACA area rule by Whitcomb brought about the realization that the slender-body drag formula applied to wing-body combinations near sonic speed.

Whitcomb[32] enunciated his well-known *NACA area rule*. Whitcomb was testing wing-body combinations in a slotted-throat wind tunnel near a Mach number of unity. He observed the shocks standing normal to the flow by schlieren pictures. He made the observation that the body of revolution and the wing-body combination having the same axial distribution of cross-sectional area had essentially the same shock-wave patterns. On the basis that the pressure drag is represented by the shock waves of the schlieren pictures, Whitcomb concluded that the drag of a slender wing-body combination was equal to that of the *equivalent body of revolution*. The equivalent body of revolution is that body of revolution having the same area distribution as the wing-body combination. An experimental verification of the equality of drag between a wing-body combination and its equivalent body of revolution is shown in Fig. 9-17. The comparison is based on drag rise, ΔC_{D_0}, which is the drag coefficient minus the constant valve at low subsonic speeds.

Once an experimental verification was made of the NACA area rule, its theoretical basis in the drag formula of Ward and the earlier work of Hayes,[43] as well as the work of others, was recognized. It was now possible to design wing-body combinations of least thickness drag using the known results for a Sears-Haack body or a Kármán ogive. For instance, to design a minimum drag wing-body combination near a Mach number of unity for a combination of zero base area and of given length and volume, the area distribution of the wing-body combination should be that for the equivalent Sears-Haack body. One way in which this can be accomplished is to start with a full Sears-Haack body as shown in Fig. 9-18. Then in the region of the

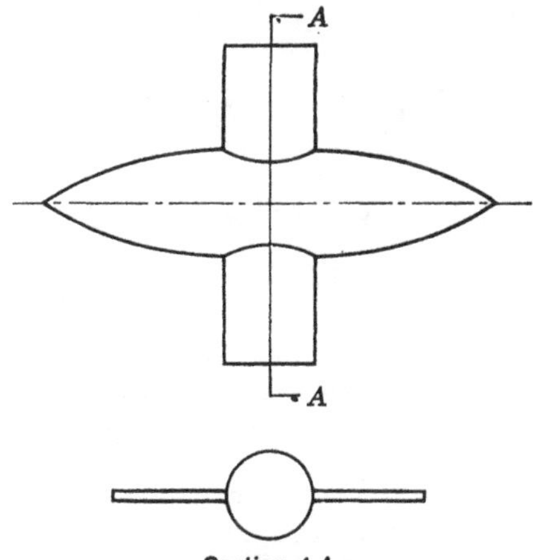

Section *AA*

FIG. 9-18. Indentation of body to minimize pressure drag at zero angle of attack according to NACA area rule.

wing-body juncture, remove as much cross section from the body in any crossflow plane as the wing contains. The wing-body combination will then have the same thickness drag as the Sears-Haack body.

Another use of Whitcomb's equivalent-body concept is its application for determining the thickness drag of a configuration which is not optimum. To do this the configuration should be sliced by crossflow planes of the kind shown in Fig. 9-18 and the cross-sectional area intercepted by the planes determined. This procedure will establish the cross-sectional area $S(x)$ as a function of axial distance. The coefficients b_n in the Fourier series for $S'(x)$ can then be determined numerically and the drag calculated from Eq. (9-41). If $S'(l)$ is not zero, the additional terms exhibited by Eq. (9-16) must be included.

Any rule as general as the NACA area rule must have its limitations. Since the rule is shown to have a theoretical basis in slender-body theory, it might be expected to be subject to the same kind of limitations as that theory. The rule is most accurate for slender configurations lying near the center of the Mach cone. For a Mach number of unity, the rule works well even for wings with unswept leading edges. If the leading edges are highly swept, then the rule will hold into the supersonic Mach-number range, since the configuration will be near the center of the Mach cone from the wing-body juncture. However, for a fixed configuration there will be an upper limit in Mach number, beyond which the NACA area rule cannot be accurately applied. A scheme to raise the upper limit to which the equivalent-body concept extends has been advanced by Whitcomb and Jones. The scheme will be termed the *supersonic area rule*. Actually the connections in which we will use the rule will be one of area only. In a more accurate sense, the rule is one of *source strength* rather than area, but its use in this connection is beyond our contemplated scope.

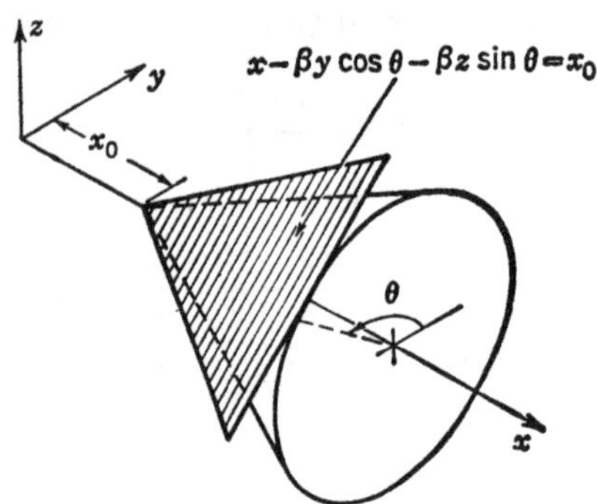

Fig. 9-19. Oblique tangent plane characterized by the parameters x_0 and θ.

The supersonic area rule utilizes fairly simple geometric construction as described by Jones[31] and Lomax and Heaslet.[30] The cutting planes are no longer crossflow planes as in the NACA area rule, but are oblique planes tangent to Mach cones as shown in Fig. 9-19. The plane shown in the figure depends on the x intercept x_0 and the line of tangency on the cone corresponding to the angle θ. The equation of the oblique plane shown is

$$x - \beta y \cos \theta - \beta z \sin \theta = x_0 \qquad (9\text{-}65)$$

The oblique plane corresponding to x_0 and θ will intercept an area $S(x_0,\theta)$ from the wing-body combination as shown in Fig. 9-20. Let $S_n(x_0,\theta)$ be the projection of this area on any crossflow plane normal to the x axis; then

$$S_n(x_0,\theta) = \frac{1}{M_0} S(x_0,\theta) \qquad (9\text{-}66)$$

The drag due to thickness of the combination is then

$$\frac{D}{q_0} = \frac{1}{2\pi} \int_0^{2\pi} \left[-\frac{1}{2\pi} \int_0^l \int_0^l S_n''(x,\theta) S_n''(\xi,\theta) \cdot \log |x - \xi| \, dx \, d\xi \right] d\theta \qquad (9\text{-}67)$$

The analogy to the first term of the drag formula of Ward, Eq. (9-16), is clear. For any value of θ the inner double integral gives the drag of the equivalent body of revolution for that value of θ. The drag of the equivalent bodies is then averaged over θ. For a Mach number of unity,

$S_n(x,\theta)$ and $S(x,\theta)$ are the same, and there is no variation with respect to θ. We therefore get the first term of Eq. (9-16) again. The formula Eq. (9-67) is limited to the case $S'(l) = 0$, a pointed base or a tangent-cylindrical base. The application of the supersonic area rule to any but simple configurations involves a large amount of work, and is frequently best accomplished numerically.

As applied in the previous paragraph the supersonic area rule is a slender-body rule. Its application as a *source strength* rule has been investigated by Lomax.[33] Briefly, an oblique-plane construction can be used to determine the axial distribution of sources equivalent to a given wing-body combination from a drag point of view. Also, the axial distributions for higher-order solutions such as quadripoles are obtained.

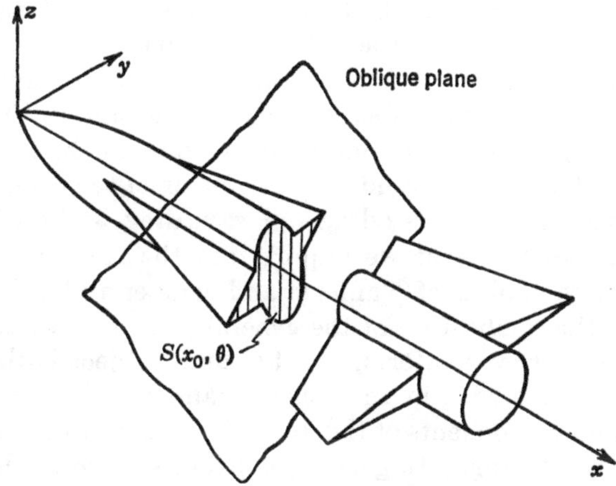

Fig. 9-20. Oblique cutting planes as used in the supersonic area rule.

Subject to certain constraints, the axial distributions are modified to minimize the drag. Then the body shape is calculated. This later step is usually very laborious if the full accuracy of linear theory is retained, but it can be simplified by descending to slender-body theory to calculate the shape. Another method of minimizing the thickness drag of wing-body combinations, not necessarily slender, has been used by Nielsen,[34] utilizing quasi-cylindrical theory. The difficulty of finding the body shape is circumvented by applying the body boundary conditions on a circular cylinder in the usual fashion of quasi-cylindrical theory. However, if the minimum drag wing-body combination does not have a quasi-cylindrical body, an accurate solution will not be obtained. Perhaps the power of these two methods lies in their ability to handle changes in body cross-section shape which are not significant in the supersonic area rule.

An important respect in which the supersonic area rule, Eq. (9-67), is incomplete has been pointed out by Lomax and Heaslet.[30] Specifically, if there exists a resultant force on the oblique area cut from the missile by any oblique plane of the supersonic area rule (Fig. 9-20) and acting in that plane, then the rule must be modified to include the effect of this force. The modification of the rule is easily made since the resultant force on the oblique area has the effect of changing the oblique area used in the supersonic area rule in a simple way. The mathematical details of this extension of the supersonic area rule together with several examples are given by Lomax and Heaslet.[30]

9-9. Minimizing Pressure Drag of Wings and Wing-Body Combinations beyond That Due to Thickness

A number of investigators have probed methods for reducing the drag due to lift of wings alone at supersonic speeds. Such methods include changes in planform and the use of camber and twist. It is useful to approach the subject of wing-body combinations of least drag due to lift in two independent steps at the risk of some possible loss in generality. In the first step we consider minimizing the drag (exclusive of thickness drag) of the main lifting member, the wing alone, and in the second step we take up the problem of adding useful volume in the form of a body. The first main item on the agenda is a discussion of the components of the drag of a lifting surface, vortex drag and wave drag, and the lower bounds for each component. Next we inquire into the methods for achieving low drag through choices of planform and camber and twist. The next item involves the application of the general principles to lifting surfaces of triangular or arrow planform, and the final subject is the addition of useful volume to the wing in an efficient manner.

The two main components of the drag of a lifting surface are the vortex drag and the wave drag. In general, a lifting surface discharges a trailing-vortex system, and the kinetic energy per unit streamwise length of the system is equal to a drag force. Also, as the surface changes angle of attack, the shock-wave configuration changes with the shocks becoming stronger. The result is an increase of wave drag. The minimization of these two components of the drag requires certain changes in planform, and for a fixed planform requires camber and twist. However, before we look at the separate components, let us examine Jones's criterion for least drag due to lift of a lifting surface similar to his criterion for least thickness drag. It is convenient to illustrate the criterion in this instance in the same way we did for the thickness drag. Consider a lifting surface as shown in Fig. 9-21. Let us suppose that a given distribution of lift over the planform is the optimum distribution yielding least drag due to lift. (In specifying any distribution of lift over a planform we suppose it to be the result of angle of attack, camber, and twist of the planform.) The Jones criterion is simply a test of whether this supposed optimum distribution is in fact optimum. Let the shape of the lifting surface in section AA to support the given load distribution correspond to a downwash velocity w_F along the section for forward flight. Similarly, let the wing reverse flight direction, maintaining the same lift distribution. Let the downwash along section AA be w_R to support the lift distribution in reverse flow. If the sum of the downwash velocities $w_F + w_R$ is constant over the wing planform, then the lift distribution is optimum. The Jones criterion is a test of a lift distribution which for *a given planform and total lift* allegedly is optimum. The criterion does not tell how to find the optimum lift distribution, nor does it guarantee the existence of such a distribution.

Let us now consider lower bounds on the vortex drag and the wave drag separately, turning first to the vortex drag. At subsonic speeds the drag due to lift is solely vortex drag. On the basis of lifting-line theory, the drag due to lift depends only on the shape of the span-load distribution and is independent of how the load is distributed chordwise. In fact, the minimum drag of a lifting surface

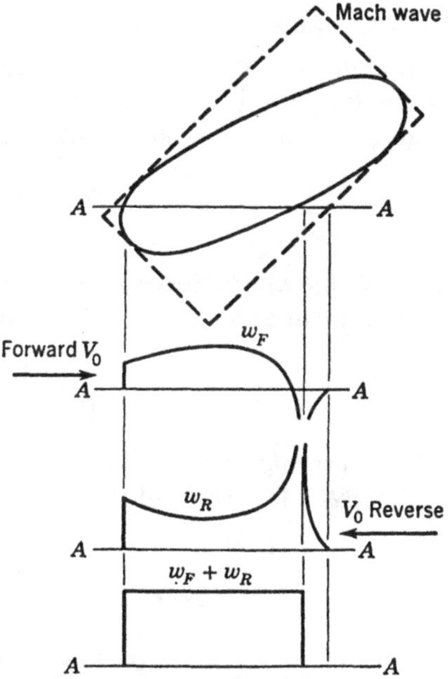

FIG. 9-21. Example illustrating Jones's criterion for minimum drag due to lift for lifting surface of given planform and total lift.

or lifting line for subsonic flow is achieved when the span loading is elliptical and is given by the well-known formula

$$\left(\frac{C_{D_i}}{\Delta C_L{}^2}\right)_{\text{vor}} \geq \frac{1}{\pi A} \qquad (9\text{-}68)$$

The reason for recounting the situation at subsonic speeds is that it is unchanged at supersonic speeds. It will be recalled that the vortex drag can be determined by considering the trailing-vortex system to trail backward in a rectilinear fashion to infinity, that is, to a region beyond the influence of the bound vortices. The kinetic energy of the vortex system is evaluated, using incompressible potential theory in the crossflow plane at infinity, the so-called Trefftz plane. At supersonic speed precisely the same theoretical model is used. The horseshoe vortices making up the trailing-vortex system are supersonic horseshoe vortices. By the time the Trefftz plane is reached, the flow field corresponds to that at the center of the lifting-surface Mach cone. At the center of the Mach cone the velocity field is independent of Mach number. In fact, at the Trefftz plane the velocity field created by supersonic horseshoe vortices is identical with that created by incompressible horseshoe vortices of equal strength and shape. These considerations explain why the vortex drag associated with a given span-load distribution is independent of Mach number and why streamwise loading as such does not influence the vortex drag but only the wave drag.

Let us now examine the lower bound established by Jones[37] for the wave drag of a lifting surface carrying a *specified lift*. The wave drag for a given lift is bounded as follows:

$$D_{\text{wave}} \geq \frac{M_0{}^2 - 1}{2} \frac{L^2}{\pi q_0 \overline{l^2}} \tag{9-69}$$

Here $\overline{l^2}$ is a characteristic mean-squared length of the surface depending on planform and Mach number and given by

$$\frac{1}{\overline{l^2}} = \frac{1}{\pi} \int_0^{2\pi} \frac{\sin^2 \theta}{[l(\theta)]^2} \, d\theta \tag{9-70}$$

The interpretation of (θ) can readily be made with the help of Figs. 9-19 and 9-22. Fix the value of θ, and thereby specify a series of parallel

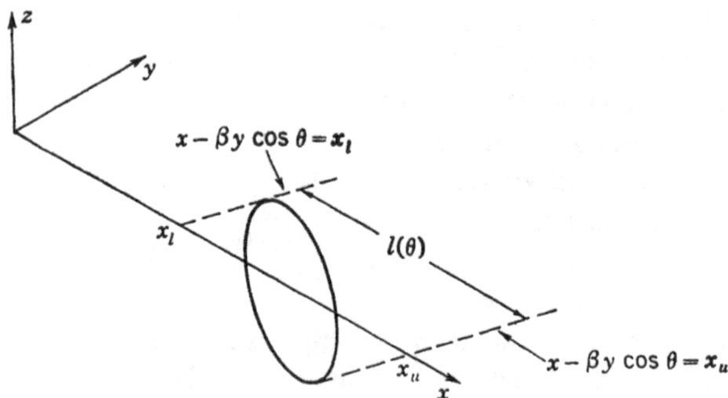

Fig. 9-22. Method of determining $l(\theta)$.

planes with x_0 as the distinguishing parameter. Let the first plane, which is just tangent to the wing planform, correspond to $x_0 = x_l$. In the xy plane the equation of the trace of this plane is from Eq. (9-65)

$$x - \beta y \cos \theta = x_l \tag{9-71}$$

The corresponding plane moving upstream from behind the wing has the trace

$$x - \beta y \cos \theta = x_u \tag{9-72}$$

The value of $l(\theta)$ is

$$l(\theta) = x_u - x_l \tag{9-73}$$

If the lifting surface were a line in the streamwise direction, then $l(\theta)$ would be the length of the line for all θ. To put Eq. (9-69) on the basis of a drag-rise factor, let us introduce some characteristic streamwise length l_0 of the configuration such as wing mean aerodynamic chord, body length, etc., and define a factor K^*

$$K^* = \frac{l_0{}^2}{\bar{l}^2} \tag{9-74}$$

If the coefficients are based on a reference area S_R, then we have the *wave drag-rise factor* of the lifting surface bounded as follows:

$$\left(\frac{C_{D_i}}{\Delta C_L{}^2}\right)_{\text{wave}} \geq K^* \frac{M_0{}^2 - 1}{2\pi} \frac{S_R}{l_0{}^2} \tag{9-75}$$

The meaning of the present lower bound should be made clear. It is the bound attained if the wing is elliptically loaded when viewed from any direction. In particular, the span loading and streamwise loading will both be elliptical. To approach or achieve the lower bound for a given planform requires optimization of camber and twist. For a given planform it is not necessarily attainable. For instance, consider a triangular lifting surface with sonic leading edges. From Eq. (9-68) such a surface has a value of $\frac{1}{\beta}\left(\dfrac{C_{D_i}}{\Delta C_L{}^2}\right)_{\text{vor}}$ of $\frac{1}{4}\pi$ and a value of $\frac{1}{\beta}\left(\dfrac{C_{D_i}}{\Delta C_L{}^2}\right)_{\text{wave}}$ of 0.087 by Eq. (9-79). The value of $\frac{1}{\beta}\dfrac{C_{D_i}}{\Delta C_L{}^2}$ is thus 0.166 in contrast to an exact lower bound of 0.222 calculated by Germain,[65] specifically for a triangular planform with sonic leading edges. Thus such a planform does not approach the Jones bound as closely as some other planform might.

Having established lower bounds on vortex drag and wave drag of the lifting surface, we are in a position to examine the possible effect of planform change on these drag components. An examination of Eq. (9-68) brings to mind the well-known fact that minimization of vortex drag requires a large aspect ratio, and this requirement is unchanged at supersonic speeds. Now in Eq. (9-75), for "wave drag-rise factor," we can change K^* to a certain extent, but we have infinite control over $S_R/l_0{}^2$. The quantity $l_0{}^2/S_R$ is what Jones has termed a "longitudinal aspect ratio." To minimize the wave drag-rise factor we must maximize the longitudinal aspect ratio. By yawing a rectangular wing behind the Mach line and decreasing its chord, the value of the drag-rise factor in Eq. (9-68) or (9-75) can be reduced to as low a value as desired. However, if this operation is carried out subject to the constraint that a constant lift be carried, the chord can be decreased only to a certain point before the boundary layer of the wing will surely separate. Viscosity thus provides the factor which limits the reduction in drag-rise factor obtainable through change in planform.

Suppose that an acceptable planform has been found and that we are now faced with the problem of trying to attain the lower bounds of vortex and wave drag-rise factors given by Eqs. (9-68) and (9-75). Generally speaking, no specific design can be carried out to insure that both minima will be attained.   We do, however, have recourse to the Jones criterion to see whether a proposed design is optimum.   Consider a flat lifting surface as the first approximation.   Usually a flat surface will not have an elliptical span loading to insure minimum vortex drag in accordance with Eq. (9-68).   (The triangular wing with subsonic leading edges is the well-known exception.)   To obtain an elliptical span loading we will have to twist and/or camber the surface.   There will probably be a number of ways in which this can be accomplished, and out of the number it is hoped that one fulfills Jones's criterion.   A practical way of testing how close a given lifting-surface design is to optimum is to evaluate $C_{D_i}/\Delta C_L^2$ and compare it with the sum of lower bounds of $(C_{D_i}/\Delta C_L^2)_{\text{vor}}$ and $(C_{D_i}/\Delta C_L^2)_{\text{wave}}$.   A specific design for a given design lift coefficient consists of cambering and twisting the surface to obtain a given lift distribution, or of computing the lift distribution resulting from a given camber and twist.   In any event, knowing the lift distribution and the camber and twist, all at the design lift coefficient, permits an evaluation of $C_{D_i}/\Delta C_L^2$ for the lifting surface at the design point.   The $C_{D_i}/\Delta C_L^2$ of the design can be compared with the lower bound to assess the excellence of the design *at the design point*.   At this time it is well to recall in connection with Eq. (9-60) that the total wing drag is due to thickness, camber, and angle of attack.   At the design point the *sum* of the drags due to camber and angle of attack equals that of the lifting surface and is determined by the preceding procedure, although the *individual components* are not determined.   The wing drag is then the drag of the lifting surface plus the thickness drag since the thickness drag is not coupled with that due to camber or angle of attack.   We thus know the lift-drag ratio of the *wing* at the design lift coefficient.   Since we have minimized the drag at fixed lift, it is clear that we have maximized the lift-drag ratio for the design lift coefficient.   We have not determined the complete drag curve, however, since it takes one other point besides C_D at the design C_L to establish the drag parabola.

Triangular Lifting Surfaces

Let us now examine the lower bounds on the vortex drag and wave drag of a triangular lifting surface.   First, with regard to vortex drag, it will be recalled that the span-load distribution is elliptical for minimum vortex drag, and that for a triangular wing with subsonic leading edges the span loading is elliptical.   The vortex drag is already a minimum, and $(C_{D_i}/\Delta C_L^2)_{\text{vor}}$ is given by Eq. (9-68).   This component of the drag is plotted against wing aspect ratio in Fig. 9-23, where it is labeled "optimum vortex drag."

It is of interest now to try to establish the lower bound of the wave drag as given by Eqs. (9-70) and (9-75).   The traces of the Mach cones

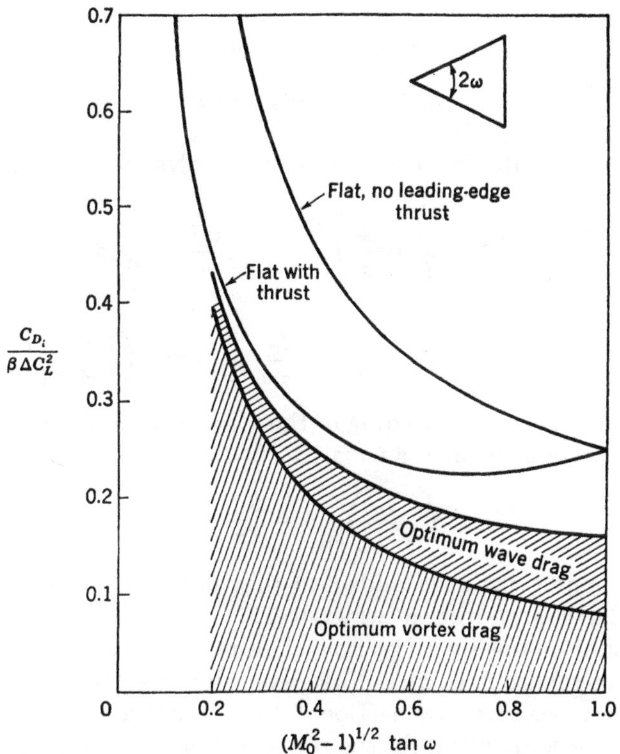

FIG. 9-23. Drag due to lift for triangular lifting surfaces.

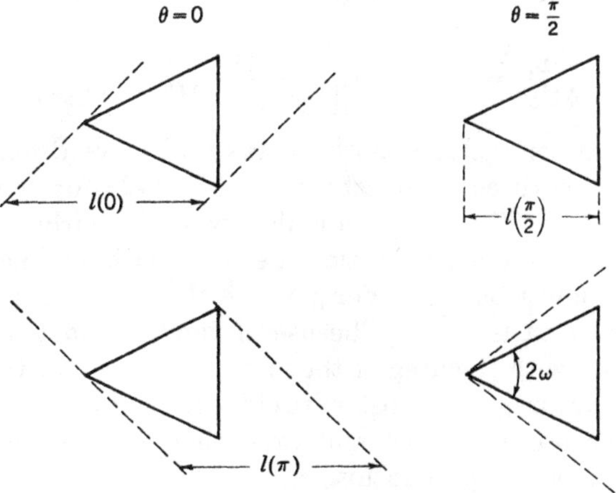

FIG. 9-24. Values of $l(\theta)$ for triangular wings.

which establish $l(\theta)$ are shown in Fig. 9-24 for $\theta = 0$, $\pi/2$, and π. It is easy to show that

$$l(\theta) = c(1 + m \cos \theta)$$
$$m = \frac{\beta A}{4} = \beta \tan \omega \tag{9-76}$$

Because of the symmetry of the problem, the value of K^* given in Eq. (9-74) is

$$K^* = \frac{c^2}{\bar{l}^2} = \frac{4}{\pi} \int_0^{\pi/2} \frac{\sin^2 \theta}{(1 + m \cos \theta)^2} \, d\theta \tag{9-77}$$

which yields

$$K^* = \frac{4}{\pi} \left[\frac{1}{m} - \frac{\pi}{2m^2} + \frac{2}{m^2(1 - m^2)^{1/2}} \tan^{-1}\left(\frac{1 - m}{1 + m}\right)^{1/2} \right] \tag{9-78}$$

The lower bound of the wave drag of the lifting surface based on the planform area as reference area is from Eq. (9-75)

$$\left(\frac{C_{D_i}}{\Delta C_L^2}\right)_{\text{wave}} = \frac{K^*}{2\pi} \beta m \tag{9-79}$$

This lower bound has been added to that for vortex drag in Fig. 9-23 where it is labeled "optimum wave drag."

The question naturally arises how close to the lower bound known triangular wing solutions come. First, consider the flat triangular lifting surface with no leading-edge section. In accordance with Eq. (9-15) the drag-rise factor is then merely the reciprocal of the lift-curve slope.

$$\frac{C_{D_i}}{\Delta C_L^2} = \frac{1}{C_{L\alpha}} = \frac{\beta E(\pi/2, k)}{2\pi m} \qquad k^2 = 1 - m^2 \tag{9-80}$$

With leading-edge suction the drag-rise factor from Eqs. (9-15) and (9-63) is

$$\frac{C_{D_i}}{\Delta C_L^2} = \frac{1 - \mu}{C_{L\alpha}} = \left[1 - \frac{(1 - m^2)^{1/2}}{2E} \right] \frac{\beta E}{2\pi m} \tag{9-81}$$

The drag-rise factor includes both vortex and wave drag. The values of $C_{D_i}/\Delta C_L^2$ for both cases are shown in Fig. 9-23 for comparison with the lower bound. The flat triangular wing is fairly far above the lower bound. At low aspect ratios the wing with leading-edge suction approaches the lower bound. For $\beta A = 4$ the leading edge is sonic, and leading-edge suction is zero. The use of camber and twist offers some gain if a solution with the drag of the lower bound can be found. At low aspect ratios the drag is almost entirely vortex drag, which is already optimum. The use of camber and twist therefore does not offer much potential gain at low aspect ratios.

To achieve the lower bound requires a triangular lifting surface fulfilling the Jones criterion for minimum drag of a lifting surface carrying a given total lift. Several efforts to achieve this lower bound have been made. S. H. Tsien[38] has attempted to obtain the least drag within the limitation of a conical lifting surface. His results are interesting. For instance, he finds that, with full leading-edge suction, there are negligible benefits of camber and twist compared to those of a flat wing. On the other hand, with no leading-edge suction, camber and twist can bring the drag down to that of the flat lifting surface with full leading-edge suction. This latter result is important if the required camber and twist also alleviate leading-edge separation, which acts to invalidate the theory for a flat triangular surface. However, it is clear that the absolute minimum drag for a given lift is not necessarily found within the limitations of conical lifting surfaces. In fact, Cohen,[39] using a different approach, has achieved a lower drag than Tsien. She superimposes a number of known solutions for cambered and twisted triangular wings in a search for the surface of optimum camber and twist. Whether such a scheme will be successful depends on whether a linear combination of known solutions can approximate closely the solution for the optimum shape. An a priori answer to this question would be difficult to give. However, as noted in connection with Eq. (9-75), the lower bound of $\dfrac{1}{\beta}\dfrac{C_{D_i}}{\Delta C_L{}^2}$ for triangular wings has been found by Germain.[65] His value of 0.222 for a sonic leading edge is closely approached by the wings of Cohen. The lower value of 0.166 on the basis of Eqs. (9-68) and (9-75) only shows that it is not possible to camber and twist a triangular lifting surface with sonic leading edges so that the loading is elliptical when viewed from any direction.

Arrow Lifting Surfaces

One of the efficient types of planforms indicated in the discussion following Eq. (9-75) is the wing of large aspect ratio with subsonic leading edges—one maximizing "lateral" and "longitudinal" aspect ratios simultaneously. One class of planforms falling in this general category is arrow wings with subsonic leading edges. Let us examine the lower bounds of vortex and wave drag for the class of arrow wings formed by cutting out part of a triangular wing as shown in Fig. 9-25. Let the wing trailing edge remain supersonic. If the arrow wing is cambered and twisted to support an elliptical span loading, then its vortex drag-rise factor is given by Eq. (9-68). If the subscript A refers to the arrow wing formed from a triangular wing denoted by subscript T, then

$$A_A = \frac{A_T}{a} \tag{9-82}$$

so that

$$\left(\frac{C_{D_i}}{\Delta C_L{}^2}\right)_{vor_A} = a\left(\frac{C_{D_i}}{\Delta C_L{}^2}\right)_{vor_T} \tag{9-83}$$

It is clear that cutting away the triangular wing does not alter the value of $\bar{l}^2$ used to calculate the lower bound for wave drag. The lower bound by Eq. (9-75) is then simply proportional to planform area S_R. Therefore

$$\left(\frac{C_{D_i}}{\Delta C_L{}^2}\right)_{\text{wave}_A} = a \left(\frac{C_{D_i}}{\Delta C_L{}^2}\right)_{\text{wave}_T}$$

Let us see how these lower bounds compare with the drag-rise factors for a flat-arrow wing with and without leading-edge suction.

Investigations which concern triangular wings are applicable in many instances to arrow wings. In particular, the value of $C_{D_i}/\Delta C_L{}^2$ for a flat lifting surface of arrow planform with a supersonic trailing edge can be obtained from the observation that the leading-edge thrust is the same

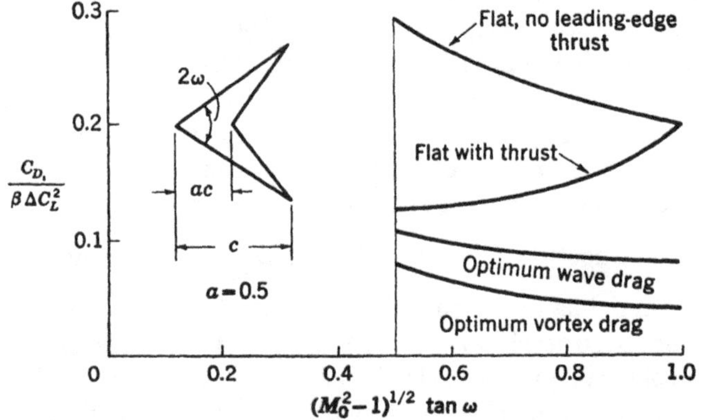

Fig. 9-25. Drag due to lift of arrow-shaped lifting surfaces.

as that for a triangular wing having the same leading edges. If the subscript A is used to denote an arrow wing, and T is the triangular wing with the same leading edges, the drag of the flat-arrow wing is

$$D_A = \alpha L_A - T = \alpha L_A - \mu \alpha L_T \tag{9-84}$$

where T is the leading-edge thrust and μ is the leading-edge suction factor for the triangular wing. With reference to Fig. 9-25 for the definition of a, the drag-rise factor for the arrow wing is

$$\frac{C_{D_i}}{\Delta C_L{}^2} = \frac{1}{(dC_L/d\alpha)_A} \left[1 - \frac{\mu_T}{a} \frac{(dC_L/d\alpha)_T}{(dC_L/d\alpha)_A} \right] \tag{9-85}$$

We thus require only the lift-curve slope of the arrow wing and the values of μ and $dC_L/d\alpha$ for the triangular wing in order to obtain the drag-rise factor for the flat-arrow wing with leading-edge suction.

The sum of the lower bounds for vortex drag and wave drag are shown in Fig. 9-25 as a function of leading-edge sweep angle for the particular family of arrow wings with $a = 0.5$. Also shown is the drag-rise factor for the lifting surface calculated from Eq. (9-85) with and without leading-edge suction.   Large gains are indicated if through the use of camber and twist the lower bound can be closely approached.   It is to be noted that the drag-rise factor for the arrow planform is potentially lower than that for the triangular wing by the factor a.   As the factor a approaches zero, so does the drag-rise factor corresponding to the sum of the optimum vortex and wave drags.   The arrow wing is approaching oblique panels of infinite aspect ratio swept behind the Mach waves.   Although the lower bound can in principle be made arbitrarily small for such wings, the mechanism of viscosity is a limiting factor.   If the total lift is fixed, the wing loading goes up as the chord goes down.   At some point the loading is so great that boundary-layer separation must occur, limiting any further reduction in drag for a constant lift through reductions in a. Tucker[40] has presented an engineering method for approximating the optimum camber and twist for arrow wings.   The use of optimum camber and twist can have a beneficial effect in controlling boundary-layer separation since the tips are usually washed out to avoid high tip loadings.

Addition of Usable Volume

We have in reality confined ourselves so far to lifting surfaces with no volume.   The addition of volume in the form of *symmetrical* wing thickness can easily be made, since the drag due to such thickness is additive to that of the lifting surface and is not coupled to it.   The drag of the lifting surface is therefore increased by the thickness drag of the wing alone, and the lift is unaltered.   The lift-drag ratio is reduced.

Now what we would like to do is add volume without reducing the lift-drag ratios of the lifting surface.   One interesting approach to this problem has been proposed by Ferri.[41]   If a wedge is mounted on the lower wing surface, the positive pressure field due to the wedge can be utilized to produce interference lift on the under surface of the wing. The lift-drag ratio can thus be greater than it would be if the volume were added as a symmetrical body.

If the volume is added in the form of a body of revolution, the body upwash will have the same effect as introduction of twist into the wing. If the wing alone already has optimum twist, it will no longer be optimum in the presence of the body.   The span loading for least vortex drag of the wing-body combination is that given by slender-body theory (see ref. 4, Chap. 5).   This span loading is closely elliptical, as shown in Table 6-1.

BASE DRAG

9-10. Physical Features of Flow at a Blunt Base; Types of Flow

The second general component of the total drag of a missile, the base drag, is not amenable to analysis solely by potential theory because it is controlled largely by interaction between the boundary layer leaving the blunt base and the external flow. Also, the theory of such interactions is far from complete, so that we must rely for engineering calculations on semiempirical correlations of base-pressure measurements. We will be concerned with two-dimensional airfoils and bodies of revolution.

The physical model of the viscous flow in the neighborhood of a blunt base is sketched in Fig. 9-26. Directly behind the base is a circulating region of fluid known as the *dead water region* of pressure p_b. Enclosing the dead water region is the boundary layer from the blunt base, and enclosing the boundary layer is the outer potential flow. As the boundary layer leaves the base, it mixes with air from the dead water region and the outer flow, and increases in thickness. The boundary layer converges toward a point on the centerline known as the *reattachment point* and straightens out in the streamwise direction further downstream.

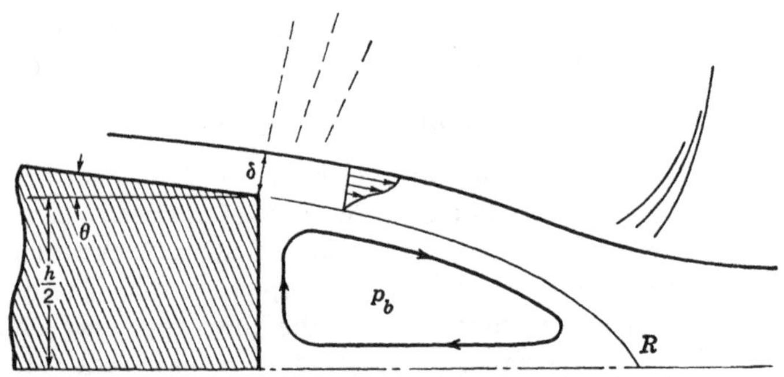

FIG. 9-26. Theoretical model of flow behind a blunt base.

Three main types of flow in the region of the base have been discussed by Chapman, Kuehn, and Larson.[44] These are the *purely laminar* type, the *transitional* type, and the *purely turbulent* type. The basis for the classification is the location of the transition point relative to the boundary-layer separation point at the body base and the reattachment point. If the transition point is downstream of the reattachment point and does not influence the base pressure, the purely laminar type prevails. If the transition point lies between the separation point and the reattachment point, the transitional type prevails. If the transition point lies upstream of the separation point so that the boundary layer at the base is turbulent, the fully turbulent type prevails.

The purely laminar type is characterized by the fact that the base pressure is independent of Reynolds number for very thin boundary-layer thickness at the separation point. This type, which occurs at very low Reynolds numbers, can be treated analytically for $\delta = 0$ as discussed by Chapman et al.[44] The transitional type is not frequently encountered at low supersonic Mach numbers. However, with the increased stability of the boundary layer accompanying increases in Mach number,[47] and with a premium on laminar boundary layers at high speeds to reduce skin friction and aerodynamic heating, its engineering importance is bound to increase. The transitional type is plagued by the general lack of understanding concerning the factors controlling the location of the transition point, and for this reason is the most difficult to treat both theoretically and experimentally. The purely turbulent type is very important from the engineering point of view, and fortunately is amenable to semiempirical treatment. Some selected references on transition and separation are given at the end of the chapter.

9-11. Basis for Correlation of Base-pressure Measurements

A number of variables are known to influence base pressure. The following list includes several of the important variables.

(1) Type of flow: laminar, transitional, or turbulent
(2) Flow dimensionality: two-dimensional or axially symmetric
(3) Angle of attack
(4) Body shape, particularly base configuration
(5) Mach number
(6) Reynolds number
(7) Heating and cooling of body

At the present time the first two variables are considered to be specified by the problem at hand, and they must be independently varied in experiments. For the time being and until Sec. 9-13, consider the angle of attack to be zero, and ignore heating and cooling effects. Within those limitations, variables (4), (5), and (6) will now be treated in the manner of Chapman,[50] as used by him to correlate extensive base-pressure measurements.

With reference to Fig. 9-26, let us postulate how the base pressure is determined. First, the general pressure level in the outer flow enclosing the boundary layer and the dead water region has a direct influence on the base pressure. The pressure change from the outer flow to the base depends on the mixing process between the boundary layer and the air on each side of it. This process depends on the boundary-layer thickness δ just before separation, and also on the velocity and density profiles of the boundary layer at separation. On the basis of this hypothesis, the body shape is important in two ways. The configuration at the base will be important in determining the average pressure and Mach number of the outer inviscid flow enclosing the boundary layer and dead water region. The general body shape will also be significant to the extent that it controls the boundary-layer thickness at the base through the pressure distribution. The Reynolds number based on body length will also influence the boundary-layer thickness. The Mach number will be significant (1) through the influence it exerts on the average pressure of the outer flow enclosing the boundary layer and dead water region, (2) through its effect on the density and velocity profiles of the boundary layer at separation, and (3) through the influence it may have on the mixing process between the boundary layer and the air on each side of it. Let us now see how variables (4), (5), and (6) might be treated on the basis of the foregoing hypothesis.

It is possible to eliminate base configuration as a variable in the experimental correlation of the base-pressure measurements if we calculate its

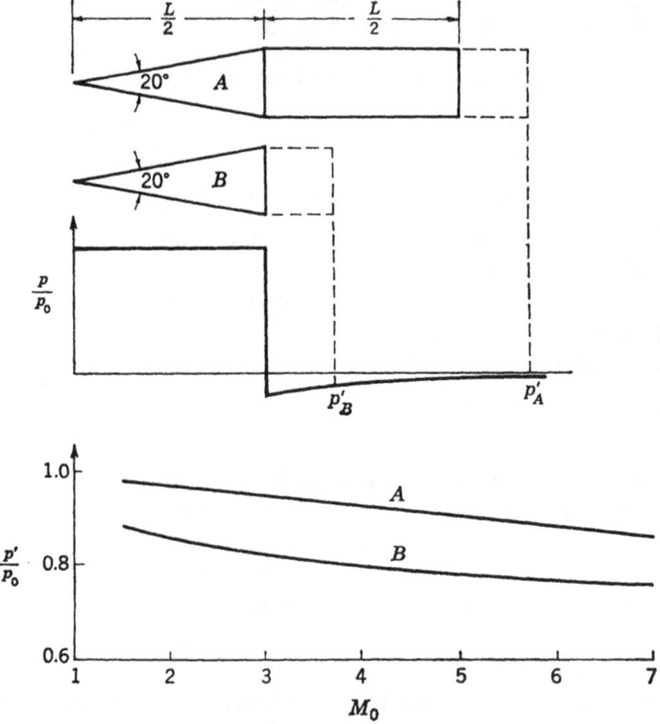

Fig. 9-27. Method of evaluating p' together with some specific values.

influence on the mean pressure of the outer flow, using inviscid fluid theory, and refer the base pressure to the calculated mean value in all measurements. (By base configuration we mean the shape of the body in front of the base as far forward as it can significantly influence the base pressure, and in particular we think of the *boattail angle* θ shown in Fig. 9-26.) We, thus, need some pressure representative of the mean pressure of the outer flow. Chapman[50] has used a quantity p' for this purpose, defined to be the average surface pressure that would prevail over the length of the dead water region if the body were prolonged by a hypothetical cylindrical extension. The model for calculating p' is shown in Fig. 9-27 for a cone and cone-cylinder together with some values of p' based on the theory of characteristics. The value of p' is taken one diameter behind the base. The base pressure p_b is now formed into the ratio p_b/p', which can be used to correlate experimental measurements of base pressure for different base configurations into a single correlation curve. The procedure to account for the influence of base configuration on base pressure is thus broken down into two steps in accordance with the following equation:

$$\frac{p_b}{p_0} = \frac{p'}{p_0}\frac{p_b}{p'}$$ (9-86)

The first factor p'/p_0 is calculated for the particular base configuration under consideration by inviscid flow theory. The second factor is taken from experimental correlation curves of a form shortly to be discussed.

Some remarks on the calculation of p' and its influence on base pressure are convenient at this point. For bodies of revolution at supersonic speeds the base configuration for three or four diameters in front of the base can influence p'. If the body is cylindrical for three or four diameters we can take p' equal to p_0. More specifically, the cylindrical length should be several multiples of the diameter times $(M_0{}^2 - 1)^{1/2}$. It is a property of two-dimensional supersonic inviscid flow that the static pressure directly behind the airfoil is free-stream static pressure. We thus have p' equal to p_0 for two-dimensional airfoils. What this means is that there is no boattail effect on the base pressure of blunt trailing-edge airfoils. One possible exception is detached flow at low Mach numbers. Actually, boattail angle can be varied on a body of revolution to reduce the total drag. Increasing the boattail angle will increase the pressure drag of the body in front of the base. However, it can raise p' above free-stream pressure, so that the base drag decreases. The least total drag usually occurs for nonzero boattail angle.

Turning now to variable (5), the Mach number, it is not immediately evident which number we should choose since the Mach number can potentially influence the base pressure in at least three ways, as previously mentioned. If the effect of Mach number were principally felt through its influence on the mixing process, then the average Mach number of the outer flow over the mixing length would be a useful one for correlation purposes. A Mach number that has proved helpful in correlating data is M', corresponding to the pressure p'. This Mach number, which is used henceforth, also helps to eliminate the effect of base configuration on the Mach number over the wake region in the same way that p' minimizes the effect of base configuration on the mean pressure of the outer flow over this region.

The final variable which we are considering, the Reynolds number, exerts its primary influence for a constant type of flow, i.e., laminar, transitional, or turbulent, through its effect on δ/h. To be sure, the boundary-layer thickness δ of the boundary layer just at separation is dependent also on Mach number and on over-all body shape. As for the Mach number, we retain it as a parameter in the correlation of base-pressure measurements, but we neglect any influence the body shape through pressure gradients may exert on the thickness of the boundary layer at separation. For a fixed value of M' we thus have that the boundary-layer thickness depends on the Reynolds number as for a flat plate. For a laminar boundary layer of length L

$$\frac{\delta}{L} \propto (\text{Re})^{-\frac{1}{2}} \tag{9-87}$$

where Re is the Reynolds number based on length L. For a turbulent boundary layer

$$\frac{\delta}{L} \propto (\text{Re})^{-\frac{1}{5}} \tag{9-88}$$

This completes the discussion of how the three variables—body shape, Mach number, and Reynolds number—determine the average pressure of the outer flow, the ratio of boundary-layer thickness to base height at separation, and the density and velocity profiles—three parameters which in the hypothesis determine base pressure.

We are now in a position to write the form of the correlation equation for variables (4), (5), and (6) with variables (1), (2), (3), and (7) held constant. In fact, a correlation in the following form is indicated on the basis of the preceding discussion:

$$\frac{p_b}{p'} = f\left(M', \frac{\delta}{h}\right) \tag{9-89}$$

The functional relationship indicates that the ratio p_b/p' should be a unique function of δ/h for constant values of M'. It is frequently convenient in engineering practice to use the Reynolds number in lieu of the boundary-layer thickness. On this basis the correlation has the following analytical form for laminar boundary layers at separation,

$$\frac{p_b}{p'} = f_1\left(M', \frac{L}{h\,\text{Re}^{\frac{1}{2}}}\right) \tag{9-90}$$

and the next form for turbulent boundary layer at separation,

$$\frac{p_b}{p'} = f_2\left(M', \frac{L}{h\,\text{Re}^{\frac{1}{5}}}\right) \tag{9-91}$$

The subscripts on f_1 and f_2 are merely used to indicate that the functions differ from f of Eq. (9-89) and from each other. The validity of the hypothesis leading to the form of the correlation is, of course, to be judged by the accuracy with which it correlates data from systematic tests.

With correlation curves of the functional forms given by Eqs. (9-90) and (9-91) we can calculate the base pressure in two steps in accordance with Eq. (9-86).   The first parameter p'/p_0 is calculated from inviscid flow theory.   The second factor p_b/p' is obtained from correlation curves of the type just discussed.

9-12. Correlation of Base-pressure Measurements for Blunt-trailing-edge Airfoils and Blunt-base Bodies of Revolution

Systematic base-pressure measurements have been made by a number of investigators.   For bodies of revolution, those of Chapman[50] are fairly extensive for Mach numbers up to 2, covering as they do the fully turbulent case and the transitional case.   For blunt-trailing-edge airfoils the data of Chapman, Wimbrow, and Kester[51] are available for both cases for Mach numbers up to 3.1, and the data of Syvertson and Gloria[49] are available for the transitional case for Mach numbers from 2.7 to 5.0. Before presenting correlation of these and other data let us note the difference in symbols between airfoils and bodies.   The base pressure for airfoils is referred to p_0, and for bodies to p', in accordance with the discussion of the previous section.   The Mach number of correlation is M_0 for the airfoils and M' for the bodies.   The over-all length is the chord c for the airfoils and length L for the bodies.   The common symbol h is the trailing-edge thickness for the airfoils and the base diameter for the bodies of revolution.   The base drag is proportional to $1 - p_b/p_0$.   Correlation curves of base pressures are presented in Figs. (9-28) to (9-32), inclusively, for use in engineering calculations.

In discussing the correlation curves, let us first consider the fully turbulent case and then the transitional case.   Under each case let us first discuss airfoils, and then bodies of revolution.   The discussion of airfoils for the fully turbulent case revolves around Eq. (9-91).   First, consider the influence of Mach number as the basic strong effect on base pressure. This basic effect would be manifest by a correlation of data for wings with thin boundary layers at the trailing edge since we would not expect much dependence of base pressure on δ/h for thin boundary layers.   Such a correlation is presented in Fig. 9-28.   Functionally, this curve can be thought of with reference to Eq. (9-91) as

$$\frac{p_b}{p_0} = f_2(M_0, 0) \tag{9-92}$$

A large decrease in base pressure accompanies increases in Mach number, in accordance with the attempt of the base pressure to approach a vacuum.   The basic effect of Mach number can conceivably be caused by changes in the shape of the boundary-layer density and velocity profiles at separation, by changes in the mixing process behind the airfoil, etc.   The second factor appears to be more important than the first.

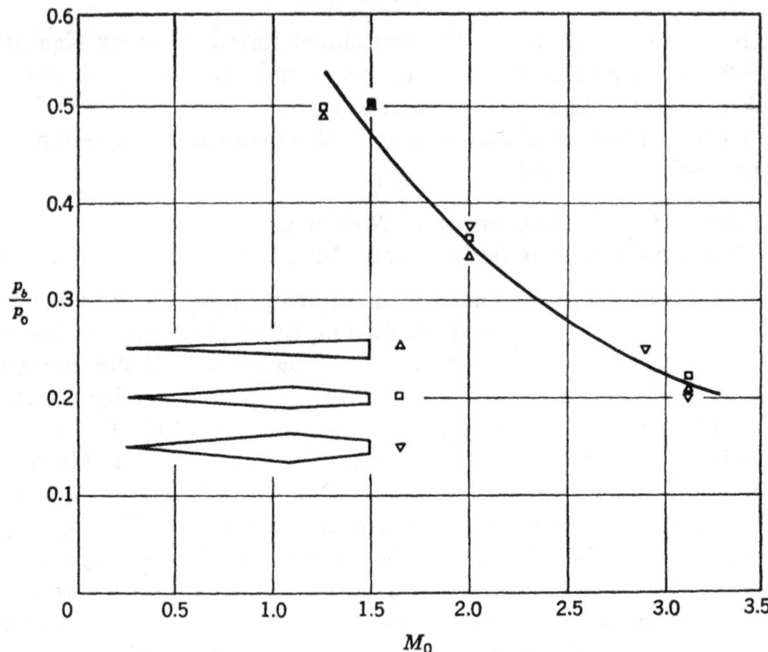

FIG. 9-28. Base-pressure correlation for airfoils with relatively thin turbulent boundary layers.

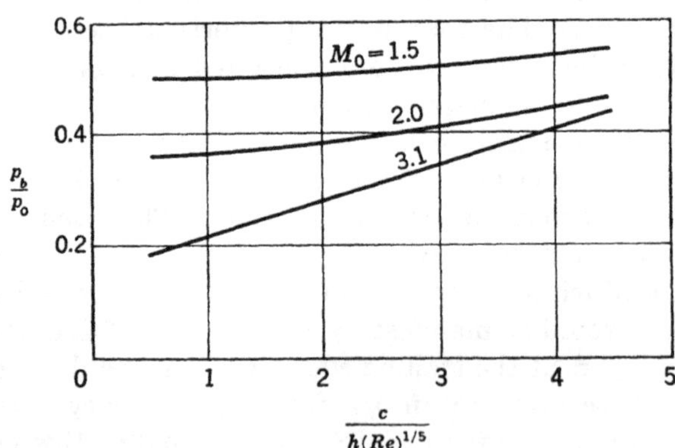

FIG. 9-29. Effect of Reynolds number on base pressure of airfoils with turbulent boundary layers.

We can imagine the changes in base pressure, superimposed on top of the basic Mach-number effect, as the boundary layer changes from thin to thick. The changes are represented by the variations of p_b/p_0 with the boundary-layer thickness parameter, $c/h\,\mathrm{Re}^{1/5}$, shown in Fig. 9-29. For the two lower Mach numbers the influence of the boundary-layer thickness on the base pressure is not large. In fact, the over-all change in base pressure is small when we consider that the airfoil boundary-layer thickness becomes as great as or greater than the trailing-edge thickness. At the Mach number of 3.1, the variation of base pressure with boundary-layer thickness is, however, larger than for 1.5 and 2.0

An examination of the correlation curves for bodies of revolution with fully turbulent boundary layers shown in Figs. 9-30 and 9-31 reveals the same qualitative effects of Mach number on base pressure for bodies as

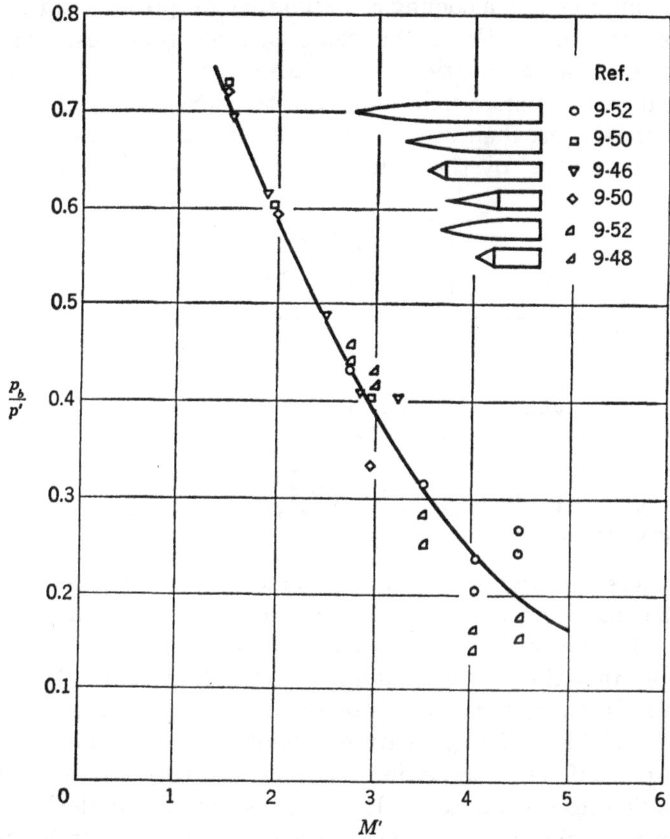

FIG. 9-30. Base-pressure correlation for bodies of revolution with relatively thin turbulent boundary layers.

for airfoils. However, the base pressure is generally higher than for airfoils at the same Mach number. This means that it is harder to maintain an "axially symmetric vacuum" than a "two-dimensional vacuum." The variation of the base pressure with boundary-layer thickness shown by Fig. 9-31 is nil. One reason for this might be that the base diameter for a body of revolution is much greater than the boundary-layer thickness; that is, the ratio δ/h is certainly much smaller for a body of revolution than for an airfoil, as evidenced by the range of $c/h \, \mathrm{Re}^{1/5}$ for airfoils in Fig. 9-29 compared to $L/h \, \mathrm{Re}^{1/5}$ for bodies of revolution in Fig. 9-31.

While the fully turbulent case is important in engineering missile applications, the transitional case is important for high speeds where heat transfer dictates a laminar boundary layer. In addition, increases in Mach number under certain circumstances have a stabilizing effect on the laminar boundary layer. See, for instance, the work of Czarnecki and Sinclair.[47] They found that cooling a parabolic body of revolution below the equilibrium temperature increased the length of laminar flow, and at high Mach numbers such cooling is mandatory for preserving the strength of missile structure. From the same sources previously mentioned, mean base-pressure correlation curves are presented in Fig. 9-32 for the transitional case. Before a discussion of the curves, a *word of caution* is necessary concerning their use. In the transitional case the base pressure is strongly influenced by the distance of the transition point behind the

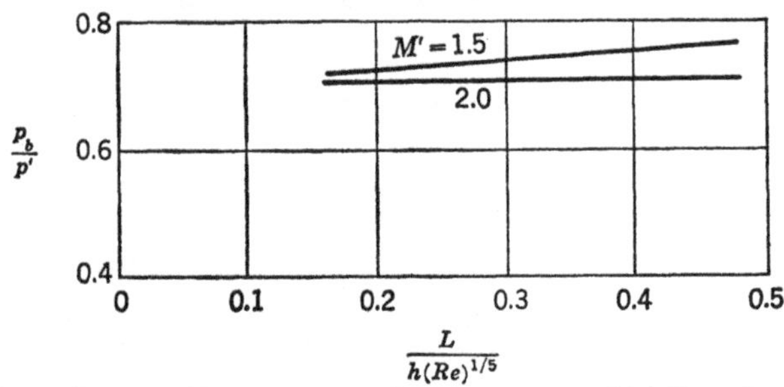

Fig. 9-31. Effect of Reynolds number on base pressure of bodies of revolution with turbulent boundary layers.

base. As a result, any of the numerous factors that can change the location of the transition point becomes a primary variable influencing base pressure. In the wind tunnel where these results were measured, the location of transition turned out to be dependent primarily on M' and δ/h. Happily then, the base-pressure data correlated on the basis of these two variables. In applications where other than the two foregoing variables can influence transition location, the correlations of Fig. 9-32 are only a first approximation. In the transitional case, the base pressure varies between the limits of base pressure for the purely laminar case, when the transition point is near the reattachment point, and base pressure for the turbulent case, when the transition point is near the separation point. These limits remain unchanged when new variables other than M' and δ/h influence transition, but the path between the limits is altered.

For the two-dimensional case the base pressure shows a rapid rise as the correlating parameter $c/h\ \mathrm{Re}^{1/2}$ increases. For small Reynolds numbers and large values of the correlating parameter the transition point is near the reattachment point, and the base pressure has the high value characteristic of the wholly laminar case.

The Reynolds number can be increased up to a critical value without moving the transition point. However, further increases in the Reynolds number cause the transition point to move toward the base and bring about the large depression noted in the base pressure. For bodies of revolution a similar result is observed. In this instance the base pressure

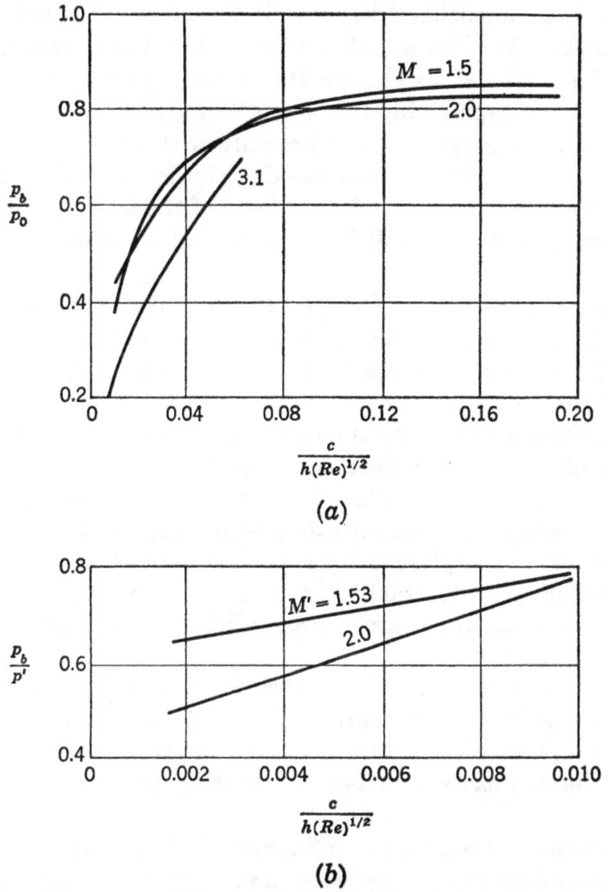

FIG. 9-32. Base-pressure correlations for (a) airfoils and (b) bodies of revolution with boundary layers turning turbulent behind the base.

decreases gradually as the Reynolds number is increased. The fundamental differences between the effect of Reynolds number for airfoils and bodies of revolution in the transitional case may be related to fundamental differences in the transition process in two-dimensional and axially symmetric flows.

9-13. Other Variables Influencing Base Pressure

Because we can discuss only qualitatively the influences of angle of attack, tail fins, and heating or cooling on base pressure, we have deferred consideration of these variables until now. First, for the variation of airfoil base pressure with angle of attack, data are available from Chapman et al.[51] For the purely turbulent case, little variation in base pressure occurs up to 5° angle of attack, the limit of the tests, for Mach numbers of 1.5 and 3.1. Changes in angle of attack can conceivably influence the base pressure through changes in p' or δ. However, it is a property of two-dimensional supersonic flow that p' is not sensitive to angle of attack. Also, for a turbulent boundary layer, pressure gradients would not be expected to change the boundary-layer thickness at the wing trailing edge appreciably for an angle of attack of 5°. These results explain the small changes in base pressure with angle of attack. If for higher angles of attack the trailing-edge shock wave should succeed in separating the upper boundary layer ahead of the trailing edge, the entire model shown in Fig. 9-26 will be altered, and changes in base pressure could result.

In contrast to its influence for the fully turbulent case, angle of attack can induce large changes in base pressure for the transitional case. If the transition point remains close to the reattachment point, it might be anticipated that the base pressure will remain constant. As a matter of observation the base pressure in some instances remains constant up to a small angle of attack, and then suddenly jumps to a higher value at a sharply defined angle of attack. If the angle is now decreased, the base pressure will again fall suddenly but sometimes with hysteresis. The phenomenon can be explained by a sudden shift in the transition point from a location near reattachment point to a position near the base. Such transition phenomena are, however, beyond the scope of engineering prediction at this time.

For bodies of revolution with a turbulent boundary layer, there is a *gradual* decrease in base pressure as the angle of attack increases. The decrease for a given change in angle of attack will become smaller as the Mach number increases because the limiting pressure of zero is being approached.

Some systematic tests on the influence of tail fins on base pressure for the fully turbulent case have been presented by Spahr and Dickey.[45] Tail fins change the general pressure level in the region of the outer flow around the dead water region but not in an axially symmetric fashion. It might be expected that their influence can be qualitatively treated like that of boattail, by taking into account the wing thickness pressure distribution. For the particular rectangular tail panels of Spahr and Dickey, negative pressure was induced behind the body base by the tail thickness pressure distribution when the trailing edge was at the base of the body. As a result, a large increase in the base drag occurred. Moving the tail forward about 1 chord length at $M_0 = 1.5$ and 2.0 eliminated the increase in base drag. By control of the airfoil section and the planform of the tail panels to induce positive pressure increments at the base, drag reduction can in principle be realized.

Heating or cooling of the boundary layer by heat transfer from the wing or body can affect the base pressure in a predictable manner. If heat is transferred from the body to the boundary layer, for instance by heating a test model, the boundary-layer temperature and speed of sound will be increased, and its Mach number will be lowered. With reference to Fig. 9-28, it is seen that the base pressure p_b will thus tend to rise. If the boundary layer is cooled by absorbing heat in the wing or body, the opposite effect will occur. Kurzweg[46] presents some systematic measurements for the effects of heat and cooling on base pressure of cone-cylinder combinations for Mach numbers from 2.5 to 5.0. He finds, as predicted, that addition of heat from the body to the air does increase the base pressure over that for no heat transfer, and cooling of the air by the body decreases the base pressure. In high-speed flight it will be necessary to cool the wing or body, that is, to lower the boundary-layer temperature below that for the adiabatic case, so that a decrease in base pressure will occur. Changes in boundary-layer thickness and changes in density and velocity profiles can also contribute to the net effect of heating or cooling on base pressure.

SKIN FRICTION

9-14. General Considerations of Skin Friction at Supersonic Speeds

The third general component of the drag is the skin friction. By the skin friction τ we mean the shearing force per unit area acting tangentially to a surface in motion relative to the viscous fluid adjacent to it. Skin friction and base drag, both being manifestations of viscosity, have much in common. For instance, we distinguish the same three cases for skin friction as for base pressure: laminar, transitional, and turbulent. The problems of skin friction and heat transfer in high-speed boundary layers are inseparable because the differential equations governing the boundary-layer velocity and temperature gradients are strongly coupled. It is a simple matter to determine the heat-transfer coefficient from the skin-friction coefficient if Reynolds analogy applies as it frequently does. However, we will not consider any heat-transfer calculations but will confine the discussion to drag. First, we describe the fundamental boundary-layer phenomena underlying skin friction in high-speed boundary layers, together with common terms used in that connection. We next present the "mean-enthalpy method" for calculating laminar skin friction illustrated by a calculative example, and then take up the same subject matter in connection with turbulent skin friction. Finally, we consider such matters as boundary layers with transition and application of flat-plate results to bodies of revolution.

The notation and units for the calculation of skin friction of high-speed boundary layers can be confusing since heat transfer and skin friction are simultaneously involved. Since this is the only part of the book where such notation is used, it seems desirable to list the notation together with an engineering set of units at this point.

SYMBOLS FOR SKIN FRICTION

c_F	local skin-friction coefficient, Eq. (9-101)
c_p	specific heat of air at constant pressure, Btu/(lb)(°R)
$\bar{c}_F$	average skin-friction coefficient over interval 0 to x
D_C	drag of cone due to skin friction, lb
D_P	drag of flat plate due to skin friction, lb
g	acceleration due to gravity of earth, 32.2 ft/sec²
h	enthalpy of air, Btu/lb
$h_0,\ h_r,\ h_s$	enthalpy of air at temperatures T_0, T_R, T_S, respectively, Btu/lb
h_i	zero of enthalpy scale, internal energy of perfect gas at absolute zero
h_t	enthalpy corresponding to stagnation temperature (and pressure), Btu/lb
J	mechanical equivalent of heat, 778 ft-lb/Btu
k	thermal conductivity of air, Btu/(ft)(sec)(°R)
L	length of boundary-layer run on cone, ft
M_0	free-stream Mach number
p	static pressure, lb/ft²
Pr	Prandtl number, $g\mu c_p/k$
q_0	free-stream dynamic pressure, lb/ft²
r	recovery factor for temperature, $(T_R - T_0)/(T_S - T_0)$
r_h	recovery factor for enthalpy, $(h_r - h_0)/(h_s - h_0)$
R	gas constant for air, 1718 ft²/(sec²)(°R)
Re	Reynolds number $V_0\rho x/\mu$
s	distance along slant surface of cone, ft
S_c	cone area, ft²
T	static temperature, °R
T_0	free-stream static temperature, °R
T_i	491.7°R
T_R	recovery temperature of insulated surface, °R
T_S	free-stream total (stagnation) temperature, °R
T_t	total temperature, °R
T_W	wall static temperature, °R
T^*	reference static temperature, °R
u	velocity parallel to plate, ft/sec
V_0	free-stream velocity, ft/sec
$x,\ y$	plate coordinates, Fig. 9-33, ft

$\gamma, \bar{\gamma}$ ratio of specific heat at constant pressure to that at constant volume, average value between temperatures T_0 and T_S when barred

δ boundary-layer thickness, ft

ϵ semiapex angle of cone, degrees

θ momentum thickness of boundary layer, ft

μ absolute viscosity, slugs/(ft)(sec)

μ_i 3.58×10^{-6} slug/(sec)(ft), reference viscosity used in Fig. 9-34 for 491.7°R and atmospheric pressure

ν kinematic viscosity, ft²/sec

ρ mass density of air, slugs/ft³

τ skin friction, lb/ft²

τ_c skin friction with compressible flow (with aerodynamic heating), lb/ft²

τ_i skin friction with incompressible flow (no aerodynamic heating), lb/ft²

$\bar{\tau}$ average skin friction between 0 and x

Superscripts and Subscripts:

0   referring to free-stream conditions or evaluated at T_0 as μ_0, ρ_0, V_0, γ_0

W evaluated at T_W as h_W, ν_W, ρ_W, or at wall as τ_W

*   evaluated at T^* as Re*, Pr*, ρ^*, c_p^*, μ^*, h^*

In the following sections we consider boundary layers which are *purely laminar, transitional,* and *purely turbulent.* Some preliminary knowledge on the part of the reader concerning boundary layers is assumed. Certain of the physical concepts and definitions pertaining to boundary layers are common to all three cases. It is our purpose to discuss at this time such of these as we shall require. To this end Fig. 9-33 has been constructed, showing in its upper part the boundary layer formed on a flat plate mounted at zero incidence as in a wind tunnel with a free stream of uniform velocity, temperature, and Mach number. The first quantity which describes the boundary layer is its thickness δ as a function of x. No sharp outer edge of the boundary layer can be discerned, so that some arbitrary definition is necessary. One such definition states that the thickness δ of the boundary layer corresponds to that position where the velocity parallel to the plate has reached 99 per cent of the free-stream velocity. The velocity u parallel to the plate can then be expressed in nondimensional form.

$$\frac{u}{V_0} = f\left(\frac{y}{\delta}\right) \approx \left(\frac{y}{\delta}\right)^n \tag{9-93}$$

The function f describes the *velocity profile* shown in the figure. For a low-speed laminar boundary layer we have an approximately parabolic velocity profile. For turbulent flow $n = \frac{1}{5}$ at low Reynolds numbers and $n = \frac{1}{9}$ at high Reynolds numbers except for a laminar sublayer.

Corresponding to the velocity profile there is also a *static temperature profile* as well as a *total temperature profile*. The static temperature is the temperature a thermometer would register if moving along with the local fluid velocity; the total temperature is the temperature of the fluid if brought to rest with respect to the plate with no energy transfer. If the velocity parallel to the plate at any position in the boundary layer is

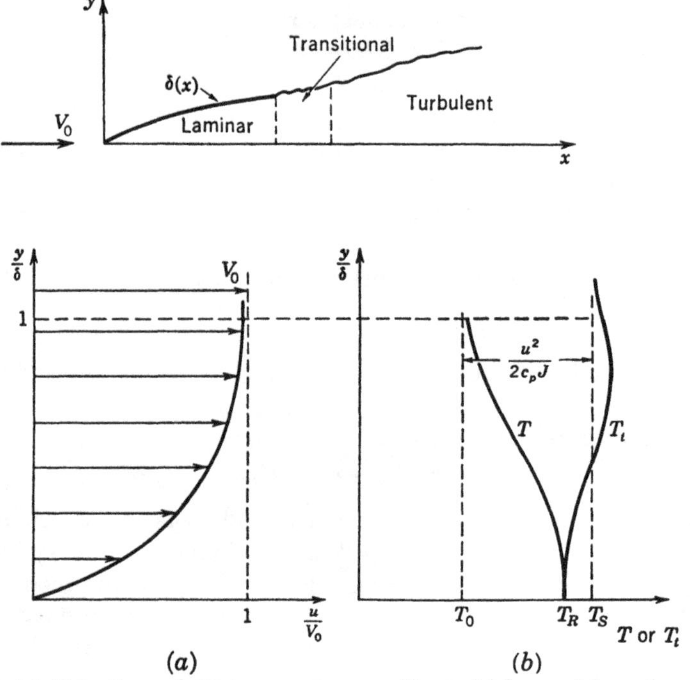

Fig. 9-33. (a) Velocity and (b) temperature profiles in high-speed boundary layer of insulated plate.

represented by u, then the total temperature and static temperature are related by

$$T_t = T + \frac{u^2}{2\bar{c}_p J} \tag{9-94}$$

Here $\bar{c}_p$ is the specific heat at constant pressure at its average value between T and T_t. In cases where the specific heat c_p is too variable to be replaced conveniently by its average value (although this can in principle always be done), we can say that the total and static temperatures are related through their corresponding enthalpies (a function only of pressure and temperature)

$$h_t = h + \frac{u^2}{2J} \tag{9-95}$$

The total temperature corresponds to that enthalpy which the local fluid mass would have if it were brought to rest with respect to the plate without any net change in work or energy such as viscous work, heat conduction, or radiation crossing the surface that contains it.

Consider now the case of an insulated plate. As the viscous layers shear one over the other, they do mechanical work on the layers between them and the plate. Since the plate is insulated, the temperature of the inner layers is thereby raised. To maintain an energy balance, the energy supplied to the inner layers by viscous work must be conducted outward again by heat conduction. It is clear that the temperature of the insulated plate will rise until the heat transferred outward from the inner layers is in balance with the viscous work done on them. The plate equilibrium temperature is called the *recovery temperature*. Let us examine the static temperature variation through the boundary layer during this physical process. At the wall we have no heat transfer, so that $\partial T/\partial y$ is zero as shown in Fig. 9-33. However, away from the wall the static temperature falls in the outward direction, and the heat conduction is away from the plate. Near the edge of the boundary layer the gradient is again small, since the shearing force is small, together with the rate work is being done on the fluid between the edge of the boundary layer and the flat plate.

The total temperature profile across the boundary layer of the insulated plate is of interest. At the wall the static and total temperatures are equal and have the common value called the recovery temperature. Since T_t is a measure of the total energy per unit mass of fluid, it must have an average value across the boundary layer equal to T_S, the free-stream total temperature. As a consequence there are regions in the boundary layer where T_t is greater than the free-stream stagnation temperatures. For the insulated plate we have $\partial T_t/\partial y$ is zero at the plate (as well as $\partial T/\partial y$) by direct application of Eq. (9-94) with $u = 0$ at the wall.

The idea of recovery temperature has been explained. For air the recovery temperature lies somewhere between T_0 and T_S. Such behavior is typical of a fluid having a Prandtl number less than unity. The *recovery factor r* is a quantity used to specify the recovery temperature

$$r = \frac{T_R - T_0}{T_S - T_0} \tag{9-96}$$

The recovery factor is thus a measure of how close the recovery temperature approaches the free-stream stagnation temperature. It is frequently convenient to define a recovery factor based on enthalpy

$$r_h = \frac{h_r - h_0}{h_s - h_0} \tag{9-97}$$

This factor is convenient when the specific heat is varying rapidly with temperature over the range of interest, as when dissociation or ionization occurs.

Before considering how the temperature variation through the boundary layer affects the skin friction, let us consider what happens when the plate temperature is not T_R.   If, by means of internal cooling of the plate, its temperature is dropped below T_R, there will be heat conduction to the plate.   The slope of the static temperature profile at the wall will not be zero, as shown in Fig. 9-33, but will be positive.   Also, the average value of T through the boundary layer will be lower.   The same comments apply to T_t.   We will consider skin friction under circumstances of cooling and heating of the plate with the plate temperature T_W less than or greater than the recovery temperature T_R.

Until now we have made no distinction between the flat plate in the wind tunnel and a flat plate flying through still air.   So far as the present analytical representation of the temperature and velocity profiles is concerned, there is no essential difference.   However, there are certain differences as far as energy transfer is concerned.   In the wind tunnel, air in a reservoir at stagnation temperature T_S is expanded to some velocity V_0 and a static temperature T_0 less than T_S.   The *free-stream air in motion does work* on the boundary layer of the plate, and thereby raises the boundary-layer static temperature.   The static temperature difference between the boundary layer and the free stream conducts heat back into the free stream.   In flight in still air at static temperature T_0, the plate moves through the still air at high speed.   In so doing, the *plate does work* on the boundary layer, raising its static temperature.   The static temperature difference between the boundary layer and the free stream sets up heat conduction into the free stream.   The direction of heat conduction is still the same.   However, in the wind tunnel the work to heat the boundary layer comes out of the free-stream flow, but in flight the work comes from the plate.

Let us now examine how the temperature and velocity profiles enter into determination of the skin friction.   The skin friction is related to the velocity gradient for small gradients through the absolute viscosity by definition

$$\tau = \mu \frac{\partial u}{\partial y} \tag{9-98}$$

We assume that μ does not depend on the gradient $\partial u/\partial y$ for the magnitudes of the gradients we are considering.   The absolute viscosity μ is dependent only on T, but through division by density it becomes the kinematic viscosity ν which is dependent on temperature and pressure.

$$\nu = \mu/\rho \tag{9-99}$$

By applying Eq. (9-98) at the wall to obtain the skin friction there, we have

$$\tau_W = \mu_W \left(\frac{\partial u}{\partial y}\right)_W \tag{9-100}$$

The value of μ_W depends on the temperature at the wall, and the value of $(\partial u/\partial y)_W$ depends on the velocity profile.

It is interesting to see how aerodynamic heating influences skin friction through opposing effects in the two terms of Eq. (9-100). What is meant by the effect of aerodynamic heating on skin friction? Let the free-stream flow conditions approaching an insulated nonradiating plate be fixed for the discussion. If we ignore aerodynamic heating, the plate will not heat up. The skin friction can then be calculated from incompressible-flow theory or correlations strictly valid for $M = 0$. However, if we consider aerodynamic heating, the plate will heat up. The skin friction must then be calculated by a method which accounts for the fact that the Mach number is not essentially zero, and the calculated skin friction will be lower than for no aerodynamic heating. This reduction in τ_W is what we term the effect of aerodynamic heating on skin friction. Specifically, the increased plate temperature has the direct effect of increasing μ_W in Eq. (9-100), and thereby increasing τ_W. However, the increased boundary-layer temperatures have a diminishing influence on $(\partial u/\partial y)_W$, which is conveniently thought of as a Reynolds-number effect. The increased temperatures reduce the densities and increase the viscosities in the constant-pressure boundary layer of the plate. The resulting decrease in Reynolds number is known to increase the boundary-layer thickness δ for both laminar and turbulent boundary layers. Since $(\partial u/\partial y)_W$ is inversely proportional to δ, aerodynamic heating has brought about a decrease in $(\partial u/\partial y)_W$. In fact, this influence of aerodynamic heating on $(\partial u/\partial y)_W$ more than offsets the increase in μ_W, so that τ_W is reduced.

If we know the velocity profile and the surface temperature, we can calculate the skin friction from Eq. (9-100). More frequently the skin friction is obtained from experimental correlations of the skin-friction coefficient. The local skin-friction coefficient is defined by

$$c_F = \frac{\tau}{\frac{1}{2}\rho^* V_0^2} \tag{9-101}$$

where ρ^* is the density evaluated at some convenient reference temperature. For incompressible flow, ρ^* is taken as the free-stream density ρ_0. We define also the local Reynolds number based on the distance x from the plate leading edge and a reference temperature T^*.

$$\text{Re}^* = \frac{V_0 x \rho^*}{\mu^*} \tag{9-102}$$

The Prandtl number based on T^* is

$$\text{Pr}^* = \frac{g\mu^* c_p{}^*}{k^*} \qquad (9\text{-}103)$$

Sometimes the average skin friction $\bar{\tau}$ between 0 and x is desired rather than the local values.   An average skin-friction coefficient based on $\bar{\tau}$ can be defined

$$\bar{c}_F = \frac{\bar{\tau}}{\tfrac{1}{2}\rho^* V_0{}^2} \qquad (9\text{-}104)$$

For low-speed flow, T^* is usually free-stream temperature.   Determining a proper value for T^* in high-speed boundary layers is a problem we will discuss shortly.

9-15. Laminar Skin Friction, Mean-enthalpy Method

The general mechanisms whereby aerodynamic heating influences skin friction have been conveyed in the previous section, and in this section an engineering method will be discussed for the calculation of laminar skin friction.   Several methods are to be found in the literature for the calculation of heat transfer and skin friction in high-speed boundary layers, notably the mean-enthalpy method used by Rubesin and Johnson,[54] and subsequently by Eckert.[55]   The mean-enthalpy method, applied by Rubesin and Johnson to laminar boundary layers, was applied to turbulent boundary layers by Sommer and Short.[66]   The essential point of these methods is to find some reference temperature which will give the skin friction of the high-speed boundary layer if used to evaluate the temperature-dependent quantities in the well-known solution for incompressible laminar boundary layers on a flat plate (Blasius solution).   If such a reference temperature can be specified, the problem of the high-speed laminar layer is reduced to an equivalent low-speed problem.   We are in the fortunate position of being able to test any particular scheme for finding such a reference temperature.   Numerical solutions are available for laminar boundary layers which take into account all the temperature-dependent physical properties such as c_p, k, and μ.   Comparison of any prospective engineering method for calculating laminar skin friction of high-speed boundary layers with the exact numerical theory discloses the accuracy of such a method.   On the other hand, exact solutions can also be used to determine what reference temperature would give the high-speed laminar skin friction if used in the low-speed theory.

The temperature profile in a high-speed boundary layer is dependent on the free-stream temperature T_0, the plate temperature T_W, and the free-stream Mach number M_0.   Let us replace the free-stream Mach number by an independent temperature parameter, the stagnation temperature T_S of the free-stream flow

$$T_S = T_0\left(1 + \frac{\bar{\gamma} - 1}{2}\,M_0{}^2\right) \tag{9-105}$$

Eckert gives two methods of determining the reference temperature T^* in terms of these three independent temperatures: T_0, T_W, and T_S. The first method is useful when the variation in specific heat c_p is not large. In this case Eckert gives the following empirical result for T^*.

$$T^* = T_0 + 0.5(T_W - T_0) + 0.22r\,\frac{\bar{\gamma} - 1}{2}\,M_0{}^2T_0$$
$$T^* = T_0 + 0.5(T_W - T_0) + 0.22r(T_S - T_0) \tag{9-106}$$

The temperature recovery factor r depends on the Prandtl number Pr^* evaluated at T^*.

$$r = (\mathrm{Pr}^*)^{\frac12} \tag{9-107}$$

The Prandtl number is not sensitive to T^*. To obtain T^*, first assume a value of T^*, obtain r from Eq. (9-107), and compute a new value from Eq. (9-106). The second method for obtaining T^* based on enthalpy has essentially the same form as the first method.

$$h^* = h_0 + 0.5(h_w - h_0) + 0.22r_h(h_s - h_0) \tag{9-108}$$

The enthalpy recovery factor r_h is also given by Eq. (9-107). If the specific heat c_p is constant, the two methods give identical results. For rapidly changing c_p as in a dissociating boundary layer, the second method is preferable.

The definitions of the skin-friction coefficient and Reynolds number, Eqs. (9-101) and (9-102), have been presented in such a fashion that the skin friction can be calculated once the reference temperature T^* is known.

$$c_F = \frac{0.664}{(\mathrm{Re}^*)^{\frac12}} = 0.664\left(\frac{V_0 x \rho^*}{\mu^*}\right)^{-\frac12}$$
$$\tau_W = c_F\,\frac{\rho^* V_0{}^2}{2} \tag{9-109}$$

To carry out the calculation we need the values of μ, h, and Pr as a function of temperature. The value of ρ is given with sufficient accuracy for undissociated air by the gas law.

$$\rho = \frac{P}{RT} \tag{9-110}$$

where $R = 1718$ for the units of p, ρ, and T given in the list of symbols in the previous section. Small plots of the temperature-dependent physical quantities are given in Fig. 9-34 for ordinary engineering calculations. For precise calculations the tables of Hilsenrath et al.[57] are available.

The reference values for Fig. 9-34 are

$$\mu_i = 3.58 \times 10^{-7} \text{ slug}/(\text{sec})(\text{ft})$$
$$T_i = 491.7°\text{R}$$
$$R = 1718 \text{ ft}^2/(\text{sec}^2)(°\text{R})$$
$$h_i = \text{internal energy of perfect gas at temperature}$$
$$\text{of absolute zero}$$

$(9\text{-}111)$

Illustrative Example

Determine the reference temperature T^*, the recovery temperature T_R, and the local skin-friction coefficient (laminar) a distance 1.0 ft behind the

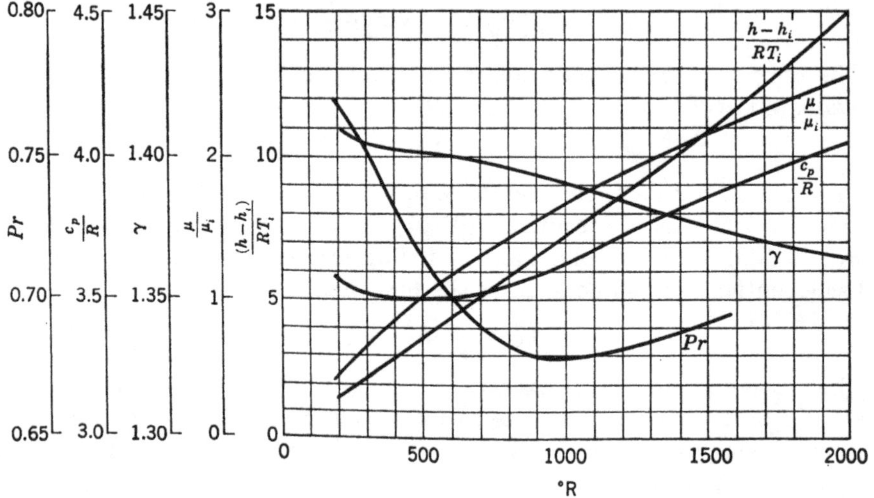

FIG. 9-34. Variation with temperature of certain physical constants for air.

leading edge of an insulated plate in a stream of static temperature 400°R, a Mach number of 3, and a pressure of 500 lb/ft². Use the constant specific-heat method. Neglect radiation.

As a trial value of T^* use 800°R. From Fig. 9-34 Pr* is 0.684, and from Eq. (9-107)

$$r = (0.684)^{1/2} = 0.827$$

Using a value of $\bar{\gamma} = 1.4$, we obtain for the stagnation temperature by Eq. (9-105)

$$T_S = 400 \left(1 + \frac{1.4 - 1}{2} 3^2 \right)$$
$$= 1120°\text{R}$$

The recovery temperature by Eq. (9-96) is

$$T_R = T_0 + r(T_S - T_0)$$
$$= 400 + (1120 - 400)(0.827)$$
$$= 995°\text{R}$$

We can now check the assumed value of T^* by Eq. (9-106). For an insulated plate $T_W = T_R$, neglecting radiation,

$$T^* = 400 + 0.5(995 - 400) + 0.22(0.827)(1120 - 400)$$
$$= 829°R$$

The values of T^* could be further improved by repeating the process with $T^* = 829°R$ as a trial value. The change in Pr^* with T^* is not large enough to warrant another approximation in this instance.

Let us now calculate the skin friction and skin-friction coefficients for $x = 1.0$ ft with the help of Eq. (9-109). To obtain c_F we need V_0, ρ^*, and μ^*. Since the speed of sound is $(\gamma R T)^{1/2}$, we have

$$V_0 = (\gamma_0 R T_0)^{1/2} M_0$$
$$V_0 = [1.403(1718)400]^{1/2}(3) = 2950 \text{ ft/sec}$$

The gas law, Eq. (9-110), yields

$$\rho^* = \frac{p}{RT^*} = \frac{500}{1718(829)} = 0.000351 \text{ slug/ft}^3$$

The viscosity ratio from Fig. 9-34 is 1.46

$$\frac{\mu^*}{\mu_i} = 1.46$$
$$\mu^* = 1.46(3.58 \times 10^{-7})$$
$$= 5.22 \times 10^{-7} \text{ slug/(ft)(sec)}$$

The Reynolds number based on the reference temperature is

$$\text{Re}^* = \frac{V_0 x \rho^*}{\mu^*} = \frac{2950(1.0)(3.51)10^{-4}}{5.22 \times 10^{-7}}$$
$$= 1.98 \times 10^6$$

The local skin-friction coefficient is

$$c_F = 0.664(\text{Re}^*)^{-1/2} = 0.664(1.98)^{-1/2}(10^{-3})$$
$$= 0.00047$$

The skin friction is

$$\tau = \frac{(4.7 \times 10^{-4})(2950)^2(3.51 \times 10^{-4})}{2}$$
$$= 0.72 \text{ lb/ft}^2$$

It is of interest to see how much the influence of T^* on the skin friction is. For no aerodynamic heating, but for the same V_0, the value of T^* would have been 400°R rather than 829°R. Let us call τ_c the skin friction taking in account aerodynamic heating, and let τ_i be the skin friction totally ignoring it. Thus, τ_i corresponds to $T^* = 400°R$, and τ_c corresponds to $T^* = 829°R$. If the quantities V_0 and x are held constant, we see from Eq. (9-109) that

$$\tau_W \propto (\mu^*\rho^*)^{\frac{1}{2}}$$

$$T^* = 400 \qquad \frac{\mu^*}{\mu_i} = 0.85$$

$$T^* = 829 \qquad \frac{\mu^*}{\mu_i} = 1.46$$

For constant pressure ρ^* is inversely proportional to T^*; thus

$$\frac{\tau_c}{\tau_i} = \left(\frac{1.46}{0.85}\right)^{\frac{1}{2}} \left(\frac{400}{829}\right)^{\frac{1}{2}}$$
$$= 0.91$$

Actually, there is not much change in skin friction at $M = 3$ due to aerodynamic heating—a decrease of 9 per cent. If the viscosity increased directly with temperature, then the decrease of ρ with temperature increase would directly offset the tendency for the viscosity to increase the skin friction. That the laminar skin friction decreases slightly as the Mach number increases can be ascribed to the fact that the rate of change of viscosity with absolute temperature is slightly less than linear. For turbulent skin friction we will find a different state of affairs.

9-16. Turbulent Skin Friction

How aerodynamic heating changes the skin friction for a turbulent boundary layer cannot be investigated along the same theoretical lines as for a laminar boundary layer. The difference arises in the fact that, whereas the physical processes in laminar boundary layers are well represented by the Navier-Stokes equations, the physical aspects of turbulent boundary layers are not well understood. We must therefore check engineering methods for calculating turbulent skin friction against experiment since we have no exact solutions. One of the first things to try might be the process that has been described in the previous section for laminar skin friction. This Eckert has done and has checked the results against experiment. His conclusion is that the general process for calculating laminar skin friction applies to turbulent skin friction with the sole change that the recovery factor now is

$$r = (\text{Pr}^*)^{\frac{1}{3}} \tag{9-112}$$

The local skin friction for turbulent flow on the basis of the Schulz-Grunow formula is

$$c_F = \frac{0.370}{(\log_{10} \text{Re}^*)^{2.584}} \tag{9-113}$$

and the average skin-friction coefficient is given by the relationship of Prandtl-Schlichting

$$\bar{c}_F = \frac{0.455}{(\log_{10} \text{Re}^*)^{2.58}} \tag{9-114}$$

Illustrative Example

Recalculate the example of the preceding section for a turbulent boundary layer.

As in the preceding example, let the trial value of T^* be 800°R. The recovery factor from Eq. (9-112) is

$$r = (\text{Pr}^*)^{\frac{1}{3}} = (0.684)^{\frac{1}{3}} = 0.881$$

The value of the recovery temperature is from Eq. (9-96)

$$T_R = 400 + 0.881(1120 - 400)$$
$$= 1035°R$$

The recovery temperature of 1035°R for the turbulent layer compares with 995°R for the laminar layer. The reference temperature now is

$$T^* = 400 + 0.5(1035 - 400) + 0.22(0.881)(1120 - 400)$$
$$= 857°R$$

A further approximation will not be attempted.

To obtain c_F we must obtain μ^* and ρ^*. From Fig. 9-34, we have

$$\frac{\mu^*}{\mu_i} = 1.50$$
$$\mu^* = 1.50(3.58)10^{-7}$$
$$= 5.37 \times 10^{-7} \text{ slug/(ft)(sec)}$$

From the gas law

$$\rho^* = \frac{500}{1718(857)} = 3.39 \times 10^{-4} \text{ slug/ft}^3$$
$$\text{Re}^* = \frac{2950(1.0)(3.39)10^{-4}}{5.37 \times 10^{-7}} = 1.86 \times 10^6$$

The local skin-friction coefficient by Eq. (9-113) is

$$c_F = \frac{0.370}{(6.270)^{2.584}}$$
$$= 0.00322$$

The skin friction by Eq. (9-109) is

$$\tau_W = \frac{0.00322(3.39 \times 10^{-4})(2.950)^2 \times 10^6}{2}$$
$$= 4.75 \text{ lb/ft}^2$$

It is of interest to see how much the skin friction has been changed as a result of aerodynamic heating. We therefore calculate the skin friction as if the reference temperature were 400°R.   Then

$$\frac{\mu^*}{\mu_i} = 0.85$$

$$\mu^* = 0.85(3.58 \times 10^{-7}) = 3.04 \times 10^{-7} \text{ slug/(ft)(sec)}$$

$$\rho^* = \frac{500}{1718(400)} = 7.28 \times 10^{-4} \text{ slug/ft}^3$$

$$\text{Re}^* = \frac{2950(1.0)(7.28 \times 10^{-4})}{3.04 \times 10^{-7}} = 7.05 \times 10^6$$

$$c_F = \frac{0.370}{(6.848)^{2.584}} = 0.00257$$

$$\tau_W = \frac{0.00257(7.28 \times 10^{-4})(2.95)^2 10^6}{2}$$

$$= 8.10 \text{ lb/ft}^2$$

The ratio of skin friction with and without aerodynamic heating is

$$\frac{\tau_c}{\tau_i} = \frac{4.75}{8.10} = 0.587$$

In this instance the skin friction of the turbulent layer has been reduced over 40 per cent as the result of aerodynamic heating, compared with only about 10 per cent for laminar flow under the same conditions.

The general effect of aerodynamic heating on skin friction is of interest. The Mach number is the primary variable, but the air temperature and plate temperature also enter as parameters. For a given air temperature and a plate of fixed thermal insulation, we can plot τ_c/τ_i against M_0. With regard to the thermal insulation, let us take the case of a perfectly insulated nonradiating plate. The variation in τ_c/τ_i is shown versus M_0 in Fig. 9-35 for ordinary air temperatures. The very considerable decrease in skin friction due to aerodynamic heating for a turbulent boundary layer is noteworthy. This decrease is much greater than for a laminar layer. Data confirming the general trend shown by this curve are to be found in Chapman and Kester.[58]

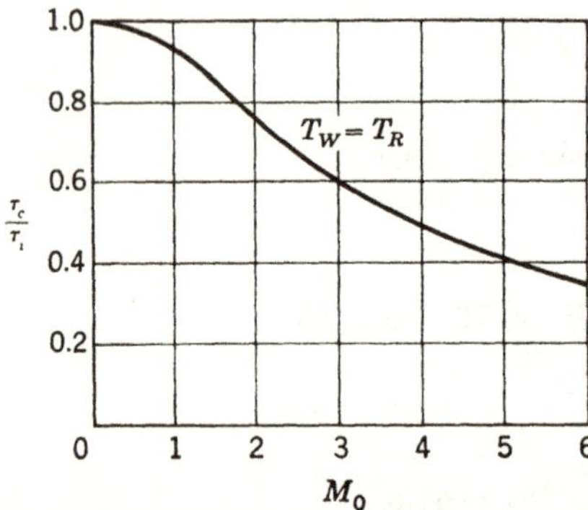

FIG. 9-35. Effect of Mach number on the local skin-friction coefficient of an insulated flat plate in air at room stagnation temperatures.

9-17. Other Variables Influencing Skin Friction

The methods of computing skin friction covered in the two previous sections apply to flat plates with no pressure gradients and at uniform temperature with completely laminar or completely turbulent flow. In practice, it is necessary to apply flat-plate results to wings and bodies, to regions of nonuniform pressure and temperature, and to boundary layers that are partially laminar and partially turbulent. Let us first consider boundary layers that are neither totally laminar nor totally turbulent.

Determining the location of the transition zone is one of the obstacles to successful prediction of skin friction of a missile under flight conditions. A few observations can be made concerning transition for particular

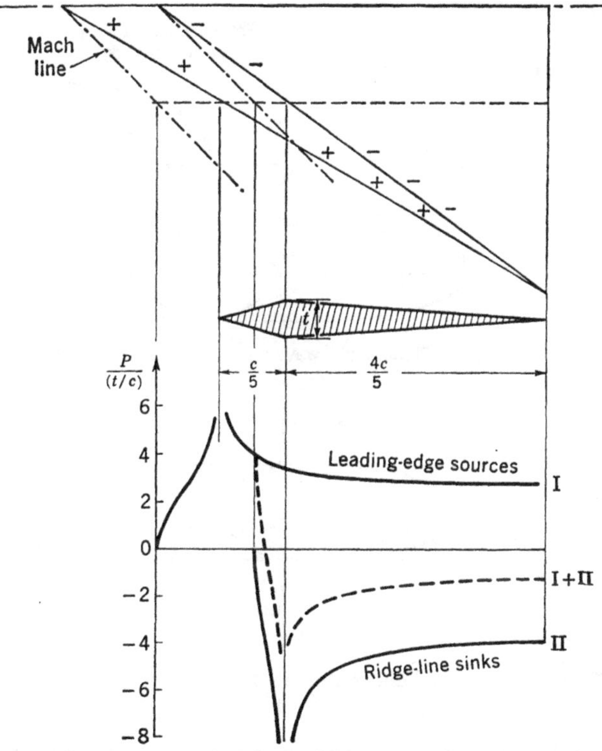

FIG. 9-36. Pressure distribution due to thickness on double-wedge triangular wing with subsonic ridge line.

bodies and wings, at least as observed in a particular wind tunnel. Because these results illustrate principles, they are of interest here. Some studies of transition have been made in connection with triangular wings of double-wedge section using the liquid-film technique as described by Vincenti.[59] One case is illustrated in Fig. 9-36 for the wing at zero angle of attack. The question to be investigated is whether the thickness pressure distribution has sharp rises which might induce transition. The thickness pressure distribution for double-wedge wings can easily be constructed by adding the pressure distribution for a pair of leading-edge sources to that for a pair of ridge-line sinks of the Jones type (Sec. 2-5).

Two cases are differentiated; case 1 of Fig. 9-36 for subsonic ridge lines, and case 2 for supersonic ridge lines. The total pressure distribution I + II for subsonic ridge lines shows a rapid increase in pressure directly behind the ridge line. This pressure rise was found[59] to induce transition at the ridge line. For a supersonic ridge line the pressure rise is delayed to the Mach lines associated with the ridge lines, and transition occurs further back on the wing. The drag measurements confirmed greater laminar flow area for case 2. Under conditions of angle of attack, the lifting pressure distribution further complicates the problem.

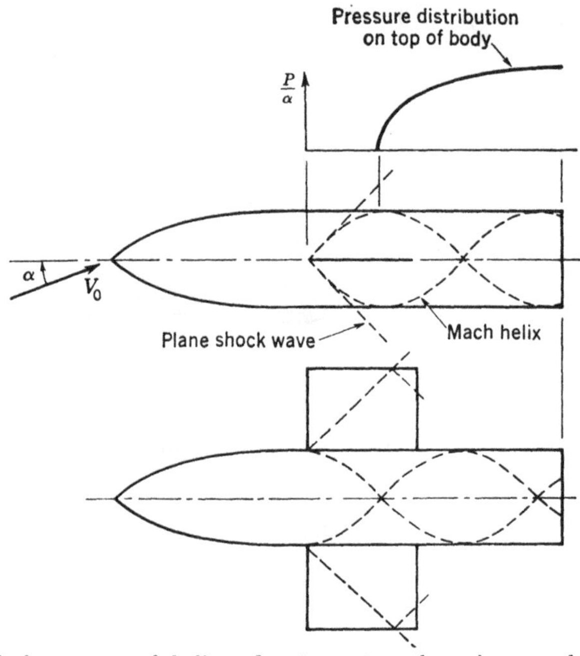

FIG. 9-37. Mach waves and helices due to rectangular wing panels of wing-body combination.

Wind-tunnel tests also show that positive pressure waves arising from the leading edge of a wing-body juncture can cause transition on the body. The boundary-layer conditions on the top and the bottom of a body in the neighborhood of a rectangular wing centrally mounted on the body have been reported by Pitts et al.[60] The general leading-edge wave pattern for such a wing-body combination is shown in Fig. 9-37. The combination at zero angle of attack produces a positive wave intersecting the body in a pair of Mach helices. The pressure distribution at the top of the body is shown to have a sharp pressure rise at the intersection of the helices which tends to induce transition. If the body angle of attack is increased, the pressure rise may be replaced by the pressure decrease of a Prandtl-Meyer fan. In this event transition would be inhibited.

Having discussed several specific examples of how pressure distribution fixes the transition zone, let us now consider the problem of calculating the skin friction if the location of the transition zone is known and if the zone is of small breadth. With reference to Fig. 9-38, the skin friction up to the transition point T can be calculated on the basis of a laminar boundary layer. However, beyond T the results for the purely turbulent boundary layer cannot be applied directly, since the turbulent boundary layer starts with finite rather than zero thickness. Some scheme is required for joining the laminar results to the turbulent results. This can be accomplished in several ways. It is assumed that the state of the turbulent boundary layer right after transition is the same as if the boundary layer had been purely turbulent from some virtual origin. The origin is located on the basis that the total skin friction up to point T is the same

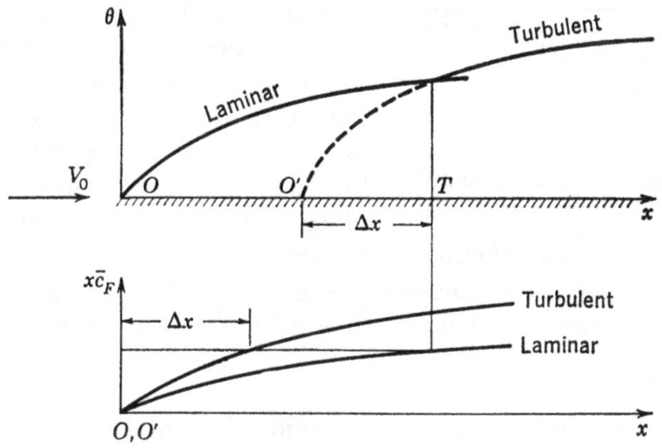

FIG. 9-38. Example illustrating method of locating virtual origin of turbulent boundary layer for transition on flat plate; narrow transition zone.

for a purely turbulent boundary layer originating at O' as for a purely laminar boundary layer originating at O. Such a condition is equivalent to equal momentum thickness θ of the laminar and turbulent boundary layers at transition. The equivalence is easily seen from the equality between the average skin-friction coefficient $\bar{c}_F$ between 0 and x, and the momentum thickness at distance x.

$$\bar{c}_F = \frac{2\theta}{x}$$

Locating the virtual origin O' requires only methods for calculating the skin-friction coefficient for purely laminar flow and purely turbulent flow, methods presented in the preceding two sections. The method of locating the virtual origin is illustrated graphically in Fig. 9-38. Curves of $x\bar{c}_F$ are constructed as functions of x, using the method for purely turbulent and purely laminar boundary layers, and the distance is Δx determined as shown.

We now consider the problem of the application of flat-plate skin friction to a nonplanar configuration such as a body of revolution. The usual general method is used, namely, the laminar boundary-layer equations are solved for a flat plate and body of revolution and are compared. This procedure has been applied by Mangler[61] to a cone, as well as by Hantsche and Wendt.[62] In the first case the cone acts in a pressure field higher than free-stream pressure, so that the reference quantities just outside the boundary layer are different from those of a flat plate in the same stream. This difference is taken into account by a simple change in reference quantities for the skin friction. Perhaps the essential difference between the cone and the flat plate is that the boundary layer is spread out as it progresses downstream. Thinning the boundary layer tends to increase the velocity gradients through it, and thereby to increase the skin friction. The theoretical analyses show that the local skin-friction coefficient on a cone is $3^{1/2}$ *greater* than the local skin-friction coefficients on a flat plate for the same boundary-layer length. Another way of stating the same result is that the local skin-friction coefficient for a cone corresponds to those for a flat plate at one-third the Reynolds number. This result applies solely to laminar flow.

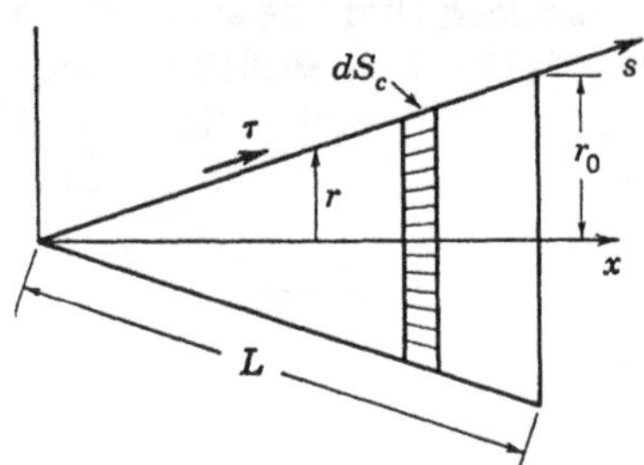

Fig. 9-39. Notation for evaluating laminar skin friction on cone.

Illustrative Example

Compare the total laminar skin-friction drag of a cone with that of a flat plate of equal area and of a length equal to the slant height of the cone for the same dynamic pressure outside the boundary layer. See Fig. 9-39.

The local skin-friction coefficient for the cone is

$$c_F = \frac{3^{1/2}(0.664)}{\mathrm{Re}^{1/2}}$$

where the local Reynolds number is

$$\mathrm{Re} = \frac{V_0 \rho_0 s}{\mu_0}$$

The drag of the cone due to skin friction is

$$D_C = \cos \epsilon \int_{S_c} c_F q_0 \, dS_C$$

with the differential area dS_C

$$dS_C = 2\pi r \, ds$$

The skin-friction drag of the cone is

$$\frac{D_C}{q_0} = \frac{4}{3^{1/2}} \frac{0.664\pi L^2 \sin \epsilon \cos \epsilon}{(V_0\rho_0 L/\mu_0)^{1/2}}$$

The average skin-friction coefficient for the plate is twice the local value for the Reynolds number based on length L.

$$\bar{c}_F = \frac{2(0.664)}{(V_0\rho_0 L/\mu_0)^{1/2}}$$

As a result the total skin-friction drag for the plate D_P is

$$\frac{D_P}{q_0} = \frac{2(0.664)\pi L^2 \sin \epsilon}{(V_0\rho_0 L/\mu_0)^{1/2}}$$

The ratio of cone skin-friction drag to "equivalent-plate" skin-friction drag is

$$\frac{D_C}{D_P} = \frac{2}{3^{1/2}} \cos \epsilon = 1.15 \cos \epsilon$$

In words, the total laminar skin-friction drag of a slender cone is 15 per cent greater than the plate of equal area and of length equal to the cone slant height. It is clear that other "equivalent plates" can be set up which will give different percentages.

To conclude our discussion of the application of flat-plate skin-friction data to nonplanar bodies, let us consider the turbulent boundary layer for cones and some results for circular cylinders. Van Driest[63] finds that the local skin-friction coefficients on cones with turbulent boundary layers correspond to those for a flat plate at half the local Reynolds number, rather than one-third the local Reynolds number as for a laminar layer. Eckert[55] concludes that the skin-friction coefficients for flat plates can be applied directly to cylinders provided the ratio of boundary-layer displacement thickness to cylinder radius does not exceed 0.01 or 0.02.

The influence of nonuniform surface temperature on the skin-friction and heat transfer for laminar flow over a flat plate has been treated by several authors, notably Chapman and Rubesin.[64]

SYMBOLS

a_0	coefficient of log term in Eq. (9-18)
A	aspect ratio of wing
A_A	aspect ratio of arrow wing
A_n, B_n	coefficients in Eq. (9-18)
A_T	aspect ratio of triangular wing
b_0	coefficient in Eq. (9-18)
b_n	Fourier coefficients in Eq. (9-39)

c	chord of two-dimensional airfoil
c_d	section drag coefficient of two-dimensional airfoil
c_{d_c}	crossflow drag coefficient, Eq. (9-34)
$c_{d\text{cam}}$	camber drag coefficient
c_F	skin-friction coefficient, Eq. (9-101)
C_A	chord-force coefficient in absence of leading-edge thrust and skin friction
C_{A_0}	value of C_A at $\alpha = 0$
C_C	chord-force coefficient including leading-edge thrust and skin friction
C_{C_0}	value of C_C at $\alpha = 0$
C_D	drag coefficient
C_{D_0}	minimum drag coefficient
C_{D_i}	"induced" drag, $C_D - C_{D_0}$
$C_{D_i}/\Delta C_L^2$	drag-rise factor
$(C_{D_i}/\Delta C_L^2)_{\text{vor}},$	
$(C_{D_i}/\Delta C_L^2)_{\text{wave}}$	drag-rise factors due to vortex drag and wave drag
C_L	lift coefficient
C_{L_0}	lift coefficient for minimum drag
$C_{L\text{opt}}$	lift coefficient for maximum lift-drag ratio
ΔC_L	$C_L - C_{L_0}$
C_N	normal-force coefficient
d	maximum diameter of body cross section
D	drag force
D_B	drag of body alone
$D_{B(W)}$	drag of body in presence of wing
D_c	crossflow drag force
D_C	drag of complete configuration
D_P	pressure drag
D_V	viscous drag
D_W	drag of wing alone
$D_{W(B)}$	drag of wing panels in presence of body
D_{wave}	wave drag
E	elliptic integral of second kind
$f_0, f_2, f_4, \ldots$	Fourier coefficients in Eq. (9-21)
g	quantity given by Eq. (9-40)
h	trailing-edge thickness; base diameter
k	drag-rise factor $C_{D_i}/\Delta C_L^2$; also modulus of elliptic integral
K	hypersonic similarity parameter, M_0 divided by fineness ratio
K^*	$l_0^2/\overline{l^2}$
l	length of body
l_0	characteristic streamwise length of configuration
$l(\theta)$	length given by Eq. (9-73)

$\overline{l^2}$	mean-squared length given by Eq. (9-70)
L	lift force; also length of boundary-layer run
$(L/D)_{max}$	maximum lift-drag ratio
L_A	lift of arrow wing
L_T	lift of triangular wing
m	$\beta \tan \omega$ for triangular wing
M_0	free-stream Mach number
M'	reference Mach number used to correlate base-pressure data
n	outward normal to missile surface; also exponent specifying body shape
p	static pressure
p_0	free-stream static pressure
p_b	base static pressure
p'	reference pressure used to correlate base-pressure data
P_B	base-pressure coefficient, $(p_b - p_0)/q_0$
P_C	$P_F - P_R$
P_F	pressure coefficient for direct flow
P_R	pressure coefficient for reverse flow
q_0	free-stream dynamic pressure
r	radial distance from body longitudinal axis
r_0	radius of base of body of revolution
S	cross-sectional area of body
S_1, S_2, S_3	surfaces of control area enclosing missile
S_c	planform area subject to crossflow
S_m	surface area of missile
S_n	projection on crossflow plane of body cross-sectional area intercepted by oblique plane
S_R	reference area
t	reciprocal of body fineness ratio; also tangent to missile surface in the τ direction; also thickness of airfoil section
t_m	maximum thickness of airfoil section
T	leading-edge thrust
V_0	free-stream velocity
Vol.	volume
$W(\mathfrak{z})$	complex potential of body
x, y, z	principal missile axes, x streamwise, y positive to right, z positive upward
x_0	coordinate of intersection of oblique plane with x axis
x_l	least value of x_0 for which oblique plane intersects wing planform
x_u	greatest value of x_0 for which oblique plane intersects wing planform

z_c	see Eq. (9-55)
z_u, z_l	distances of upper and lower airfoil surfaces, respectively, measured from chord joining leading and trailing edges
$\bar{z}$	see Eq. (9-53)
$\mathfrak{z}$	$y + iz$
$\mathfrak{z}_0$	$\mathfrak{z}$ coordinate of centroid of base of body
α	angle of attack
β	$(M_0{}^2 - 1)^{1/2}$
δ	tangent to body surface in streamwise direction; boundary-layer displacement thickness
θ	polar angle in crossflow plane with $\theta = 0$ plane horizontal
θ, ϕ	angular parameters used in Eq. (9-38)
λ	taper ratio of wing
Λ_{le}	sweep angle of wing leading edge
μ	leading-edge suction factor defined by Eq. (9-11)
μ_T	value of μ for triangular wing
ν	outward normal to base contour in crossflow plane
ξ	dummy variable of integration
σ	tangent to body surface in crossflow plane of base
τ	skin friction; also wing thickness ratio
ϕ	potential function
ϕ_0	potential function for crossflow plane of base at zero angle of attack
ϕ_α	potential function for crossflow plane of base due to angle of attack
ω	semiapex angle of triangular wing

Subscripts:

l	lower surface of missile
u	upper surface of missile

REFERENCES

1. Brown, Clinton E.: Aerodynamics of Bodies at High Speeds, sec. B in "Aerodynamic Components of Aircraft at High Speeds," vol. VII of "High-speed Aerodynamics and Jet Propulsion," Princeton University Press, Princeton, 1957.

2. von Kármán, Theodore: On the Foundations of High-speed Aerodynamics, sec. A, p. 12, in "General Theory of High-speed Aerodynamics," vol. VI of "High-speed Aerodynamics and Jet Propulsion," Princeton University Press, Princeton, 1956.

3. von Kármán, Theodore: The Problem of Resistance in Compressible Fluids, *Atti V convegno fondazione Alessandro Volta, Rome,* 1935.

4. Massachusetts Institute of Technology, Department of Electrical Engineering: "Tables of Supersonic Flow around Cones," by the Staff of the Computing Section, Center of Analysis, under direction of Zdenek Kopal, *MIT Tech. Rept.* 1, Cambridge, 1947.

5. Taylor, G. I., and J. W. Maccoll: *Proc. Roy. Soc. London A,* vol. 139, pp. 278–311, 1933.

6. Ames Research Staff: Equations, Tables, and Charts for Compressible Flow, *NACA Tech. Repts.* 1135, 1953.

7. Ehret, D. M., V. J. Rossow, and V. I. Stevens: An Analysis of the Applicability of the Hypersonic Similarity Law to the Study of Flow about Bodies of Revolution at Zero Angle of Attack, *NACA Tech. Notes* 2250, December, 1950.

8. Rossow, V. J.: Applicability of the Hypersonic Similarity Rule to Pressure Distributions Which Include the Effects of Rotation for Bodies of Revolution at Zero Angle of Attack, *NACA Tech. Notes* 2399, June, 1951.

9. Tsien, Hsue-Shen: Similarity Laws of Hypersonic Flow, *J. Math. and Phys.*, vol. 25, no. 3, 1946.

10. Newton, Sir Isaac: "Principia," "Motte's Translation Revised," pp. 657–661, University of California Press, Berkeley, 1946.

11. Todhunter, I.: "Researches in the Calculus of Variations," pp. 167–173, reprint, Stechert-Hafner, Inc., New York, 1924.

12. Eggers, A. J., Jr., M. M. Resnikoff, and D. H. Dennis: Bodies of Revolution for Minimum Drag at High Supersonic Airspeeds, *NACA Tech. Repts.* 1306, 1958.

13. Jorgensen, L. H.: Nose Shapes for Minimum Pressure Drag at Supersonic Mach Numbers, *J. Aeronaut. Sci.*, vol. 21, no. 4, pp. 276–279, Readers' Forum, 1954.

14. Ehret, Dorris M.: Accuracy of Approximate Methods for Predicting Pressures on Pointed Nonlifting Bodies of Revolution in Supersonic Flow, *NACA Tech. Notes* 2764, August, 1952.

15. Van Dyke, Milton D.: Practical Calculation of Second Order Supersonic Flow past Nonlifting Bodies of Revolution, *NACA Tech. Notes* 2744, 1952.

16. Ferri, Antonio: "Elements of Aerodynamics of Supersonic Flow," The Macmillan Company, New York, 1949.

17. Haack, W.: Geschossenformen kleinsten Wellenwiderstandes, *Ber. Lilienthal-Ges. Luftfahrt*, vol. 139.

18. Sears, William R.: On Projectiles of Minimum Wave Drag, *Quart. Appl. Math.*, vol. 14, no. 4, 1947.

19. Eggers, A. J., Jr., and Raymond C. Savin: Approximate Methods for Calculating the Flow about Nonlifting Bodies of Revolution at High Supersonic Airspeeds, *NACA Tech. Notes* 2579, 1951.

20. von Kármán, Theodore, and Norton B. Moore: Resistance of Slender Bodies Moving with Supersonic Velocities with Special Reference to Projectiles, *Trans. ASME*, vol. 54, no. 23, pp. 303–310, 1932.

21. Grant, Frederick C., and Morton Cooper: Tables for the Computation of Wave Drag of Arrow Wings of Arbitrary Airfoil Section, *NACA Tech. Notes* 3185, June, 1954.

22. Puckett, Allen E.: Supersonic Wave Drag of Thin Airfoils, *J. Aeronaut. Sci.*, vol. 13, no. 9, pp. 475–484, 1946.

23. Laurence, T.: Charts of the Wave Drag of Wings at Zero Lift, *RAE Tech. Note Aero.* 2139, revised, November, 1952.

24. Nielsen, Jack N.: Effect of Aspect Ratio and Taper on the Pressure Drag at Supersonic Speeds of Unswept Wings at Zero Lift, *NACA Tech. Notes* 1487, November, 1947.

25. Puckett, A. E., and H. T. Stewart: Aerodynamic Performance of Delta Wings at Supersonic Speeds, *J. Aeronaut. Sci.*, vol. 14, no. 10, pp. 567–578, 1947.

26. Katzen, E. D., and G. E. Kaattari: Drag Interference between a Pointed Cylindrical Body and Triangular Wings of Various Aspect Ratios at Mach Numbers of 1.50 and 2.02, *NACA Tech. Notes* 3794, November, 1956.

27. Heaslet, Max A., and Harvard Lomax: Supersonic and Transonic Small Perturbation Theory, sec. D, pp. 219–221, in "General Theory of High-speed Aerodynamics," vol. VI of "High-speed Aerodynamics and Jet Propulsion," Princeton University Press, Princeton, 1954.

28. Nielsen, Jack N.: Quasi-cylindrical Theory of Wing-Body Interference at Supersonic Speeds and Comparison with Experiment, *NACA Tech. Repts.* 1252, 1955.

29. Nielsen, Jack N., and Frederick H. Matteson: Calculative Method for Estimating the Interference Pressure Field at Zero Lift on a Symmetrical Swept-back Wing Mounted on a Circular Cylindrical Body, *NACA Research Mem.* A9E19, 1949.

30. Lomax, Harvard, and Max A. Heaslet: Recent Developments in the Theory of Wing-Body Wave Drag, *J. Aeronaut. Sci.*, vol. 23, no. 12, pp. 1061–1074, 1956.

31. Jones, Robert T.: Theory of Wing-Body Drag at Supersonic Speeds, *NACA Tech. Repts.* 1284, 1956.

32. Whitcomb, Richard T.: A Study of the Zero-lift Drag-rise Characteristics of Wing-Body Combinations near the Speed of Sound, *NACA Tech. Repts.* 1273, 1956.

33. Lomax, Harvard: Nonlifting Wing-Body Combinations with Certain Geometric Restraints Having Minimum Wave Drag at Low Supersonic Speeds, *NACA Tech. Repts.* 1297, 1957.

34. Nielsen, Jack N.: General Theory of Wave-drag Reduction for Combinations Employing Quasicylindrical Bodies with an Application to Swept Wing and Body Combinations, *NACA Tech. Notes* 3722, 1956.

35. Jones, Robert T.: The Minimum Drag of Thin Wings in Frictionless Flow, *J. Aeronaut. Sci.*, vol. 18, no. 2, pp. 75–81, 1951.

36. Jones, Robert T.: Theoretical Determination of the Minimum Drag of Airfoils at Supersonic Speeds, *J. Aeronaut. Sci.*, vol. 19, no. 12, pp. 813–822, 1952.

37. Jones, Robert T.: Minimum Wave Drag for Arbitrary Arrangements of Wings and Bodies, *NACA Tech. Repts.* 1335, 1957.

38. Tsien, S. H.: The Supersonic Conical Wing of Minimum Drag, *J. Aeronaut. Sci.*, vol. 22, no. 12, pp. 805–817, 1955.

39. Cohen, Doris: The Warping of Triangular Wings for Minimum Drag at Supersonic Speeds, *J. Aeronaut. Sci.*, vol. 24, no. 1, pp. 67–68, Readers' Forum, 1957.

40. Tucker, Warren A.: A Method for the Design of Sweptback Wings Warped to Produce Specific Flight Characteristics at Supersonic Speeds, *NACA Tech. Repts.*, 1226, 1955.

41. Ferri, Antonio: On the Use of Interfering Flow Fields for the Reduction of Drag at Supersonic Speeds, *J. Aeronaut. Sci.*, vol. 24, no. 1, pp. 1-18, 1957.

42. Jones, Robert T., and Doris Cohen: Aerodynamics of Wings at High Speeds, sec. A, pp. 221–223, in "Aerodynamic Components of Aircraft at High Speeds," vol. VII of "High-speed Aerodynamics and Jet Propulsion," Princeton University Press, Princeton, 1957.

43. Hayes, Wallace D.: Linearized Supersonic Flow, *North Am. Aviation Rept.* AL-222, June 18, 1947.

44. Chapman, Dean R., Donald M. Kuehn, and Howard K. Larson: Investigation of Separated Flows in Supersonic and Subsonic Streams with Emphasis on the Effect of Transition, *NACA Tech. Notes* 3869, March, 1957.

45. Spahr, J. Richard, and Robert R. Dickey: Effect of Tail Surfaces on the Base Drag of a Body of Revolution at Mach Numbers of 1.5 and 2.0, *NACA Tech. Notes* 2360, April, 1951.

46. Kurzweg, H. H.: The Base Pressure Measurements of Heated, Cooled, and Boat-tailed Models at Mach Numbers 1.5 to 5.0, *U.S. Naval Ordnance Lab. Symposium on Aeroballistics*, University of Texas, Nov. 16-17, 1950.

47. Czarnecki, K. R., and Archibald R. Sinclair: An Investigation of the Effects of Heat Transfer on Boundary-layer Transition on a Parabolic Body of Revolution (NACA RM-10) at a Mach Number of 1.61, *NACA Tech. Repts.* 1240, 1955.

48. Bogdonoff, Seymour M.: A Preliminary Study of Reynolds Number Effects on Base Pressure at $M = 2.95$, *Princeton Aeronaut. Eng. Lab. Rept.* 182, June 12, 1950.

49. Syvertson, Clarence A., and Hermilo R. Gloria: An Experimental Investigation of the Zero-lift-drag Characteristics of Symmetrical Blunt-trailing-edge Airfoils at Mach Numbers from 2.7 to 5.0, *NACA Research Mem.* A53B02, April, 1953.

50. Chapman, Dean R.: An Analysis of Base Pressures at Supersonic Velocities and Comparison with Experiment, *NACA Tech. Repts.* 1051, 1951.

51. Chapman, Dean R., William R. Wimbrow, and Robert H. Kester: Experimental Investigation of Base Pressure on Blunt-trailing-edge Wings at Supersonic Velocities, *NACA Tech. Repts.* 1109, 1952.

52. Reller, John O., and Frank M. Hamaker: An Experimental Investigation of the Base Pressure Characteristics of Nonlifting Bodies of Revolution at Mach Numbers from 2.73 to 4.98, *NACA Tech. Notes* 3393, 1955.

53. Korst, H. H.: A Theory of Base Pressure in Transonic and Supersonic Flow, *J. Appl. Mechanics*, December, 1956.

54. Rubesin, M. W., and H. A. Johnson: A Critical Review of Skin-friction and Heat-transfer Solutions of the Laminar Boundary Layer of a Flat Plate, *Trans. ASME*, vol. 71, no. 4, pp. 385–388, 1949.

55. Eckert, Ernst R. G.: Survey of Heat Transfer at High Speeds, *WADC Tech. Rept.* 54–70, Wright-Patterson Air Force Base, Ohio.

56. Eckert, Ernst R. G.: Engineering Relations for Friction and Heat Transfer to Supersonic High Velocity Flow, *J. Aeronaut. Sci.*, vol. 22, no. 8, pp. 385–387, Readers' Forum, 1955.

57. Hilsenrath, Joseph, et al.: Tables of the Thermal Properties of Gases, *Natl. Bur. Standards Circ.* 564, November, 1955.

58. Chapman, Dean R., and Robert H. Kester: Measurements of Turbulent Skin Friction on Cylinders in Axial Flow at Subsonic and Supersonic Velocities, *J. Aeronaut. Sci.*, vol. 20, no. 7, pp. 441–448, 1953.

59. Vincenti, Walter G.: Comparison between Theory and Experiment for Wings at Supersonic Speeds, *NACA Tech. Repts.* 1033, 1951.

60. Pitts, William C., Jack N. Nielsen, and Maurice P. Gionfriddo: Comparison between Theory and Experiment for Interference Pressure Field between Wing and Body at Supersonic Speeds, *NACA Tech. Notes* 3128, 1954.

61. Mangler, W.: Boundary Layers on Bodies of Revolution in Symmetrical Flow, *Aerodynamische Versuchsanstalt Göttingen*, E. V., Rept. 45-A-17, as translated by M. S. Medvedeff, Goodyear Aircraft Corp., Akron, Ohio, Mar. 6, 1946.

62. Hantsche, W., and H. Wendt: The Laminar Boundary Layer and Circular Cone at Zero Incidence in a Supersonic Stream, *Brit. MAP Rept. and Transl.* 276, 1946.

63. Van Driest, E. R.: Turbulent Boundary Layer on a Cone in Supersonic Flow at Zero Angle of Attack, *J. Aeronaut. Sci.*, vol. 19, no. 1, pp. 55–57, 1952.

64. Chapman, D. R., and M. W. Rubesin: Temperature and Velocity Profiles in the Compressible Laminar Boundary Layer with Arbitrary Distribution of Surface Temperature, *J. Aeronaut. Sci.*, vol. 16, no. 9, pp. 547–565, 1949.

65. Germain, P.: Sur le Minimum de traînée d'une aile de forme en plan donnée, *Compt. rend.*, vol. 244, pp. 1135–1138, Feb. 25, 1957.

66. Sommer, S. C., and B. J. Short: Free-flight Measurements of Turbulent Boundary-layer Skin Friction in the Presence of Severe Aerodynamic Heating at Mach Numbers from 2.8 to 7.0, *NACA Tech. Notes* 3391, March, 1955.

67. Graham, Ernest W.: The Pressure on a Slender Body of Non-uniform Cross-sectional Shape in Axial Supersonic Flow, *Douglas Aircraft Co. Rept.* SM 13346-A, July 20, 1949.

68. Oswatitsch, K., and F. Keune: Aequivalenzsatz, Ahnlichkeitssätze für schallnahe Geschwindigkeiten und Widerstand nicht angestellter Körper kleiner Spannweite, *Z. angew. Math. u. Phys.*, Zurich, vol. 6, 1955.

69. Legendre, Robert: "Limite sonique de la résistance d'ondes d'un aeronef," *Compt. rend.*, vol. 236, pp. 2480–2479, June 29, 1953.

Transition

70. Jones, Robert A.: An Experimental Study at a Mach Number of 3 of the Effect of Turbulence Level and Sandpaper-type Roughness on Transition on a Flat Plate, *NASA Mem.* 2-9-L, March, 1959.

71. Locktenbert, B. H.: Transition in a Separated Laminar Boundary Layer, *ARC* 19,007, *FM* 2495, January, 1957.

72. Jack, John R., Richard J. Wisniewski, and N. S. Diaconis: Effects of Extreme Surface Cooling on Boundary Layer Transition, *NACA Tech. Notes* 4094, 1957.

73. Jack, John R., and N. S. Diaconis: Variation of Boundary Layer Transition with Heat Transfer on Two Bodies of Revolution at a Mach Number of 3.12, *NACA Tech. Notes* 3562, 1955.

74. Czarnecki, K. R., and Archibald R. Sinclair: An Extension of the Effects of Heat Transfer on Boundary Layer Transition on a Parabolic Body of Revolution (NACA RM-10) at a Mach Number of 1.61, *NACA Tech. Notes* 3166, 1954.

75. Jedlicka, James R., Max E. Wilkins, and Alvin Seiff: Experimental Determination of Boundary Layer Transition on a Body of Revolution at $M = 3.5$, *NACA Tech. Notes* 3342, 1954.

76. Diaconis, N. S., Richard J. Wisniewski, and John R. Jack: Heat Transfer and Boundary Layer Transition on Two Blunt Bodies at Mach Number 3.12, *NACA Tech. Notes* 4099, 1957.

Separation

77. Chapman, Dean R., Donald M. Kuehn, and Howard K. Larson: Investigation of Separated Flows in Supersonic and Subsonic Streams with Emphasis on the Effects of Transition, *NACA Tech. Notes* 3869, 1957.

78. Kuehn, Donald M.: Experimental Investigation of the Pressure Rise Required for the Incipient Separation of Turbulent Boundary Layers in Two-dimensional Supersonic Flow, *NASA Mem.* 1-21-59A, February, 1959.

79. Czarnecki, K. R., and Archibald R. Sinclair: A Note on the Effect of Heat Transfer on Peak Pressure Rise Associated with Separation of Turbulent Boundary Layer on a Body of Revolution (NACA RM-10) at a Mach Number of 1.61, *NACA Tech. Notes* 3997, April, 1957.

80. Gadd, G. E., D. W. Holder, and J. D. Regan: An Experimental Investigation of the Interaction between Shock Waves and Boundary Layer, *Proc. Roy. Soc. London A*, vol. 226, no. 1165, pp. 227–253, 1954.

81. Hakkinen, R. J., I. Greber, L. Trilling, and S. S. Abarbanel: The Interaction of an Oblique Shock Wave with a Laminar Boundary Layer, *NASA Mem.* 2-18-59W, March, 1959.

82. Gadd, G. E.: The Effects of Convex Surface Curvature on Boundary Layer Separation in Supersonic Flow, *ARC* 18,038, *FM* 2335, November, 1955.

83. Holder, D. W., and G. E. Gadd: The Interaction between Shock Waves and Boundary Layer and Its Relation to Base Pressure in Supersonic Flow, *Natl. Phys. Lab. Symposium on Boundary Layer Effects in Aerodynamics*, Paper 8, Teddington, England, April, 1955.

84. Vas, I.E., and S. M. Bogdonoff: Interaction of a Turbulent Boundary Layer with a Step at $M = 3.85$, *Princeton Univ. Dept. Aeronaut. Eng. Rept.* 295, April, 1955.

85. Bogdonoff, Seymour M.: Some Experimental Studies of the Separation of Supersonic Turbulent Boundary Layers, *Princeton Univ. Dept. Aeronaut. Eng. Rept.* 336, June, 1955.

CHAPTER 10

STABILITY DERIVATIVES

In the previous chapters of the book we have been concerned mainly with the aerodynamics of component parts of the missile and particular types of interference between components.   In this chapter we take the broad point of view and consider all forces and moments as functions of all linear and angular velocities.   The rates of change of any force or moment coefficient with respect to linear or angular velocity components of the missile or time derivatives thereof are called *stability derivatives*. Stability derivatives are in reality partial derivatives; they can be of any degree and include any number of the velocity components as independent variables as well as time.   These stability derivatives are the usual aerodynamic inputs in dynamical analyses.   Again, the feature that probably distinguishes this chapter from previous ones is the general approach of treating all stability derivatives rather than the specialized approach of treating a few derivatives intensively that characterizes earlier chapters.

Before embarking on general methods of evaluating stability derivatives, we must give careful consideration to notation and to reference quantities.   It is to be noted that the axis system to be used will represent a departure from the previous usage in earlier chapters in accordance with the discussion in Sec. 10-1.   Section 10-2 is concerned with the general nature of aerodynamic forces and the assumptions which lead to the concept of a stability derivative.   In Secs. 10-3 and 10-4 the powerful Maple-Synge method is brought into play systematically to extract as much information as possible on stability derivatives from the rotational and mirror symmetries of the missile.   The Bryson analysis is used in Sec. 10-5 to show how most of the stability derivatives for certain classes of slender missiles can be calculated by means of apparent mass coefficients, and the analysis is applied to a slender triangular wing in Sec. 10-6.   General methods of evaluating apparent mass quantities using complex variable theory are considered in Sec. 10-7, and a table of apparent masses is compiled in Table 10-3.   A number of illustrative examples to explain the use of the table of apparent masses are given in Secs. 10-8 and 10-9.   In Sec. 10-10 the variations with aspect ratio of the stability derivatives of a triangular wing are discussed.   The information considered up to this point deals largely with missiles having no empennage behind the wing. When the empennage lies behind the wing, the fore-going methods and others can be used to determine the empennage contribution to the stability derivatives. These matters are discussed in Sec. 10-11 and are illustrated by a calculative example.

10-1. Reference Axes; Notation

Perhaps the first problem arising in the study of stability derivatives is the choice of a system of reference axes. This choice is not an obvious one since the systems used in stability analyses include body axes, wind axes, stability axes, Eulerian axes, and pseudo-Eulerian axes, and no one set of axes will meet all requirements. From the point of view of notational uniformity, it would be desirable to retain the same set of axes used in the previous chapters. However, this procedure would lead to a system with the positive longitudinal axis rearward and the positive vertical axis upward in direct opposition to most of the foregoing systems of axes. Also, for such an axis system, the usual positive directions of φ, ψ, θ and p, q, r would not correspond to the positive right-hand rotations of the system. For these and other reasons, it was decided to *standardize the reference axes for stability derivatives to a set of body axes* coinciding in direction with the principal axes of inertia of the missile. (Any axis of symmetry will be a principal axis of inertia.) The positive directions are taken as shown in Fig. 10-1. This choice of reference axes allows us to invoke directly the symmetry properties of the missile in studying their effects on the stability derivatives without an intermediate transformation from one system of axes to another. Once the stability derivatives have been determined with respect to a standard system of body axes, they can, however, be transferred at will to any other axis system. It should be borne in mind that a system of axes fixed in the body also has the advantage in dynamical analysis that the moments of inertia are not functions of time.

With reference to Fig. 10-1, the reference axes X, Y, Z constitute a set of axes fixed in the missile with its origin coinciding with the missile center of gravity. The capital letters are used so that no confusion with the axes x, y, z need arise. The components along X, Y, and Z of the missile force, moment, translational velocity of the missile center of gravity, and its angular displacement are given in Fig. 10-1 in symbol and sign. The positive moment directions, angular velocities, and angular displacements all correspond to positive rotations by the right-hand rule for the positive axis directions. The translational velocity components of the missile center of mass, u, v, and w, are not to be confused with the components of the local fluid velocity along x, y, and z as used, for instance, in Eq. (6-1).

The angular displacements θ, ψ, and φ are to be given special attention. They are to be differentiated strictly from the angles of attack, sideslip, and bank as defined in Sec. 1-4.   The angles of attack and sideslip have a *kinematic definition* based on the components of the free-stream velocity along the body axes of the missile.   The angular displacements θ, ψ, and φ, on the other hand, are used to measure the *missile attitude* with respect to a fixed set of axes, and in no way require motion of the missile relative to the surrounding air for their definition.   Let X_0, Y_0, and Z_0 be stationary axes fixed in space, and consider a missile moving with respect to these fixed axes.   Let us now describe one of many possible ways of specifying the angular position of the missile at any particular instant of

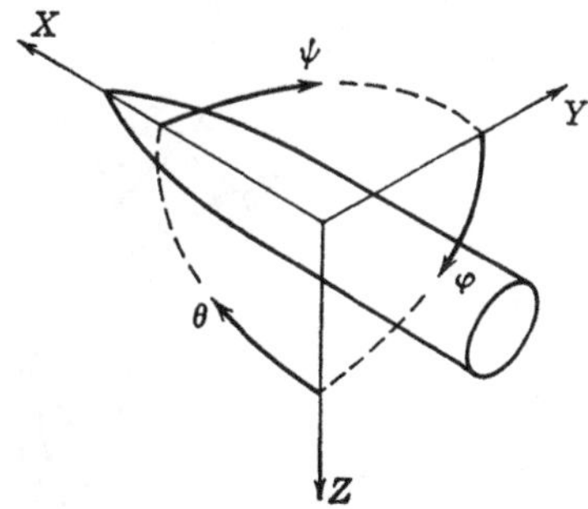

FIG. 10-1. Standard conventions and symbols.

Axes			Moments about axis			Angles		Velocities	
Designation	Symbol	Force along axis	Designation	Symbol	Positive direction	Designation	Symbol	Linear	Angular
Longitudinal	X	X	Roll	L	$Y \rightarrow Z$	Roll	φ	u	p
Lateral	Y	Y	Pitch	M	$Z \rightarrow X$	Pitch	θ	v	q
Normal	Z	Z	Yaw	N	$X \rightarrow Y$	Yaw	ψ	w	r

time.   We shall do this by successively yawing, pitching, and rolling the X_0, Y_0, Z_0 axes until they coincide in direction with the axes X, Y, and Z fixed in the missile as shown in Fig. 10-2.   The angles of yaw, pitch, and roll, ψ, θ, and φ, then describe uniquely the missile attitude.   First, yaw the missile by an angular displacement ψ around OZ_0 so that X_0 goes into X_1 and Y_0 into Y_1.   Then pitch the missile by an angle θ about the OY_1 axis so that X_1 moves to X_2 and Z_0 to Z_2.   Finally, roll the missile by angle φ around the OX_2 axis (or OX axis) so that the point Y_2 moves to Y and Z_2 to Z.   It is to be noted that the angles ψ, θ, and φ are not about mutually perpendicular axes.   The operations of yaw, pitch, and roll are always to be performed in that order since angular displacements do not follow the ordinary law of vector addition but, in fact, follow a noncom-

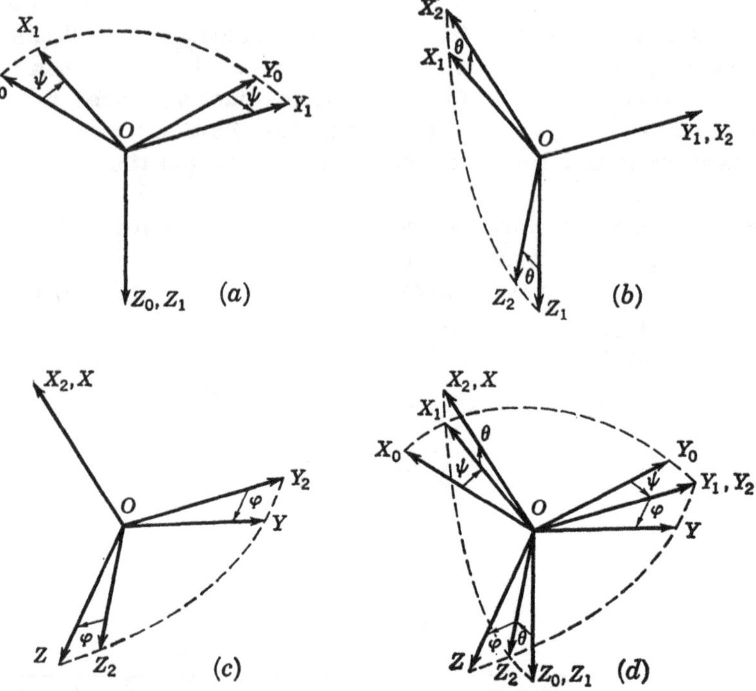

Fig. 10-2. System of angular displacements. (a) Yaw about OZ_0 by ψ; (b) pitch about OY_1 by θ; (c) roll about OX_2 by ϕ; (d) composite diagram.

mutative law. Under the foregoing system the direction cosines of the final missile body axes X, Y, and Z to the fixed axes X_0, Y_0, and Z_0 are given by the Table 10-1. The angles of yaw, pitch, and roll are thus a

TABLE 10-1. DIRECTION COSINES OF BODY AXES

Body axes \ Fixed axes	OX_0	OY_0	OZ_0
OX	$\cos\theta\cos\psi$	$\cos\theta\sin\psi$	$-\sin\theta$
OY	$-\cos\varphi\sin\psi +$ $\sin\varphi\sin\theta\cos\psi$	$\cos\varphi\cos\psi +$ $\sin\varphi\sin\theta\sin\psi$	$\sin\varphi\cos\theta$
OZ	$\sin\varphi\sin\psi +$ $\cos\varphi\sin\theta\cos\psi$	$-\sin\varphi\cos\psi +$ $\cos\varphi\sin\theta\sin\psi$	$\cos\varphi\cos\theta$

system of three angular displacements that specify the angular orientation of any missile in space with reference to a fixed set of axes. As such, these angles are pure geometric quantities independent of the kinematics of the missile. For small values of θ, ψ, and φ, these quantities can be considered as taken about the missile body axes. Under these circumstances the direction cosines become those shown in Table 10-2.

TABLE 10-2. DIRECTION COSINES OF BODY AXES FOR SMALL
ANGULAR DISPLACEMENTS

Body axes \ Fixed axes	OX_0	OY_0	OZ_0
OX	1	ψ	$-\theta$
OY	$-\psi$	1	φ
OZ	θ	$-\varphi$	1

10-2. General Nature of Aerodynamic Forces; Stability Derivatives

The forces and moments acting on a missile result mainly from the missile propulsive system, gravitational attraction, and from the reaction of the air on the missile as a result of its motion. (This is not to say that other types of forces cannot be involved.) In this chapter we are concerned only with the reaction of the air on the missile by virtue of its motion. Consider a missile which has been flying for some time in air which is at rest at great distances from the missile. The forces on the missile at any particular instant depend in general on the entire history of its motion in the air. This result is generally true for subsonic velocities since the missile wake can be "felt" by the missile from all points in the wake at all times. At supersonic speeds, the pressure disturbances from any point are confined to its downstream Mach cone. As a consequence, in steady supersonic flow only a small length of the wake can influence the missile. The functional dependence of any particular force or moment on the *complete dynamical history of the missile* can be written

$$F = f[u(t),\, v(t),\, w(t),\, p(t),\, q(t),\, r(t)] \tag{10-1}$$

Obviously the dependence of the force on the complete history of the motion is too complicated a relationship to be of much use in analysis. We must therefore simplify the relationship on the basis of some plausible assumption. By suitably restricting the types of missile motion, we can accomplish such a simplification. The forces on a missile resulting from a sudden change in angle of attack would depend on the past history of the motion for a definite period after the sudden change. The forces acting on a missile which is undergoing sinusoidal oscillations of high frequency will certainly depend on more than the instantaneous dynamical state of the missile. The aerodynamic forces associated with boundary-layer separation such as hysteresis in lift near maximum lift certainly depend on more than just the instantaneous values of u, v, w, p, q, and r.

A missile becoming rapidly hot as a result of aerodynamic heating develops forces which depend on the history of its motion. However, for missile motions which take place sufficiently slowly and for which effects of heating and viscosity are not significant, it is reasonable to assume that the missile forces and moments depend primarily on the *instantaneous dynamical state* specified by the linear and angular velocity components. (If we were also to include the linear and angular acceleration components, we could also take into account in part the immediate past history of the missile to the degree of approximation that $\dot{u}$, $\dot{v}$, $\dot{w}$, $\dot{p}$, $\dot{q}$, and $\dot{r}$ are independent variables in Taylor series for the forces and moments.) Let us formulate the consequences of the simplifying assumption mathematically. Let X, Y, Z, L, M, N be the components of the missile force and moment corresponding to a dynamical state described by u, v, w, p, q, and r. Under the assumption that the force and moment components depend only on the instantaneous values of u, v, w, p, q, and r, we can write more specifically than Eq. (10-1) that

$$X = X(u,v,w,p,q,r) \tag{10-2}$$

with similar equations for the other components. If further we assume that the function given in Eq. (10-2) is analytic—for instance, there are no hysteresis effects that make it double-valued—we can then expand it in a Taylor series about some point u_0,v_0,w_0,p_0,q_0,r_0. Let us expand about the point $(0,0,0,0,0,0)$ so that

$$X = \sum_{ijklmn=0,\,1,\,2,\,\ldots} x_{ijklmn}(u_0,v_0,w_0,p_0,q_0,r_0)u^i v^j w^k p^l\, q^m r^n \tag{10-3}$$
$$u_0 = v_0 = w_0 = p_0 = q_0 = r_0 = 0$$

where x_{ijklmn} is in general a function which depends on u_0,v_0,w_0,p_0,q_0,r_0 but which is a constant in the present case. By the theory of Taylor expansions, it is known that the function x_{ijklmn} is related to the partial derivative $\partial x^g / \partial u^i \partial v^j \partial w^k \partial p^l \partial q^m \partial r^n$ where $g = i + j + k + l + m + n$. This partial derivative is called a *stability derivative* with the possible application of a multiplicative constant depending on the exact definition of the stability derivative. It is thus clear why the stability derivative depends on the particular values of p_0, q_0, r_0, u_0, v_0, and w_0 for its value. Let us write out just the constant and first-degree terms of the general expansion of Eq. (10-3) about the point u_0,v_0,w_0,p_0,q_0, and r_0.

$$X = x_{000000}(u_0,v_0,w_0,p_0,q_0,r_0) + x_{100000}(u - u_0) + x_{010000}(v - v_0)$$
$$+ x_{001000}(w - w_0) + x_{000100}(p - p_0) + x_{000010}(q - q_0)$$
$$+ x_{000001}(r - r_0) \tag{10-4}$$
$$X = X_0 + \frac{\partial X}{\partial u}\delta u + \frac{\partial X}{\partial v}\delta v + \frac{\partial X}{\partial w}\delta w + \frac{\partial X}{\partial p}\delta p + \frac{\partial X}{\partial q}\delta q + \frac{\partial X}{\partial r}\delta r \tag{10-5}$$

We have six derivatives in X including terms of first degree in the general expansion, and we obtain six more each for Y, Z, L, M, and N. Of these 36 derivatives, those due to the components of the linear velocity, u, v, and w, are termed *resistance derivatives*.

$$X_u \quad Y_u \quad Z_u \quad L_u \quad M_u \quad N_u$$
$$X_v \quad Y_v \quad Z_v \quad L_v \quad M_v \quad N_v$$
$$X_w \quad Y_w \quad Z_w \quad L_w \quad M_w \quad N_w$$

The 18 derivatives depending on components of the angular velocity are termed *rotary derivatives*.

$$X_p \quad Y_p \quad Z_p \quad L_p \quad M_p \quad N_p$$
$$X_q \quad Y_q \quad Z_q \quad L_q \quad M_q \quad N_q$$
$$X_r \quad Y_r \quad Z_r \quad L_r \quad M_r \quad N_r$$

If we had included the independent variables $\dot{u}$, $\dot{v}$, $\dot{w}$, $\dot{p}$, $\dot{q}$, and $\dot{r}$ in the general expansion given by Eq. (10-3), we would have obtained 36 more *acceleration derivatives* including terms of first degree. Of these acceleration derivatives, experience has shown that certain ones can be important. The ones with which we will be concerned include

$$Z_{\dot{w}}, \ M_{\dot{w}}, \ Y_{\dot{v}}, \ N_{\dot{v}}$$

By including terms of degree higher than the first, *higher-order derivatives* without limit can be generated. Certain derivatives of this kind with which we will be concerned include

$$L_{\alpha\beta}, \ L_{\alpha p}, \ L_{\beta p}, \ N_{\alpha p}$$

Again the assumptions underlying stability derivatives as they are used in practice are that the missile forces and moments depend only on the instantaneous values of u, v, w, p, q, r and possible $\dot{w}$ and $\dot{v}$, and that the functional relationship between forces (and moments) and these independent variables is a Taylor series. It must also be borne in mind that the stability derivative is a function which depends on particular values of u_0, v_0, w_0, p_0, q_0, r_0, $\dot{w}_0$, and $\dot{v}_0$ for its value. Luckily, however, the functional dependence is usually simple.

The stability derivatives as defined above are dimensional, and some consistent scheme of making them nondimensional must now be introduced. A reference area is needed as well as a reference length. It is frequent practice to use different reference lengths for different purposes. For instance, pitching-moment coefficient is usually based on the wing mean aerodynamic chord whereas rolling-moment coefficient is usually based on the wing span. For the purpose of general treatments, it is desirable to use only one reference length λ and one reference area S_R. Conversion to other reference quantities can readily be made for specific cases. If q_0 is the free-stream dynamic pressure, the force coefficients are taken to be

$$C_X = \frac{X}{q_0 S_R} \qquad C_Y = \frac{Y}{q_0 S_R} \qquad C_Z = \frac{Z}{q_0 S_R} \qquad (10\text{-}6)$$

and the moment coefficients are taken to be

$$C_m = \frac{M}{q_0 S_R \lambda} \qquad C_n = \frac{N}{q_0 S_R \lambda} \qquad C_l = \frac{L}{q_0 S_R \lambda} \qquad (10\text{-}7)$$

The velocity components u, v, and w are made nondimensional by division by V_0, yielding u/V_0 and the angles of attack and sideslip.

$$\frac{u}{V_0} \qquad \beta = \frac{v}{V_0} \qquad \alpha = \frac{w}{V_0} \qquad (10\text{-}8)$$

The approximations to the angles of attack and sideslip α and β as given by Eq. (10-8) are valid only if α and β are small compared to unity, as discussed in Sec. 1-4, but this range will be wide enough for the purposes of this chapter, which is based almost exclusively on linear theory. The angular velocities are made nondimensional as follows:

$$\frac{p\lambda}{2V_0}, \frac{q\lambda}{2V_0}, \frac{r\lambda}{2V_0}$$

The use of the factor 2 makes $p\lambda/2V_0$ the helix angle of the wing tips in case λ is the wing span. The accelerations may also be nondimensionalized as follows:

$$\frac{\dot{u}\lambda}{2V_0^2} \qquad \frac{\dot{v}\lambda}{2V_0^2} = \frac{\dot{\beta}\lambda}{2V_0} \qquad \frac{\dot{w}\lambda}{2V_0^2} = \frac{\dot{\alpha}\lambda}{2V_0} \qquad \frac{\dot{p}\lambda^2}{2V_0^2} \qquad \frac{\dot{q}\lambda^2}{2V_0^2} \qquad \frac{\dot{r}\lambda^2}{2V_0^2}$$

(The use of the factor 2 in the acceleration derivatives is convenient because combinations such as $M_q + M_{\dot{\alpha}}$ occur in many problems.)

We have completely nondimensionalized the force, moments, velocities, and accelerations. Let us now specify the notation for the *resistance stability* derivatives in terms of the nondimensional component parts.

$$C_{X_u} = \frac{\partial C_X}{\partial(u/V_0)} \qquad C_{X_\alpha} = \frac{\partial C_X}{\partial\alpha} \qquad C_{X_\beta} = \frac{\partial C_X}{\partial\beta}$$

$$C_{Y_u} = \frac{\partial C_Y}{\partial(u/V_0)} \qquad C_{Y_\alpha} = \frac{\partial C_Y}{\partial\alpha} \qquad C_{Y_\beta} = \frac{\partial C_Y}{\partial\beta}$$

$$C_{Z_u} = \frac{\partial C_Z}{\partial(u/V_0)} \qquad C_{Z_\alpha} = \frac{\partial C_Z}{\partial\alpha} \qquad C_{Z_\beta} = \frac{\partial C_Z}{\partial\beta}$$

$$C_{l_u} = \frac{\partial C_l}{\partial(u/V_0)} \qquad C_{l_\alpha} = \frac{\partial C_l}{\partial\alpha} \qquad C_{l_\beta} = \frac{\partial C_l}{\partial\beta}$$

$$C_{m_u} = \frac{\partial C_m}{\partial(u/V_0)} \qquad C_{m_\alpha} = \frac{\partial C_m}{\partial\alpha} \qquad C_{m_\beta} = \frac{\partial C_m}{\partial\beta}$$

$$C_{n_u} = \frac{\partial C_n}{(u/V_0)} \qquad C_{n_\alpha} = \frac{\partial C_n}{\partial\alpha} \qquad C_{n_\beta} = \frac{\partial C_n}{\partial\beta}$$

The notation used for the *rotary stability derivatives* is

$$C_{X_p} = \frac{\partial C_X}{\partial(p\lambda/2V_0)} \qquad C_{X_q} = \frac{\partial C_X}{\partial(q\lambda/2V_0)} \qquad C_{X_r} = \frac{\partial C_X}{\partial(r\lambda/2V_0)}$$

$$C_{Y_p} = \frac{\partial C_Y}{\partial(p\lambda/2V_0)} \qquad C_{Y_q} = \frac{\partial C_Y}{\partial(q\lambda/2V_0)} \qquad C_{Y_r} = \frac{\partial C_Y}{\partial(r\lambda/2V_0)}$$

$$C_{Z_p} = \frac{\partial C_Z}{\partial(p\lambda/2V_0)} \qquad C_{Z_q} = \frac{\partial C_Z}{\partial(q\lambda/2V_0)} \qquad C_{Z_r} = \frac{\partial C_Z}{\partial(r\lambda/2V_0)}$$

$$C_{l_p} = \frac{\partial C_l}{\partial(p\lambda/2V_0)} \qquad C_{l_q} = \frac{\partial C_l}{\partial(q\lambda/2V_0)} \qquad C_{l_r} = \frac{\partial C_l}{\partial(r\lambda/2V_0)}$$

$$C_{m_p} = \frac{\partial C_m}{\partial(p\lambda/2V_0)} \qquad C_{m_q} = \frac{\partial C_m}{\partial(q\lambda/2V_0)} \qquad C_{m_r} = \frac{\partial C_m}{\partial(r\lambda/2V_0)}$$

$$C_{n_p} = \frac{\partial C_n}{\partial(p\lambda/2V_0)} \qquad C_{n_q} = \frac{\partial C_n}{\partial(q\lambda/2V_0)} \qquad C_{n_r} = \frac{\partial C_n}{\partial(r\lambda/2V_0)}$$

The notation used for the *acceleration derivatives* is

$$C_{X_{\dot\alpha}} = \frac{\partial C_X}{\partial(\dot\alpha\lambda/2V_0)} \qquad C_{X_{\dot\beta}} = \frac{\partial C_X}{\partial(\dot\beta\lambda/2V_0)} \qquad C_{X_{\dot p}} = \frac{\partial C_X}{\partial(\dot p\lambda^2/2V_0{}^2)}$$

$$C_{Y_{\dot\alpha}} = \frac{\partial C_Y}{\partial(\dot\alpha\lambda/2V_0)} \qquad C_{Y_{\dot\beta}} = \frac{\partial C_Y}{\partial(\dot\beta\lambda/2V_0)} \qquad C_{Y_{\dot p}} = \frac{\partial C_Y}{\partial(\dot p\lambda^2/2V_0{}^2)}$$

$$C_{Z_{\dot\alpha}} = \frac{\partial C_Z}{\partial(\dot\alpha\lambda/2V_0)} \qquad C_{Z_{\dot\beta}} = \frac{\partial C_Z}{\partial(\dot\beta\lambda/2V_0)} \qquad C_{Z_{\dot p}} = \frac{\partial C_Z}{\partial(\dot p\lambda^2/2V_0{}^2)}$$

$$C_{l_{\dot\alpha}} = \frac{\partial C_l}{\partial(\dot\alpha\lambda/2V_0)} \qquad C_{l_{\dot\beta}} = \frac{\partial C_l}{\partial(\dot\beta\lambda/2V_0)} \qquad C_{l_{\dot p}} = \frac{\partial C_l}{\partial(\dot p\lambda^2/2V_0{}^2)}$$

$$C_{m_{\dot\alpha}} = \frac{\partial C_m}{\partial(\dot\alpha\lambda/2V_0)} \qquad C_{m_{\dot\beta}} = \frac{\partial C_m}{\partial(\dot\beta\lambda/2V_0)} \qquad C_{m_{\dot p}} = \frac{\partial C_m}{\partial(\dot p\lambda^2/2V_0{}^2)}$$

$$C_{n_{\dot\alpha}} = \frac{\partial C_n}{\partial(\dot\alpha\lambda/2V_0)} \qquad C_{n_{\dot\beta}} = \frac{\partial C_n}{\partial(\dot\beta\lambda/2V_0)} \qquad C_{n_{\dot p}} = \frac{\partial C_n}{\partial(\dot p\lambda^2/2V_0{}^2)}$$

with $\dot q$ and $\dot r$ derivatives similar.

The higher-order derivatives are specified in the same manner as the derivatives of first degree:

$$C_{Y_{\alpha\dot p}} = \frac{\partial^2 C_Y}{\partial\alpha\,\partial(\dot p\lambda^2/2V_0{}^2)} \tag{10-9}$$

We will sometimes call the resistance derivatives, which depend on the translational velocity components u, v, and w, together with the rotary derivatives, which depend on the angular velocity components p, q, and r, jointly the *velocity stability derivatives* in contrast to the *acceleration stability derivatives*.

A number of the derivatives have special importance or special names:

Static stability:

$$C_{m_\alpha} \qquad \text{static longitudinal stability}$$
$$C_{n_\beta} \qquad \text{directional stability; weathercock stability}$$

Damping derivatives:

$$
\begin{array}{ll}
C_{m_q},\ C_{m_{\dot\alpha}} & \text{damping in pitch} \\
C_{n_r},\ C_{n_{\dot\beta}} & \text{damping in yaw} \\
C_{l_p} & \text{damping in roll}
\end{array}
$$

Dihedral effect:

$$
-C_{l_\beta}
$$

The significance of C_{m_α} and C_{n_β} is that they are the "spring constants" for pitching and yawing motions and largely determine the natural frequencies of the modes. For stability, C_{m_α} is negative, and C_{n_β} is positive. The damping derivatives act effectively as the "damper" in a spring-mass-damper system and control the rate at which oscillations are damped. The reasons for two terms for damping in pitch and yaw are discussed in Sec. 10-10. The dihedral effect is a measure of the rolling moment developed by the missile as a result of sideslip. If the rolling moment is positive (right wing down) for negative sideslip, the dihedral effect is "stable," and the missile rolls into the turn.

10-3. Properties of Stability Derivatives Resulting from Missile Symmetries; Maple-Synge Analysis for Cruciform Missiles

Before we concern ourselves with methods for evaluating stability derivatives, it is desirable to deduce what general information we can concerning stability derivatives from the symmetry properties of the missile. However, the reader who is interested at this time only in final results can go to Sec. 10-5, in which the apparent-mass method of evaluating stability derivatives is treated. The elegant Maple-Synge analysis[25] systematically deduces the consequences of the several types of symmetry possessed by missiles, and it is the basis of this section. As pointed out in the previous section, the stability derivatives depend for their values on the values of u, v, w, p, q, and r. We will consider several important cases in this connection.

$$
\begin{array}{llllll}
\textit{Case 1: } \text{Roll and pitch} & u \neq 0 & w \neq 0 & p \neq 0 & v = q = r = 0 \\
\textit{Case 2: } \text{Pitch and no roll} & u \neq 0 & w \neq 0 & p = 0 & v = q = r = 0 \\
\textit{Case 3: } \text{Roll and no pitch} & u \neq 0 & w = 0 & p \neq 0 & v = q = r = 0 \\
\textit{Case 4: } \text{No roll and no pitch} & u \neq 0 & w = 0 & p = 0 & v = q = r = 0
\end{array}
$$

Two distinct types of symmetry are important in so far as stability derivatives are concerned: *rotational symmetry* and *mirror symmetry*. The rotational symmetry has been specified in terms of *covering operations*. If by rotation through a particular angle about its longitudinal axis a missile can be brought from one orientation to another indistinguishable from the first, a covering operation is said to have been performed. If by successive rotations in the amount of $2\pi/n$ radians the missile undergoes a succession of covering operations, it is said by Maple and Synge to possess *n-gonal symmetry*. Mirror symmetry, on the other hand, is intuitively apparent. When we say a missile possesses a vertical plane of symmetry, we mean that it possesses mirror symmetry from one side of the plane to the other; that is, the missile part to the left of the plane is the mirror image of the missile part to the right of the plane. Missiles frequently possess several planes of mirror symmetry. In Fig. 10-3 missiles possessing 1-, 2-, 3-, and 4-gonal symmetry but no mirror symmetry are contrasted with missiles possessing 1-, 2-, 3-, and 4-gonal symmetry and also mirror symmetry. We shall call a missile having three planes of mirror symmetry and 3-gonal symmetry a *triform missile* and one with four planes of mirror symmetry and 4-gonal symmetry a *cruciform missile*. The two symmetry properties together yield general information on the analytical form of stability derivatives, and also specify which derivatives are necessarily zero.

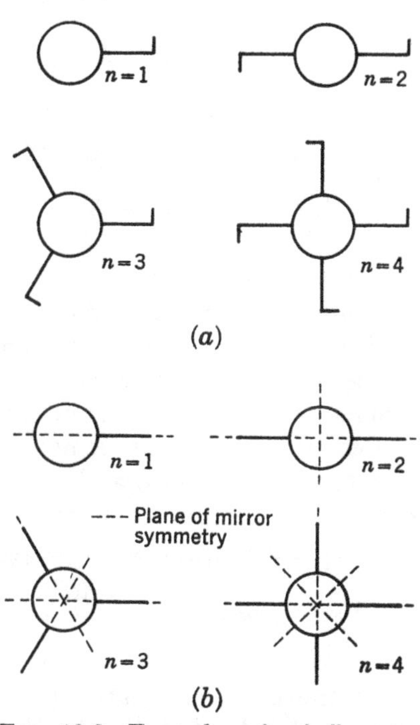

FIG. 10-3. Examples of missile symmetries. (a) *N*-gonal symmetry, no mirror symmetry; (b) *N*-gonal and mirror symmetries.

In the analysis which follows it is convenient to specify certain complex combinations of quantities as follows:

$$
\begin{aligned}
F &= Z + iY & T &= N + iM \\
\bar{F} &= Z - iY & \bar{T} &= N - iM \\
\nu &= w + iv & \omega &= r + iq \\
\bar{\nu} &= w - iv & \bar{\omega} &= r - iq
\end{aligned}
\tag{10-10}
$$

The symbols are those specified by Fig. 10-1. The first assumption in the analysis is one concerning the general nature of the aerodynamic forces. In accordance with the preceding section and within its limitations we assume that the aerodynamic forces and moments of the missile depend only on u, v, w, p, q, and r. We further assume that these forces and moments are given by a Taylor series in v, w, q, and r with the coefficients functions depending on u and p. Since the coefficients are functions of u and p, we have lost no generality in comparison with Eq. (10-3). Also any quadratic or cubic dependence of forces and moments on v or p

can be considered without involving terms of second and third degree in
the general expansion. In terms of the complex quantities we thus have

$$F = Z + iY = \sum_{ijkl} f_{ijkl}(u,p)\, \nu^i \bar{\nu}^j \omega^k \bar{\omega}^l$$

$$T = N + iM = \sum_{ijkl} t_{ijkl}(u,p)\, \nu^i \bar{\nu}^j \omega^k \bar{\omega}^l$$

$$X = \sum_{ijkl} x_{ijkl}(u,p)\, \nu^i \bar{\nu}^j \omega^k \bar{\omega}^l \tag{10-11}$$

$$L = \sum_{ijkl} l_{ijkl}(u,p)\, \nu^i \bar{\nu}^j \omega^k \bar{\omega}^l$$

The coefficients f_{ijkl}, t_{ijkl}, x_{ijkl}, l_{ijkl} are complex-valued functions of u and p.
In Appendix A at the end of the chapter, the consequences of rotational
and mirror symmetry are systematically deduced in so far as the coefficients in Eq. (10-11) are concerned. We will concern ourselves with the
results here, and refer the interested reader to Appendix A at the end of
the chapter for the mathematical details.

From Eqs. (10A-24) and (10A-25) the general terms of the series for the
drag and rolling moment are

$$X = x_{0000}^{(E)} + x_{0011}^{(E)}(q^2 + r^2) + 2x_{1001}^{(O)}(wr + qv) + 2x_{1001}^{(E)}(wq - vr)$$
$$+ x_{1100}^{(E)}(w^2 + v^2) + \text{terms of fourth degree} \tag{10-12}$$

$$L = l_{0000}^{(O)} + l_{0011}^{(O)}(q^2 + r^2) + 2l_{1001}^{(E)}(wr + qv) + 2l_{1001}^{(O)}(qw - vr)$$
$$+ l_{1100}^{(O)}(w^2 + v^2) + \text{terms of fourth degree} \tag{10-13}$$

Here the coefficients are functions of u and p. The superscript (E)
denotes that the function is even in p and the superscript (O) denotes an
odd function in p. Similarly, we have results for the forces Y and Z.

$$Z = f_{0010}^{(O)}r - f_{0010}^{(E)}q + f_{1000}^{(E)}w - f_{1000}^{(O)}v$$
$$Y = f_{0010}^{(E)}r + f_{0010}^{(O)}q + f_{1000}^{(O)}w + f_{1000}^{(E)}v \tag{10-14}$$

The expansions for N and M are analogous to those for Z and Y, respectively, with the superscripts (E) and (O) interchanged.

$$N = t_{0010}^{(E)}r - t_{0010}^{(O)}q + t_{1000}^{(O)}w - t_{1000}^{(E)}v$$
$$M = t_{0010}^{(O)}r + t_{0010}^{(E)}q + t_{1000}^{(E)}w + t_{1000}^{(O)}v \tag{10-15}$$

The expansions for Y, Z, M, and N contain no terms of second degree.

Equations (10-12) to (10-15) inclusive give the Maple-Synge expansions for all six forces and moments in powers of w, v, q, and r with coefficients which are functions of u and p. The stability derivatives are
formed by differentiating the forces and moments with respect to u, v, w,
p, q, and r. When the roll rate is zero, the following relationships help to
reduce the number of stability derivatives which are nonzero.

$$x_{ijkl}^{(O)} = l_{ijkl}^{(O)} = f_{ijkl}^{(O)} = t_{ijkl}^{(O)} = 0 \qquad\qquad \text{if } p = 0$$

$$\frac{\partial}{\partial p} x_{ijkl}^{(E)} = \frac{\partial}{\partial p} l_{ijkl}^{(E)} = \frac{\partial}{\partial p} f_{ijkl}^{(E)} = \frac{\partial}{\partial p} t_{ijkl}^{(E)} = 0 \qquad \text{if } p = 0 \tag{10-16}$$

The derivatives which exist for the four cases are summarized in Fig. 10-4.

It is interesting to determine the number of independent stability derivatives for each of the four cases listed in Fig. 10-4. However,

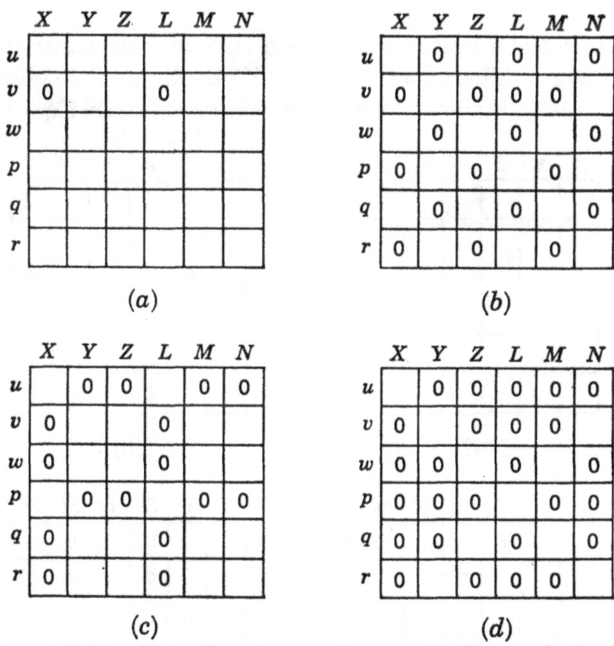

FIG. 10-4. Zero stability derivatives for cruciform and triform missiles. (a) Case 1: $u \neq 0, w \neq 0, p \neq 0, v = q = r = 0$; (b) case 2: $u \neq 0, w \neq 0, p = 0, v = q = r = 0$; (c) case 3: $u \neq 0, w = 0, p \neq 0, v = q = r = 0$; (d) case 4: $u \neq 0, w = 0, p = 0, v = q = r = 0$.

before doing so, let us note that certain equalities prevail among the derivatives by virtue of Eqs. (10-14) and (10-15), namely,

$$Z_r = Y_q \qquad Z_q = -Y_r \qquad Z_w = Y_v \qquad Z_v = -Y_w$$
$$N_r = M_q \qquad N_q = -M_r \qquad N_w = M_v \qquad N_v = -M_w \tag{10-17}$$

When these equalities are taken into consideration, it is clear that, of the 34 nonzero derivatives, for case 1, 26 are independent; for case 2, 13 are independent; for case 3, 12; and, for case 4, only 6. Since the total number of derivatives without considerations of symmetry is 36, a large reduction in the number has been made by means of the Maple-Synge analysis. While the analysis establishes which derivatives are zero by virtue of symmetry, other derivatives may be zero by virtue of special aerodynamic reasons. We will consider methods for evaluating the stability derivatives later, but will first carry out the Maple-Synge analysis for missiles with 2- and 3-gonal symmetry and mirror symmetry.

10-4. Maple-Synge Analysis for Triform Missiles and Other Missiles

It is interesting to examine the results of the Maple-Synge analysis for missiles with 3-gonal and 2-gonal symmetry as well as mirror symmetry. The actual analyses for these two cases are carried out in Appendix B at the end of the chapter. Only the results of the analyses will be discussed in this section.

The triform missile presents an interesting case in comparison with a cruciform missile. The expansions for Y, Z, M, and N given by Eqs. (10B-3) to (10B-6), inclusive, are to be compared with Eqs. (10-14) and

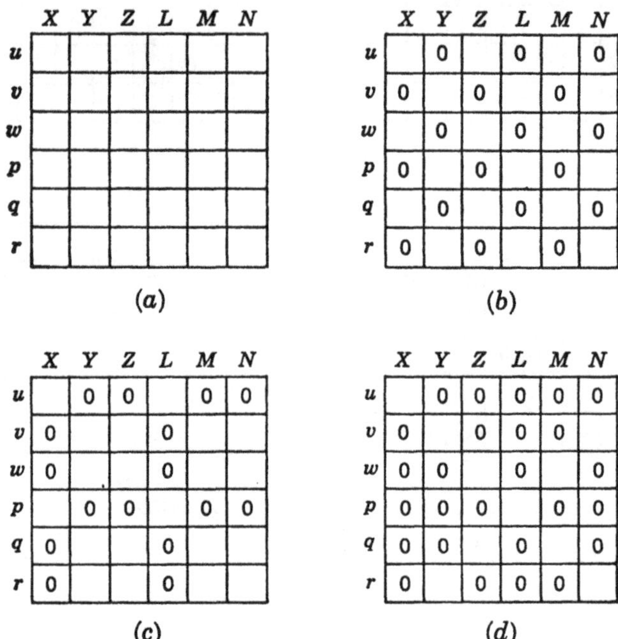

(a)

	X	Y	Z	L	M	N
u						
v						
w						
p						
q						
r						

(b)

	X	Y	Z	L	M	N
u		0		0		0
v	0		0		0	
w		0		0		0
p	0		0		0	
q		0		0		0
r	0		0		0	

(c)

	X	Y	Z	L	M	N
u		0	0		0	0
v	0			0		
w	0			0		
p		0	0		0	0
q	0			0		
r	0			0		

(d)

	X	Y	Z	L	M	N
u		0	0	0	0	0
v	0		0	0	0	
w	0	0		0		0
p	0	0	0		0	0
q	0	0		0		0
r	0		0	0	0	

FIG. 10-5. Zero stability derivatives for missile with 2-gonal and mirror symmetries. (a) Case 1: $u \neq 0$, $w \neq 0$, $p \neq 0$, $v = q = r = 0$; (b) case 2: $u \neq 0$, $w \neq 0$, $p = 0$, $v = q = r = 0$; (c) case 3: $u \neq 0$, $w = 0$, $p \neq 0$, $v = q = r = 0$; (d) case 4: $u \neq 0$, $w = 0$, $p = 0$, $v = q = r = 0$.

(10-15), which give the corresponding quantities for a cruciform missile. It turns out that the first-degree terms in each case are identical in form, but the triform missile has many terms of second degree where the cruciform missile has none. For the triform missile the X and L forces are given by Eqs. (10B-9) and (10B-10). These results compared with those for cruciform missiles given by Eqs. (10-12) and (10-13) reveal that the forms of the equation are identical for the two cases through terms of second degree.

With the series for the forces and moments explicitly determined, we can now obtain the stability derivatives by direct differentiation with respect to u, v, w, p, q, and r. The differentiations will not be carried out. However, the derivatives which are not identically zero from symmetry conditions are precisely those listed in Fig. 10-4 for cruciform missiles. It also turns out that the eight equalities between stability derivatives for a cruciform missile given by Eq. (10-17) are also true for cases 3 and 4 for a triform missile. It is to be noted that, even though the triform missile has many of the stability derivative properties of a cruciform missile, the numerical values of its stability derivatives are generally different from those of a cruciform missile.

The general Maple-Synge expansions for Y, Z, M, N, X, and L are derived in Appendix B at the end of the chapter for a missile with 2-gonal symmetry and mirror symmetry. The stability derivatives based on the results are summarized in Fig. 10-5. It is interesting to compare the derivatives which are zero for the present case with those which are zero for the cruciform-triform case, as listed in Fig. 10-4. For case 1, X_v and L_v are not zero in the present circumstances; and, for case 2, L_v is not zero. For cases 3 and 4, the derivatives which are zero by virtue of symmetry are identical for missiles with *2-gonal* and mirror symmetry, for triform missiles, and for cruciform missiles.

All derivatives listed in cases 1 and 2 are not independent. In fact, the following equalities hold for these cases:

$$Z_u = w \frac{\partial}{\partial u} (Z_w) \qquad N_p = w \frac{\partial}{\partial p} (N_w)$$

$$Y_p = w \frac{\partial}{\partial p} (Y_w) \qquad M_u = w \frac{\partial}{\partial u} (M_w) \tag{10-18}$$

For case 1 the additional equalities hold:

$$Z_p = w \frac{\partial}{\partial p} (Z_w) \qquad N_u = w \frac{\partial}{\partial u} (N_w)$$

$$Y_u = w \frac{\partial}{\partial u} (Y_w) \qquad M_p = w \frac{\partial}{\partial p} (M_w) \tag{10-19}$$

10-5. General Expression for Stability Derivatives in Terms of Inertia Coefficients; Method of Bryson

Hitherto we have been concerned only with the general properties of stability derivatives derivable from the symmetry properties of the missile. Now we will be concerned with methods for actual evaluation of the derivatives. A number of approaches for evaluating the derivatives are possible for slender configurations. There is the direct approach used by Nonweiler[1] of determining the potential, calculating the pressure distribution by Bernoulli's equation, and integrating the pressure distribution to obtain the force or moment concerned. If the square terms in Bernoulli's equation are included, the integrations can become very complicated in many cases, i.e., Sec. 5-5. Also, special account must be taken of leading-edge suction. A second method used by Ward and extended by Sacks[2] considers the gross forces and moments evaluated from a consideration of the pressures acting on the control surface enclosing the missile, together with the momentum flux through the surface. This second method makes extensive use of residue theory and conformal mapping, giving the stability derivatives in terms of the coefficients of the Laurent series for transforming the missile cross sections into a circle. A third approach which will be used here is the apparent-mass method used by Bryson.[3] This method is a direct one if the apparent-mass coefficients of the missile cross section are known. It automatically takes into account effects of leading-edge suction.

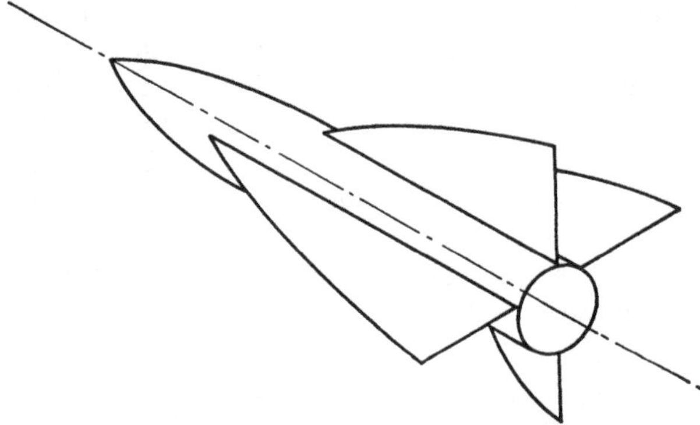

FIG. 10-6. Missile type readily amenable to analysis by apparent-mass methods.

Before embarking on the *method of apparent masses* or *method of inertial coefficients* as we will variously term it, let us consider the general class of configurations to which the method applies. Generally speaking, the wing wake must not influence the empennage, or the tail wake must not influence the wing for a canard missile. Missiles of the type shown in Fig. 10-6 are readily handled by apparent-mass methods. The influence of wing wake on the empennage is treated in Chap. 7 and in Sec. 10-11. It is probably important to realize that the method of apparent masses gives stability derivatives, not gross forces or moments. If the force or moment in question is zero when v, w, p, q, or r is zero, then the derivative with respect to any one of these independent variables also automatically gives the forces or moment for nonzero values of these variables. However, this would not be true, for instance, for lift or pitching moment associated with wing camber or wing twist. In such cases it is probably better to calculate the force or moment acting when v, w, p, q, or r is zero by special methods rather than the apparent-mass methods.

Bryson's method of apparent masses is based on certain results of Lamb[4] which will be quoted here without proof. Consider a missile moving through an infinite expanse of fluid stationary at infinity, and let the system of body axes X, Y, Z have its origin fixed at the center of gravity of the missile as shown in Fig. 10-7.   Consider a crossflow plane fixed in the fluid perpendicular to the X axis.   The potential in this plane depends (except for a function of X, Sec. 3-4, which cannot influence the stability derivatives considered herein) only on the normal velocities of the missile cross section in the plane at the instant under consideration. Let ξ, η, and ζ be parallel to X, Y, and Z, and let v_1 and v_2 be the linear velocities of the missile cross section in the plane along the η and ζ axes, respectively.   Also designate the angular velocity of the missile cross

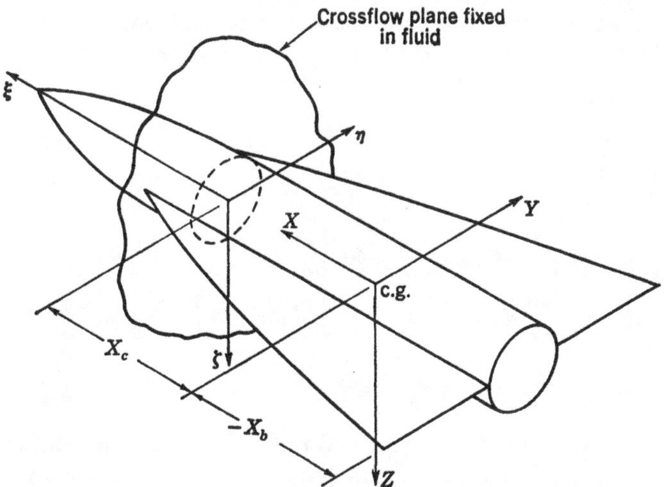

Fig. 10-7. Coordinate systems for apparent-mass analysis.

section about the ξ axis as p.   If we designate the potentials due to unit values of v_1, v_2, and p as ϕ_1, ϕ_2, and ϕ_3, respectively, we have for the complete potential

$$\phi = v_1\phi_1 + v_2\phi_2 + p\phi_3 \tag{10-20}$$

(We are neglecting any influence of the log term proportional to the rate of change of missile cross-sectional area.   Its influence on the stability derivatives is nil for a missile with a horizontal or vertical plane of symmetry except in so far as the drag is concerned.)   The kinetic energy of the fluid per unit length along ξ can be expressed by the well-known integral[5]

$$T = -\tfrac{1}{2}\rho \oint_C \phi \frac{\partial \phi}{\partial n} \, ds$$

where the contour C is the periphery of the missile cross section in the crossflow plane, and n is the outward normal.   The kinetic energy can be expressed with the help of Eq. (10-20) as

$$- \frac{T}{\frac{1}{2}\rho S_R} = \frac{v_1{}^2}{S_R} \oint_C \phi_1 \frac{\partial \phi_1}{\partial n} \, ds + \frac{v_1 v_2}{S_R} \oint_C \phi_1 \frac{\partial \phi_2}{\partial n} \, ds$$

$$+ \frac{v_1(\lambda p)}{\lambda S_R} \oint_C \phi_1 \frac{\partial \phi_3}{\partial n} \, ds + \frac{v_1 v_2}{S_R} \oint_C \phi_2 \frac{\partial \phi_1}{\partial n} \, ds$$

$$+ \frac{v_2{}^2}{S_R} \oint_C \phi_2 \frac{\partial \phi_2}{\partial n} \, ds + \frac{v_2(\lambda p)}{\lambda S_R} \oint_C \phi_2 \frac{\partial \phi_3}{\partial n} \, ds$$

$$+ \frac{v_1(\lambda p)}{\lambda S_R} \oint_C \phi_3 \frac{\partial \phi_1}{\partial n} \, ds + \frac{v_2(\lambda p)}{\lambda S_R} \oint_C \phi_3 \frac{\partial \phi_2}{\partial n} \, ds$$

$$+ \frac{(\lambda p)^2}{\lambda^2 S_R} \oint_C \phi_3 \frac{\partial \phi_3}{\partial n} \, ds \quad (10\text{-}21)$$

The reference length λ has been introduced together with a reference area S_R. The nine integrals are called the inertia coefficients of the cross section, and are given the notation A_{ij} in accordance with the following array:

$$\begin{vmatrix} A_{11} & A_{12} & A_{13} \\ A_{21} & A_{22} & A_{23} \\ A_{31} & A_{32} & A_{33} \end{vmatrix}$$

$$= - \begin{vmatrix} \dfrac{1}{S_R} \oint_C \phi_1 \dfrac{\partial \phi_1}{\partial n} \, ds & \dfrac{1}{S_R} \oint_C \phi_1 \dfrac{\partial \phi_2}{\partial n} \, ds & \dfrac{1}{\lambda S_R} \oint_C \phi_1 \dfrac{\partial \phi_3}{\partial n} \, ds \\[2ex] \dfrac{1}{S_R} \oint_C \phi_2 \dfrac{\partial \phi_1}{\partial n} \, ds & \dfrac{1}{S_R} \oint_C \phi_2 \dfrac{\partial \phi_2}{\partial n} \, ds & \dfrac{1}{\lambda S_R} \oint_C \phi_2 \dfrac{\partial \phi_3}{\partial n} \, ds \\[2ex] \dfrac{1}{\lambda S_R} \oint_C \phi_3 \dfrac{\partial \phi_1}{\partial n} \, ds & \dfrac{1}{\lambda S_R} \oint_C \phi_3 \dfrac{\partial \phi_2}{\partial n} \, ds & \dfrac{1}{\lambda^2 S_R} \oint_C \phi_3 \dfrac{\partial \phi_3}{\partial n} \, ds \end{vmatrix} \quad (10\text{-}22)$$

It is of interest to note a reciprocal relationship for inertial coefficients. This relationship is based on the particular form of Green's[6] theorem valid for potential functions ϕ_1, ϕ_2, and ϕ_3 which follows:

$$\oint_C \phi_i \frac{\partial \phi_j}{\partial n} \, ds = \oint_C \phi_j \frac{\partial \phi_i}{\partial n} \, ds \quad (10\text{-}23)$$

Thus, we have

$$A_{ij} = A_{ji} \quad (10\text{-}24)$$

and the kinetic energy of the fluid per unit length becomes

$$T = \tfrac{1}{2}\rho S_R[v_1{}^2 A_{11} + v_2{}^2 A_{22} + (\lambda p)^2 A_{33} + 2v_1 v_2 A_{12} + 2v_1(\lambda p) A_{13}$$
$$+ 2v_2(\lambda p) A_{23}] \quad (10\text{-}25)$$

It is convenient at this point to relate the velocities v_1 and v_2 to linear and angular velocities v, w, q, and r, but with the substitution of α and β as independent variables for v and w. Thus, we obtain

$$v_1 = v + rX = \beta V_0 + rX$$
$$v_2 = w - qX = \alpha V_0 - qX \quad (10\text{-}26)$$

The power of the method now is that the forces and moments Y, Z, L, M, and N can be simply determined by differentiation of the kinetic energy given by Eq. (10-25). (We are not including the thrust force X, because the method of apparent masses is not suited to its determination.) The formulas for obtaining the force dY/dX and dZ/dX per unit axial distance and the rolling moment dL/dX per unit axial distance are taken from Lamb,[4] and are presented in their particular form for the present application without proof.

$$\frac{dY}{dX} = -\frac{d}{dt}\left(\frac{\partial T}{\partial v_1}\right) + p\frac{\partial T}{\partial v_2}$$

$$\frac{dZ}{dX} = -\frac{d}{dt}\left(\frac{\partial T}{\partial v_2}\right) - p\frac{\partial T}{\partial v_1} \qquad (10\text{-}27)$$

$$\frac{dL}{dX} = -\frac{d}{dt}\left(\frac{\partial T}{\partial p}\right) + v_2\frac{\partial T}{\partial v_1} - v_1\frac{\partial T}{\partial v_2}$$

The differentiation is in the fixed crossflow plane, and the total derivative d/dt must reflect the changing coordinate X_c of the crossflow plane with time. Thus

$$\frac{d}{dt} = \frac{\partial}{\partial t} + \frac{dX_c}{dt}\frac{\partial}{\partial X} = \frac{\partial}{\partial t} - V_0\frac{\partial}{\partial X} \qquad (10\text{-}28)$$

If we carry out the differentiations indicated by Eq. (10-27), we obtain the forces and rolling moment per unit length.

$$\frac{dY}{dX} = -\rho S_R[A_{11}\dot{v}_1 + A_{12}\dot{v}_2 + A_{13}(\lambda\dot{p})]$$

$$+ \rho S_R V_0\frac{\partial}{\partial X}[A_{11}v_1 + A_{12}v_2 + A_{13}(\lambda p)]$$

$$+ \rho S_R p[A_{12}v_1 + A_{22}v_2 + A_{23}(\lambda p)] \qquad (10\text{-}29)$$

$$\frac{dZ}{dX} = -\rho S_R[A_{12}\dot{v}_1 + A_{22}\dot{v}_2 + A_{23}(\lambda\dot{p})]$$

$$+ \rho S_R V_0\frac{\partial}{\partial X}[A_{12}v_1 + A_{22}v_2 + A_{23}(\lambda p)]$$

$$- \rho S_R p[A_{11}v_1 + A_{12}v_2 + A_{13}(\lambda p)] \qquad (10\text{-}30)$$

$$\frac{dL}{dX} = -\rho S_R\lambda[A_{13}\dot{v}_1 + A_{23}\dot{v}_2 + A_{33}(\lambda\dot{p})]$$

$$+ \lambda\rho S_R V_0\frac{\partial}{\partial X}[A_{13}v_1 + A_{23}v_2 + A_{33}(\lambda p)]$$

$$+ \rho S_R v_2[A_{11}v_1 + A_{12}v_2 + A_{13}(\lambda p)]$$

$$- \rho S_R v_1[A_{12}v_1 + A_{22}v_2 + A_{23}(\lambda p)] \qquad (10\text{-}31)$$

Since the axial distributions of sideforce Y, normal force $-Z$, and rolling moment L are known along the body, direct integrations from missile base to missile apex will yield the Y, Z, L, and M and N. Let us first put Eqs. (10-29), (10-30), and (10-31) into appropriate nondimensional form by dividing all forces by $\rho V_0^2 S_R/2$ and all moments by $\rho V_0^2 S_R\lambda/2$, where S_R and λ are the reference area and reference length, respectively. We also introduce the parameters α, β, $\lambda p/2V_0$, $\lambda q/2V_0$, and $\lambda r/2V_0$ as the independent variables. By these means we obtain

$$\frac{dC_Y}{d(X/\lambda)} = -4\left\{ A_{11}\left[\left(\frac{\lambda\dot{\beta}}{2V_0}\right) + \beta\left(\frac{\dot{V}_0\lambda}{2V_0{}^2}\right) + \frac{X}{\lambda}\left(\frac{\dot{r}\lambda^2}{2V_0{}^2}\right)\right]\right.$$

$$+ A_{12}\left[\left(\frac{\lambda\dot{\alpha}}{2V_0}\right) + \alpha\left(\frac{\dot{V}_0\lambda}{2V_0{}^2}\right) - \frac{X}{\lambda}\left(\frac{\dot{q}\lambda^2}{2V_0{}^2}\right)\right] + A_{13}\left(\frac{\lambda^2\dot{p}}{2V_0{}^2}\right)\right\}$$

$$+ 2\frac{d}{d(X/\lambda)}\left\{ A_{11}\left[\beta + 2\left(\frac{r\lambda}{2V_0}\right)\frac{X}{\lambda}\right]\right.$$

$$+ A_{12}\left[\alpha - 2\left(\frac{q\lambda}{2V_0}\right)\frac{X}{\lambda}\right] + 2A_{13}\left(\frac{\lambda p}{2V_0}\right)\right\}$$

$$+ 4\left(\frac{\lambda p}{2V_0}\right)\left\{ A_{12}\left[\beta + 2\left(\frac{r\lambda}{2V_0}\right)\frac{X}{\lambda}\right]\right.$$

$$\left. + A_{22}\left[\alpha - 2\left(\frac{q\lambda}{2V_0}\right)\frac{X}{\lambda}\right] + 2A_{23}\left(\frac{\lambda p}{2V_0}\right)\right\} \quad (10\text{-}32)$$

$$\frac{dC_Z}{d(X/\lambda)} = -4\left\{ A_{12}\left[\left(\frac{\lambda\dot{\beta}}{2V_0}\right) + \beta\left(\frac{\dot{V}_0\lambda}{2V_0{}^2}\right) + \frac{X}{\lambda}\left(\frac{\dot{r}\lambda^2}{2V_0{}^2}\right)\right]\right.$$

$$+ A_{22}\left[\left(\frac{\lambda\dot{\alpha}}{2V_0}\right) + \alpha\left(\frac{\dot{V}_0\lambda}{2V_0{}^2}\right) - \frac{X}{\lambda}\left(\frac{\dot{q}\lambda^2}{2V_0{}^2}\right)\right] + A_{23}\left(\frac{\lambda^2\dot{p}}{2V_0{}^2}\right)\right\}$$

$$+ 2\frac{d}{d(X/\lambda)}\left\{ A_{12}\left[\beta + 2\left(\frac{r\lambda}{2V_0}\right)\frac{X}{\lambda}\right]\right.$$

$$+ A_{22}\left[\alpha - 2\left(\frac{q\lambda}{2V_0}\right)\frac{X}{\lambda}\right] + 2A_{23}\left(\frac{\lambda p}{2V_0}\right)\right\}$$

$$- 4\left(\frac{\lambda p}{2V_0}\right)\left\{ A_{11}\left[\beta + 2\left(\frac{r\lambda}{2V_0}\right)\frac{X}{\lambda}\right]\right.$$

$$\left. + A_{12}\left[\alpha - 2\left(\frac{q\lambda}{2V_0}\right)\frac{X}{\lambda}\right] + 2A_{13}\left(\frac{\lambda p}{2V_0}\right)\right\} \quad (10\text{-}33)$$

$$\frac{dC_l}{d(X/\lambda)} = -4\left\{ A_{13}\left[\left(\frac{\lambda\dot{\beta}}{2V_0}\right) + \beta\left(\frac{\dot{V}_0\lambda}{2V_0{}^2}\right) + \frac{X}{\lambda}\left(\frac{\dot{r}\lambda^2}{2V_0{}^2}\right)\right]\right.$$

$$+ A_{23}\left[\left(\frac{\lambda\dot{\alpha}}{2V_0}\right) + \alpha\left(\frac{\dot{V}_0\lambda}{2V_0{}^2}\right) - \frac{X}{\lambda}\left(\frac{\dot{q}\lambda^2}{2V_0{}^2}\right)\right] + A_{33}\left(\frac{\lambda^2\dot{p}}{2V_0{}^2}\right)\right\}$$

$$+ 2\frac{d}{d(X/\lambda)}\left\{ A_{13}\left[\beta + 2\left(\frac{r\lambda}{2V_0}\right)\frac{X}{\lambda}\right]\right.$$

$$+ A_{23}\left[\alpha - 2\left(\frac{q\lambda}{2V_0}\right)\frac{X}{\lambda}\right] + 2A_{33}\left(\frac{\lambda p}{2V_0}\right)\right\}$$

$$+ 2\left[\alpha - 2\left(\frac{q\lambda}{2V_0}\right)\frac{X}{\lambda}\right]\left\{ A_{11}\left[\beta + 2\left(\frac{r\lambda}{2V_0}\right)\frac{X}{\lambda}\right]\right.$$

$$+ A_{12}\left[\alpha - 2\left(\frac{q\lambda}{2V_0}\right)\frac{X}{\lambda}\right] + 2A_{13}\left(\frac{\lambda p}{2V_0}\right)\right\}$$

$$- 2\left[\beta + 2\left(\frac{r\lambda}{2V_0}\right)\frac{X}{\lambda}\right]\left\{ A_{12}\left[\beta + 2\left(\frac{r\lambda}{2V_0}\right)\frac{X}{\lambda}\right]\right.$$

$$\left. + A_{22}\left[\beta - 2\left(\frac{q\lambda}{2V_0}\right)\frac{X}{\lambda}\right] + 2A_{23}\left(\frac{\lambda p}{2V_0}\right)\right\} \quad (10\text{-}34)$$

We will now obtain the specific formulas for the derivatives of C_Y, C_Z, C_l, C_m, and C_n by α, β, $p\lambda/2V_0$, $\lambda q/2V_0$, and $r\lambda/2V_0$—25 derivatives in all. Considering first the derivatives of C_Y, we obtain from Eq. (10-32)

$$\frac{dC_{Y_\alpha}}{d(X/\lambda)} = -4A_{12}\left(\frac{\lambda\dot{V}_0}{2V_0{}^2}\right) + 2\frac{dA_{12}}{d(X/\lambda)} + 4\left(\frac{\lambda p}{2V_0}\right)A_{22}$$

$$\frac{dC_{Y_\beta}}{d(X/\lambda)} = -4A_{11}\left(\frac{\lambda\dot{V}_0}{2V_0{}^2}\right) + 2\frac{dA_{11}}{d(X/\lambda)} + 4\left(\frac{\lambda p}{2V_0}\right)A_{12}$$

$$\frac{dC_{Y_p}}{d(X/\lambda)} = 4\frac{dA_{13}}{d(X/\lambda)} + 4A_{12}\left[\beta + 2\frac{X}{\lambda}\left(\frac{r\lambda}{2V_0}\right)\right]$$

$$+ 4A_{22}\left[\alpha - 2\frac{X}{\lambda}\left(\frac{q\lambda}{2V_0}\right)\right] + 16A_{23}\left(\frac{\lambda p}{2V_0}\right) \qquad (10\text{-}35)$$

$$\frac{dC_{Y_q}}{d(X/\lambda)} = -4\frac{d}{d(X/\lambda)}\left[A_{12}\left(\frac{X}{\lambda}\right)\right] - 8\left(\frac{\lambda p}{2V_0}\right)\left(\frac{X}{\lambda}\right)A_{22}$$

$$\frac{dC_{Y_r}}{d(X/\lambda)} = 4\frac{d}{d(X/\lambda)}\left[A_{11}\left(\frac{X}{\lambda}\right)\right] + 8\left(\frac{\lambda p}{2V_0}\right)\left(\frac{X}{\lambda}\right)A_{12}$$

To obtain the gross forces and moments, it is necessary to integrate from the missile base at X_b on the negative X axis (Fig. 10-7) to X_n at the missile apex. We denote the *value of the inertial coefficients at the missile base by a bar* as in $\bar{A}_{11}$, $\bar{A}_{22}$, etc. We furthermore indicate X integrals of the inertial coefficients A_{ij} as follows:

$$B_{ij} = \int_{(X/\lambda)_b}^{(X/\lambda)_n} A_{ij}\, d\left(\frac{X}{\lambda}\right)$$

$$C_{ij} = \int_{(X/\lambda)_b}^{(X/\lambda)_n} A_{ij}\left(\frac{X}{\lambda}\right) d\left(\frac{X}{\lambda}\right) \qquad (10\text{-}36)$$

$$D_{ij} = \int_{(X/\lambda)_b}^{(X/\lambda)_n} A_{ij}\left(\frac{X}{\lambda}\right)^2 d\left(\frac{X}{\lambda}\right)$$

In terms of A_{ij}, B_{ij}, and C_{ij} the integration of Eq. (10-35) yields

$$C_{Y_\alpha} = -4\left(\frac{\lambda\dot{V}_0}{2V_0{}^2}\right)B_{12} - 2\bar{A}_{12} + 4\left(\frac{\lambda p}{2V_0}\right)B_{22}$$

$$C_{Y_\beta} = -4\left(\frac{\lambda\dot{V}_0}{2V_0{}^2}\right)B_{11} - 2\bar{A}_{11} + 4\left(\frac{\lambda p}{2V_0}\right)B_{12}$$

$$C_{Y_p} = -4\bar{A}_{13} + 4\alpha B_{22} + 4\beta B_{12} + 16\left(\frac{\lambda p}{2V_0}\right)B_{23}$$

$$- 8\left(\frac{\lambda q}{2V_0}\right)C_{22} + 8\left(\frac{\lambda r}{2V_0}\right)C_{12} \qquad (10\text{-}37)$$

$$C_{Y_q} = 4\bar{A}_{12}\left(\frac{X}{\lambda}\right)_b - 8\left(\frac{\lambda p}{2V_0}\right)C_{22}$$

$$C_{Y_r} = -4\bar{A}_{11}\left(\frac{X}{\lambda}\right)_b + 8\left(\frac{\lambda p}{2V_0}\right)C_{12}$$

In similar fashion the derivatives for C_Z and C_l can be obtained, and only the results are quoted here.

$$C_{z_\alpha} = -4\left(\frac{\lambda \dot{V}_0}{2V_0{}^2}\right) B_{22} - 2\bar{A}_{22} - 4\left(\frac{\lambda p}{2V_0}\right) B_{12}$$

$$C_{z_\beta} = -4\left(\frac{\lambda \dot{V}_0}{2V_0{}^2}\right) B_{12} - 2\bar{A}_{12} - 4\left(\frac{\lambda p}{2V_0}\right) B_{11}$$

$$C_{z_p} = -4\bar{A}_{23} - 4\alpha B_{12} - 4\beta B_{11} - 16\left(\frac{\lambda p}{2V_0}\right) B_{13}$$

$$+ 8\left(\frac{\lambda q}{2V_0}\right) C_{12} - 8\left(\frac{\lambda r}{2V_0}\right) C_{11} \qquad (10\text{-}38)$$

$$C_{z_q} = 4\bar{A}_{22}\left(\frac{X}{\lambda}\right)_b + 8\left(\frac{\lambda p}{2V_0}\right) C_{12}$$

$$C_{z_r} = -4\bar{A}_{12}\left(\frac{X}{\lambda}\right)_b - 8\left(\frac{\lambda p}{2V_0}\right) C_{11}$$

$$C_{l_\alpha} = -4\left(\frac{\dot{V}_0\lambda}{2V_0{}^2}\right) B_{23} - 2\bar{A}_{23} + 4\alpha B_{12} + 2\beta(B_{11} - B_{22})$$

$$+ 4\left(\frac{\lambda p}{2V_0}\right) B_{13} - 8\left(\frac{\lambda q}{2V_0}\right) C_{12} + 4\left(\frac{\lambda r}{2V_0}\right) (C_{11} - C_{22})$$

$$C_{l_\beta} = -4\left(\frac{\dot{V}_0\lambda}{2V_0{}^2}\right) B_{13} - 2\bar{A}_{13} + 2\alpha(B_{11} - B_{22}) - 4\beta B_{12}$$

$$- 4\left(\frac{\lambda p}{2V_0}\right) B_{23} - 4\left(\frac{\lambda q}{2V_0}\right) (C_{11} - C_{22}) - 8\left(\frac{\lambda r}{2V_0}\right) C_{12}$$

$$C_{l_p} = -4\bar{A}_{33} + 4\alpha B_{13} - 4\beta B_{23} - 8\left(\frac{\lambda q}{2V_0}\right) C_{13} - 8\left(\frac{\lambda r}{2V_0}\right) C_{23} \qquad (10\text{-}39)$$

$$C_{l_q} = 4\bar{A}_{23}\left(\frac{X}{\lambda}\right)_b - 8\alpha C_{12} - 4\beta(C_{11} - C_{22}) - 8\left(\frac{\lambda p}{2V_0}\right) C_{13}$$

$$+ 16\left(\frac{\lambda q}{2V_0}\right) D_{12} - 8\left(\frac{\lambda r}{2V_0}\right) (D_{11} - D_{22})$$

$$C_{l_r} = -4\bar{A}_{13}\left(\frac{X}{\lambda}\right)_b + 4\alpha(C_{11} - C_{22}) - 8\beta C_{12} - 8\left(\frac{\lambda p}{2V_0}\right) C_{23}$$

$$- 8\left(\frac{\lambda q}{2V_0}\right) (D_{11} - D_{22}) - 16\left(\frac{\lambda r}{2V_0}\right) D_{12}$$

The pitching-moment and yawing-moment derivatives are obtained by taking the moment of the C_Z and C_Y distributions about the origin of the X, Y, Z axes, which was taken at the center of gravity of the missile.

$$C_{m_\alpha} = 4\left(\frac{\lambda \dot{V}_0}{2V_0{}^2}\right)C_{22} + 2\left[B_{22} + \bar{A}_{22}\left(\frac{X}{\lambda}\right)_b\right] + 4\left(\frac{\lambda p}{2V_0}\right)C_{12}$$

$$C_{m_\beta} = 4\left(\frac{\lambda \dot{V}_0}{2V_0{}^2}\right)C_{12} + 2\left[B_{12} + \bar{A}_{12}\left(\frac{X}{\lambda}\right)_b\right] + 4\left(\frac{\lambda p}{2V_0}\right)C_{11}$$

$$C_{m_p} = 4\left[B_{23} + \bar{A}_{23}\left(\frac{X}{\lambda}\right)_b\right] + 4\alpha C_{12} + 4\beta C_{11} + 16\left(\frac{\lambda p}{2V_0}\right)C_{13}$$
$$- 8\left(\frac{\lambda q}{2V_0}\right)D_{12} + 8\left(\frac{\lambda r}{2V_0}\right)D_{11} \tag{10-40}$$

$$C_{m_q} = -4\left[\bar{A}_{22}\left(\frac{X}{\lambda}\right)_b^2 + C_{22}\right] - 8\left(\frac{\lambda p}{2V_0}\right)D_{12}$$

$$C_{m_r} = 4\left[\bar{A}_{12}\left(\frac{X}{\lambda}\right)_b^2 + C_{12}\right] + 8\left(\frac{\lambda p}{2V_0}\right)D_{11}$$

$$C_{n_\alpha} = -4\left(\frac{\lambda \dot{V}_0}{2V_0{}^2}\right)C_{12} - 2\left[B_{12} + \bar{A}_{12}\left(\frac{X}{\lambda}\right)_b\right] + 4\left(\frac{\lambda p}{2V_0}\right)C_{22}$$

$$C_{n_\beta} = -4\left(\frac{\lambda \dot{V}_0}{2V_0{}^2}\right)C_{11} - 2\left[B_{11} + \bar{A}_{11}\left(\frac{X}{\lambda}\right)_b\right] + 4\left(\frac{\lambda p}{2V_0}\right)C_{12}$$

$$C_{n_p} = -4\left[B_{13} + \bar{A}_{13}\left(\frac{X}{\lambda}\right)_b\right] + 4\alpha C_{22} + 4\beta C_{12}$$
$$+ 16\left(\frac{\lambda p}{2V_0}\right)C_{23} - 8\left(\frac{\lambda q}{2V_0}\right)D_{22} + 8\left(\frac{\lambda r}{2V_0}\right)D_{12} \tag{10-41}$$

$$C_{n_q} = 4\left[\bar{A}_{12}\left(\frac{X}{\lambda}\right)_b^2 + C_{12}\right] - 8\left(\frac{\lambda p}{2V_0}\right)D_{22}$$

$$C_{n_r} = -4\left[\bar{A}_{11}\left(\frac{X}{\lambda}\right)_b^2 + C_{11}\right] + 8\left(\frac{\lambda p}{2V_0}\right)D_{12}$$

Equations (10-37) to (10-41) inclusive give *25 velocity derivatives* in terms of the inertia coefficients which can be obtained from the apparent-mass coefficients presented in Table 10-3. By use of these formulas we can systematically calculate the stability derivatives for slender missiles typified by that pictured in Fig. 10-6. It is of interest to note that the damping-in-roll derivative C_{l_p} is the only one involving $\bar{A}_{33}$, and $\bar{A}_{33}$ is frequently the most difficult inertia coefficient to obtain.

TABLE 10-3. APPARENT MASS COEFFICIENTS

A. Line:

$$m_{11} = 0$$
$$m_{12} = 0$$
$$m_{13} = 0$$
$$m_{22} = \frac{\pi\rho b^2}{4}$$
$$m_{23} = 0$$
$$m_{33} = \frac{\pi\rho b^4}{128}$$

TABLE 10-3. APPARENT MASS COEFFICIENTS (*Continued*)

B. Circle:

$m_{11} = \pi\rho a^2$

$m_{12} = 0$

$m_{13} = 0$

$m_{22} = \pi\rho a^2$

$m_{23} = 0$

$m_{33} = 0$

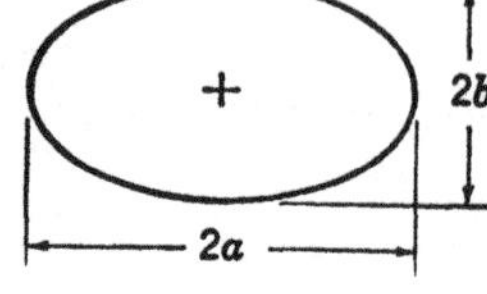

C. Ellipse:

$m_{11} = \pi\rho a^2$

$m_{12} = 0$

$m_{13} = 0$

$m_{22} = \pi\rho b^2$

$m_{23} = 0$

$m_{33} = \dfrac{\pi\rho(a^2 - b^2)^2}{8}$

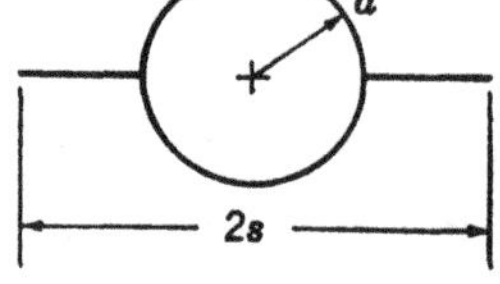

D. Planar midwing, circular body:

$m_{11} = \pi\rho a^2$

$m_{12} = 0$

$m_{13} = 0$

$m_{22} = \pi\rho s^2 \left(1 - \dfrac{a^2}{s^2} + \dfrac{a^4}{s^4} \right)$

$m_{23} = 0$

$m_{33} = \dfrac{\pi\rho s^4}{8} \qquad \text{if } a = 0$

$$m_{33} = \frac{\pi\rho s^4}{8} \left\{ \left[(1 + R^2)^2 \tan^{-1}\frac{1}{R} \right]^2 + 2R(1 - R^2)(R^4 - 6R^2 + 1) \tan^{-1}\frac{1}{R} - \pi^2 R^4 + R^2(1 - R^2)^2 \right\} \qquad \text{where } R = \frac{a}{s}$$

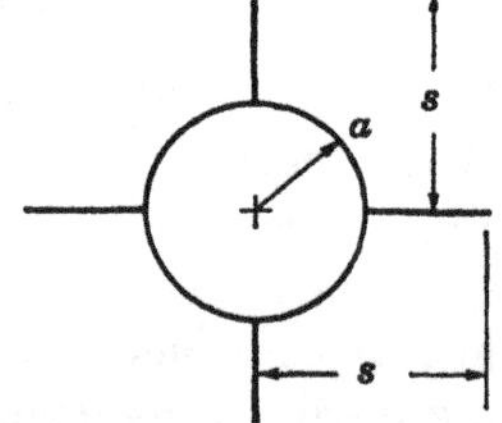

E. Cruciform wing, circular body:

$m_{11} = \pi\rho s^2 \left(1 - \dfrac{a^2}{s^2} + \dfrac{a^4}{s^4} \right)$

$m_{12} = 0$

$m_{13} = 0$

$m_{22} = \pi\rho s^2 \left(1 - \dfrac{a^2}{s^2} + \dfrac{a^4}{s^4} \right)$

$m_{23} = 0$

$m_{33} = \dfrac{2\rho s^4}{\pi} \qquad \text{if } a = 0$

m_{33}: Fig. 10-16   if $a \neq 0$

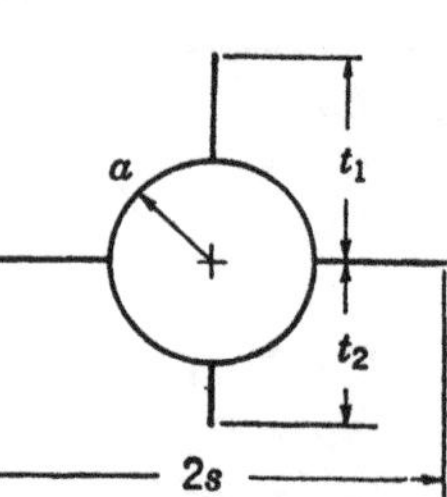

F. Midtail empennage with circular body:

$$m_{11} = \frac{\pi\rho s^2}{4} \left\{ \frac{t_1^2}{s^2} \left(1 + \frac{a^4}{t_1^4} \right) + \frac{t_2^2}{s^2} \left(1 + \frac{a^4}{t_2^4} \right) - 2 \left(1 + \frac{a^2}{s^2} \right)^2 + 2 \left[\left(1 + \frac{a^4}{s^2 t_1^2} \right) \left(1 + \frac{t_1^2}{s^2} \right) \left(1 + \frac{a^4}{s^2 t_2^2} \right) \left(1 + \frac{t_2^2}{s^2} \right) \right]^{\frac{1}{2}} \right\}$$

$m_{12} = 0$

m_{13}: See Ref. 13.

$m_{22} = \pi\rho s^2 \left(1 - \dfrac{a^2}{s^2} + \dfrac{a^4}{s^4} \right)$

$m_{23} = 0$

TABLE 10-3. APPARENT MASS COEFFICIENTS (*Continued*)

G. Multifinned body, three or more fins:

$$m_{11} = m_{22}$$
$$= 2\pi\rho s^2 \left\{ \left[\frac{1 + (a^2/s^2)^{n/2}}{2} \right]^{4/n} - \frac{1}{2}\left(\frac{a}{s}\right)^2 \right\}$$

$$m_{12} = 0$$
$$m_{13} = 0$$
$$m_{23} = 0$$
$$m_{33} = 0.533\rho s^4 \qquad n = 3 \qquad a = 0$$
$$m_{33} = \frac{2\rho s^4}{\pi} \qquad n = 4 \qquad a = 0$$
$$m_{33} = \frac{\pi\rho s^4}{2} \qquad n = \infty \qquad a = 0$$

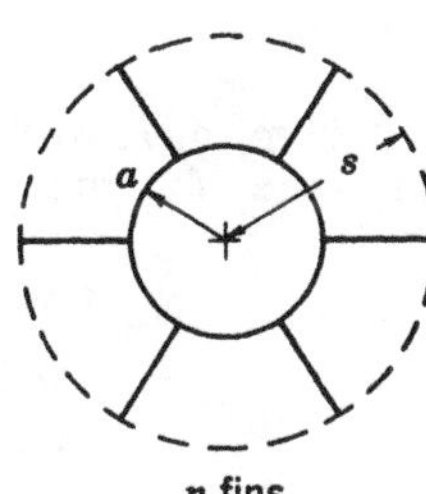

H. Regular inscribed polygon:

$$m_{11} = m_{22} = 0.654\pi\rho a^2 \qquad n = 3$$
$$= 0.787\pi\rho a^2 \qquad n = 4$$
$$= 0.823\pi\rho a^2 \qquad n = 5$$
$$= 0.867\pi\rho a^2 \qquad n = 6$$

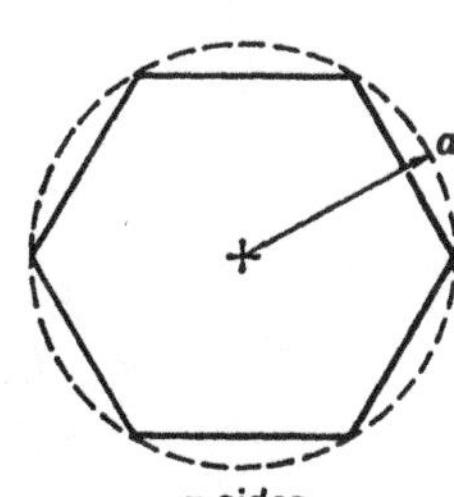

I. Tangent-tail empennage, circular body:

$$m_{11} = 2\pi\rho \left\{ c^2 - \frac{a^2}{2} + \frac{4c^2 \sin\lambda \cos^2(\lambda/2)}{3(\lambda + \sin\lambda)} \left[\sin^2\frac{\lambda}{2} - \frac{3\lambda \cos^2(\lambda/2)}{\lambda + \sin\lambda} \right] + 2(r^2 - c^2) \right\}$$

$$m_{22} = 2\pi\rho \left\{ c^2 - \frac{a^2}{2} - \frac{4c^2 \sin\lambda \cos^2(\lambda/2)}{3(\lambda + \sin\lambda)} \left[\sin^2\frac{\lambda}{2} - \frac{3\lambda \cos^2(\lambda/2)}{\lambda + \sin\lambda} \right] \right\}$$

$$m_{12} = 0$$
$$m_{23} = 0$$

where

$$\frac{a}{s} = \frac{1}{\pi} \left\{ \sinh^{-1}\left(\frac{\lambda}{2}\tan\frac{\lambda}{2}\right)^{1/2} + \left[\frac{\lambda}{2}\tan\frac{\lambda}{2} + \left(\frac{\lambda}{2}\right)^2 \tan^2\frac{\lambda}{2} \right]^{1/2} \right\}$$

$$\frac{c}{a} = \frac{\pi}{\lambda + \sin\lambda}$$

$$2 + \frac{t_1}{a} = \frac{\pi}{\dfrac{\lambda}{h/c + 1} + \tan^{-1}\dfrac{\sin\lambda}{h/c - \cos\lambda}}$$

$$\frac{t_2}{a} = \frac{\pi}{\dfrac{\lambda}{f/c - 1} + \tan^{-1}\dfrac{\sin\lambda}{f/c + \cos\lambda}}$$

$$r = \frac{1}{4}\left(h + \frac{c^2}{h} + f + \frac{c^2}{f} \right)$$

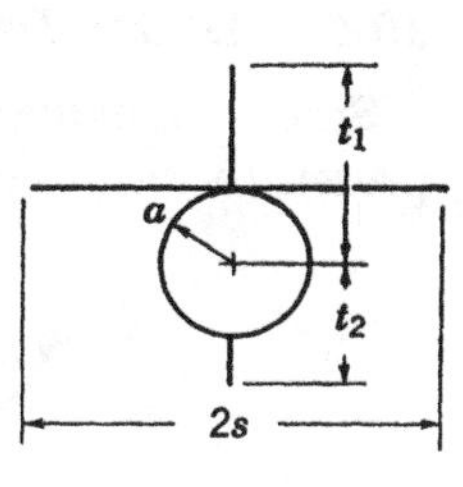

J. Midtail empennage, elliptical body:

$$m_{11} = \pi\rho(4c^2 - k^2 - 2ab - b^2)$$

$$m_{22} = \frac{\pi\rho}{(a-b)^2} \left[s^2(a^2 + b^2) + 2ab^2(a - b) - 2abs(s^2 - a^2 + b^2)^{1/2} \right]$$

$$k = \frac{as - b(s^2 + a^2 - b^2)^{1/2}}{a - b}$$

$$c = \frac{f_1 + f_2}{4}$$

$$f_{1,2}^2 = k^2 + \left[\tau_{1,2} + \frac{(a+b)^2}{4\tau_{1,2}} \right]^2$$

$$\tau_{1,2} = \frac{1}{2}\left[t_{1,2} + (t_{1,2} - a^2 + b^2)^{1/2} \right]$$

$$m_{12} = 0$$
$$m_{13} = 0$$

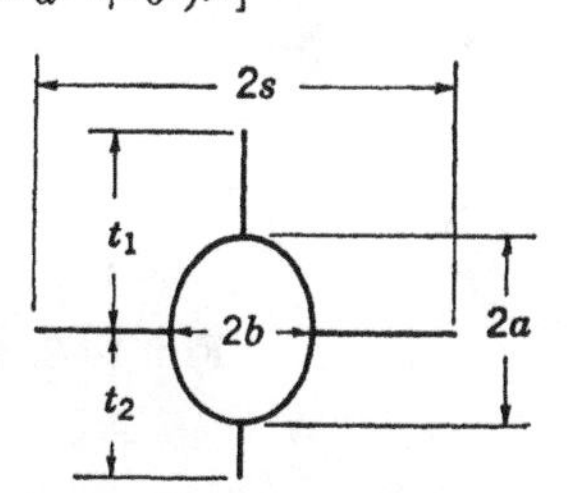

Certain acceleration derivatives also follow easily from Eqs. (10-32), (10-33), and (10-34). With $\dot{\alpha}$ derivatives given by

$$\frac{\partial}{\partial \dot{\alpha}} = \frac{\partial}{\partial(\dot{\alpha}\lambda/2V_0)}$$

and $\dot{\beta}$, $\dot{p}$, $\dot{q}$, $\dot{r}$, and $\dot{V}_0$ derivatives given by $\dot{\beta}\lambda/2V_0$, $\dot{p}\lambda^2/2V_0^2$, $\dot{q}\lambda^2/2V_0^2$, $\dot{r}\lambda^2/2V_0^2$, $\dot{V}_0\lambda/2V_0^2$, we obtain

$$
\begin{aligned}
C_{Y\dot{V}_0} &= -4\alpha B_{12} - 4\beta B_{11} & C_{Y\dot{\alpha}} &= -4B_{12} \\
C_{Y\dot{\beta}} &= -4B_{11} & C_{Y\dot{p}} &= -4B_{13} \\
C_{Y\dot{q}} &= 4C_{12} & C_{Y\dot{r}} &= -4C_{11}
\end{aligned}
\tag{10-42}
$$

$$
\begin{aligned}
C_{Z\dot{V}_0} &= -4\alpha B_{22} - 4\beta B_{12} & C_{Z\dot{\alpha}} &= -4B_{22} \\
C_{Z\dot{\beta}} &= -4B_{12} & C_{Z\dot{p}} &= -4B_{23} \\
C_{Z\dot{q}} &= 4C_{22} & C_{Z\dot{r}} &= -4C_{12}
\end{aligned}
\tag{10-43}
$$

$$
\begin{aligned}
C_{l\dot{V}_0} &= -4\alpha B_{23} - 4\beta B_{13} & C_{l\dot{\alpha}} &= -4B_{23} \\
C_{l\dot{\beta}} &= -4B_{13} & C_{l\dot{p}} &= -4B_{33} \\
C_{l\dot{q}} &= 4C_{23} & C_{l\dot{r}} &= -4C_{13}
\end{aligned}
\tag{10-44}
$$

$$
\begin{aligned}
C_{m\dot{V}_0} &= 4\alpha C_{22} + 4\beta C_{12} & C_{m\dot{\alpha}} &= 4C_{22} \\
C_{m\dot{\beta}} &= 4C_{12} & C_{m\dot{p}} &= 4C_{23} \\
C_{m\dot{q}} &= -4D_{22} & C_{m\dot{r}} &= 4D_{12}
\end{aligned}
\tag{10-45}
$$

$$
\begin{aligned}
C_{n\dot{V}_0} &= -4\alpha C_{12} - 4\beta C_{11} & C_{n\dot{\alpha}} &= -4C_{12} \\
C_{n\dot{\beta}} &= -4C_{11} & C_{n\dot{p}} &= -4C_{13} \\
C_{n\dot{q}} &= 4D_{12} & C_{n\dot{r}} &= -4D_{11}
\end{aligned}
\tag{10-46}
$$

10-6. Stability Derivatives of Slender Flat Triangular Wing

Since triangular wings are of particular importance to missiles, they present an appropriate means of illustrating the power of the method of

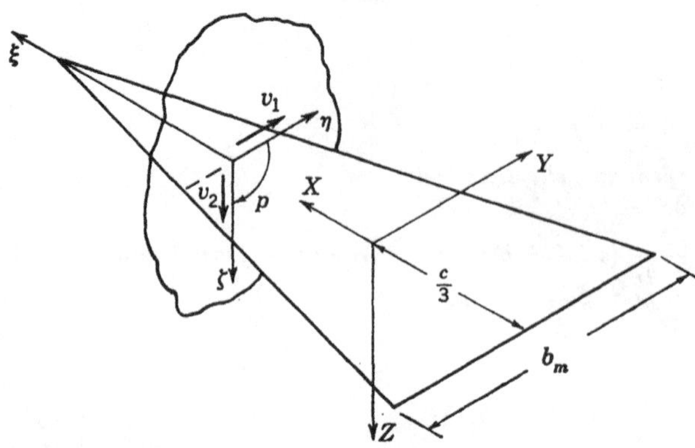

FIG. 10-8. Axes and notation for slender triangular wing.

inertia coefficients for evaluating stability derivatives. In fact, we will now systematically deduce the *velocity stability derivatives* for a slender flat triangular wing, using Eqs. (10-37) to (10-41). The derivatives of the drag force or those relating to axial velocity u are not given.

Consider the slender triangular wing shown in Fig. 10-8. Let the center of moments be at the centroid of wing area, let the wing planform be the reference area, and *let the total wing span b_m be the reference length* λ of the equations of the previous section. The inertia coefficients A_{ij} are evaluated using Eq. (10-22) wherein ϕ_1, ϕ_2, and ϕ_3 are due to unit velocities v_1, v_2, and p, as indicated in Fig. 10-9. Since v_1 of the wing produces no flow, we have $\phi_1 = 0$. It immediately follows that

$$\bar{A}_{11} = \bar{A}_{12} = \bar{A}_{13} = 0 \quad (10\text{-}47)$$

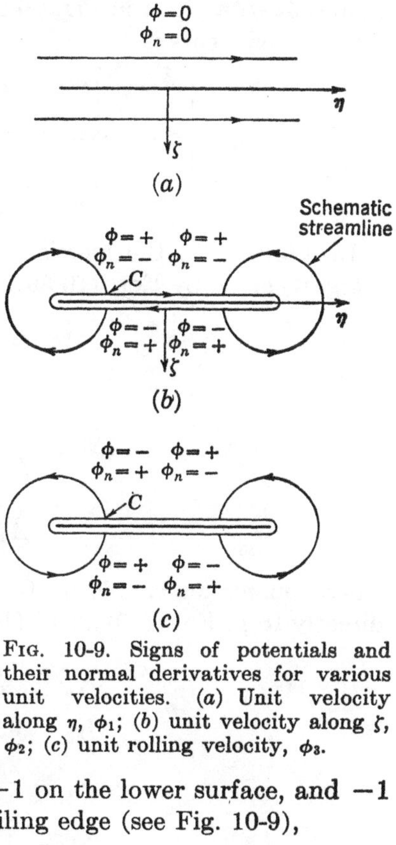

The potential ϕ_2 is that for unit v_2 of the wing or unit angle of attack. It is well known that the potential distribution across the span of a slender triangular wing is elliptical,[7] and that its lift-curve slope is 2, based on the area of a circle of diameter equal to the span b_m. These facts are sufficient to establish that

$$\phi_2 = \pm(s^2 - \eta^2)^{\frac{1}{2}} \quad (10\text{-}48)$$

where the plus sign refers to the upper surface, and the negative sign to the lower surface. Also $\partial\phi_2/\partial n = +1$ on the lower surface, and -1 on the upper surface. Thus, at the trailing edge (see Fig. 10-9),

FIG. 10-9. Signs of potentials and their normal derivatives for various unit velocities. (a) Unit velocity along η, ϕ_1; (b) unit velocity along ζ, ϕ_2; (c) unit rolling velocity, ϕ_3.

$$\bar{A}_{22} = -\frac{1}{S_R}\oint_C \phi_2\frac{\partial\phi_2}{\partial n}\,ds = -\frac{4}{S_R}\int_0^{s_m}[-(s_m^2 - \eta^2)^{\frac{1}{2}}]\,d\eta$$

$$= \frac{\pi b_m^2}{4S_R} \quad (10\text{-}49)$$

The coefficient $\bar{A}_{33}$ is determined with the help of the potential for unit p taken from Lamb:[8]

$$\phi_3 = \pm\tfrac{1}{2}\eta(s^2 - \eta^2)^{\frac{1}{2}} \quad (10\text{-}50)$$

The signs are chosen in accordance with Fig. 10-9. Thus, at the trailing edge

$$\bar{A}_{33} = -\frac{1}{\lambda^2 S_R} \oint \phi_3 \frac{\partial \phi_3}{\partial n}\, ds$$

$$= -\frac{4}{\lambda^2 S_R} \int_0^{s_m} [-\tfrac{1}{2} p \eta (s^2 - \eta^2)^{1/2}] p \eta\, d\eta$$

$$= \frac{\pi b_m^4}{128 S_R \lambda^2} \tag{10-51}$$

The only independent inertia coefficient left, $\bar{A}_{23}$, is zero, as the signs of ϕ_2 and $\partial \phi_3/\partial n$ given in Fig. 10-9 readily show. The complete matrix of inertia coefficients is

$$\begin{vmatrix} \bar{A}_{11} & \bar{A}_{12} & \bar{A}_{13} \\ \bar{A}_{21} & \bar{A}_{22} & \bar{A}_{23} \\ \bar{A}_{31} & \bar{A}_{32} & \bar{A}_{33} \end{vmatrix} = \begin{vmatrix} 0 & 0 & 0 \\ 0 & \dfrac{\pi b_m^2}{4 S_R} & 0 \\ 0 & 0 & \dfrac{\pi b_m^4}{128 S_R \lambda^2} \end{vmatrix} \tag{10-52}$$

In addition to the inertial coefficients A_{ij} we need certain of their integrals given by Eqs. (10-36). These quantities are easily found to be

$$B_{22} = \frac{\pi}{6}$$

$$C_{22} = -\frac{\pi}{36A}$$

$$D_{22} = \frac{4\pi}{135A^2} \tag{10-53}$$

$$\bar{A}_{22} = \frac{\pi A}{4}$$

These quantities enable us to write down the 25 velocity derivatives directly from Eqs. (10-37) to (10-41).

$$C_{Y_\alpha} = \tfrac{2}{3}\pi \frac{p b_m}{2 V_0}$$

$$C_{Y_\beta} = 0$$

$$C_{Y_p} = \frac{2\pi \alpha}{3} + \frac{2\pi}{9A} \frac{q b_m}{2 V_0} \tag{10-54}$$

$$C_{Y_q} = \frac{2\pi}{9A} \frac{p b_m}{2 V_0}$$

$$C_{Y_r} = 0$$

$$C_{Z_\alpha} = -\frac{\pi A}{2} - \frac{2\pi}{3} \frac{b_m \dot{V}_0}{2 V_0^2}$$

$$C_{Z_\beta} = 0$$

$$C_{Z_p} = 0 \tag{10-55}$$

$$C_{Z_q} = \frac{-2\pi}{3}$$

$$C_{Z_r} = 0$$

$$C_{l_\alpha} = -\frac{\pi\beta}{3} + \frac{\pi}{9A}\frac{rb_m}{2V_0}$$

$$C_{l_\beta} = -\frac{\pi\alpha}{3} - \frac{\pi}{9A}\frac{qb_m}{2V_0}$$

$$C_{l_p} = -\frac{\pi A}{32}$$

$$C_{l_q} = -\frac{\pi\beta}{9A} + \frac{32\pi}{135A^2}\frac{rb_m}{2V_0}$$

$$C_{l_r} = \frac{\pi\alpha}{9A} + \frac{32\pi}{135A^2}\frac{qb_m}{2V_0}$$

(10-56)

$$C_{m_\alpha} = -\frac{\pi}{9A}\frac{b_m\dot{V}_0}{2V_0^2}$$

$$C_{m_\beta} = 0$$

$$C_{m_p} = 0$$

$$C_{m_q} = -\frac{\pi}{3A}$$

$$C_{m_r} = 0$$

(10-57)

$$C_{n_\alpha} = -\frac{\pi}{9A}\frac{pb_m}{2V_0}$$

$$C_{n_\beta} = 0$$

$$C_{n_p} = -\frac{\pi\alpha}{9A} - \frac{32\pi}{135A^2}\frac{qb_m}{2V_0}$$

$$C_{n_q} = -\frac{32\pi}{135A^2}\frac{pb_m}{2V_0}$$

$$C_{n_r} = 0$$

(10-58)

The foregoing derivatives include a number of kinds of forces and moments: static, damping, Magnus, etc. Some discussion of these types of forces and moments will be given in Sec. 10-10 when we examine the effect of aspect ratio on the foregoing results. The noteworthy feature of the foregoing analysis is the powerful manner in which it yields results. It is known that dihedral introduced geometrically into the wing can have an important influence on certain of the foregoing derivatives. Ribner and Malvestuto[15] have included the effects of geometric dihedral in their study of the stability derivatives of slender triangular wings. The appearance of the aspect ratio in the denominator of certain of the stability derivatives is due to the particular choice of reference area and length in this case, and does not indicate that the derivatives are particularly important for low aspect ratios.

The *acceleration derivatives* can be easily written from Eqs. (10-42) to (10-46) inclusive. The only new coefficient appearing is B_{33}, which is found to be

$$B_{33} = \frac{\pi}{320} \qquad (10\text{-}59)$$

Of the 30 acceleration derivatives given by the equations, only the following seven are not zero.

$$C_{z\dot{V}_0} = -\frac{2\pi\alpha}{3} \qquad C_{z\dot{\alpha}} = -\frac{2\pi}{3} \qquad C_{z\dot{q}} = -\frac{\pi}{9A}$$

$$C_{l\dot{p}} = -\frac{\pi}{80} \qquad\qquad (10\text{-}60)$$

$$C_{m\dot{V}_0} = -\frac{\pi\alpha}{9A} \qquad C_{m\dot{\alpha}} = -\frac{\pi}{9A} \qquad C_{m\dot{q}} = -\frac{16\pi}{135A^2}$$

It is interesting to interpret the results for the derivatives of C_m and C_z with respect to α, q, $\dot{\alpha}$, $\dot{q}$, and $\dot{V}_0$ in terms of the center of pressure of the forces involved.   Dividing C_m by C_z yields the center-of-pressure position in fractions of the reference length from the wing centroid (two-thirds root chord position).   Converting these results to fractions of the root chord c, we obtain

$$\left(\frac{\bar{X}}{c}\right)_\alpha = 0 \qquad \left(\frac{\bar{X}}{c}\right)_q = -\frac{1}{4} \qquad \left(\frac{\bar{X}}{c}\right)_{\dot{\alpha}} = -\frac{1}{12}$$

$$\left(\frac{\bar{X}}{c}\right)_{\dot{q}} = -\frac{8}{15} \qquad \left(\frac{\bar{X}}{c}\right)_{\dot{V}_0} = -\frac{1}{12} \qquad\qquad (10\text{-}61)$$

The minus signs indicate that the centers of pressure are behind the centroids in each case, except the center of pressure associated with α, which is at the centroid.   Increasing aspect ratio to the point where the triangular wing is no longer slender will cause certain of the centers of pressure to move, as discussed in Sec. 10-10.

It is also of interest to compare the zero and nonzero terms as determined in this section with the zero and nonzero derivatives given in Fig. 10-5, which applies to a triangular wing.   It will be seen that all the zero terms deduced on the basis of the Maple-Synge analysis do in fact turn out to be zero.   However, a large number of additional terms are also zero, by virtue of the particular aerodynamic properties of a slender triangular wing.

10-7. General Method of Evaluating Inertia Coefficients and Apparent Masses

Several methods are available for evaluating the inertia coefficients. There is, first, the method of evaluating directly the integrals given by Eq. (10-22), which was utilized in the preceding section in determining the apparent-mass coefficients for a triangular wing.   However, a more powerful method exists based on the theory of residues.   This requires only a knowledge of the transformation that maps the missile cross section conformally onto a circle of radius c, with no distortion at infinity. From this method we can find the inertia coefficients without any difficulty except when they may require summing an intractable infinite series. It has been used by a number of authors, including Ward,[9] Bryson,[10] Summers,[11] and Sacks.[2] The treatment of Bryson is the basis of this section. It is our primary purpose here to derive simple formulas for the apparent-mass coefficients in terms of the transformation which turns the missile cross section into the circle of radius c. The reader who is content with the apparent-mass results of Table 10-3 may proceed to those results directly. In Eq. (10-22) we have already defined the *inertia coefficients* in terms of the potentials ϕ_1, ϕ_2, and ϕ_3 for two translations and one rotation of a given missile cross section. We now define the *apparent-mass coefficients* as

$$m_{ij} = m_{ji} = -\rho \oint_C \phi_i \frac{\partial \phi_j}{\partial n}\, ds \qquad i, j = 1, 2, 3 \tag{10-62}$$

The apparent-mass coefficients so defined are usually called "additional" apparent-mass coefficients since they induce on a body in a fluid dynamical effects additional to those due to the mass of the body itself. Because such a distinction is unnecessary for our purposes, we shall dispense with the adjective "additional." The apparent-mass coefficients do not actually have the dimensions of mass, but have dimensions that are readily apparent from their relationships to the truly nondimensional inertia coefficients.

$$A_{11} = \frac{m_{11}}{\rho S_R} \qquad A_{12} = A_{21} = \frac{m_{12}}{\rho S_R} \qquad A_{22} = \frac{m_{22}}{\rho S_R}$$

$$A_{13} = A_{31} = \frac{m_{13}}{\rho \lambda S_R} \qquad A_{23} = A_{32} = \frac{m_{23}}{\rho \lambda S_R} \tag{10-63}$$

$$A_{33} = \frac{m_{33}}{\rho \lambda^2 S_R}$$

The quantities λ and S_R are the reference length and area, respectively. Although the quantities m_{ij} do not have the dimensions of mass, we will call them apparent masses for short. It might be asked why a table of apparent masses rather than dimensionless inertia coefficients is being presented. The reason is that the inertia coefficients depend on reference quantities λ and S_R which are not usually properties of the cross section whereas the apparent-mass coefficients do not depend on λ and S_R.

It is well now to consider the crossflow plane $\mathfrak{z}$ of a given missile cross section, as shown in Fig. 10-10, together with the transformed plane ζ in which the missile becomes a circle of radius c. Because we require the flow fields at infinity in the physical plane to be undistorted in the trans-

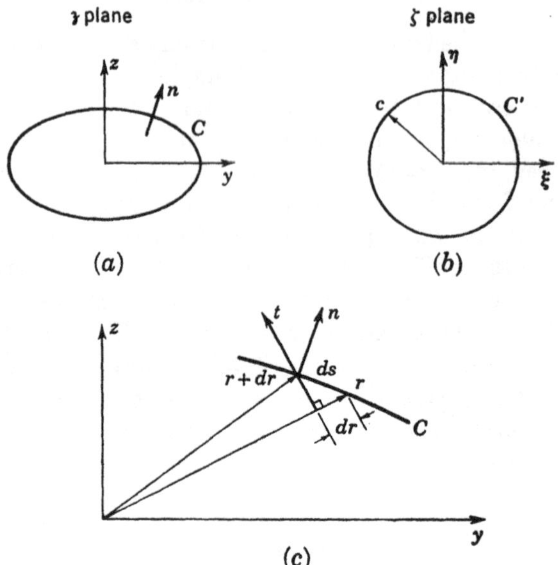

Fig. 10-10. Notation in determination of apparent-mass coefficients. (a) Physical plane; (b) transformed plane; (c) boundary conditions for ϕ_3.

formed plane, the general mapping of the $\jmath$ plane into the ζ plane is given by

$$\jmath = \zeta + \sum_{n=0}^{\infty} \frac{a_n}{\zeta^n} \tag{10-64}$$

Let us now examine the normal derivatives of ϕ_1, ϕ_2, and ϕ_3 on the boundary C in the physical plane, as shown in Fig. 10-10. Since ϕ_1 corresponds to unit velocity along y, we have

$$\frac{\partial \phi_1}{\partial n} = \cos(n,y) = \frac{dz}{ds} \tag{10-65}$$

where $\cos(n,y)$ is the cosine of the angle between n and the y axis. Similarly, for ϕ_2 we have for unit vertical velocity

$$\frac{\partial \phi_2}{\partial n} = \cos(n,z) = -\frac{dy}{ds} \tag{10-66}$$

Also, for unit angular velocity we have

$$\frac{\partial \phi_3}{\partial n} = r\cos(n,t) = -r\frac{dr}{ds} = -\frac{1}{2}\frac{dr^2}{ds} = -\frac{1}{2}\frac{d(\jmath\bar{\jmath})}{ds} \tag{10-67}$$

(Note that p is now taken positive when y rotates to z since we are using the axes x, y, z rather than X, Y, and Z in this derivation.) The stratagem now brought into play to allow the use of residue theory is to form complex combinations of the apparent masses as follows:

$$m_{11} + im_{21} = -\rho \oint_C \phi_1\left(\frac{\partial \phi_1}{\partial n} + i\frac{\partial \phi_2}{\partial n}\right) ds$$

$$= i\rho \oint_C \phi_1\, d\mathfrak{z} = i\rho \oint_C (W_1 - i\psi_1)\, d\mathfrak{z}$$

$$m_{12} + im_{22} = -\rho \oint_C \phi_2\left(\frac{\partial \phi_1}{\partial n} + i\frac{\partial \phi_2}{\partial n}\right) ds$$

$$= i\rho \oint_C \phi_2\, d\mathfrak{z} = i\rho \oint_C (W_2 - i\psi_2)\, d\mathfrak{z} \qquad (10\text{-}68)$$

$$m_{13} + im_{23} = -\rho \oint_C \phi_3\left(\frac{\partial \phi_1}{\partial n} + i\frac{\partial \phi_2}{\partial n}\right) ds$$

$$= i\rho \oint_C \phi_3\, d\mathfrak{z} = i\rho \oint_C (W_3 - i\psi_3)\, d\mathfrak{z}$$

The apparent mass m_{33} has its own special formula with the help of Eq. (10-67):

$$m_{33} = -\rho \oint_C \phi_3\left[-\frac{d(\mathfrak{z}\bar{\mathfrak{z}})}{2}\right] = \frac{\rho}{2} \oint_C (W_3 - i\psi_3)\, d(\mathfrak{z}\bar{\mathfrak{z}}) \qquad (10\text{-}69)$$

It is clear that the integrals with the exception of that for m_{33} have analytic integrands to which the theory of residues is applicable. The parts of the integrals involving the stream function can be expressed in terms of the geometric properties of the missile cross section. Integrating by parts

$$\oint_C \psi_i\, d\mathfrak{z} = \oint_C d(\mathfrak{z}\psi_i) - \oint_C \mathfrak{z}\, d\psi_i \qquad (10\text{-}70)$$

and using the Cauchy-Riemann equation

$$\frac{\partial \psi}{\partial s} = \frac{\partial \phi}{\partial n} \qquad (10\text{-}71)$$

we obtain

$$\oint_C \psi_i\, d\mathfrak{z} = \oint_C d(\mathfrak{z}\psi_i) - \oint_C \mathfrak{z}\frac{\partial \phi_i}{\partial n}\, ds \qquad (10\text{-}72)$$

For the motions involved here ψ is a single-valued continuous function on the boundary so that the perfect differential $d(\mathfrak{z}\psi_i)$ is zero taken around the boundary. We thus have with the help of Eqs. (10-65), (10-66), and (10-67)

$$\oint_C \psi_1\, d\mathfrak{z} = -\oint_C \mathfrak{z}\, dz = -S_C$$

$$\oint_C \psi_2\, d\mathfrak{z} = \oint_C \mathfrak{z}\, dy = -iS_C \qquad (10\text{-}73)$$

$$\oint_C \psi_3\, d\mathfrak{z} = \frac{1}{2}\oint_C \mathfrak{z}\, d(\mathfrak{z}\bar{\mathfrak{z}}) = -i\mathfrak{z}_c S_C$$

where S_C is the cross-sectional area, and $\mathfrak{z}_c$ is the complex coordinate of the centroid of the missile cross section. The part of the integrals of

Eqs. (10-68) and (10-69) involving the complex potentials $W_i(\mathfrak{z})$ will be evaluated in the ζ plane by the use of residue theory. To do this we must determine the expansions for W_1, W_2, W_3 valid in the region exterior to the circle in the ζ plane and isolate the coefficient of the ζ^{-1} term.

First we will derive the expansions for W_1 and W_2, which are similar, and then the expansion for W_3. If $W_1(\mathfrak{z})$ is the complex potential for the flow in the $\mathfrak{z}$ plane for translation of the body with unit velocity along the positive y axis with the fluid stationary at infinity, then $W_1(\mathfrak{z}) - \mathfrak{z}$ describes the flow for the body stationary with the flow velocity at

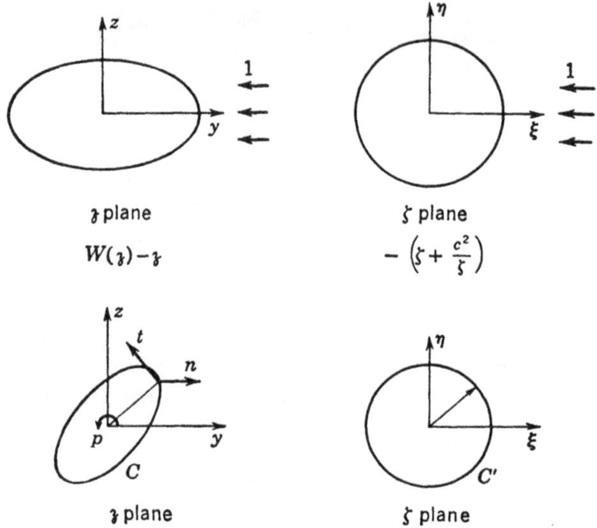

Fig. 10-11. Further notation for use in determination of apparent-mass coefficients.

infinity in the negative y direction with unit speed, as shown in Fig. 10-11. The complex potential for the flow in the ζ plane is formed by making the substitution $\mathfrak{z} = \mathfrak{z}(\zeta)$ into $W_1(\mathfrak{z}) - \mathfrak{z}$ so that

$$W_1(\mathfrak{z}(\zeta)) - \mathfrak{z}(\zeta) = -\left(\zeta + \frac{c^2}{\zeta}\right) \tag{10-74}$$

where we have equated the transformed complex potential to the known complex potential for flow past a circular cylinder. With the help of Eq. (10-64) we have the final result for $W_1(\mathfrak{z}(\zeta))$

$$W_1(\mathfrak{z}(\zeta)) = -\frac{c^2}{\zeta} + \sum_{n=0}^{\infty} \frac{a_n}{\zeta^n} \tag{10-75}$$

The same technique serves to determine the complex potential for unit velocity in the upward direction

$$W_2(\mathfrak{z}(\zeta)) = i(\zeta - \mathfrak{z}) - i\frac{c^2}{\zeta} = -i\frac{c^2}{\zeta} - i\sum_{r=0}^{\infty} \frac{a_n}{\zeta^n} \tag{10-76}$$

To obtain $W_3(\mathfrak{z}(\zeta))$ we make use of a clever result given by Milne-Thompson[26] for two-dimensional motion of an arbitrary body translating and rotating in an infinite fluid with no circulation. The function $2i\psi$ on the circular cylinder into which the body is transformed by Eq. (10-64) is called the *boundary function*. The function $2i\psi_3$ has the value obtained from Eqs. (10-67) and (10-71) on the boundary of the circle

$$2i\psi_3 = -i(\mathfrak{z}\bar{\mathfrak{z}}) \tag{10-77}$$

According to Milne-Thompson if this boundary function can be expanded into a series of positive powers of ζ and a series of negative powers, the complex potential is equal to the series of negative powers of ζ.

$$\mathfrak{z} = \zeta + \sum_{n=0}^{\infty} \frac{a_n}{\zeta^n} = f(\zeta)$$

$$\bar{\mathfrak{z}} = \bar{\zeta} + \sum_{n=0}^{\infty} \frac{\bar{a}_n}{\bar{\zeta}^n} = \bar{f}(\bar{\zeta}) = \bar{f}\left(\frac{c^2}{\bar{\zeta}}\right) \tag{10-78}$$

so that

$$2i\psi_3 = -if(\zeta)\bar{f}\left(\frac{c^2}{\bar{\zeta}}\right) \tag{10-79}$$

on the circle. In accordance with the result of Milne-Thompson, $W_3(\mathfrak{z}(\zeta))$ is the series of negative powers extracted from Eq. (10-79), which series we will denote as **PP**, the principal part. Thus

$$W_3(\mathfrak{z}(\zeta)) = -i\,\mathbf{PP}\left[\bar{f}(\zeta)f\left(\frac{c^2}{\bar{\zeta}}\right)\right]$$

$$= -i\sum_{n=1}^{\infty} \frac{b_n}{\zeta^n} \tag{10-80}$$

where

$$b_n = \sum_{m=-1}^{\infty} \frac{a_{m+n}\bar{a}_n}{c^{2m}} \qquad \begin{array}{ll} a_n = 1 & n = -1 \\ a_{-n} = 0 & n > 1 \end{array} \tag{10-81}$$

Having now determined the series expansions for W_1, W_2, and W_3, we can now return to the evaluation of Eq. (10-68). Since the term of degree ζ^{-1} is the only one contributing to the integrals of W_i we have by Cauchy's theorem of residues

$$\oint W_1\,d\mathfrak{z} = 2\pi i(a_1 - c^2)$$

$$\oint W_2\,d\mathfrak{z} = 2\pi i(-ia_1 - ic^2) \tag{10-82}$$

$$\oint W_3\,d\mathfrak{z} = 2\pi i(-ib_1)$$

The results for the apparent masses from Eq. (10-68) are now, with the help of Eqs. (10-73) and (10-82),

$$\begin{aligned}
m_{11} + im_{12} &= i\rho[2\pi i(a_1 - c^2) + iS_C] \\
m_{12} + im_{22} &= i\rho[2\pi(a_1 + c^2) - S_C] \\
m_{13} + im_{23} &= i\rho[2\pi b_1 - \bar{z}_c S_C]
\end{aligned} \qquad (10\text{-}83)$$

These results give a simple means of evaluating all the apparent masses of the missile cross section except m_{33}, if the transformation of the missile cross section into the circle of radius c is known. It is to be noted that all the quantities in Eq. (10-83) are then known. However, b_1 is an infinite series given by Eq. (10-81), which may or may not be readily summed.

The immediately preceding equation gives general formulas for all apparent masses except m_{33}. This apparent mass requires the following special treatment for its evaluation because it is represented by a non-analytic integral.

$$m_{33} = \frac{\rho}{2} \oint_C W_3 \, d(z\bar{z}) - i\frac{\rho}{2} \oint_C \psi_3 \, d(z\bar{z}) \qquad (10\text{-}84)$$

Now, integrating by parts, we have

$$\oint_C \psi_3 \, d(z\bar{z}) = \oint_C d(\psi_3 z\bar{z}) - \oint_C z\bar{z} \frac{\partial \phi_3}{\partial n} \, ds \qquad (10\text{-}85)$$

and from Eq. (10-67)

$$\oint_C \psi_3 \, d(z\bar{z}) = \oint_C d(\psi_3 z\bar{z}) + \frac{1}{2} \oint_C (z\bar{z}) \, d(z\bar{z}) = 0 \qquad (10\text{-}86)$$

so that we are left with

$$m_{33} = \frac{\rho}{2} \oint_C W_3 \, d(z\bar{z}) \qquad (10\text{-}87)$$

The stratagem for evaluating this nonanalytic integral is to find some function analytic outside the circle, which is numerically equal to $z\bar{z}$ on the circle. By substituting this analytic expression for $z\bar{z}$ into the integrand, we do not change the numerical value of the integral, but we do make it analytic so that it becomes amenable to treatment by the calculus of residues. The key then is the analytic expression equal to $z\bar{z}$ on the circle C' (Fig. 10-11). On C' we have

$$z\bar{z} = f(\zeta)\bar{f}(\bar{\zeta}) = f(\zeta)\bar{f}\left(\frac{c^2}{\zeta}\right)$$

$$= \left(\sum_{n=-1}^{\infty} \frac{a_n}{\zeta^n} \right)\left(\sum_{m=-1}^{\infty} \frac{\bar{a}_m \zeta^m}{c^{2m}} \right) \qquad (10\text{-}88)$$

with
$$a_{-1} = \bar{a}_{-1} = 1$$

or
$$\bar{\bar{\bar{\jmath}}} = \sum_{n=-\infty}^{\infty} \frac{b_n}{\zeta^n} \qquad \text{on } C' \tag{10-89}$$

where b_n is given by Eq. (10-81). We then have found the desired analytic function. Making use of Eqs. (10-80) and (10-89), we find that Eq. (10-87) becomes

$$m_{33} = -\frac{i\rho}{2} \oint_{C'} \left(\sum_{n=1}^{\infty} \frac{b_n}{\zeta^n} \right) \left(\sum_{m=-\infty}^{\infty} \frac{m b_m}{\zeta^{m+1}} \right) d\zeta \tag{10-90}$$

Only for those terms with $m = -n$ do we get a contribution so that

$$m_{33} = -\frac{i\rho}{2} (2\pi i) \sum_{n=1}^{\infty} n b_n b_{-n} \tag{10-91}$$

However, it can be seen from Eq. (10-81) that

$$b_{-n} = \frac{\bar{b}_n}{c^{2n}} \tag{10-92}$$

The final result for m_{33} is

$$m_{33} = \pi\rho \sum_{n=1}^{\infty} \frac{n b_n \bar{b}_n}{c^{2n}} \tag{10-93}$$

The results for the apparent masses are now collected.

$$m_{11} = 2\pi\rho \left[c^2 - \frac{S_C}{2\pi} - \mathbf{R}(a_1) \right]$$
$$m_{12} = m_{21} = -2\pi\rho\mathbf{I}(a_1)$$
$$m_{22} = 2\pi\rho \left[c^2 - \frac{S_C}{2\pi} + \mathbf{R}(a_1) \right]$$
$$m_{13} = m_{31} = -2\pi\rho\mathbf{I}\left(b_1 - \tfrac{1}{3}c\,\frac{S_C}{2\pi} \right) \tag{10-94}$$
$$m_{23} = 2\pi\rho\mathbf{R}\left(b_1 - \tfrac{1}{3}c\,\frac{S_C}{2\pi} \right)$$
$$m_{33} = \pi\rho \sum_{n=1}^{\infty} \frac{n b_n \bar{b}_n}{c^{2n}}$$

Illustrative Example

Calculate the apparent masses and inertia coefficients for a slender triangular wing.

The transformation that takes the line of span b_m in the $\mathfrak{z}$ plane into the circle of radius c in the ζ plane is

$$\mathfrak{z} = \zeta + \frac{c^2}{\zeta} = \zeta + \frac{b_m{}^2}{16\zeta} = \zeta + \sum_{n=1}^{\infty} \frac{a_n}{\zeta^n}$$

where we have identified the transformation with Eq. (10-64). The values of the coefficients a_n determine all the apparent masses in accordance with Eq. (10-94). These coefficients are

$$a_1 = \frac{b_m{}^2}{16}$$
$$a_{-1} = 1 \qquad \text{Eq. (10-81)}$$
$$a_0 = a_2 = a_3 = a_4 = \cdots = 0$$

The coefficients b_n from Eq. (10-81) are

$$b_0 = b_1 = 0$$
$$b_2 = a_1 \bar{a}_1 c^2 = \frac{b_m{}^4}{256}$$

Forming the apparent masses from Eq. (10-94), we get

$$m_{11} = 2\pi\rho \left(\frac{b_m{}^2}{16} - 0 - \frac{b_m{}^2}{16} \right) = 0$$
$$m_{12} = 0$$
$$m_{22} = 2\pi\rho \left(\frac{b_m{}^2}{16} + 0 + \frac{b_m{}^2}{16} \right) = \pi\rho \frac{b_m{}^2}{4}$$
$$m_{13} = 0$$
$$m_{23} = 0$$
$$m_{33} = 2\pi\rho \left(\frac{b_m{}^4}{256} \right)^2 \left(\frac{16}{b_m{}^2} \right)^2 = \frac{\pi\rho b_m{}^4}{128}$$

The nonzero coefficients from Eq. (10-63) are

$$A_{22} = \frac{m_{22}}{\rho S_R} = \frac{b_m{}^2}{4 S_R}$$
$$A_{33} = \frac{m_{33}}{\rho \lambda^2 S_R} = \frac{\pi b_m{}^4}{128 \lambda^2 S_R}$$

These results for A_{22} and A_{33} are in accordance with the values given in Eqs. (10-49) and (10-51) and obtained by different means.

10-8. Table of Apparent Masses with Application to the Stability Derivatives of Cruciform Triangular Wings

The apparent-mass coefficients are known for a large number of typical missile cross sections in whole or in part. The apparent masses for a number of such cross sections have been collected and are presented in

Table 10-3. A number of interesting and useful points arise in connection with the table. Perhaps the first point of interest is that the coefficients m_{11} and m_{22} for an ellipse, be it the special case of a circle or a straight line, depend only on the span normal to the direction of motion. For a multifinned body of three or more equally spaced equal-span fins, m_{11} and m_{22} are equal. If for any cross section the coefficients m_{11} and m_{22} are equal, then the coefficients for translation in all directions in the plane are equal. A circle has a greater apparent mass than any regular inscribed polygon. For a cross section with a vertical plane of symmetry the coefficients m_{12} and m_{23} are zero. For a horizontal plane of mirror symmetry, m_{12} and m_{13} are zero.

To illustrate the use of the table, let us apply it to the calculation of the stability derivatives of the slender cruciform missile shown in Fig. 10-12. In the next section we will consider a number of other examples. Let the reference area S_R be the wing planform area $\frac{1}{2}bc$, and let the reference length λ be the maximum span b. The origin of the system of axes is taken a distance X_n behind the wing apex. The stability derivatives are given in terms of the inertia coefficients A_{ij} in Eqs. (10-37) to (10-41), inclusive. The inertia coefficients are obtained from the following apparent-mass coefficients from Table 10-3:

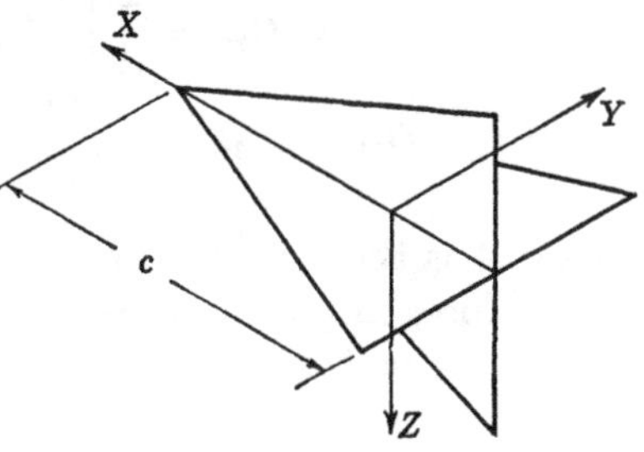

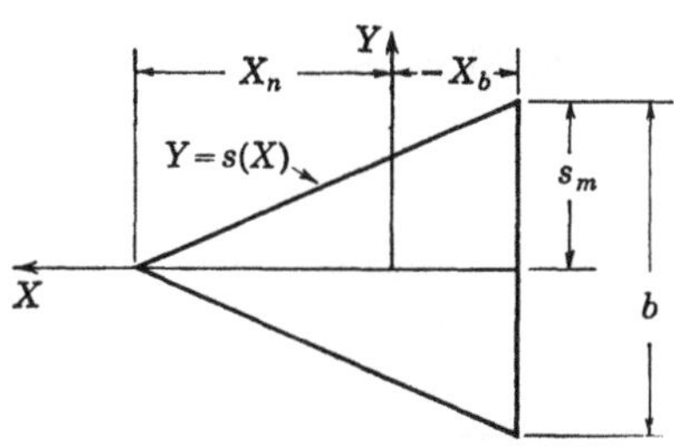

FIG. 10-12. Slender cruciform wing.

$$m_{11} = m_{22} = \pi \rho s^2$$
$$m_{33} = \frac{2\rho s^4}{\pi}$$
$$m_{12} = m_{13} = m_{23} = 0$$

(10-95)

The nonzero inertia coefficients are thus

$$A_{11} = A_{22} = \frac{m_{11}}{\rho S_R} = \frac{\pi s^2}{S_R}$$
$$A_{33} = \frac{m_{33}}{\rho \lambda^2 S_R} = \frac{2s^4}{\pi b^2 S_R}$$

(10-96)

The integrals of the inertia coefficients B_{11}, C_{11}, and D_{11} given by Eq. (10-36) are required:

$$B_{11} = \int_{(X/b)_b}^{(X/b)_n} A_{11}\, d\left(\frac{X}{b}\right)$$

(10-97)

With reference to the notation of Fig. 10-12,

$$A_{11} = \frac{\pi s_m{}^2}{S_R}\left(\frac{X_n - X}{c}\right)^2$$

with the result that

$$B_{11} = B_{22} = \frac{\pi}{6}$$

Similarly,

$$C_{11} = C_{22} = \frac{\pi}{A}\left[\frac{1}{3}\left(1 + \frac{X_b}{c}\right) - \frac{1}{4}\right]$$
$$D_{11} = D_{22} = \frac{\pi}{A^2}\left[\frac{2}{5} - \left(1 + \frac{X_b}{c}\right) + \frac{2}{3}\left(1 + \frac{X_b}{c}\right)^2\right]$$

where A is the aspect ratio.

The foregoing results apply to any position of the origin of the body axes.   Let us take the origin at the wing centroid of area so that

$$\frac{X_b}{c} = -\frac{1}{3}$$

The coefficients then become

$$\bar{A}_{11} = \bar{A}_{22} = \frac{\pi A}{4}$$

$$\bar{A}_{33} = \frac{A}{8\pi}$$

$$B_{11} = B_{22} = \frac{\pi}{6} \tag{10-98}$$

$$C_{11} = C_{22} = -\frac{\pi}{36A}$$

$$D_{11} = D_{22} = \frac{4\pi}{135A^2}$$

It will be noted that these coefficients are simply related to those for a triangular wing with the exception of $\bar{A}_{33}$.   The results for the stability derivatives follow immediately.

$$C_{Y_\alpha} = \frac{2\pi}{3}\frac{pb}{2V_0}$$

$$C_{Y_\beta} = -\frac{\pi}{2}A - \frac{2\pi}{3}\frac{b\dot{V}_0}{2V_0{}^2}$$

$$C_{Y_p} = \frac{2\pi}{3}\alpha + \frac{2\pi}{9A}\frac{qb}{2V_0} \tag{10-99}$$

$$C_{Y_q} = \frac{2\pi}{9A}\frac{pb}{2V_0}$$

$$C_{Y_r} = \frac{2\pi}{3}$$

$$C_{z_\alpha} = -\frac{\pi}{2} A - \frac{2\pi}{3} \frac{b\dot{V}_0}{2V_0^2}$$

$$C_{z_\beta} = -\frac{2\pi}{3} \frac{pb}{2V_0}$$

$$C_{z_p} = -\frac{2\pi}{3} \beta + \frac{2\pi}{9A} \frac{rb}{2V_0} \qquad (10\text{-}100)$$

$$C_{z_q} = -\frac{2\pi}{3}$$

$$C_{z_r} = \frac{2\pi}{9A} \frac{pb}{2V_0}$$

$$C_{l_\alpha} = C_{l_\beta} = C_{l_q} = C_{l_r} = 0$$

$$C_{l_p} = -\frac{A}{2\pi} \qquad (10\text{-}101)$$

$$C_{m_\alpha} = -\frac{\pi}{9A} \frac{b\dot{V}_0}{2V_0^2}$$

$$C_{m_\beta} = -\frac{\pi}{9A} \frac{pb}{2V_0}$$

$$C_{m_p} = -\frac{\pi}{9A} \beta + \frac{32\pi}{135A^2} \frac{rb}{2V_0} \qquad (10\text{-}102)$$

$$C_{m_q} = -\frac{\pi}{3A}$$

$$C_{m_r} = \frac{32\pi}{135A^2} \frac{pb}{2V_0}$$

$$C_{n_\alpha} = -\frac{\pi}{9A} \frac{pb}{2V_0}$$

$$C_{n_\beta} = \frac{\pi}{9A} \frac{b\dot{V}_0}{2V_0^2}$$

$$C_{n_p} = -\frac{\pi}{9A} \alpha - \frac{32\pi}{135A^2} \frac{qb}{2V_0} \qquad (10\text{-}103)$$

$$C_{n_q} = -\frac{32\pi}{135A^2} \frac{pb}{2V_0}$$

$$C_{n_r} = -\frac{\pi}{3A}$$

This example illustrates the utility of Table 10-3 for evaluating stability derivatives. The effect of adding a round or elliptical body to a cruciform wing can be readily determined.

Let us see how the results obtained above correspond to the forces and moments arising out of the Maple-Synge analysis. With references to Fig. 10-4 it can be seen that the following derivatives are zero besides those predicted to be zero on the basis of the Maple-Synge analysis:

Case 1: $M_\alpha, N_\beta, L_r, L_q, L_\alpha, Z_p, M_p$
Case 2: M_α, N_β, L_r
Case 3: M_α, N_β
Case 4: M_α, N_β

The derivatives M_α and N_β are zero because of the conical flow field associated with a cruciform wing and the particular choice of moment center (at the wing centroid).

	Y	Z	L	M	N
α		$-\frac{\pi}{2}A$		0	
β	$-\frac{\pi}{2}A$				0
p			$-\frac{A}{2\pi}$		
q		$-\frac{2\pi}{3}$		$-\frac{\pi}{3A}$	
r	$\frac{2\pi}{3}$				$-\frac{\pi}{3A}$

(a)

	Y	Z	L	M	N
α	$\frac{2\pi}{3}\left(\frac{pb}{2V_0}\right)$				$-\frac{\pi}{9A}\left(\frac{pb}{2V_0}\right)$
β		$-\frac{2\pi}{3}\left(\frac{pb}{2V_0}\right)$		$-\frac{\pi}{9A}\left(\frac{pb}{2V_0}\right)$	
p	$\frac{2\pi}{3}\alpha$	$-\frac{2\pi}{3}\beta$		$-\frac{\pi}{9A}\beta$	$-\frac{\pi}{9A}\alpha$

(b)

	Y	Z	L	M	N
p	$\frac{2\pi}{9A}\left(\frac{qb}{2V_0}\right)$	$\frac{2\pi}{9A}\left(\frac{rb}{2V_0}\right)$		$\frac{32\pi}{135A^2}\left(\frac{rb}{2V_0}\right)$	$\frac{-32\pi}{135A^2}\left(\frac{qb}{2V_0}\right)$
q	$\frac{2\pi}{9A}\left(\frac{pb}{2V_0}\right)$				$\frac{-32\pi}{135A^2}\left(\frac{pb}{2V_0}\right)$
r		$\frac{2\pi}{9A}\left(\frac{pb}{2V_0}\right)$		$\frac{32\pi}{135A^2}\left(\frac{pb}{2V_0}\right)$	

(c)

Fig. 10-13. Classes of derivatives for slender cruciform wing. (a) "Ordinary"; (b) Magnus; and (c) gyroscopic derivatives.

It is informative to try to classify the various types of forces and moments arising for the cruciform wing. The classification is divided into *Magnus terms, gyroscopic terms,* and *other terms.* By Magnus forces and moments we mean those forces and moments developing as a result of roll at angle of attack or at angle of sideslip. Such terms here are proportional to βp or αp and are listed in Fig. 10-13. It is easy to show that the Magnus forces in this case act a distance $c/12$ *behind* the wing centroid.

The second class of forces and moments are gyroscopic. A gyroscopic force or moment is taken to be one proportional to the product of two of the three angular velocities: p, q, and r. For instance, the term

$$C_Z = \frac{2\pi}{9A} \frac{rb}{2V_0} \frac{pb}{2V_0}$$

gives rise to two of the derivatives shown in Fig. 10-13. A missile rotating about two axes will tend to act like a gyroscope, as a result of the gyroscopic terms. It can readily be shown that the gyroscopic forces act a distance $\frac{8}{15}c$ *behind* the centroid of wing area in a position off the wing planform.

The other aerodynamic terms listed in Fig. 10-13 are the static terms in pitch, Z_α and M_α, and the static terms in sideslip, Y_β and N_β. Likewise we have the damping terms due to pitching velocity, Z_q and M_q, and those due to yawing velocity, Y_r and N_r. It can be seen that the damping forces act a distance $c/4$ *behind* the wing centroid. The term L_p is the damping in roll. Certain miscellaneous terms associated with $b\dot{V}_0/2V_0^2$ are due to axial acceleration of the missile.

10-9. Further Examples of the Use of Apparent-mass Table

A number of stability derivative problems involving complicated interference effects can frequently be solved, using the apparent-mass coefficients of Table 10-3. The examples selected here are just a few of many possible. As a first example let us determine the lift-curve slope of a cruciform wing and body combination.

Example 1

The lift in the plane of the body axis and the wind direction will now be determined, using Table 10-3. The included angle α_c, between the body axis and the wind direction, and the angle of bank φ are both considered arbitrary. From Eqs. (10-37) and (10-38) and Fig. 10-15

$$\begin{aligned} C_{Z_\alpha} &= -2\bar{A}_{22} & C_Z &= -2\bar{A}_{22}\alpha_c \cos\varphi \\ C_{Y_\beta} &= -2\bar{A}_{11} & C_Y &= -2\bar{A}_{11}\alpha_c \sin\varphi \end{aligned} \tag{10-104}$$

Let us take the lift equal to the normal force to the degree of approximation of this calculation:

$$\begin{aligned} C_L &= -C_Y \sin\varphi - C_Z \cos\varphi \\ &= 2\alpha_c(\bar{A}_{11}\sin^2\varphi + \bar{A}_{22}\cos^2\varphi) \\ &= 2\alpha_c\bar{A}_{11} \end{aligned}$$

since $\qquad\qquad \bar{A}_{11} = \bar{A}_{22}$

The value of $\bar{A}_{11}$ from Table 10-3 is

$$\bar{A}_{11} = \frac{\bar{m}_{11}}{\rho S_R} = \frac{\pi \rho s^2}{\rho S_R}\left(1 - \frac{a^2}{s^2} + \frac{a^4}{s^4}\right)$$

so that

$$C_L = \frac{2\pi s^2}{S_R}\left(1 - \frac{a^2}{s^2} + \frac{a^4}{s^4}\right)\alpha_c \tag{10-105}$$

This result is to be compared with Eq. (5-35), with which it is in agreement. We note that the lift-curve slope does not depend on φ for a slender cruciform wing-body combination, nor does it depend on the precise planform of the wings. It depends only on the missile cross section at the maximum span.

Example 2

Consider a missile of n equally spaced equal-span fins as shown in configuration G of Table 10-3. Let us calculate how the damping in roll C_{l_p} is affected by the number of fins. With reference to Eq. (10-39), we see that, for $\alpha = \beta = q = r = 0$, we have

$$C_{l_p} = -4\bar{A}_{33}$$

Since C_{l_p} is directly proportional to $\bar{A}_{33}$, and since we can let the reference area and length be constant as n changes, we can write

$$\frac{(C_{l_p})_n}{(C_{l_p})_2} = \frac{(m_{33})_n}{(m_{33})_2} \tag{10-106}$$

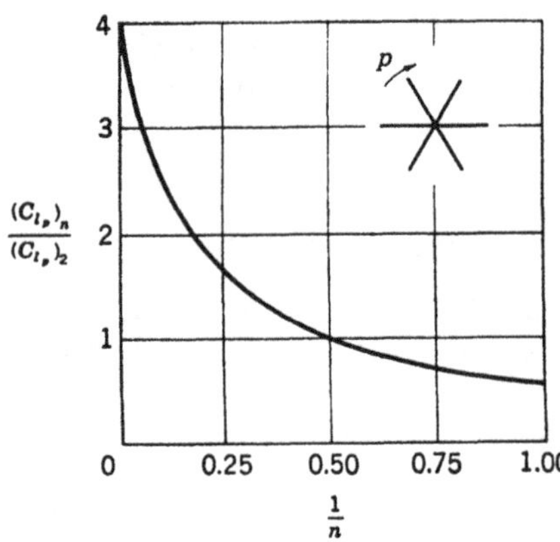

FIG. 10-14. Effect of number of fins on damping in roll.

This ratio has been calculated from the numerical results given in Table 10-3 and the results are shown in Fig. 10-14. It is seen that the addition of fins to a missile adds to the damping in roll at a decreasing rate, as would be expected. The influence of the body is treated in the next example.

Example 3

Consider a planar or cruciform missile of fixed span, and permit the body radius to vary. Let us determine how the damping in roll is affected by changes in body radius a, as shown in Fig. 10-15. According to Eq. (10-39), the damping in roll is

$$C_{l_p} = -4\bar{A}_{33} + 4\alpha B_{13} - 8\frac{q\lambda}{2V_0}C_{13} - 4\beta B_{23} - 8\frac{r\lambda}{2V_0}C_{23}$$

Since both planar and cruciform missiles considered have horizontal and vertical planes of symmetry, the following inertia coefficients are zero,

$$A_{12} = A_{13} = A_{23} = 0$$

and we are left with

$$C_{l_p} = -4\bar{A}_{33}$$

If we base C_{l_p} on total span and total panel area, including that blanketed by the body, the reference quantities will be constant as body radius varies. We can then write

$$\frac{C_{l_p}}{(C_{l_p})_{a=0}} = \frac{m_{33}}{(m_{33})_{a=0}} \quad (10\text{-}107)$$

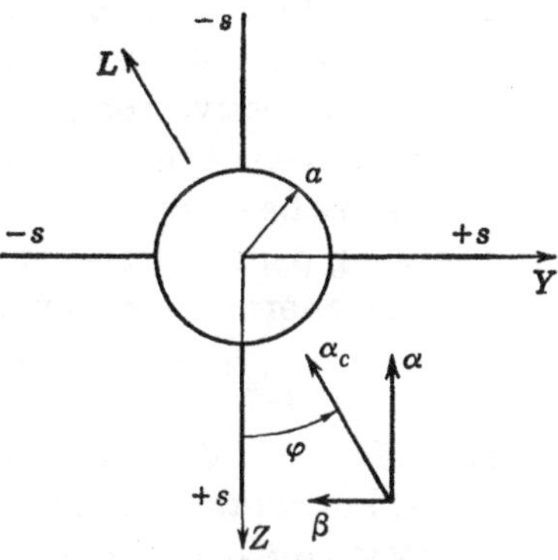

FIG. 10-15. Crossflow plane at wing trailing edges of slender cruciform missile.

Known values of m_{33} can be used in this equation to obtain the change in damping, or known values of C_{l_p} can be used to obtain m_{33}. Numer-

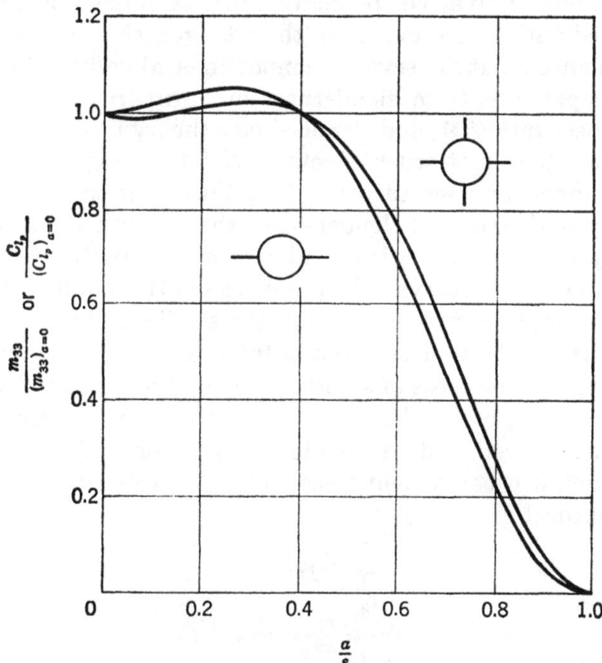

FIG. 10-16. Effect of body radius on damping in roll for fixed span.

ical results are available for C_{l_p} for both the planar and cruciform cases given by Adams and Dugan.[27] Analytical results are given for the planar case by Lomax and Heaslet.[29] These results, plotted in Fig. 10-16 as a function of a/s, can be thought of as yielding damping-in-roll coefficients or apparent-mass coefficients. In any event, it is clear that the addition of a body with a value of a/s up to 0.4 causes very little change in the damping in roll. Actually the angle of attack on the inboard stations is relatively ineffective compared to that on the outboard stations. The loss of effectiveness due to blanketing of the inboard stations by the body can therefore be easily compensated by favorable interference of the body on the wing panels. The small difference between the ratios for planar and for cruciform wings is notable.

Further Example

The effect of interference among the various parts of an empennage, fuselage, horizontal tail, upper vertical tail, and ventral fin on the stability derivative C_{n_β} is treated in Sec. 10-11, on the basis of the apparent-mass coefficients of Table 10-3.

10-10. Effect of Aspect Ratio on Stability Derivatives of Triangular Wings

Although slender-body theory proves to be a powerful tool for calculating the stability derivatives of many types of missile configurations, it must be used with discretion when the configurations are not slender, as we have pointed out in several connections already. The first-order effects of departures from slenderness are primarily a function of the effective aspect ratio BA, and slender-body theory is in a sense the theory for $BA = 0$. The first-order effects of BA for wings are well approximated by supersonic wing theory. It is thus clear that a comparison of the stability derivatives of slender-body theory with those of supersonic wing theory for triangular wings will give much insight into the application of slender-body theory to the prediction of the stability derivatives of nonslender complete missile configurations. This comparison will now be made with the help of the results for the stability derivatives of triangular wings on the basis of supersonic wing theory as collected by Ribner and Malvestuto.[15] The comparison will bring to light significant phenomena not predicted by slender-body theory. For the purpose of the discussion, it is convenient to consider the stability derivatives in the following natural groupings:

Static stability:	$C_{L_\alpha},\ C_{Y_\beta},\ C_{m_\alpha},\ C_{n_\beta}$
Roll damping:	C_{l_p}
Pitch damping:	$C_{m_q},\ C_{m_{\dot\alpha}},\ C_{L_q},\ C_{L_{\dot\alpha}}$
Dihedral effect:	$-C_{l_\beta}$
Magnus forces:	$C_{Y_{\alpha p}},\ C_{n_{\alpha p}}$

The reference area is taken as the wing planform area, and the reference length is taken as the wing span. If the semiapex angle of the wing is ω, then the primary independent variable for the discussion, the effective aspect ratio, is

$$BA = 4B \tan \omega$$
$$B = (M_0{}^2 - 1)^{1/2}$$

Static Stability Derivatives

The results of slender-body theory and of supersonic wing theory[15] for the variation with BA of C_{L_α} and C_{m_α} are as follows:

Slender-body theory:

$$C_{L_\alpha} = \frac{\pi A}{2}$$
$$C_{m_\alpha} = 0 \tag{10-108}$$

Supersonic wing theory:

$$C_{L_\alpha} = \frac{\pi A}{2E(\pi/2,\, k)} \qquad BA < 4$$
$$= \frac{4}{B}; \qquad BA > 4 \tag{10-109}$$
$$C_{m_\alpha} = 0; \qquad BA > 4 \qquad \text{or} \qquad BA < 4$$

where,
$$k^2 = 1 - \left(\frac{BA}{4}\right)^2 \tag{10-110}$$

The ratio of C_{L_α} calculated by supersonic wing theory to that calculated by slender-body theory is designated as $C_{L_\alpha}^*$ and is plotted in Fig. 10-17 against BA. It is clear that the slender-body theory is about 35 per cent in error for $BA = 4$ where the leading edge becomes sonic. The results

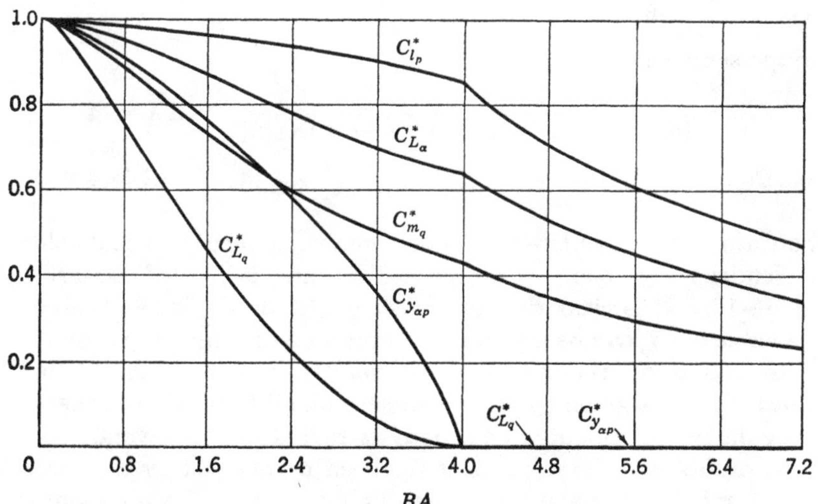

FIG. 10-17. Effect of aspect ratio on stability derivatives of triangular wing.

for C_{m_α} for both theories are identically zero. (The wing centroid is the moment center.) This result is a straightforward result of the fact that both theories must give conical flow fields and conical pressure fields for a conical configuration. (See Sec. 2-4.)

The lateral derivatives C_{Y_β} and C_{n_β} are zero for a flat triangular wing on the basis of slender-body theory. It is clear that the normal forces on the wing can have no components along the lateral body axis. Any side forces or yawing moments must therefore result from forces on the side edges of the wing, so-called *leading-edge suction forces* arising from the high flow velocities around a sharp leading edge.[15] An asymmetric sideslip condition is necessary to produce C_Y or C_n, which can occur only if α and β are both not zero. For triangular wings with supersonic leading edges, these forces are zero because of the absence of leading-edge suction. [The leading-edge pressure coefficients corresponding to oblique shock-expansion theory are applied in a plane normal to the leading edge.) The extent to which leading-edge suction produces C_Y and C_n for triangular wings with subsonic leading edges depends to a considerable degree on the physical condition of the leading edges. The sharp leading edge of the theory on which infinite suction pressure acts is a mathematical idealization. Only by some degree of leading-edge rounding will any appreciable fraction of the upper theoretical limit be achieved.

Damping in Roll

The values of C_{l_p} on the basis of supersonic wing theory have been given by Brown and Adams.[16]

Slender-body theory:

$$C_{l_p} = \frac{-\pi A}{32} \tag{10-111}$$

Supersonic wing theory:

$$C_{l_p} = \frac{-\pi A k^2}{16[(1 + k^2)E(\pi/2\ k) - (1 - k^2)K(k)]} \qquad BA < 4$$

$$C_{l_p} = -\frac{1}{3B} \qquad BA > 4 \tag{10-112}$$

The ratio of C_{l_p} calculated by supersonic wing theory to that calculated by slender-body theory is designated $C_{l_p}^*$ and is plotted against BA in Fig. 10-17. It is clear that the effect of BA on C_{l_p} is less than on C_{L_α}. The value of $C_{l_p}^*$ can be assumed to apply well to wing-body combinations up to values of a/s of about 0.4, on the basis of Fig. 10-16. The values of C_{l_p} are given by Brown and Heinke,[19] from wind-tunnel tests of triangular wings mounted on a body of revolution for several supersonic Mach numbers. These values have been normalized by the theoretical value for $BA = 0$ given by slender-body theory, and are presented as a function of BA in Fig. 10-18. The general variation with BA of the damping in roll is the same theoretically and experimentally, and the agreement between experiment and theory is fair on an absolute basis. The difference is greatest in the region of $BA = 4$ where the leading edge becomes sonic. It is known that the disagreement of C_{L_α} on the basis of theory and of experiment is similar to that shown[28] in Fig. 10-18. For

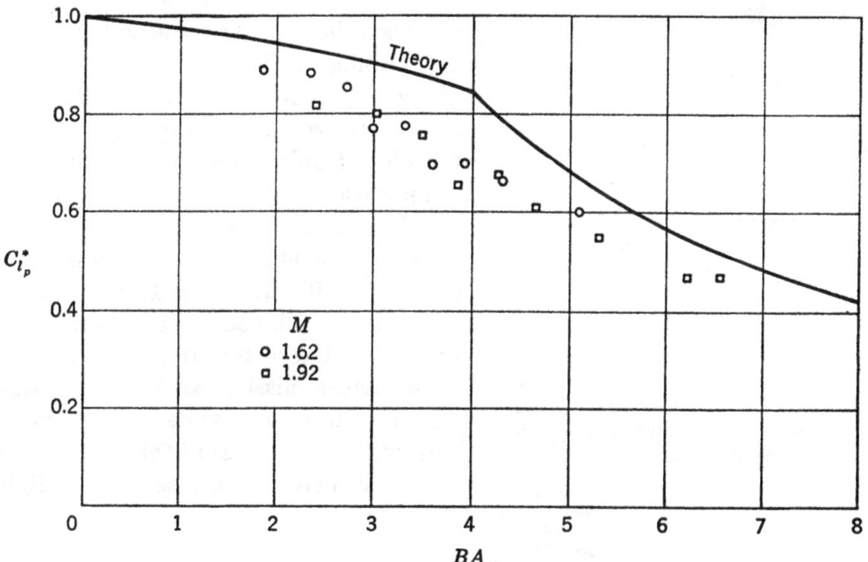

Fig. 10-18. Comparison between experiment and theory for damping in roll of triangular wings.

C_{L_α} the disagreement around $BA = 4$ is known to result from the transonic-type flow resulting from the sonic velocity normal to the leading edge of the wing, and undoubtedly similar effects prevail for C_{l_p}.

Damping in Pitch

The damping-in-pitch derivatives are bothersome, in that they include the combined effects of $\dot{\alpha}$ and q, and the effects of these two independent variables have quite different behavior with changing BA. To obtain a proper understanding of the term C_{L_q}, C_{m_q}, $C_{L_{\dot{\alpha}}}$, and $C_{m_{\dot{\alpha}}}$ it is vital to understand the differences in the types of motion characterized by the two conditions $\dot{\alpha} = 0$, $q \neq 0$, and $\dot{\alpha} \neq 0$, $q = 0$. To illustrate these two types of motion, Fig. 10-19 has been prepared. The angle θ as shown is the angle between a *fixed direction* and the wing chord, and α is the angle between the *instantaneous flight direction* and the wing chord. The instantaneous flight direction is the instantaneous direction of the velocity of the center of gravity. Consider now *uniform motion* with q constant and $\dot{\alpha} = 0$. This is seen clearly to be characterized by perfect loops in a

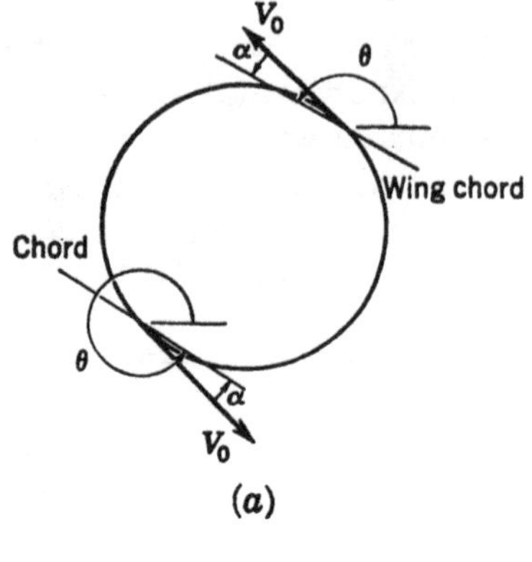

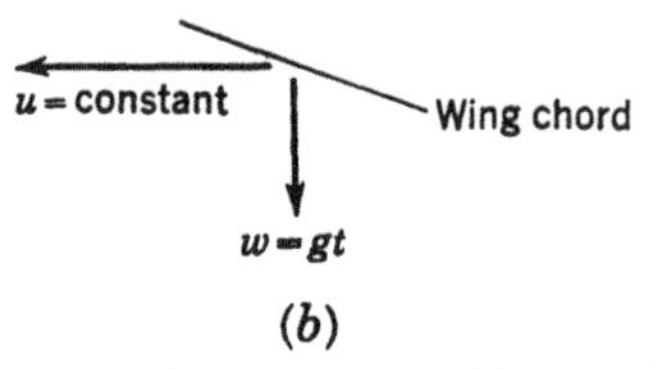

FIG. 10-19. Types of uniform motion involved in damping in pitch. (a) $\dot{\alpha} = 0$, q = constant; (b) $q = 0$, $\dot{\alpha}$ = constant.

vertical plane, since $\theta = q$ is a constant and α is also constant, so that $\dot{\alpha} = 0$. The second type of *uniform motion* characterized by $q = 0$ and $\dot{\alpha}$ = constant corresponds to a wing of fixed attitude undergoing a uniform vertical acceleration as if freely falling. If one of the foregoing cases of uniform motion prevails, then the appropriate stability derivative applies.

A type of motion which prevails probably more frequently than the former examples of uniform motion is sinusoidal pitching oscillation, which is a combination of the two former motions. Several types of sinusoidal motion are illustrated in Fig. 10-20. In case 1, the missile axis is always aligned in the flight direction, so that $\alpha = 0$ and q is sinusoidal. In case 2, the missile axis has a constant direction in space while $\dot{\alpha}$ is varying sinusoidally as a result of changes in vertical velocity. In case 3, the flight

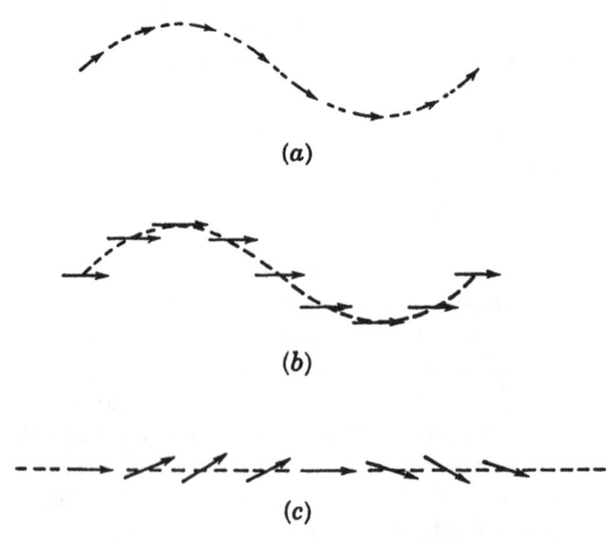

FIG. 10-20. Types of sinusoidal pitching motions. (a) $\dot{\alpha} = 0$, $q = \theta = \sin \omega t$; (b) $\dot{\alpha} = \sin \omega t$, $q = \theta = 0$; (c) $q = \dot{\alpha} = \sin \omega t$.

path is essentially straight while the missile axis is changing direction around it in a sinusoidal manner. It is clear that θ and α are both equal and in phase such that $\dot{\alpha} = \theta = q$. The significant damping derivatives are in this case $(C_{L_{\dot{\alpha}}} + C_{L_q})$ and $(C_{m_{\dot{\alpha}}} + C_{m_q})$.

Let us inspect the effects of BA on C_{L_q} and C_{m_q} first. These results are available in the work of Miles,[17] as well as results for $\dot{\alpha}$.

Slender-body theory:

$$C_{L_q} = \frac{2\pi}{3}$$

$$C_{m_q} = \frac{-\pi}{3A} \tag{10-113}$$

Supersonic wing theory:

$$C_{L_q} = \frac{2\pi}{3}\left[\frac{3k^2}{(2k^2 - 1)E(\pi/2, k) + (1 - k^2)K(k)} - \frac{2}{E(\pi/2, k)} \right]$$

$$C_{m_q} = \frac{-\pi k^2}{3A[(2k^2 - 1)E + (1 - k^2)K]} \qquad \text{for } BA < 4 \tag{10-114}$$

$$C_{L_q} = 0$$

$$C_{m_q} = -\frac{16}{9BA^2} \qquad \text{for } BA > 4 \tag{10-115}$$

These results are *based on the wing span as reference length*, and are for rotation about the wing centroid which coincides with the center of moments. Different positions of the center of rotation and center of moments are discussed subsequently. The ratios of C_{L_q} and C_{m_q} on the basis of supersonic wing theory to those on the basis of slender-body theory are designated $C_{L_q}^*$ and $C_{m_q}^*$, and are presented as a function of BA in Fig. 10-17. For small values of BA the force C_{L_q} acts a distance $c/4$ behind the wing centroid. For $BA > 4$ the value of C_{L_q} is zero; the value of C_{m_q}, however, is not zero but negative. As a result, the center of pressure has moved an infinite distance behind the wing. Thus, C_{m_q} will have a stabilizing influence for any axis of rotation in front of the $11/12$ root-chord position for all BA values.

Let us inspect the effects of BA on $C_{L_{\dot{\alpha}}}$ and $C_{m_{\dot{\alpha}}}$.

Slender-body theory:

$$C_{L_{\dot{\alpha}}} = \frac{2\pi}{3}$$

$$C_{m_{\dot{\alpha}}} = \frac{-\pi}{9A} \tag{10-116}$$

Supersonic wing theory:

$$C_{L_{\dot{\alpha}}} = \frac{2\pi}{3B^2}\left[\frac{3k^2(B^2 + 1)}{(2k^2 - 1)E(\pi/2, k) + (1 - k^2)K(k)} - \frac{(2B^2 + 3)}{E(\pi/2, k)} \right]$$

$$C_{m_{\dot{\alpha}}} = \frac{-\pi}{9B^2A}\left[\frac{3k^2(B^2 + 1)}{(2k^2 - 1)E + (1 - k^2)K} - \frac{2B^2 + 3}{E} \right] \qquad \text{for } BA < 4 \tag{10-117}$$

$$C_{L_{\dot\alpha}} = \frac{-16}{3B^3 A}$$

$$C_{m_{\dot\alpha}} = \frac{8}{9B^3 A^2} \qquad \text{for } BA > 4 \tag{10-118}$$

The values of $C_{L_{\dot\alpha}}^*$ and $C_{m_{\dot\alpha}}^*$ do not depend solely on the parameter BA, as do C_{L_q} and C_{m_q}, but rather on both B and A. The values of these parameters are shown in Fig. 10-21 as a function of BA for $B = 1$. The values of $C_{L_{\dot\alpha}}^*$ and $C_{m_{\dot\alpha}}^*$ have the same numerical values, a fact indicating a uniform center of pressure for $\dot\alpha$ motion in distinct contrast to q motion. As previously mentioned, this center of pressure occurs a distance $c/12$ behind the centroid. Thus, as long as $C_{L_{\dot\alpha}}^*$ is positive, $C_{m_{\dot\alpha}}$ will tend to damp $\dot\alpha$ motions for rotations about axes in front of the three-fourths root-chord position behind the wing apex. However, as shown in Fig. 10-21, for some value of A depending on B, the value of $C_{L_{\dot\alpha}}$ becomes negative and therefore destabilizing for positions of the axis in front of the three-fourths root-chord position.

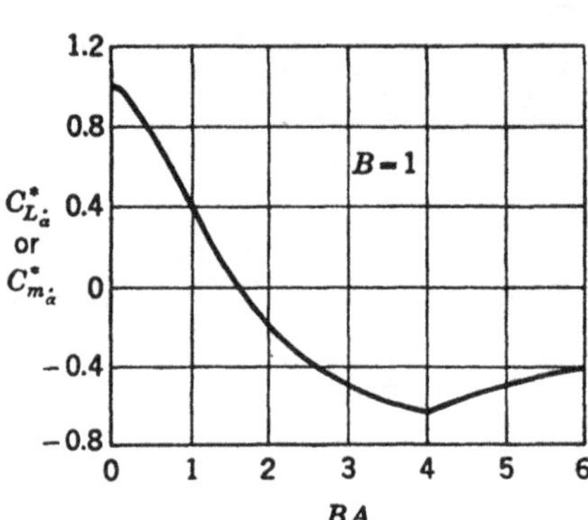

FIG. 10-21. Variation with aspect ratio of certain pitch damping derivatives of triangular wings.

For changes in center of rotation the values of C_{L_q} and C_{m_q} will be altered in a manner different from the usual moment axis transformation, because the downwash distribution along the wing will be substantially altered. Let $C'_{L_{\dot\alpha}}$, $C'_{m_{\dot\alpha}}$, C'_{L_q}, C'_{m_q} be values of these derivatives for given centers of rotation and moments which may be different. Let $C_{L_{\dot\alpha}}$, $C_{m_{\dot\alpha}}$, C_{L_q}, and C_{m_q} be the values for a new center of moments a distance l_1 *behind* the old center of moments, and for a new center of rotation a distance l_2 *behind* the old center of rotation. The quantities are related as follows:

$$C_{L_{\dot\alpha}} = C'_{L_{\dot\alpha}}$$

$$C_{m_{\dot\alpha}} = C'_{m_{\dot\alpha}} + \frac{l_1}{\lambda} C'_{L_{\dot\alpha}}$$

$$C_{L_q} = C'_{L_q} - 2C_{L_\alpha} \frac{l_2}{\lambda} \tag{10-119}$$

$$C_{m_q} = C'_m - 2C'_{m_\alpha} \frac{l_2}{\lambda} + C'_{L_q} \frac{l_1}{\lambda} - 2 \frac{l_1}{\lambda} \frac{l_2}{\lambda} C_{L_\alpha}$$

The quantity λ is the reference length, and the factor 2 is a result of the fact that the derivatives are based on $q\lambda/2V_0$ and $\dot\alpha\lambda/2V_0$. It is to be noted that the $\dot\alpha$ derivatives transform exactly as α derivatives as, indeed, also do the q derivatives if l_2 is zero. However, with a change in the center of q rotation, the redistribution of downwash along the wing chord introduces the terms proportional to l_2.

Dihedral Effect

The dihedral effect $-C_{l_\beta}$ is the negative of the rolling moment due to sideslip.   If the missile is in a positive sideslip attitude with the windward side on the right, facing forward, and if the rolling moment is negative, tending to roll the missile into a left turn with the left wing moving

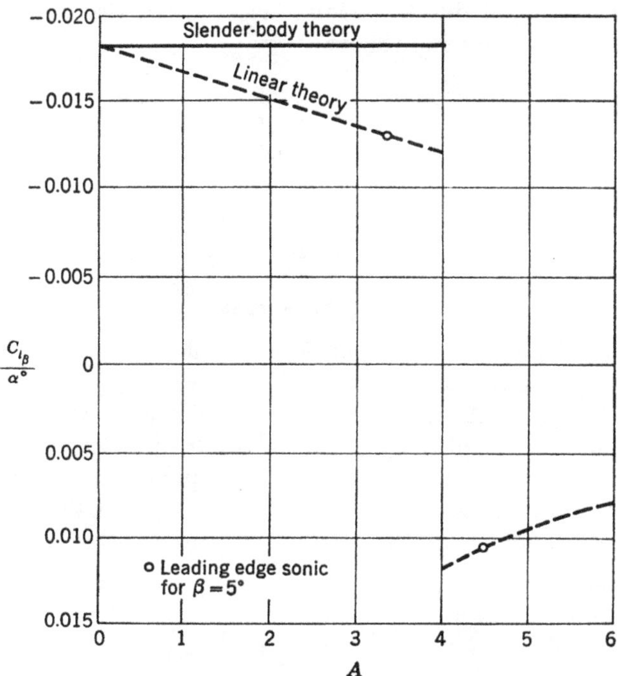

Fig. 10-22. Rolling moment due to sideslip of triangular wings; $M_0{}^2 = 2$.

downward, then the dihedral effect is stabilizing or positive.   Thus C_{l_β} negative is stabilizing.   The value of C_{l_β} has been determined for a large number of wing planforms on the basis of supersonic wing theory by Jones, Spreiter, and Alksne.[18]   For angle of sideslip greater than zero, there is an asymmetry in the sweep of the leading edges, and hence of the wing planform, as viewed in the streamwise direction.   This change in planform is significant in supersonic wing theory, and in effect causes the dependence of C_l on β for a fixed value of α to be slightly nonlinear. Actually, as long as either leading edge does not change from subsonic to supersonic, or conversely, as a result of sideslip, the dependence of C_l on β can be taken as linear.   It is apparent that the value of C_{l_β}, besides depending on Mach number and aspect ratio, will also depend on sideslip angle β.

Figure 10-22 has been prepared to show the qualitative dependence of C_{l_β} for a triangular wing on A for $M_0{}^2 = 2$. Up to the point where the leading edge becomes sonic, slender-body theory gives a fair estimate of the dihedral effect. However, when the leading edge of the wing becomes sonic, BA approximately equal to 4, the dihedral effect changes from stable to unstable. The two circles on the curve show the precise values of aspect ratio for which the leading edges are sonic for a sideslip angle of 5°. The effects of thickness will influence the value of A for which the leading-edge shock wave detaches.

The rolling moment due to yawing velocity can also be calculated on the basis of supersonic wing theory.[15]

Slender-body theory:

$$C_{l_r} = \frac{\pi\alpha}{9A} \tag{10-120}$$

Supersonic wing theory:

$$C_{l_r} = \frac{\pi(1 + 9A^2/16)\alpha}{9AE(\pi/2,\ k)} \tag{10-121}$$

Magnus Forces

By Magnus forces we mean the force C_Y developing as the result of rolling velocity at angle of attack (or C_L as a result of roll at sideslip) and proportional to αp. At angle of attack α it is clear that no force can develop along the lateral body axis as a result of pressure forces normal to the wing planform. Any side force or yawing moment must arise as a result of leading-edge suction forces. An analysis of the ideal leading-edge suction forces[15] yields the following results for the Magnus forces:

Slender-body theory:

$$C_{Y_{\alpha p}} = \frac{2\pi}{3}$$

$$C_{n_{\alpha p}} = \frac{-\pi}{9A} \tag{10-122}$$

Supersonic wing theory:

$$C_{Y_{\alpha p}} = \frac{4\pi k^3}{3E(\pi/2,\ k)[(k^2 + 1)E(\pi/2,\ k) - (1 - k^2)K(k)]}$$

$$C_{n_{\alpha p}} = \frac{-2\pi k^3(A/16 + 1/9A)}{E[(k^2 + 1)E - (1 - k^2)K]} \tag{10-123}$$

The value of $C_{Y_{\alpha p}}$ as given by this equation decreases continuously with BA as shown in Fig. 10-17 to a value of zero at $BA = 4$. For $BA > 4$ there is no leading-edge suction, so that $C_{Y_{\alpha p}}$ and $C_{n_{\alpha p}}$ are both zero.

10-11. Contribution of the Empennage to Certain Stability Derivatives; Empennage Interference Effects

Up to this point in the present chapter, we have concerned ourselves with the stability of what are called *tailless configurations*. This is not to say that the configuration has no tail. We mean by a *tailless missile* one for which the stabilizing and control surfaces are contained entirely, or for the most part, between the crossflow planes that bracket the wing. This definition is a functional one, in order to separate missiles into one class, which can be wholly treated by apparent-mass methods, and the opposite class, which requires a consideration of wing-tail interference as described in Chap. 7. For a missile which consists essentially of a winged part plus an empennage in tandem, the contributions of the separate parts to the stability derivatives can be calculated by the foregoing methods. But, in addition, account must be taken of the wing-tail interference. In this section we will be concerned with wing-tail interference phenomena not treated in Chap. 7, and with interference effects between the various parts of the empennage. The *empennage* is comprised of *body*, a *horizontal tail*, an *upper vertical tail*, and a *lower vertical tail* or *ventral fin*. As in the preceding section, it is convenient to consider the derivatives in the following natural groups:

Static stability:	C_{L_α}, C_{m_α}, C_{Y_β}, C_{n_β}
Damping in roll:	C_{l_p}
Pitch damping:	C_{L_q}, C_{m_q}, $C_{L_{\dot\alpha}}$, $C_{m_{\dot\alpha}}$
Yaw damping:	C_{Y_r}, C_{n_r}, $C_{Y_{\dot\beta}}$, $C_{n_{\dot\beta}}$

Static Stability Derivatives

The static stability derivatives of the empennage are influenced by interference between the various parts of the empennage, and between body and wing vortices and the empennage. In so far as C_{L_α} and C_{m_α} are concerned, both these influences have already been treated at some length for the condition of zero sideslip. In principle, the values of C_{Y_β} and C_{n_β} could be similarly treated, except for the fact that the upper and lower vertical tails differ in size and shape, unlike the left and right horizontal tail panels. (The cruciform missile is a notable exception.) In this section we confine our attention to the effects of β, and consider successively the empennage interference effects and the body and wing vortices. It is desirable to have a generalized scheme for analyzing the special empennage interference effects arising from the inequality of upper and lower vertical tail spans. The scheme we will now outline is based on slender-body theory and is generalized to nonslender missiles. It applies equally to low, mid, or high horizontal tail positions and is valid over that range of α and β for which the empennage sideforce and yawing-moment characteristics are linear.

The general scheme for analyzing empennage interference effects is based on systematically building up the empennage from its component parts, as shown in Fig. 10-23. Starting first with the quantity on the left side of the equation, the sideforce on the body is subtracted from the sideforce represented by the cross section $BHUL$ since the sideforce on the body alone is developed essentially on the body nose and does not represent a contribution of the empennage.   The notation is as follows:

$$
\begin{array}{ll}
B & \text{body alone} \\
H & \text{horizontal tail alone} \\
U & \text{upper tail alone} \\
L & \text{lower tail alone} \\
E & \text{empennage } (BHUL - B)
\end{array}
$$

The first term on the right side of the equation in Fig. 10-23 represents the effect on the sideforce of adding the horizontal tail to the body, and the second and third terms represent the effects of adding successively the upper vertical tail and then the lower vertical tail.   These three terms are to be "normalized" into sideforce ratios that can be applied to nonslender configurations.   It is logical to normalize the sideforce due to the addition of any given component by the sideforce of the component

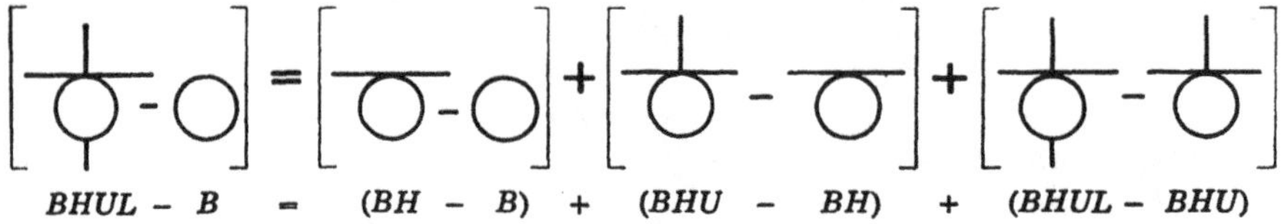

$$BHUL - B \;=\; (BH - B) \;+\; (BHU - BH) \;+\; (BHUL - BHU)$$

Fig. 10-23. Decomposition of empennage.

itself calculated on the same basis.   Any tendency of the means of calculation to underpredict or overpredict would be minimized by the formation of such a ratio.   Furthermore, by proper choice of the definitions of the components alone, the ratios can be given the direct physical significance of interference effects.   Let us now write the equation for the sideforce on the basis of the build-up shown in Fig. 10-23, and then form the sideforce ratios.

$$
\begin{aligned}
Y_E = Y_{BHUL} - Y_B &= (Y_{BH} - Y_B) + (Y_{BHU} - Y_{BH}) \\
&\quad + (Y_{BHUL} - Y_{BHU}) \\
&= \left(\frac{Y_{BH} - Y_B}{Y_B}\right) Y_B + \left(\frac{Y_{BHU} - Y_{BH}}{Y_U}\right) Y_U \\
&\quad + \left(\frac{Y_{BHUL} - Y_{BHU}}{Y_L}\right) Y_L \qquad (10\text{-}124)
\end{aligned}
$$

It is to be noted that the sideforce due to the addition of the horizontal tail to the body has been normalized by the sideforce of the body alone, rather than that of the horizontal tail which is zero.   The above scheme for the sideforce applies equally to any other forces or moments due to the empennage, although a different order of build-up might be desirable in some cases.   We have not yet defined precisely what we mean by the various components.   The *body alone* is a pointed body of revolution with the same base cross section as the body cross section at the empennage location. The *upper tail alone* corresponds to the upper tail panel mounted on a perfect reflection plane; that is, the sideforce due to the upper tail alone is one half the sideforce on a surface composed of two panels similar to the upper tail panel. A corresponding definition holds for the *lower vertical tail alone* or the ventral fin. The horizontal tail alone is the wing formed by joining the horizontal tail panels together.

The three sideforce ratios shown in parenthesis in Eq. (10-124) are given the following notation, and at the same time are specified in terms of apparent mass ratios

$$(K_{11})_B = \frac{Y_{BH} - Y_B}{Y_B} = \frac{(m_{11})_{BH} - (m_{11})_B}{(m_{11})_B} \tag{10-125}$$

$$(K_{11})_U = \frac{Y_{BHU} - Y_{BH}}{Y_U} = \frac{(m_{11})_{BHU} - (m_{11})_{BH}}{(m_{11})_U} \tag{10-126}$$

$$(K_{11})_L = \frac{Y_{BHUL} - Y_{BHU}}{Y_L} = \frac{(m_{11})_{BHUL} - (m_{11})_{BHU}}{(m_{11})_L} \tag{10-127}$$

These sideforce ratios are analogous to the lift ratios K_B and K_W used in Chap. 5 to specify the lift interference of wing-body combinations. We can now write the final result for the empennage sideforce and yawing moment.

$$(C_{Y_\beta})_E = (C_{Y_\beta})_B(K_{11})_B + (C_{Y_\beta})_U(K_{11})_U + (C_{Y_\beta})_L(K_{11})_L \tag{10-128}$$

$$(C_{n_\beta})_E = -\frac{(l_{cm})_V(C_{Y_\beta})_E}{l_r} \tag{10-129}$$

The quantity $(l_{cm})_V$ is the distance between the center of moments and the center of pressure of the sideforce on the empennage. It is interesting to interpret each term of Eq. (10-128) physically. The ratio $(K_{11})_B$ shows how much the sideforce on the body is increased (or decreased) by the addition of the horizontal tail. The ratio $(K_{11})_U$ shows how much the sideforce of the *BH* combination is increased by the addition of the upper vertical tail in multiples of the upper vertical tail mounted on a reflection plane. It thus includes any increase in force on the upper tail due to the sidewash effects of *BH*, and it includes any sideforce on *BH* generated by the upper tail. The factor $(K_{11})_L$ has the same general interpretation as $(K_{11})_U$. However, now the sidewash effects over the ventral fin can be enhanced by the action of the upper vertical tail, and the sideforce generated by the ventral fin can conceivably be caught in part by the upper vertical tail. The physical significance of these quantities has been further discussed by Nielsen and Kaattari,[23] as well as their application to the effects of ventral fins on directional stability. The subsequent example shows the application of the analysis to a cruciform empennage. In Eqs. (10-128) and (10-129) the values $(C_{Y_\beta})_B$, $(C_{Y_\beta})_U$, and $(C_{Y_\beta})_L$ are to be obtained from experiment or the most accurate available theory.

Up to this point we have been concerned with the effects of sideslip only for that range of angles of attack and sideslip for which nonlinear effects of vortices or other causes are unimportant. For large enough angles of attack or sideslip, however, vortices discharged by the wing or body, or both, will produce significant nonlinear effects. In principle these vortex effects can be treated in essentially the same manner as they were treated in Chap. 7 for wing-tail interference at zero sideslip. How-

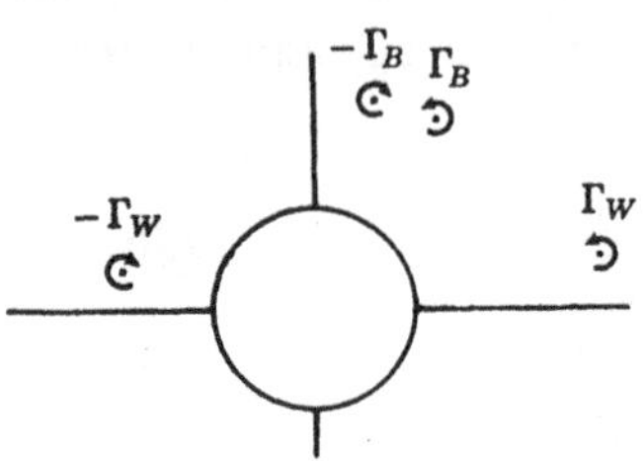

Fig. 10-24. Body and wing vortices in vicinity of empennage for combined pitch and sideslip.

ever, the qualitative effects of sideslip are different from those of pitch, and so a qualitative discussion of the effects of sideslip should prove useful. The principal qualitative difference due to the addition of sideslip is illustrated by Fig. 10-24. The body and wing vortices have been displaced laterally with respect to the tail, so that no symmetry exists about a vertical plane. The displacement laterally of the vortices is due principally to the facts that the wing vortices are discharged by the wing essentially in the streamwise direction, and the trailing-edge shock wave tends to align the body vortices in the streamwise direction. The body vortex farther from the upper vertical tail is stabilizing, tending to increase the directional stability, but the body vortex nearer the tail has a dominant destabilizing effect. Nielsen and Kaattari[23] have discussed methods for calculating C_{Y_β} and C_{n_β}, including body and wing vortex effects. A brief résumé of this discussion is now given.

The calculation of the effects of the wing and body vortices on C_{Y_β} and C_{n_β} proceeds from a knowledge of their strengths and positions at the empennage. The theoretical basis for determining the wing vortex strengths and positions has been covered in Chap. 6, and similar information for the body vortices can be obtained with the help of Chap. 4. It is essential to include the effects of image vortices inside the body if completely erroneous results are to be avoided. The external and internal vortices induce velocities normal to the horizontal tail and upper and lower vertical tails, which vary spanwise but not chordwise if calculated by the method of Sec. 6-4. These normal induced velocities can be interpreted as twisting the horizontal tail, and the upper and lower vertical tails. The resulting forces can be estimated by a strip-theory integration across the individual surfaces. Although such a strip-theory method neglects panel-panel interference, such interference can be accounted for by the more sophisticated methods of reverse flow discussed in Sec. 7-6. When the vortices are very close to the surfaces of the empennage, they will undergo large lateral movements as a result of their images in the empennage. A strong coupling will then be introduced between the effect of the vortices on the empennage and the effect of the empennage on the vortices. This strong coupling can be accounted for by the method of Sec. 4-9. However, it should be borne in mind that the boundary layer will tend to diminish this strong coupling, and that secondary vortices induced at sharp exterior corners may tend further to modify the coupling.

Up to this point we have considered the effects of sideslip and pitch to be independent, in so far as our discussion was concerned. Any coupling between angles of attack and sideslip should produce a term in C_{L_α} and C_{m_α} proportional to β or a term in C_{Y_β} and C_{n_β} proportional to α. An examination of Eqs. (10-37) to (10-40) reveals no such terms arising in the apparent-mass method. Therefore coupling of the type considered does not occur for slender missiles obeying slender-body theory. However, for extremes in angles of attack and sideslip, q_∞ or Mach-number effects proportional to the product $\alpha\beta$ frequently appear. Their calculation in some cases can be made on the basis of shock-expansion theory. Coupling between α and β also arises in the effects of wing and body vortices on the empennage, since both the vortex strength and the lateral displacements of the vortices depend on α and β.

Damping in Roll

The damping-in-roll derivative C_{l_p} is unique in that it is the sole derivative requiring a knowledge of the apparent-mass coefficient m_{33}, which is usually more difficult to obtain than the other coefficients. We will be occupied with the quantitative interference effects between the various parts of the empennage which have an influence on C_{l_p}, but will confine our consideration of wing-tail interference to a few qualitative remarks. As a starting point the equation for the roll-damping derivative based on slender-body theory, Eq. (10-39), is given

$$C_{l_p} = -4\bar{A}_{33} + 4\alpha B_{13} - 8\frac{q\lambda}{2V_0}C_{13} - 4\beta B_{23} - 8\frac{r\lambda}{2V_0}C_{23} \quad (10\text{-}130)$$

For a conventional empennage with a vertical plane of symmetry, we have

$$A_{12} = A_{23} = 0$$

with the result that

$$C_{l_p} = -4\bar{A}_{33} + 4\alpha B_{13}$$

Thus we have a damping-in-roll derivative which varies with angle of attack. However, if the empennage has also a horizontal plane of mirror symmetry, then B_{13} is zero and the term proportional to angle of attack disappears. Let us confine our discussion henceforth to C_{l_p} at zero angle of attack.

A study of empennage interference effects on C_{l_p} can be conveniently carried out using the same general method for C_{Y_β} and C_{n_β}. With reference to Fig. 10-23, we can write an equation for damping in roll similar to Eq. (10-124) for sideforce.

$$(C_{l_p})_E = [(C_{l_p})_{BH'} - (C_{l_p})_B] + [(C_{l_p})_{BH'U} - (C_{l_p})_{BH'}]$$
$$+ [(C_{l_p})_{BH'UL} - (C_{l_p})_{BH'U}] \quad (10\text{-}131)$$

In this case we are taking the *horizontal tail alone H'* to include the tail area blanketed by the body, and *not* to be just the surface formed by the exposed panels. This shift in definition of the tail alone from our hitherto invariable practice is particularly convenient for the study of C_{l_p}, and is used in this connection only.

In normalizing the contributions to the damping in roll of the successive additions of H', U, and L to the empennage, we divide by the damping in roll of H', U, and L alone, respectively, thereby specifying three damping-in-roll ratios

$$(K_{33})_{H'} = \frac{(C_{l_p})_{BH'} - (C_{l_p})_B}{(C_{l_p})_{H'}} = \frac{(m_{33})_{BH'} - (m_{33})_B}{(m_{33})_{H'}} \quad (10\text{-}132)$$

$$(K_{33})_U = \frac{(C_{l_p})_{BH'U} - (C_{l_p})_{BH'}}{(C_{l_p})_U} = \frac{(m_{33})_{BH'U} - (m_{33})_{BH'}}{(m_{33})_U} \quad (10\text{-}133)$$

$$(K_{33})_L = \frac{(C_{l_p})_{BH'UL} - (C_{l_p})_{BH'U}}{(C_{l_p})_L}$$

$$= \frac{(m_{33})_{BH'UL} - (m_{33})_{BH'U}}{(m_{33})_L} \quad (10\text{-}134)$$

The damping-in-roll derivative of the empennage is

$$(C_{l_p})_E = (K_{33})_{H'}(C_{l_p})_{H'} + (K_{33})_U(C_{l_p})_U + (K_{33})_L(C_{l_p})_L \quad (10\text{-}135)$$

These equations permit the calculation of the damping in roll in so far as the apparent-mass coefficients m_{33} are available. The ratios $(K_{33})_{H'}$, $(K_{33})_U$, and $(K_{33})_L$ have the same physical interpretations with respect to damping in roll as the K_{22} coefficients have for directional stability. This particular method of calculating C_{l_p} is instructive when the spans of the upper and lower vertical tails are unequal since it shows the relative effectiveness of the two surfaces. However, for a cruciform empennage or any empennage with a horizontal plane of symmetry, the relative effectiveness of the upper and lower vertical tails may be of no great concern. In this case a detailed decomposition of the interference effects by the foregoing method would be unnecessary, and more direct methods such as those in Sec. 10-9 may be preferable. A collection of data on C_{l_p} for triform empennages is given by Stone.[20]

The interference effects of wing and body vortices on C_{l_p} of the empennage differ. The rolling wing lays down a vortex which thereafter has little tendency to rotate, and, as the vortex moves essentially streamwise from wing to tail, the tail rolls with respect to the vortex. The angular phase difference due to such motion is directly proportional to p. From a knowledge of the vortex strengths, which differ from side to side unless $\alpha = 0$, and their positions, an estimate of their effects on C_{l_p} can be made. The body vortices remain fixed in direction if they are not entrained by the wing flow field, and the time average of the forces developed by the empennage rotating through them will depend on the precise vortex configuration that obtains. Because of body roll and boundary-layer effects, it seems probable that vortices of different strengths and radial locations may be generated, so that the time average of the empennage rolling moment is not zero. Also interaction between the body vortices and the rolling wing field can complicate the phenomenon.

Damping in Pitch (and Yaw)

In the past it has been the usual practice to assume that the contribution of the empennage to the damping in pitch overrides that of other sources which are neglected.[24] This assumption is usually justified when the centers of gravity and of moments are much closer to the wing center of pressure than to the tail center of pressure. In this analysis we will assume arbitrary positions of the centers of moments and of gravity and see how the damping depends on the actual positions of these quantities. First, let us study damping due to q, and then that due to $\dot{\alpha}$.

Consider a missile with the center of moments distinct from the center of gravity, which moves in a circular path, as shown in Fig. 10-25, with $\alpha = 0$ and $q = $ constant. Such motion could be obtained on the end of a whirling arm. There is no downwash field (that due to wing thickness is neglected) due to angle of attack, since $\alpha = 0$. However, because of the rotation of q, there is a distribution of vertical velocity of the air along the missile, as shown in Fig. 10-25. The upward velocity of the air at the horizontal tail is approximately $q(l_{cg})_H$. The local angle of attack due to the q motion at the horizontal tail is thus

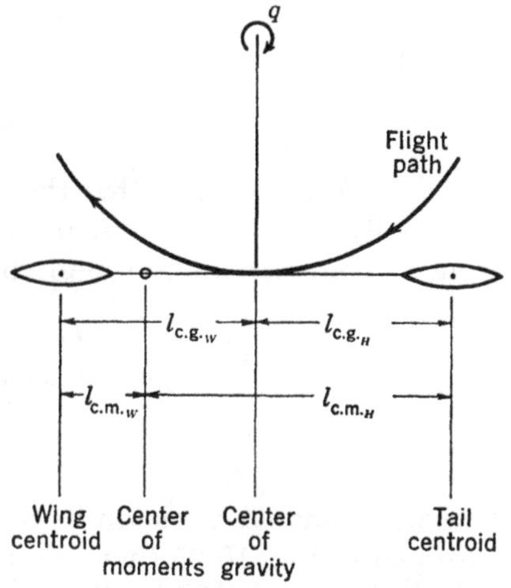

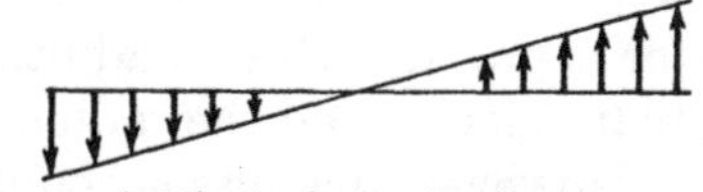

FIG. 10-25. Wing-tail combination in uniform q motion.

$$(\Delta\alpha)_H = \frac{q(l_{cg})_H}{V_0} \qquad (10\text{-}136)$$

neglecting any downwash at the tail resulting from wing lift due to q. The increase in lift of the empennage resulting from $(\Delta\alpha)_H$ is

$$(\Delta C_L)_E = \left(\frac{dC_L}{d\alpha}\right)^*_E \frac{q_H}{q_0} \frac{q(l_{cg})_H}{V_0} \tag{10-137}$$

where the asterisk indicates that the lift-curve slope of the empennage is to be evaluated at the Mach number existing at the horizontal tail location. The dynamic pressure q_H is that prevailing at the horizontal tail. The contribution of the empennage to C_{L_q} and C_{m_q} of the missile is thus

$$(C_{L_q})_E = 2\left(\frac{dC_L}{d\alpha}\right)^*_E \frac{q_H}{q_0}\left(\frac{l_{cg}}{l_r}\right)_H \tag{10-138}$$

$$(C_{m_q})_E = -2\left(\frac{dC_L}{d\alpha}\right)^*_E \frac{q_H}{q_0}\left(\frac{l_{cg}}{l_r}\right)_H\left(\frac{l_{cm}}{l_r}\right)_H \tag{10-139}$$

Similarly for the wing we have

$$(C_{L_q})_W = -2\left(\frac{dC_L}{d\alpha}\right)_W\left(\frac{l_{cg}}{l_r}\right)_W \tag{10-140}$$

$$(C_{m_q})_W = -2\left(\frac{dC_L}{d\alpha}\right)_W\left(\frac{l_{cg}}{l_r}\right)_W\left(\frac{l_{cm}}{l_r}\right)_W \tag{10-141}$$

where we have assumed that the dynamic pressure at the wing location is essentially free-stream dynamic pressure. The total contribution of wing and empennage to the pitch damping is thus

$$(C_{m_q})_{W+E} = -2\left(\frac{dC_L}{d\alpha}\right)^*_E \frac{q_H}{q_0}\left(\frac{l_{cg}}{l_r}\right)_H\left(\frac{l_{cm}}{l_r}\right)_H$$
$$-2\left(\frac{dC_L}{d\alpha}\right)_W\left(\frac{l_{cg}}{l_r}\right)_W\left(\frac{l_{cm}}{l_r}\right)_W \tag{10-142}$$

One point should be noted in connection with this equation. We have assumed in Fig. 10-25 that the missile is fixed to the rotating arm at its center of gravity. This is in accordance with the general notion that the velocity of a missile is specified by the translational velocity of the missile center of gravity, plus an additional velocity determined from the missile angular velocity and the radius vector measured from the center of gravity. However, the missile could be attached to the whirling arm at some point other than the center of gravity, but this case has been precluded in the derivation.

Equation (10-142) is of interest because it displays the roles of the wing and empennage as well as the roles of the center of gravity and center of moments in pitch damping. For flight of a missile, the center of moments and center of gravity are taken to be coincident. If the center of gravity is sufficiently close to the wing center of pressure (due to q), it is clear that the tail contribution to C_{m_q} outweighs that of the wing. However, if the wing and tail lifts are not of greatly different magnitude, the center of gravity might not lie sufficiently close to the wing center of pressure for the contribution of the tail to C_{m_q} to be of overriding importance.

A simplified analysis can also be used to calculate the contribution of the empennage to $C_{L_{\dot\alpha}}$ and $C_{m_{\dot\alpha}}$. The motion corresponding to constant $\dot\alpha$ with $q = 0$ is shown in Fig. 10-26. The motion in question is that of a wing-empennage moving downward with constant acceleration and no angular velocity. Thus, unlike the case of C_{m_q}, no question of the center of rotation (gravity) position arises in the determination of $C_{m_{\dot\alpha}}$. The essential concept which makes possible the simplified analysis is the so-called *downwash lag* concept. It is assumed that the downwash field of the wing at the empennage lags the wing angle of attack by the time it takes the wake to travel from the wing to the empennage. It is further assumed that the downwash field at the empennage is the steady-state

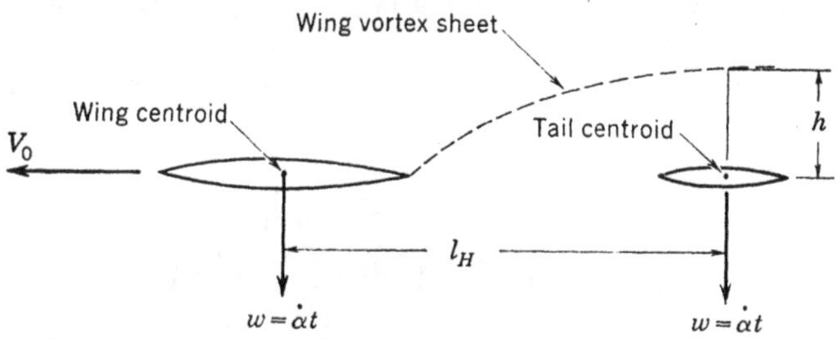

FIG. 10-26. Wing-tail combination in uniform $\dot\alpha$ motion.

downwash field corresponding to the angle of attack of the wing specified by the first assumption. (As the figure shows, the empennage has moved downward with respect to the wing vortex sheet a distance h.) If the wing angle of attack changes by an amount $\Delta\alpha_W$, the change in angle of attack of the horizontal tail is

$$\Delta\alpha_H = -\left(\frac{d\epsilon}{d\alpha}\right)_H \Delta\alpha_W \tag{10-143}$$

By the downwash lag concept

$$\Delta\alpha_W = -\frac{\dot\alpha l_H}{V_0} \tag{10-144}$$

with the result that

$$\Delta\alpha_H = \frac{\dot\alpha l_H}{V_0}\left(\frac{d\epsilon}{d\alpha}\right)_H \tag{10-145}$$

The horizontal tail length l_H will subsequently be specified. The lift developed by the empennage as a result of $\Delta\alpha_H$ is

$$(\Delta C_L)_E = \left(\frac{dC_L}{d\alpha}\right)_E^* \frac{q_H}{q_0} \frac{\dot\alpha l_H}{V_0}\left(\frac{d\epsilon}{d\alpha}\right)_H \tag{10-146}$$

or

$$(C_{L_{\dot\alpha}})_E = 2\left(\frac{dC_L}{d\alpha}\right)_E^* \frac{q_H}{q_0}\left(\frac{d\epsilon}{d\alpha}\right)_H \frac{l_H}{l_r} \tag{10-147}$$

The contribution of the empennage to $C_{m_{\dot{\alpha}}}$ is

$$(C_{m_{\dot{\alpha}}})_E = -2\left(\frac{dC_L}{d\alpha}\right)^*_E \frac{q_H}{q_0}\left(\frac{d\epsilon}{d\alpha}\right)_H \frac{l_H}{l_r}\left(\frac{l_{cm}}{l_r}\right)_H \qquad (10\text{-}148)$$

Because we have used a simplified analysis based on steady-flow quantities to analyze a complicated unsteady-flow process, the precise definition of the tail length l_H has been lost. Tobak[22] has shed some light on this matter using his unsteady-flow analysis based on the indicial-function method. Tobak finds that the downwash lag concept is essentially correct, but depends for its accuracy on the proper choice of l_H. The proper choice turns out to be approximately the length from wing centroid to tail centroid. The distance from the tail centroid to the center of rotation is not involved, since rotations are not involved in pure $\dot{\alpha}$ motions, as we have noted. The values of $C_{m_{\dot{\alpha}}}$ and $C_{L_{\dot{\alpha}}}$ of Martin et al.[21] are in accord with the simplified analysis only if the above choice is made for the tail length.

The damping-in-yaw derivatives corresponding to those for pitch are

$$\begin{aligned}
(C_{Y_r})_E &= -2\left(\frac{dC_Y}{d\beta}\right)^*_E \frac{q_V}{q_0}\left(\frac{l_{cg}}{l_r}\right)_V \\[2mm]
(C_{n_r})_E &= +2\left(\frac{dC_Y}{d\beta}\right)^*_E \frac{q_V}{q_0}\left(\frac{l_{cg}}{l_r}\right)_V\left(\frac{l_{cm}}{l_r}\right)_V \\[2mm]
(C_{Y_{\dot{\beta}}})_E &= 2\left(\frac{dC_Y}{d\beta}\right)^*_E \frac{q_V}{q_0}\left(\frac{d\sigma}{d\beta}\right)_V \frac{l_V}{l_r} \\[2mm]
(C_{n_{\dot{\beta}}})_E &= -2\left(\frac{dC_Y}{d\beta}\right)^*_E \frac{q_V}{q_0}\left(\frac{d\sigma}{d\beta}\right)_V \frac{l_V}{l_r}\left(\frac{l_{cm}}{l_r}\right)_V
\end{aligned} \qquad (10\text{-}149)$$

The asterisk now applies to slopes evaluated at the Mach number prevailing on the upper and lower vertical tails, and the subscript V indicates mean quantities over the upper and lower vertical tails. The quantity $d\sigma/d\beta$ is treated in Chap. 6. The total damping in yaw for sinusoidal oscillations is $C_{n_r} - C_{n_{\dot{\beta}}}$, in contrast to the quantity $C_{m_q} + C_{m_{\dot{\alpha}}}$ for damping of sinusoidal pitching oscillations.

Illustrative Example

Let us calculate the contributions of a cruciform empennage to the derivatives C_{Y_β}, C_{n_β}, C_{l_p}, C_{L_q}, C_{m_q}, $C_{L_{\dot{\alpha}}}$, and $C_{m_{\dot{\alpha}}}$ for the example missile of Fig. 7-9. The center of moments is taken at the missile center of gravity at the two-thirds chord location of the wing-body juncture. Let $M_0 = 2$, $\alpha = 5°$, and $\beta = 0°$. The reference area is taken to be the exposed wing area, and the reference length is the mean aerodynamic chord of the exposed wing panels. Body or wing vortex effects are to be disregarded,

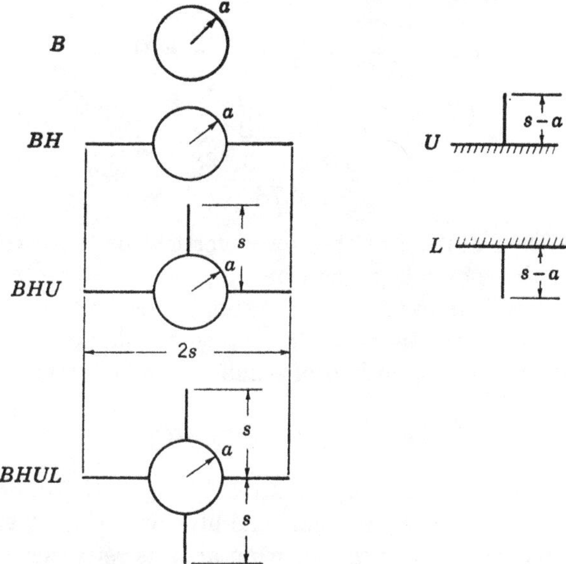

FIG. 10-27. Components in build-up of cruciform empennage.

so that the quantities calculated apply only to the range of α and β over which they are linear.

(a) C_{Y_β}, C_{n_β}:

The static derivative C_{L_α} due to the empennage was calculated in Sec. 7-5, using a method different from that to be used here to calculate C_{Y_β} and C_{n_β}. For a cruciform empennage, we have

$$C_{Y_\beta} = -C_{L_\alpha}$$
$$C_{n_\beta} = -C_{m_\alpha}$$

The first step in the present procedure is to evaluate the interference factors $(K_{11})_B$, $(K_{11})_U$, and $(K_{11})_L$ given by Eqs. (10-125), (10-126), and (10-127). To do this, we must utilize the apparent mass results of Table 10-3 with the notation of Fig. 10-27.

$$(m_{11})_B = \pi\rho a^2 = 0.316\pi\rho$$

$$(m_{11})_{BH} = \pi\rho a^2 = 0.316\pi\rho$$

$$(m_{11})_{BHUL} = \pi\rho s^2 \left(1 - \frac{a^2}{s^2} + \frac{a^4}{s^4}\right) = 3.00\pi\rho$$

$$(m_{11})_{BHU} = \frac{\pi\rho s^2}{4}\left(1 + \frac{a^2}{s^2}\right)\left[-\left(1 + \frac{a^2}{s^2}\right) + 2^{3/2}\left(1 + \frac{a^4}{s^4}\right)^{1/2}\right] = 1.568\pi\rho$$

$$(m_{11})_U = \frac{\pi\rho(s - a)^2}{2} = 0.782\pi\rho$$

$$(m_{11})_L = \frac{\pi\rho(s - a)^2}{2} = 0.782\pi\rho$$

The desired interference factors are now

$$(K_{11})_B = \frac{0.316 - 0.316}{0.316} = 0$$

$$(K_{11})_U = \frac{1.568 - 0.316}{0.782} = 1.602$$

$$(K_{11})_L = \frac{3.000 - 1.568}{0.782} = 1.832$$

We note that the addition of the lower vertical panel to the empennage with the other three panels present develops about 15 per cent more sideforce than the addition of the upper vertical panel to the empennage with only the horizontal panels present. Since the upper vertical tail and lower vertical tail correspond to one-half of whole wings, their sideforce curve slopes are

$$(C_{Y_\beta})_U = (C_{Y_\beta})_L = -\tfrac{1}{2}(C_{L_\alpha})_H$$

The horizontal tail alone is a triangular wing with supersonic leading edges, so that it has the two-dimensional lift-curve slope based on its own area. Thus, based on the exposed wing area as reference area S_R,

$$(C_{Y_\beta})_U = (C_{Y_\beta})_L = -\frac{1}{2}\frac{4}{(M_0{}^2 - 1)^{1/2}}\frac{S_H}{S_R}$$

$$= -\frac{1}{2}\frac{4}{3^{1/2}}\left(\frac{1.25}{2.25}\right)^2 = -0.356$$

Thus from Eq. (10-128)

$$(C_{Y_\beta})_E = 1.602(-0.356) + 1.832(-0.356)$$
$$= -1.221 \text{ per radian}$$

The tail length is taken as that between the centroids of the wing and tail panels.

$$(l_{cm})_V = \tfrac{1}{3}(2.25) + 3.16 + \tfrac{2}{3}(1.25) = 4.74$$
$$(l_r) = \tfrac{2}{3}(c_r)_W = 1.50$$

Thus from Eq. (10-129)

$$(C_{n_\beta})_E = -\frac{4.74}{1.50}(-1.221) = 3.86 \text{ per radian}$$

It is noted in passing that these values of C_{Y_β} and C_{n_β} do not include body or wing vortex interference.

(b) C_{l_p}:

The previous derivatives C_{Y_β} and C_{n_β} were calculated by accounting for all interference effects arising in a step-by-step composition of the empennage. Although Eqs. (10-132) to (10-135) provided for an analogous calculation for C_{l_p}, we will use an alternate method. From Fig. 10-14 we see that the addition of vertical panels to the horizontal tail increases the damping in roll in the ratio 1.625 to 1. Here we are keeping the span of all panels constant and letting the body radius vary, in particular becoming zero in this instance. Now let the body radius grow to a value yielding $a/s = 0.31$, and Fig. 10-16 shows another 2 per cent increase in damping in roll. Thus, if $(C_{l_p})_{H'}$ is the damping-in-roll derivative for the horizontal tail including the part blanketed by the body, we have

$$(C_{l_p})_E = 1.625(1.02)(C_{l_p})_{H'}$$

On the basis of Eq. (10-112), the tail-alone damping in roll is

$$(C_{l_p})_{H'} = -\tfrac{1}{3}(M_0{}^2 - 1)^{1/2} = -0.1925$$

Based on the wing area and wing span, the value of $(C_{l_p})_{H'}$ is

$$(C_{l_p})_{H'} = -(0.1925)\frac{1.812}{2.812}\left(\frac{1.812}{2.25}\right)^2 = -0.080$$

The damping in roll for the complete empennage is thus

$$(C_{l_p})_E = 1.625(1.02)(-0.080) = -0.133$$

(c) $C_{L_q}, C_{m_q}, C_{L_{\dot{\alpha}}}, C_{m_{\dot{\alpha}}}$:

The damping-in-pitch derivative C_{L_q} from Eq. (10-138) is

$$(C_{L_q})_E = 2\left(\frac{dC_L}{d\alpha}\right)_E^* \frac{q_H}{q_0}\left(\frac{l_{cg}}{l_r}\right)_H$$

the asterisk denoting that the empennage contribution to the lift-curve slope is to be evaluated at the Mach number prevailing at the horizontal tail. We will assume that this Mach number is the same as the free-stream Mach number, and that the dynamic pressure at the tail q_H is the same as the free-stream dynamic pressure. Since the distance to the center of gravity has already been evaluated as 4.74 and

$$\left(\frac{dC_L}{d\alpha}\right)_E = -(C_{Y_\beta})_E = 1.221$$

the value of $(C_{l_q})_E$ is

$$(C_{L_q})_E = 2(1.221)(1)\left(\frac{4.74}{1.50}\right) = 7.72$$

The center of moments being coincident with the center of gravity, $(C_{m_q})_E$ is

$$(C_{m_q})_E = -7.72\left(\frac{4.74}{1.50}\right) = -24.4$$

Here the derivative is based on the mean aerodynamic chord of the wing $\bar{c}_W$, the derivative being with respect to $q\bar{c}_W/2V_0$.

The derivatives $(C_{L_{\dot{\alpha}}})_E$ and $(C_{m_{\dot{\alpha}}})_E$ are given by Eqs. (10-147) and (10-148) as

$$(C_{L_{\dot{\alpha}}})_E = 2\left(\frac{dC_L}{d\alpha}\right)_E^* \frac{q_H}{q_0}\frac{l_H}{l_r}\left(\frac{d\epsilon}{d\alpha}\right)_H$$

$$(C_{m_{\dot{\alpha}}})_E = -(C_{L_{\dot{\alpha}}})_E\left(\frac{l_{\text{om}}}{l_r}\right)_H$$

where l_H is the distance from wing centroid to tail centroid. The value of $(d\epsilon/d\alpha)_H$ is obtained from the illustrative example of Sec. 7-5. Since, in the absence of Mach number and dynamic pressure changes from the free-stream value at the tail position, the horizontal tail effectiveness is

$$\eta_H = 1 - \left(\frac{d\epsilon}{d\alpha}\right)_H$$

we can readily determine $(d\epsilon/d\alpha)_H$ from the fact that $\eta_H = 0.73$. Therefore,

$$\left(\frac{d\epsilon}{d\alpha}\right)_H = 0.27$$

The values of the $\dot{\alpha}$ derivatives now are

$$(C_{L_{\dot{\alpha}}})_E = 2(1.221)(1)\left(\frac{4.74}{1.50}\right)(0.27) = 2.08$$

$$(C_{m_{\dot{\alpha}}})_E = -2.08\left(\frac{4.74}{1.50}\right) = -6.59$$

Again the derivative is with respect to $\dot{\alpha}\bar{c}_W/2V_0$. The damping-in-yaw derivatives follow readily from Eq. (10-149).

SYMBOLS OTHER THAN STABILITY DERIVATIVES

The following symbols do not include those for the stability derivatives since these are fully described in Sec. 10-2.

a	radius of body at wing trailing edges of cruciform missile (Fig. 10-15); radius of body of a cruciform empennage (Fig. 10-27)
a_n	complex coefficient in mapping function taking missile cross section into circle of radius c
A	aspect ratio of wing alone or wing panels joined together
A_{ij}	inertial coefficient of missile cross section; $i, j = 1, 2, 3$
$\bar{A}_{ij}$	value of inertial coefficients at missile base
b	span of cruciform wing
b_m	span of planar wing
b_n	complex coefficients associated with expansion for $W_3(\tilde{s})$
B	$(M_0^2 - 1)^{\frac{1}{2}}$

B_{ij} see Eq. (10-36)

c root chord of triangular wing; radius of circle into which missile cross section is mapped

C_{ij} see Eq. (10-36)

C_{L_α} lift-curve slope

$(C_{L_\alpha})^*$, $(C_{m_\alpha})^*$, etc.   ratio of lift-curve slope on basis of supersonic wing theory to that based on slender-body theory

C_{L_α}', C_{m_α}', etc.   values before change in center of moments and center of rotation

$(dC_L/d\alpha)_E{}^*$ lift-curve slope of empennage evaluated at local Mach number of horizontal tail surfaces

$(dC_Y/d\beta)_E{}^*$ sideforce curve slope evaluated at local Mach number of vertical empennage surfaces

C_{m_α} moment-curve slope

C_X, C_Y, C_Z, C_l, C_m, C_n force coefficients for X, Y, Z, L, M, and N

D degree of term in expansion for stability derivative in Maple-Synge analysis

D_{ij} see Eq. (10-36)

E elliptic integral of second kind

f_{ijkl} complex-valued function of u and p occurring as coefficients in Taylor expansion of Maple-Synge analysis

$f_{ijkl}^{(E)}$, $f_{ijkl}^{(O)}$ even and odd *real* functions of p

$f_{ijkl}^{(R)}$, $f_{ijkl}^{(I)}$ real and imaginary parts of f_{ijkl}

F $Z + iY$; also incomplete elliptic integral of first kind

F' value of F after rotation about X axis through $2\pi/n$ radians

g $i + j + k + l + m + n$

i, j, k, l exponents in Eq. (10-11)

I imaginary part of a complex-valued function

k modulus of elliptic integral, $(1 - B^2 A^2/16)^{1/4}$

K complete elliptic integral of first kind

$(K_{11})_B$, $(K_{11})_L$, $(K_{11})_U$ sideforce ratios defined by Eqs. (10-125), (10-126), and (10-127)

$(K_{33})_{H'}$, $(K_{33})_L$, $(K_{33})_U$ rolling-moment ratios defined by Eqs. (10-132), (10-133), and (10-134)

l_1 distance of new center of moments behind old center of moments

l_2 distance of new center of rotation behind old center of rotation

l_{cg} distance from centroid (wing or tail) to missile center of gravity, Fig. 10-25

l_{cm} distance from centroid (wing or tail) to center of moments, Fig. 10-25

l_r reference length

L, M, N	positive moments about X, Y, and Z
L'	value of L after rotation about X axis by $2\pi/n$ radians
m, n	integers in Maple-Synge analysis
m_{ij}	apparent-mass coefficients defined by Eq. (10-62) $i,j = 1,2,3$
M_0	free-stream Mach number
n	outward normal to missile cross section in crossflow plane; also number defining degree of missile rotational symmetry, Fig. 10-3
p, q, r	angular velocities X, Y, and Z
p'	value of p after rotation about X axis by $2\pi/n$ radians
P, Q	integers in Maple-Synge analysis; $P = i + k$, $Q = j + l$
q_0	free-stream dynamic pressure
q_H	dynamic pressure at horizontal tail
$\mathbf{R}$	real part of a complex-valued function
s	distance measured along contour of missile cross section in crossflow plane; also local semispan of triangular wing, planar wing-body combination, and cruciform wing-body combination
s_m	maximum semispan of triangular wing
S_C	cross-sectional area of missile
S_H	exposed area of horizontal tail
S_R	reference area
t	time
T	$N + iM$; also kinetic energy of flow per unit length along X axis of missile
T'	value of T after rotation about X axis through $2\pi/n$ radians
u, v, w	linear velocity components of missile center of mass along X, Y, and Z axes
$u_0, v_0, w_0, p_0, q_0, r_0$	values of u, v, w, p, q, and r about which general Taylor series for X is expanded; Eq. (10-3)
$u_r, v_r, w_r, p_r, q_r, r_r$	values of u, v, w, p, q, and r after transformation of mirror symmetry in Maple-Synge analysis
v_1, v_2	velocity components of missile cross section along η and ζ axes, respectively, Fig. 10-8
V_0	free-stream velocity
W_1, W_2, W_3	$\phi_1 + i\psi_1$, $\phi_2 + i\psi_2$, $\phi_3 + i\psi_3$
x, y, z	set of axes illustrated in Figs. 10-10 and 10-11, x positive rearward along missile longitudinal axis
x_{ijklmn}	$\partial^g X/\partial u^i\, \partial v^j\, \partial w^k \partial p^l\, \partial q^m \partial r^n$; $g = i + j + k + l + m + n$
X, Y, Z	set of axes fixed in missile, Fig. 10-1; also set of force components acting on missile along X, Y, and Z. (Context reveals which definition applies.)
X_0, Y_0, Z_0	positions of X, Y, Z for zero pitch, yaw, and roll

$X_1,\ Y_1,\ Z_1$	positions of X, Y, Z after yaw about OZ_0
$X_2,\ Y_2,\ Z_2$	positions of X, Y, Z after yaw about OZ_0 and then pitch about OY_1
X_0	value of X force accompanying u_0, v_0, w_0, p_0, q_0, and r_0
X_b	value of X at missile base
X_c	X coordinate of fixed cross-sectional plane through which missile is passing, Fig. 10-7
X_n	value of X at missile apex
$X_r,\ Y_r,\ Z_r$	values of X, Y, Z after mirror reflection
X'	X coordinate after rotation about X axis by $2\pi/n$ radians; $X' = X$
$\bar{X}$	X coordinate of center of pressure
$\mathfrak{z}$	$y + iz$
$\mathfrak{z}_C$	value of $\mathfrak{z}$ at centroid of missile cross section
α	angle of attack
α_c	angle between missile longitudinal axis and free-stream velocity
$\Delta\alpha_H$	change in local angle of attack at horizontal tail
$\Delta\alpha_W$	change in local angle of attack at wing
β	angle of sideslip
$(d\epsilon/d\alpha)_H$	rate of change of downwash angle at tail with wing angle of attack
ζ	$\xi + i\eta$; complex variable of plane in which missile cross section is circle of radius c
η	vertical axis in ζ plane; also lateral coordinate in crossflow plane of Fig. 10-8
η_H	effectiveness of horizontal tail
$\theta,\ \psi,\ \varphi$	angles of pitch, yaw, and roll describing missile attitude, Fig. 10-2
λ	general reference length used in defining stability derivatives
ν	$w + iv$
ν'	$e^{-i\varphi}$
$\xi,\ \eta,\ \zeta$	axes parallel to X, Y, and Z and fixed to crossflow plane through which missile is passing, Fig. 10-7
ρ	free-stream density
$d\sigma/d\beta$	rate of change of sidewash angle with angle of sideslip
φ	angle of roll
ϕ	potential function
$\phi_1,\ \phi_2,\ \phi_3$	potential functions due to unit values of v_1, v_2, and p
ψ	angle of yaw; also stream function
$\psi_1,\ \psi_2,\ \psi_3$	stream functions corresponding to ϕ_1, ϕ_2, and ϕ_3
ψ_i	ψ_1, ψ_2, or ψ_3
ω	$r + iq$
ω'	$\omega e^{-i\varphi}$

Subscripts:

B	due to body or due to addition of body
E	empennage
H	horizontal tail panels
H'	horizontal tail including area blanketed by body
L	lower vertical tail
r	quantity after mirror reflection
U	upper vertical tail
W	wing

REFERENCES

1. Nonweiler, T.: Theoretical Stability Derivatives of a Highly Swept Delta Wing and Slender Body Combination, *Coll. Aeronaut. (Cranfield) Rept.* 50, 1951.

2. Sacks, Alvin H.: Aerodynamic Forces, Moments, and Stability Derivatives for Slender Bodies of General Cross-section, *NACA Tech. Notes* 3283, November, 1954.

3. Bryson, Arthur E., Jr.: Stability Derivatives for a Slender Missile with Application to a Wing-Body-Vertical Tail Configuration, *J. Aeronaut. Sci.*, vol. 20, no. 5, pp. 297–308, 1953.

4. Lamb, Horace: "Hydrodynamics," 6th ed., pp. 160–168, Cambridge University Press, New York.

5. Milne-Thompson, L. M.: "Theoretical Hydrodynamics," 2d ed., sec. 3.76, The Macmillan Company, New York, 1950.

6. Milne-Thompson, L. M.: "Theoretical Hydrodynamics," 2d ed., sec. 2.62, The Macmillan Company, New York, 1950.

7. Jones, Robert T.: Properties of Low-aspect-ratio Wings at Speeds Below and Above the Speeds of Sound, *NACA Tech. Repts.* 835, 1946.

8. Lamb, Horace: "Hydrodynamics," 6th ed., p. 88, Cambridge University Press, New York.

9. Ward, G. N.: Supersonic Flow past Slender Pointed Bodies, *Quart. J. Appl. Math.*, vol. II, part I, 1949.

10. Bryson, Arthur E., Jr.: Evaluation of the Inertia Coefficients of the Cross-section of a Slender Body, *J. Aeronaut. Sci.*, vol. 21, no. 6, pp. 424–427, 1954.

11. Summers, Richard G.: On Determining the Additional Apparent Mass of a Wing-Body-Vertical Tail Configuration, *J. Aeronaut. Sci.*, vol. 20, no. 12, pp. 856–857, 1953.

12. Bryson, Arthur E., Jr.: The Aerodynamic Forces on a Slender Low (or High) Wing, Circular Body, Vertical Tail Configuration, *J. Aeronaut. Sci.*, vol. 21, no. 8, pp. 574–575, 1954.

13. Bryson, Arthur E., Jr.: Comment on the Stability Derivatives of a Wing-Body-Vertical Tail Configuration, *J. Aeronaut. Sci.*, vol. 21, no. 1, p. 59, 1954.

14. Graham, E. W.: A Limiting Case for Missile Rolling Moments, *J. Aeronaut. Sci.*, vol. 18, no. 9, pp. 624–628, 1951.

15. Ribner, Herbert S., and Frank S. Malvestuto, Jr.: Stability Derivatives of Triangular Wings at Supersonic Speeds, *NACA Tech. Repts.* 908, 1948.

16. Brown, Clinton E., and Mac C. Adams: Damping in Pitch and Roll of Triangular Wings at Supersonic Speeds, *NACA Tech. Repts.* 892, 1948.

17. Miles, J. W.: The Application of Unsteady Flow Theory to the Calculation of Dynamic Stability Derivatives, *North Am. Aviation Rept.* AL-957.

18. Jones, A. L., J. R. Spreiter, and A. Alksne: The Rolling Moment Due to Sideslip of Triangular, Trapezoidal, and Related Plan Forms in Supersonic Flow, *NACA Tech. Notes* 1700, 1948.

19. Brown, Clinton E., and Harry S. Heinke, Jr.: Preliminary Wind-tunnel Tests of Triangular and Rectangular Wings in Steady Roll at Mach Numbers of 1.69 and 1.92, *NACA Tech. Notes* 3740, 1956.

20. Stone, David G.: A Collection of Data for Zero-lift Damping in Roll of Wing-Body Combinations as Determined with Rocket-powered Models Equipped with Roll-torque Nozzles, *NACA Tech. Notes* 3455, 1957.

21. Martin, John C., Margaret S. Diederich, and Percy J. Bobbitt: A Theoretical Investigation of the Aerodynamics of Wing-Tail Combinations Performing Time Dependent Motions at Supersonic Speeds, *NACA Tech. Notes* 3072, 1954.

22. Tobak, Murray: On the Use of the Indicial Function Concept in the Analysis of Unsteady Motions of Wing and Wing-Tail Combinations, *NACA Tech. Repts.* 1188, 1954.

23. Nielsen, Jack N., and George E. Kaattari: The Effects of Vortex and Shock-expansion Fields on Pitch and Yaw Instabilities of Supersonic Airplanes, *Inst. Aeronaut. Sci. Preprint* 743, June, 1957.

24. Durand, William Frederick: "Aerodynamic Theory," vol. V, pp. 44–45, Durand Reprinting Committee, California Institute of Technology, Pasadena.

25. Maple, C. G., and J. L. Synge: Aerodynamic Symmetry of Projectiles, *Quart. Appl. Math.*, vol. 6, no. 4, 1949.

26. Milne-Thompson, L. M.: "Theoretical Hydrodynamics," 2d ed., sec. 9.63, The Macmillan Company, New York, 1950.

27. Adams, Gaynor J., and Duane W. Dugan: Theoretical Damping in Roll and Rolling Moment Due to Differential Wing Incidence for Slender Cruciform Wings and Wing-Body Combinations, *NACA Tech. Repts.* 1088, 1952.

28. Love, Eugene S.: Investigation at Supersonic Speeds of 22 Triangular Wings Representing Two Airfoil Sections for Each of 11 Apex Angles, *NACA Tech. Repts.* 1238, 1955.

29. Lomax, Harvard, and Max A. Heaslet: Damping-in-roll Calculations for Slender Swept-back Wings and Slender Wing-Body Combinations, *NACA Tech. Notes* 1950, September, 1949.

APPENDIX 10A. MAPLE-SYNGE ANALYSIS FOR CRUCIFORM MISSILE

In this appendix we will deduce the effects of rotational and mirror symmetry on the stability derivatives of a cruciform missile. Consider now a missile possessing n-gonal symmetry, and let it undergo a rotation through an angle

$$\varphi = \frac{2\pi}{n} \tag{10A-1}$$

Under this rotation the physical forces and moments do not change; that is, they are invariants of the transformation. Let the original system of notation given by Eq. (10-10) apply to the missile before rotation, and let the same symbols with primes refer to the same physical quantities described now in terms of the new coordinates. Thus,

$$
\begin{aligned}
F' &= Fe^{-i\varphi} & T' &= Te^{-i\varphi} & X' &= X & L' &= L \\
\nu' &= \nu e^{-i\varphi} & u' &= u & \omega' &= \omega e^{-i\varphi} & p' &= p
\end{aligned}
\tag{10A-2}
$$

Now, if these primed quantities are substituted into Eq. (10-11), we must obtain an equality. Furthermore the functions f_{ijkl}, t_{ijkl}, x_{ijkl}, and l_{ijkl}

remain unchanged since only the independent variables v, w, q, and r were varied in the transformation. Carrying out this substitution, we obtain

$$
\begin{aligned}
e^{i\varphi} \sum_{ijkl} f_{ijkl}(u,p)\,v^i \bar{\nu}^j \omega^k \bar{\omega}^l e^{-i\varphi(i-j+k-l)} &= \sum_{ijkl} f_{ijkl}(u,p)\,v^i \bar{\nu}^j \omega^k \bar{\omega}^l \\
e^{i\varphi} \sum_{ijkl} t_{ijkl}(u,p)\,v^i \bar{\nu}^j \omega^k \bar{\omega}^l e^{-i\varphi(i-j+k-l)} &= \sum_{ijkl} t_{ijkl}(u,p)\,v^i \bar{\nu}^j \omega^k \bar{\omega}^l \\
\sum_{ijkl} x_{ijkl}(u,p)\,v^i \bar{\nu}^j \omega^k \bar{\omega}^l e^{-i\varphi(i-j+k-l)} &= \sum_{ijkl} x_{ijkl}(u,p)\,v^i \bar{\nu}^j \omega^k \bar{\omega}^l \\
\sum_{ijkl} l_{ijkl}(u,p)\,v^i \bar{\nu}^j \omega^k \bar{\omega}^l e^{-i\varphi(i-j+k-l)} &= \sum_{ijkl} l_{ijkl}(u,p)\,v^i \bar{\nu}^j \omega^k \bar{\omega}^l
\end{aligned}
\tag{10A-3}
$$

To preserve the equality for arbitrary values of ν, $\bar{\nu}$, ω, and $\bar{\omega}$ we must have, for F and T,

$$
\exp\left[i\left(\frac{2\pi}{n}\right)(1 - i + j - k + l) \right] = 1
$$

so that

$$
i - j + k - l - 1 = mn \qquad m = 0, \pm1, \pm2, \ldots \tag{10A-4}
$$

In a like manner we obtain, for X and L,

$$
i - j + k - l = mn \qquad m = 0, \pm1, \pm2, \ldots \tag{10A-5}
$$

Equations (10A-4) and (10A-5) must be satisfied by missiles possessing n-gonal symmetry.

Let us now consider a systematic scheme for investigating the terms of Eq. (10-11) term by term to see if their retentions are compatible with Eqs. (10A-4) and (10A-5). We are interested in the degree D of the terms and the symmetry number n of the missile. Let us introduce numbers P and Q

$$
P = i + k \qquad Q = j + l \tag{10A-6}
$$

which are determined by D and n as follows,

$$
P + Q = D \qquad P - Q = mn + 1 \tag{10A-7}
$$

for the f_{ijkl} and t_{ijkl} terms. Likewise, the coefficient x_{ijkl} and l_{ijkl} are governed by

$$
P + Q = D \qquad P - Q = mn \tag{10A-8}
$$

Equations (10A-7) and (10A-8) can be considered *selection rules* for picking those terms of degree D in Eq. (10-11), the retention of which is compatible with n-gonal symmetry of the missile.

To illustrate the use of the selection rules let us apply them to ascertain admissible terms in the expansions for the forces and moments of a cruciform missile, $n = 4$. It is convenient to construct a *PQ diagram*, in which P is abscissa and Q is ordinate. Only positive values of P and Q are admissible because of analyticity. The PQ diagram for $n = 4$ is shown in Fig. 10-28, where the sets of straight lines corresponding to Eq. (10A-7) are shown. Where the curves intersect at integral values of the coordinates, allowable values of P and Q are found. For instance, no intersection is found for terms of degree $D = 0$, and only the intersection $P = 1$, $Q = 0$ is found for terms of first degree. The terms of first degree are then found from the following sets of values of i, j, k, and l, yielding $P = 1$ and $Q = 0$:

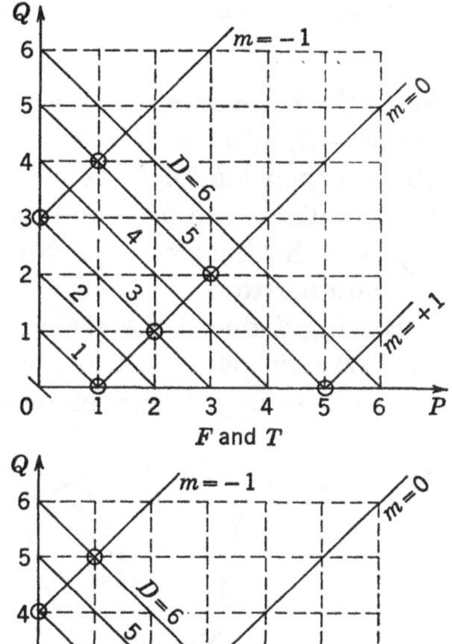

FIG. 10-28. *PQ* diagrams for cruciform missile.

$P = 1$		$Q = 0$	
i	k	j	l
1	0	0	0
0	1	0	0

The terms in question are $f_{1000}\nu$ and $f_{0010}\omega$, as well as $t_{1000}\nu$ and $t_{0010}\omega$. No second-degree terms appear, but terms of third degree arise in the set of values $P = 0$, $Q = 3$, and $P = 2$, $Q = 1$. Again the sets of values of i, j, k, and l are

$P = 0$		$Q = 3$		$P = 2$		$Q = 1$	
i	k	j	l	i	k	j	l
0	0	0	3	2	0	0	1
0	0	1	2	2	0	1	0
0	0	2	1	1	1	0	1
0	0	3	0	1	1	1	0
				0	2	0	1
				0	2	1	0

The general expansion for the force F can now be written up to but not including terms of fifth degree

$$F = f_{1000}\nu + f_{0010}\omega + f_{0003}\bar{\omega}^3 + f_{0102}\bar{\nu}\bar{\omega}^2 + f_{0201}\bar{\nu}^2\bar{\omega} + f_{0300}\bar{\nu}^3$$
$$+ f_{2001}\nu^2\bar{\omega} + f_{2100}\nu^2\bar{\nu} + f_{1011}\nu\omega\bar{\omega} + f_{1110}\nu\bar{\nu}\omega$$
$$+ f_{0021}\omega^2\bar{\omega} + f_{0120}\bar{\nu}\omega^2 + \text{terms of fifth degree} \quad (10A-9)$$

A similar expansion exists for T. The PQ diagram for the force X and the rolling moment L is included in Fig. 10-28. The series obtained from this diagram has terms as follows:

$$X = x_{0000} + x_{0011}\omega\bar{\omega} + x_{1001}\nu\bar{\omega} + x_{0110}\bar{\nu}\omega + x_{1100}\nu\bar{\nu}$$
$$+ \text{ terms of fourth degree} \quad (10A\text{-}10)$$

Since there are potentially 1 term for $D = 0$, 4 terms for $D = 1$, 10 terms for $D = 2$, 20 terms for $D = 3$, etc., we should have many terms in the above expansion for F, but rotational symmetry has reduced the number to 12. There are potentially 35 terms in the expression for X including forms of third degree, and through rotational symmetry we have reduced the number to 5.

Having deduced the general consequences of rotational symmetry, let us now turn our attention to mirror symmetry. The positive conventions of the axes, forces, moments, etc., were given in Fig. 10-1, and these

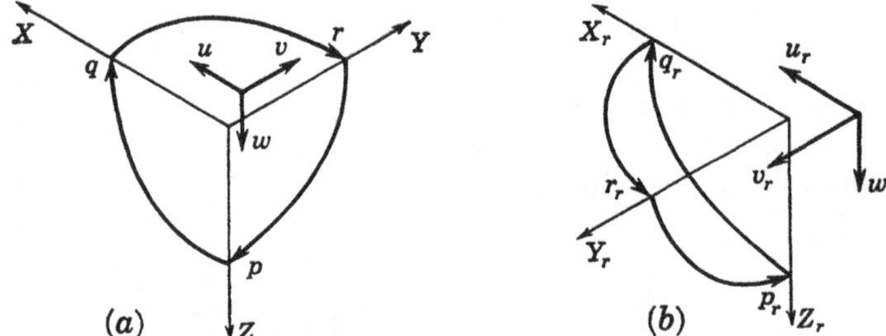

Fig. 10-29. Positive conventions involved in mirror symmetry (a) before and (b) after reflection.

conventions are again repeated in part in Fig. 10-29. We will take the plane of mirror symmetry to be the XZ plane. Let the Y axis be reflected in the plane to obtain the new axis Y_r. The reflected axis system is thus

$$X = X_r \qquad Y = -Y_r \qquad Z = Z_r \qquad (10A\text{-}11)$$

It is important now to define positive quantities in the new system in the same general manner as in the old system. Thus the linear velocity components and the forces are positive along the positive X_r, Y_r, and Z_r axes. Positive directions of the angular velocities and moments correspond to *cyclic rotation of the axes*. If $X_r \to Y_r$, then r_r is positive; and, if $Z_r \to X_r$, then q_r is positive, and similarly for the moments. We thus obtain

$$\begin{array}{llll} u_r = u & p_r = -p & X_r = X & L_r = -L \\ v_r = -v & q_r = q & Y_r = -Y & M_r = M \quad (10A\text{-}12) \\ w_r = w & r_r = -r & Z_r = Z & N_r = -N \end{array}$$

It is to be noted that the positive vectors for the angular velocities and moments do not necessarily correspond to positive axis directions because X_r, Y_r, Z_r is not a right-handed system of axes as X, Y, Z is.

The set of variables given by Eq. (10-10) now becomes

$$
\begin{aligned}
F_r &= Z_r + iY_r = Z - iY = \bar{F} \\
T_r &= N_r + iM_r = -N + iM = -\bar{T} \\
\nu_r &= w_r + iv_r = w - iv = \bar{\nu} \\
\bar{\nu}_r &= w_r - iv_r = w + iv = \nu \\
\omega_r &= r_r + iq_r = -r + iq = -\bar{\omega} \\
\bar{\omega}_r &= r_r - iq_r = -r - iq = -\omega
\end{aligned}
\tag{10A-13}
$$

We now assume that the forces and moments can be expressed in the new coordinates in precisely the same form as in Eq. (10A-3) and that the functions f_{ijkl}, t_{ijkl}, x_{ijkl}, and l_{ijkl} are the identical functions in either system of coordinates. Thus

$$
\begin{aligned}
F_r &= \sum_{ijkl} f_{ijkl}(u_r, p_r)\, \nu_r{}^i \bar{\nu}_r{}^j \omega_r{}^k \bar{\omega}_r{}^l \\
T_r &= \sum_{ijkl} t_{ijkl}(u_r, p_r)\, \nu_r{}^i \bar{\nu}_r{}^j \omega_r{}^k \bar{\omega}_r{}^l
\end{aligned}
\tag{10A-14}
$$

We already know that $F_r = \bar{F}$ and $T_r = -\bar{T}$, so that

$$
\begin{aligned}
\sum_{ijkl} f_{ijkl}(u_r, p_r)\, \nu_r{}^i \bar{\nu}_r{}^j \omega_r{}^k \bar{\omega}_r{}^l &= \sum_{ijkl} \bar{f}_{ijkl}(u, p)\, \bar{\nu}^i \nu^j \bar{\omega}^k \omega^l \\
-\sum_{ijkl} t_{ijkl}(u_r, p_r)\, \nu_r{}^i \bar{\nu}_r{}^j \omega_r{}^k \bar{\omega}_r{}^l &= \sum_{ijkl} \bar{t}_{ijkl}(u, p)\, \bar{\nu}^i \nu^j \bar{\omega}^k \omega^l
\end{aligned}
\tag{10A-15}
$$

or from Eq. (10A-13)

$$
\begin{aligned}
\sum_{ijkl} f_{ijkl}(u, -p)\, \bar{\nu}^i \nu^j (-\bar{\omega})^k (-\omega)^l &= \sum_{ijkl} \bar{f}_{ijkl}(u, p)\, \bar{\nu}^i \nu^j \bar{\omega}^k \omega^l \\
-\sum_{ijkl} t_{ijkl}(u, -p)\, \bar{\nu}^i \nu^j (-\bar{\omega})^k (-\omega)^l &= \sum_{ijkl} \bar{t}_{ijkl}(u, p)\, \bar{\nu}^i \nu^j \bar{\omega}^k \omega^l
\end{aligned}
\tag{10A-16}
$$

For the equalities given by Eq. (10A-16) to be true we must have

$$
\begin{aligned}
f_{ijkl}(u, -p) &= (-1)^{k+l} \bar{f}_{ijkl}(u, p) \\
t_{ijkl}(u, -p) &= -(-1)^{k+l} \bar{t}_{ijkl}(u, p)
\end{aligned}
\tag{10A-17}
$$

Now we must further break f_{ijkl} and t_{ijkl} down into even and odd functions of the rolling velocity p, and into real and imaginary parts. If the superscript (R) stands for real and (I) for imaginary, we write

$$
\begin{aligned}
f_{ijkl} &= f_{ijkl}^{(R)} + i f_{ijkl}^{(I)} \\
t_{ijkl} &= t_{ijkl}^{(R)} + i t_{ijkl}^{(I)}
\end{aligned}
\tag{10A-18}
$$

Substitution into Eq. (10A-17) gives

$$
f_{ijkl}^{(R)}(u, -p) + i f_{ijkl}^{(I)}(u, -p) = (-1)^{k+l}[f_{ijkl}^{(R)}(u, p) - i f_{ijkl}^{(I)}(u, p)]
$$

It follows that $f_{ijkl}^{(R)}$ and $f_{ijkl}^{(I)}$ must be even or odd functions in p according as $k + l$ is odd or even.

$$
\begin{array}{ll}
(k + l)_{\text{even}} & (k + l)_{\text{odd}} \\
f_{ijkl}^{(R)} \text{ even in } p & f_{ijkl}^{(R)} \text{ odd in } p \\
f_{ijkl}^{(I)} \text{ odd in } p & f_{ijkl}^{(I)} \text{ even in } p
\end{array}
$$

Here let the superscript (O) stand for an odd function of p and the superscript (E) for an even function; then

$$
\begin{aligned}
f_{ijkl} &= f_{ijkl}^{(E)} + i f_{ijkl}^{(O)} & k + l \text{ even} \\
f_{ijkl} &= f_{ijkl}^{(O)} + i f_{ijkl}^{(E)} & k + l \text{ odd}
\end{aligned}
\tag{10A-19}
$$

If we complete similar analyses for T_r, X_r, and L_r, we obtain the following set of relationships for the odd or even nature of the real and imaginary parts of the functions:

$k + l$ *even:*

$$
\begin{aligned}
f_{ijkl} &= f_{ijkl}^{(E)} + i f_{ijkl}^{(O)} \\
x_{ijkl} &= x_{ijkl}^{(E)} + i x_{ijkl}^{(O)} \\
t_{ijkl} &= t_{ijkl}^{(O)} + i t_{ijkl}^{(E)} \\
l_{ijkl} &= l_{ijkl}^{(O)} + i l_{ijkl}^{(E)}
\end{aligned}
\tag{10A-20}
$$

$k + l$ *odd:*

$$
\begin{aligned}
f_{ijkl} &= f_{ijkl}^{(O)} + i f_{ijkl}^{(E)} \\
x_{ijkl} &= x_{ijkl}^{(O)} + i x_{ijkl}^{(E)} \\
t_{ijkl} &= t_{ijkl}^{(E)} + i t_{ijkl}^{(O)} \\
l_{ijkl} &= l_{ijkl}^{(E)} + i l_{ijkl}^{(O)}
\end{aligned}
\tag{10A-21}
$$

We are now in a position to determine the expansions for the force and moment coefficients in real rather than in complex form. The analysis will be carried out for the four cases listed at the beginning of Sec. 10-3 up to and including terms of second degree. With reference to Eq. (10A-10) for X, we now have

$$
\begin{aligned}
X = x_{0000} + x_{0011}(r^2 + q^2) + x_{1001}(w + iv)(r - iq) \\
+ x_{0110}(w - iv)(r + iq) + x_{1100}(w^2 + v^2)
\end{aligned}
$$
$$
+ \text{ terms of fourth degree} \tag{10A-22}
$$

Since X must be real, the functions x_{ijkl} have the properties

$$
\begin{aligned}
x_{0000} &= x_{0000}^{(E)} \\
x_{0011} &= x_{0011}^{(E)} \\
x_{1001} &= x_{1001}^{(O)} + i x_{1001}^{(E)} \\
x_{0110} &= x_{1001}^{(O)} - i x_{1001}^{(E)} \\
x_{1100} &= x_{1100}^{(E)}
\end{aligned}
\tag{10A-23}
$$

From these properties the real form of X becomes

$$
X = x_{0000}^{(E)} + x_{0011}^{(E)}(q^2 + r^2) + 2x_{1001}^{(O)}(wr + qv)
$$
$$
+ 2x_{1001}^{(E)}(wq - vr) + x_{1100}^{(E)}(w^2 + v^2)
$$
$$
+ \text{ terms of fourth degree} \tag{10A-24}
$$

We can write an expression for the rolling moment L from Eq. (10A-24) by observing that the expansion of L is of the same form as that for X, see Eq. (10A-10), but that the superscripts (E) and (O) must be reversed.

$$L = l_{0000}^{(O)} + l_{0011}^{(O)}(q^2 + r^2) + 2l_{1001}^{(E)}(wr + qv)$$
$$+ 2l_{1001}^{(O)}(qw - vr) + l_{1100}^{(O)}(w^2 + v^2)$$
$$+ \text{terms of fourth degree} \quad (10A\text{-}25)$$

The equations for X and L as given above are the complete Taylor expansions for these forces for a cruciform configuration up to and including terms of second degree (there are no terms of third degree) about the point $w = v = q = r = 0$. The coefficients depend on the linear velocity along the missile longitudinal axis and the roll rate or spin about it.

The expansions for F and T given by Eq. (10A-9) are separable into real and imaginary parts to yield Y, Z, M, and N.

$$F = Z + iY = (f_{0010}^{(O)} + if_{0010}^{(E)})(r + iq)$$
$$+ (f_{1000}^{(E)} + if_{1000}^{(O)})(w + iv)$$
$$+ \text{terms of third degree} \quad (10A\text{-}26)$$

so that

$$Z = f_{0010}^{(O)}r - f_{0010}^{(E)}q + f_{1000}^{(E)}w - f_{1000}^{(O)}v$$
$$Y = f_{0010}^{(E)}r + f_{0010}^{(O)}q + f_{1000}^{(O)}w + f_{1000}^{(E)}v \quad (10A\text{-}27)$$

Similarly for M and N we obtain

$$M = t_{0010}^{(O)}r + t_{0010}^{(E)}q + t_{1000}^{(E)}w + t_{1000}^{(O)}v$$
$$N = t_{0010}^{(E)}r - t_{0010}^{(O)}q + t_{1000}^{(O)}w - t_{1000}^{(E)}v \quad (10A\text{-}28)$$

APPENDIX 10B. MAPLE-SYNGE ANALYSIS FOR TRIFORM MISSILES AND OTHER MISSILES

Let us now consider the triform missile and construct first the *PQ diagram* for complex force F and complex moment T in accordance with the selection rule, Eq. (10A-4). The *PQ* diagram is shown in Fig. 10-30. There is no term of degree zero. The terms of degree unity correspond to the values $P = 1$, $Q = 0$.

$P = 1$		$Q = 0$	
i	k	j	l
0	1	0	0
1	0	0	0

The terms of degree two correspond to $P = 0$, $Q = 2$.

$P = 0$		$Q = 2$	
i	k	j	l
0	0	0	2
0	0	2	0
0	0	1	1

Thus the expansions for F and T for terms up to and including those of second degree are

$$F = f_{0010}\omega + f_{1000}\nu + f_{0002}\bar{\omega}^2 + f_{0200}\bar{\nu}^2 + f_{0101}\bar{\nu}\bar{\omega}$$
$$+ \text{ terms of third degree} \quad (10B\text{-}1)$$

$$T = t_{0010}\omega + t_{1000}\nu + t_{0002}\bar{\omega}^2 + t_{0200}\bar{\nu}^2 + t_{0101}\bar{\nu}\bar{\omega}$$
$$+ \text{ terms of third degree} \quad (10B\text{-}2)$$

In accordance with Eqs. (10A-20) and (10A-21), the complex terms f_{ijkl} and t_{ijkl} can be decomposed into real and imaginary parts that are either

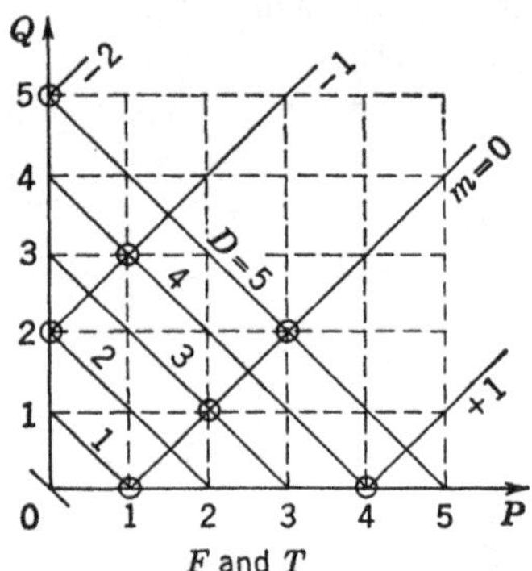

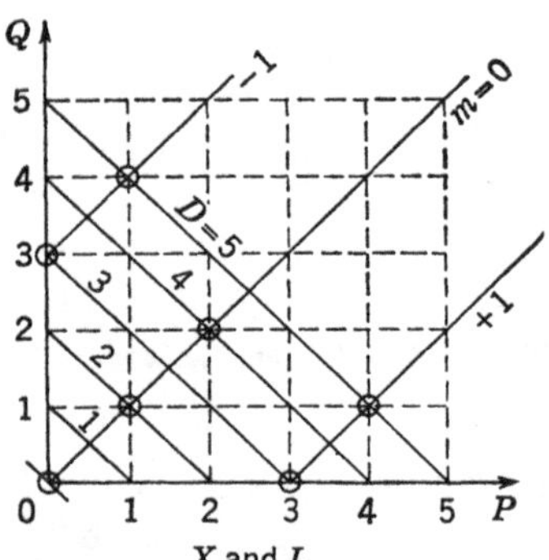

Fig. 10-30. PQ diagram for triform missile.

odd or even functions of p. When the analysis is carried out and F and T separated into their component parts, we obtain

$$Y = f^{(O)}_{1000}w + f^{(E)}_{1000}v + f^{(E)}_{0010}r + f^{(O)}_{0010}q + f^{(O)}_{0002}(r^2 - q^2)$$
$$- 2f^{(E)}_{0002}rq + f^{(O)}_{0200}(w^2 - v^2) - 2f^{(E)}_{0200}wv$$
$$+ f^{(E)}_{0101}(wr - vq) - f^{(O)}_{0101}(wq + rv)$$
$$+ \text{ terms of third degree} \quad (10B\text{-}3)$$

$$Z = f^{(E)}_{1000}w - f^{(O)}_{1000}v + f^{(O)}_{0010}r - f^{(E)}_{0010}q + f^{(E)}_{0002}(r^2 - q^2)$$
$$+ 2f^{(O)}_{0002}rq + f^{(E)}_{0200}(w^2 - v^2) + 2f^{(O)}_{0200}wv$$
$$+ f^{(O)}_{0101}(wr - vq) + f^{(E)}_{0101}(wq + vr)$$
$$+ \text{ terms of third degree} \quad (10B\text{-}4)$$

$$M = t^{(E)}_{1000}w + t^{(O)}_{1000}v + t^{(O)}_{0010}r + t^{(E)}_{0010}q + t^{(E)}_{0200}(w^2 - v^2)$$
$$- 2t^{(O)}_{0200}wv + t^{(E)}_{0002}(r^2 - q^2) - 2t^{(O)}_{0002}rq + t^{(O)}_{0101}(wr - vq)$$
$$- t^{(E)}_{0101}(wq + vr) + \text{terms of third degree} \qquad (10\text{B-}5)$$
$$N = t^{(O)}_{1000}w - t^{(E)}_{1000}v + t^{(E)}_{0010}r - t^{(O)}_{0010}q + t^{(O)}_{0200}(w^2 - v^2)$$
$$+ 2t^{(E)}_{0200}wv + t^{(O)}_{0002}(r^2 - q^2) + 2t^{(E)}_{0002}rq + t^{(E)}_{0101}(wr - vq)$$
$$+ t^{(O)}_{0101}(wq + vr) + \text{terms of third degree} \qquad (10\text{B-}6)$$

These equations for a triform missile are to be compared with Eqs. (10-14) and (10-15) for a cruciform missile. The significant difference is that the cruciform missile expansions contain no terms of second degree whereas those for the triform contain many such terms.

Turning now to the expansions for the thrust force and rolling moment X and L of a triform missile, we construct the PQ diagram in accordance with the selection rule, Eq. (10A-5). The PQ diagram is shown in Fig. 10-30. A single term of degree zero appears

$P = 0$		$Q = 0$	
i	k	j	l
0	0	0	0

The terms of degree two correspond to $P = 1$, $Q = 1$.

$P = 1$		$Q = 1$	
i	k	j	l
0	1	0	1
0	1	1	0
1	0	0	1
1	0	1	0

The general expansions for X and L are thus

$$X = x_{0000} + x_{0011}\omega\bar{\omega} + x_{1001}\nu\bar{\omega} + x_{0110}\bar{\nu}\omega + x_{1100}\nu\bar{\nu}$$
$$+ \text{terms of third degree} \qquad (10\text{B-}7)$$
$$L = l_{0000} + l_{0011}\omega\bar{\omega} + l_{1001}\nu\bar{\omega} + l_{0110}\bar{\nu}\omega + l_{1100}\nu\bar{\nu}$$
$$+ \text{terms of third degree} \qquad (10\text{B-}8)$$

Because X and L are real, the following equalities must hold:

$$x^{(E)}_{1001} = -x^{(E)}_{0110} \qquad x^{(O)}_{1100} = 0 \qquad x^{(O)}_{0000} = 0$$
$$x^{(O)}_{1001} = x^{(O)}_{0110} \qquad x^{(O)}_{0011} = 0 \qquad l^{(E)}_{0000} = 0$$
$$l^{(O)}_{1001} = l^{(O)}_{0110} \qquad l^{(E)}_{1100} = 0$$
$$l^{(E)}_{1001} = l^{(E)}_{0110} \qquad l^{(E)}_{0011} = 0$$

The final real expansions for X and L are

$$X = x^{(E)}_{0000} + 2x^{(O)}_{1001}(wr + qv) + 2x^{(E)}_{1001}(wq - vr)$$
$$+ x^{(E)}_{1100}(w^2 + v^2) + x^{(E)}_{0011}(r^2 + q^2) + \text{terms of third degree} \qquad (10\text{B-}9)$$
$$L = l^{(O)}_{0000} + 2l^{(E)}_{1001}(wr + qv) + 2l^{(O)}_{1001}(wq - vr)$$
$$+ l^{(O)}_{1100}(w^2 + v^2) + l^{(O)}_{0011}(r^2 + q^2) + \text{terms of third degree} \qquad (10\text{B-}10)$$

These equations for X and L are to be compared with Eqs. (10-12) and (10-13) for a cruciform missile. It is seen that the expansions are identical through terms of second degree.

Let us round out the analysis by consideration of missiles of 2-gonal and mirror symmetries. The PQ diagrams based on the selection rules, Eqs. (10A-4) and (10A-5), are given in Fig. 10-31. Inspection of this figure reveals that the expansions for F and T contain terms of odd degree

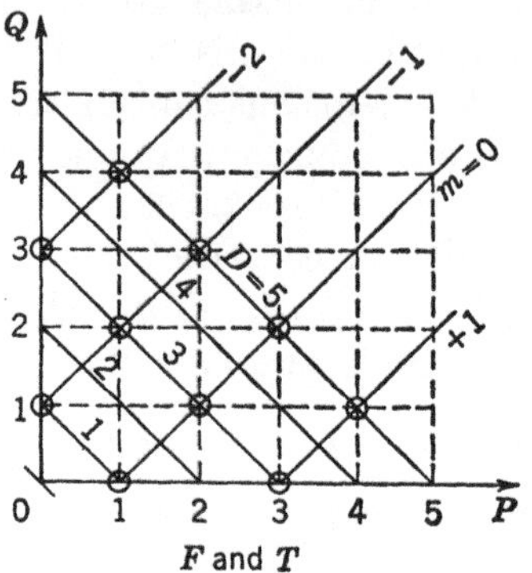

F and T

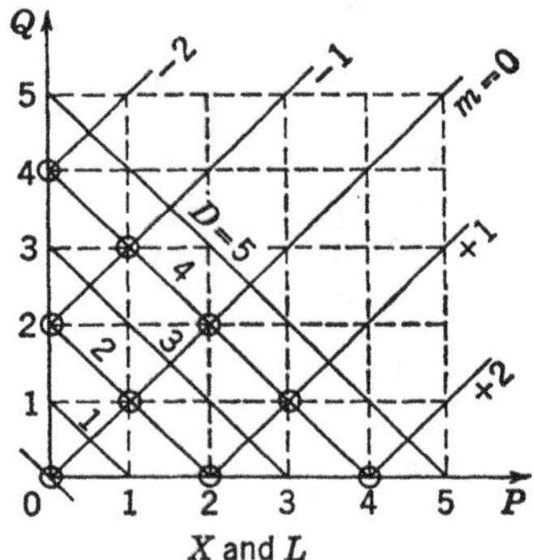

X and L

Fig. 10-31. PQ diagram for missile with 2-gonal and mirror symmetries.

only (as for a cruciform missile), in contrast to the expansions for a triform missile which contain terms of all but degree zero. For terms of first degree we have the conditions $P = 1$, $Q = 0$, and $P = 0$, $Q = 1$.

$P = 1$		$Q = 0$		$P = 0$		$Q = 1$	
i	k	j	l	i	k	j	l
0	0	1	0	0	0	0	1
1	0	0	0	0	1	0	0

The expansions for F and T in complex form are

$$F = f_{1000}\nu + f_{0100}\bar{\nu} + f_{0010}\omega + f_{0001}\bar{\omega} + \text{terms of third degree} \quad \text{(10B-11)}$$
$$T = t_{1000}\nu + t_{0100}\bar{\nu} + t_{0010}\omega + t_{0001}\bar{\omega} + \text{terms of third degree} \quad \text{(10B-12)}$$

When the complex-valued functions f_{ijkl} and t_{ijkl} are split into real and imaginary parts which are either odd or even functions of p, we obtain, for Y, Z, M, and N,

$$Y = (f^{(O)}_{0100} + f^{(O)}_{1000})w + (f^{(E)}_{1000} - f^{(E)}_{0100})v + (f^{(O)}_{0010} - f^{(O)}_{0001})q$$
$$+ (f^{(E)}_{0001} + f^{(E)}_{0010})r + \text{terms of third degree} \quad (10B\text{-}13)$$

$$Z = (f^{(E)}_{0100} + f^{(E)}_{1000})w + (f^{(O)}_{0100} - f^{(O)}_{1000})v + (f^{(E)}_{0001} - f^{(E)}_{0010})q$$
$$+ (f^{(O)}_{0001} + f^{(O)}_{0010})r + \text{terms of third degree} \quad (10B\text{-}14)$$

$$M = (t^{(E)}_{0100} + t^{(E)}_{1000})w + (t^{(O)}_{1000} - t^{(O)}_{0100})v + (t^{(E)}_{0010} - t^{(E)}_{0001})q$$
$$+ (t^{(O)}_{0001} + t^{(O)}_{0010})r + \text{terms of third degree} \quad (10B\text{-}15)$$

$$N = (t^{(O)}_{0100} + t^{(O)}_{1000})w + (t^{(E)}_{0100} - t^{(E)}_{1000})v + (t^{(O)}_{0001} - t^{(O)}_{0010})q$$
$$+ (t^{(E)}_{0001} + t^{(E)}_{0010})r + \text{terms of third degree} \quad (10B\text{-}16)$$

Consider now the thrust force X and the rolling moment L, all possible terms of zero and second degree appearing in the PQ diagram of Fig. 10-31. For the terms of second degree, we have

$P = 2$	$Q = 0$			$P = 1$	$Q = 1$			$P = 0$	$Q = 2$		
i	k	j	l	i	k	j	l	i	k	j	l
2	0	0	0	1	1	0	0	0	2	0	0
1	0	1	0	0	1	1	0	0	1	0	1
0	0	2	0	1	0	0	1	0	0	0	2
				0	0	1	1				

The complex expansions for X and L are thus

$$X = x_{0000} + x_{0011}\omega\bar{\omega} + x_{1001}\nu\bar{\omega} + x_{0110}\bar{\nu}\omega + x_{1100}\nu\bar{\nu} + x_{0020}\omega^2$$
$$+ x_{2000}\nu^2 + x_{1010}\nu\omega + x_{0002}\bar{\omega}^2 + x_{0200}\bar{\nu}^2 + x_{0101}\bar{\nu}\bar{\omega}$$
$$+ \text{terms of fourth degree} \quad (10B\text{-}17)$$

$$L = l_{0000} + l_{0011}\omega\bar{\omega} + l_{1001}\nu\bar{\omega} + l_{0110}\bar{\nu}\omega + l_{1100}\nu\bar{\nu} + l_{0020}\omega^2$$
$$+ l_{2000}\nu^2 + l_{1010}\nu\omega + l_{0002}\bar{\omega}^2 + l_{0200}\bar{\nu}^2 + l_{0101}\bar{\nu}\bar{\omega}$$
$$+ \text{terms of fourth degree} \quad (10B\text{-}18)$$

When the relationships among the functions x_{ijkl} and l_{ijkl} are taken into account, the real expansions for X and L are

$$X = x^{(E)}_{0000} + (x^{(E)}_{1100} + x^{(E)}_{2000} + x^{(E)}_{0200})w^2$$
$$+ (x^{(E)}_{1100} - x^{(E)}_{2000} - x^{(E)}_{0200})v^2 + (x^{(E)}_{0011} - x^{(E)}_{0020} - x^{(E)}_{0002})q^2$$
$$+ (x^{(E)}_{0011} + x^{(E)}_{0020} + x^{(E)}_{0002})r^2 + 2(x^{(O)}_{0002} - x^{(O)}_{0020})rq$$
$$+ 2(x^{(O)}_{0200} - x^{(O)}_{2000})vw + 2x^{(E)}_{0101}(wq + rv) - 2x^{(O)}_{1010}(wr - vq)$$
$$+ 2x^{(E)}_{1001}(wq - vr) + 2x^{(O)}_{1001}(wr + vq)$$
$$+ \text{terms of fourth degree} \quad (10B\text{-}19)$$

$$L = l^{(O)}_{0000} + (l^{(O)}_{1100} + l^{(O)}_{0020} + l^{(O)}_{0002})w^2 + (l^{(O)}_{1100} - l^{(O)}_{2000} - l^{(O)}_{0200})v^2$$
$$+ (l^{(O)}_{0011} - l^{(O)}_{0020} - l^{(O)}_{0002})q^2 + (l^{(O)}_{0011} + l^{(O)}_{0020} + l^{(O)}_{0002})r^2$$
$$+ 2(l^{(E)}_{0002} - l^{(E)}_{0020})rq + 2(l^{(E)}_{0200} - l^{(E)}_{2000})wv + 2l^{(O)}_{0101}(wq + vr)$$
$$+ 2l^{(O)}_{1001}(wq - vr) + 2l^{(E)}_{1001}(wr + vq) - 2l_{1010}(wr + vq)$$
$$+ \text{terms of fourth degree} \quad (10B\text{-}20)$$

The derivatives with respect to u, v, w, p, q, and r have been determined, and those which are zero are shown in Fig. 10-5 for four cases.

NAME INDEX

Abarbanel, S. S., 348
Ackeret, J., 15, 25, 84
Adams, G. J., 216, 221, 227, 257, 393, 421
Adams, M. C., 396, 420
Alden, H. L., 205
Alksne, A. Y., 128, 143, 177, 180, 401, 420
Allen, H. J., 89, 90, 110, 259, 260
Ames Research Staff, 16, 33, 275, 345

Bleviss, Z. O., 227, 258
Bobbitt, P. J., 412, 421
Bogdonoff, S. M., 319, 346, 348
Brown, C. E., 344, 396, 420, 421
Bryson, A. E., Jr., 111, 205, 349, 363, 364, 372, 379, 420
Busemann, A., 15, 16, 23, 33, 243, 247, 249, 254
Byrd, P., 31, 33

Chapman, D. R., 245, 258, 312–314, 317, 319, 322, 336, 341, 346–348
Churchill, R., 42, 47, 61
Cohen, D., 16, 33, 237, 258, 297, 309, 346
Cooper, M., 290, 345
Czarnecki, K. R., 312, 320, 346, 348

Dennis, D. H., 285–287, 345
DeYoung, J., 146, 180
Diaconis, N. S., 348
Dickey, R. R., 322, 346
Diederich, M. S., 412, 421
Dugan, D. W., 216, 221, 227, 242, 257–259, 393, 421
Durand, W. F., 117, 142, 409, 421

Eckert, E. R. G., 330, 331, 334, 341, 347
Eggers, A. J., Jr., 278, 279, 285–287, 345
Ehret, D., 278–280, 345
Erdélyi, A., 58, 61

Ferri, A., 311, 345, 346
Flax, A. H., 114, 142
Föppl, L., 91, 92, 94, 108–110, 167
Fraenkel, L. E., 42, 61, 74, 110
Frick, C. W., Jr., 234, 258
Friedman, M. D., 31, 33
Frost, R. C., 238, 258

Gadd, G. E., 258, 348
Gallagher, J. J., 246, 259
Garrick, I. E., 9, 33
Germain, P., 305, 309, 347
Gionfriddo, M. D., 338, 347
Gloria, H. R., 317, 347
Goin, K. L., 237, 238, 243, 244, 258
Goldstein, S., 90, 110
Graham, E. W., 296, 347, 420
Graham, M. E., 140, 143, 180, 234, 237, 258
Grant, F. C., 290, 345
Greber, I., 348

Haack, W., 270, 280, 281, 283–285, 299, 345
Haefeli, R. C., 154, 180
Hakkinen, R. J., 348
Hamaker, F. M., 319, 347
Hantsche, W., 340, 347
Hayes, W. D., 296, 299, 346
Heaslet, M. A., 38, 61, 198, 205, 222, 258, 267, 292, 300–302, 345, 346, 393, 421
Heinke, H. S., Jr., 396, 421
Hikido, K., 216, 242, 258
Hilsenrath, J., 331, 347
Holder, D. W., 258, 348

Jack, J. R., 348
Jedlicka, J. R., 348
Johnson, H. A., 330, 347
Jones, A. L., 128, 143, 401, 420

Jones, R. A., 348
Jones, R. T., 16, 17, 24, 33, 148, 180, 235,
    258, 296, 297, 300, 302–306, 308, 337,
    346, 420
Jorgensen, L. H., 85–87, 89, 110, 287, 345

Kaattari, G. E., 89, 110, 131, 136, 142,
    193, 196, 201, 205, 291, 292, 295, 345,
    405, 406, 421
Kaden, H., 149, 151, 176, 179
Kahane, A., 74, 110
Kainer, J. H., 258
Katzen, E. D., 136, 142, 291, 292, 295,
    345
Katzoff, S., 180
Kester, R. H., 317, 322, 336, 347
Keune, F., 296, 347
Kopal, Z., 275, 344
Korst, H. H., 347
Kuehn, D. M., 245, 258, 312, 346, 348
Kurzweg, H. H., 319, 323, 346

Lagerstrom, P. A., 17, 33, 132, 140, 142,
    143, 180, 234, 235, 237, 258
Lamb, H., 364, 367, 375, 420
Larson, H. K., 245, 258, 312, 346, 348
Laurence, T., 290, 345
Lawrence, H. R., 114, 142
Legendre, R., 296, 347
Lennertz, J., 117
Liepmann, H. W., 8, 33
Lin, C. C., 111
Lindsey, W. F., 89, 110
Locktenbert, B. H., 348
Lomax, H., 38, 61, 198, 205, 258, 267,
    292, 300–302, 345, 346, 393, 421
Love, E. S., 397, 421

Maccoll, J. W., 73, 110, 275, 344
Malvestuto, F. S., Jr., 377, 394–396, 402,
    420
Mangler, W., 340, 347
Maple, C. G., 349, 358, 359–363, 378,
    389, 418, 421, 427
Marte, J. E., 258
Martin, J. C., 412, 421
Matteson, F. H., 295, 346
Miles, J. W., 399, 420
Milne-Thompson, L. M., 25, 33, 92, 111,
    167, 180, 366, 383, 420, 421

Mirels, H., 154, 180, 242, 258
Moore, N. B., 278, 280, 345
Morikawa, G., 114, 142, 182, 205
Mueller, J. N., 246, 259
Munk, M., 39, 61

Newton, I., 281, 284–287, 345
Nielsen, J. N., 17, 33, 83, 89, 110, 114,
    131, 136, 142, 193, 196, 201, 205,
    248, 258, 290, 295, 301, 338, 345–347,
    405, 406, 421
Nonweiler, T., 363, 420

Oswatitsch, K., 296, 347

Perkins, E. W., 85–87, 89, 110
Pitts, W. C., 131, 136, 142, 193, 196, 205,
    338, 347
Puckett, A. E., 8, 33, 290, 292, 293, 345

Raney, D. J., 86, 89, 110
Regan, J. D., 348
Reller, J. O., 319, 347
Resnikoff, M. M., 286, 287, 345
Ribner, H. S., 377, 394–396, 402, 420
Rogers, A. W., 21, 33, 168, 180
Rossow, V. J., 278, 345
Rubesin, M. W., 330, 341, 347
Rubinow, S. I., 9, 33

Sacks, A. H., 94, 101, 111, 124, 142, 151,
    173, 176, 179, 180, 364, 379, 420
Sauer, R., 16, 33
Savin, R. C., 278, 279, 345
Schindel, L. H., 205
Schlichting, H., 334
Sears, W. R., 270, 280, 281, 283–285,
    299, 345
Seiff, A., 348
Short, B. J., 330, 347
Silverstein, A., 180
Sinclair, A. R., 312, 320, 346, 348
Solarski, A., 74, 110
Sommer, S. C., 330, 347
Spahr, J. R., 136, 143, 322, 346
Spreiter, J. R., 124, 142, 151, 173, 176,
    179, 180, 222, 258, 401, 420

Staff of Ames 1- by 3-foot Supersonic Wind Tunnel, 15, 16, 33, 241, 243, 244, 249, 258
Stevens, V. I., 278, 345
Stewart, H. J., 20, 33, 290, 292, 345
Stone, D. G., 408, 421
Summers, R. G., 379, 420
Synge, J. L., 349, 358–363, 378, 389, 418, 421, 427
Syvertson, C. A., 317, 347

Tang, K. K., 136, 142
Taylor, G. I., 73, 110, 275, 344
Tobak, M., 259, 260, 412, 421
Todhunter, I., 281, 286, 345
Trefftz, E., 117, 155, 303
Trilling, L., 348
Trockenbrodt, E., 146, 180
Tsien, H. S., 276, 345

Tsien, S. H., 309, 346
Tucker, W. A., 238, 258, 311, 346

Van Driest, E. R., 341, 347
Van Dyke, M. D., 263, 278–280, 287, 345
Vas, I. E., 348
Vincenti, W. G., 266, 337, 338, 347
von Kármán, T., 271, 278, 280, 281, 284, 299, 344, 345

Ward, G. N., 34, 40, 46, 52, 55, 61, 81, 148, 180, 182, 205, 261, 265, 269, 294, 296, 297, 299, 300, 364, 379, 420
Warren, C. H. E., 7
Wendt, H., 340, 347
Westwater, F. L., 148, 150, 168, 179
Whitcomb, R. T., 296, 298–300, 346
Wilkins, M. E., 348
Wimbrow, W. R., 317, 322, 347
Wisniewski, R. J., 348

SUBJECT INDEX

Acceleration stability derivatives, 355, 357
  general formulas in terms of apparent masses, 374
  slender triangular wing, 377, 378
Ackeret theory, classification, 15
  description, 15, 16
  swept wings, 25
Addition of volume at angle of attack, lifting surface of minimum drag, 311
Aerodynamic controls, prediction techniques, general approach, 208
Aerodynamic heating, effect, on laminar skin friction, 330–334
  on turbulent skin friction, 334–336
  physics, 326–330
Afterbody effects, planar wing-body combinations, 118, 131–134, 140
Air-jet spoilers, definition, 210
Airfoils, base-pressure correlation, 317–321
  (See also Base-drag correlation)
Airplanes versus missiles, 1
All-movable controls, coupling effects, 228–234
  boundary conditions, 229, 230
  coupling, no control, 231
    pitch controls, 231–233
    roll control, 233, 234
  loading coefficient, for horizontal panels, 230
    for vertical panels, 231
  summary of results, 232
  symmetry properties of velocity components, 229, 230
  types of couplings, 228, 231
  cruciform configurations, 225–227
    pitching effectiveness, 225, 226
      maximizing pitch control, 226
      panel-panel interference, 226, 227
    rolling effectiveness, 226–228
      characteristics feature of cruciform arrangements, 226
      effect of radius-semispan ratio, 227, 228
      numerical results, 228
      panel-panel interference, 226, 227
      reverse roll, 226
  planar configurations, 213–225
    loading distribution due to control deflection, 220, 221
      body loading coefficient, 221

All-movable controls, planar configurations, loading distribution due to control deflection, numerical example, 220
      panel loading coefficient, 221
    pitching effectiveness, 214–219, 224, 225
      body forces, 216–218
      boundary conditions, 214
      calculative example, 224, 225
      center of pressure, 219
      configuration lift, 218
      doublet solutions, 215, 216
      panel forces, 216–218
      potential function, 216
    rolling effectiveness, 221–225
      analytical solution, 222
      calculative example, 224, 225
      nonslender configurations, 224
      numerical values, 223
      physical explanation, 223
      reverse-flow methods, 221, 222
      reverse roll, 224
All-movable tip controls, definition, 209
  lift effectiveness, 239
  numerical values of effectiveness, 240
Allen's crossflow theory, center of pressure, 89, 90
  lift force, 89, 90
Angle, of attack, effect on base pressure, 322
  sine definition, 5
  small angle definition, 5
  tangent definition, 5
  of bank, 3, 4
  of sideslip, 4–6
  sine definition, 5, 6
  small angle definition, 5
  tangent definition, 5, 6
Angle-of-attack drag, supersonic airfoils, 290
Angular displacements, pitch, 350–353
  roll, 350–353
  yaw, 350–353
Apparent mass, boundary conditions for, 380
  circle, 372
  coordinate system, 365
  cruciform wing, circular body, 372
  definition, 379
  ellipse, 372

Apparent mass, examples in use of tables,
387–389, 391–393
   damping in roll, cruciform missile,
   392, 393
      multifinned missile, 392
      planar missile, 392, 393
   lift of cruciform missile, 391, 392
   general formulas, acceleration derivatives, 374
      velocity derivatives, 369–371
   general method, derivatives, 364–371
   for evaluation, 378–386
   inertia coefficients, definition, 366
   line, 371
   midtail empennage, circular body, 372
      elliptical body, 373
   multifinned body, three or more fins,
   373
   notation in theory, 380, 382
   planar midwing, circular body, 372
   regular inscribed polygon, 373
   slender cruciform wing, 386–388
   slender triangular wing, 385–386
   table, 371–374
   tangent-tail empennage, circular body,
   373
Arrow lifting surfaces, minimum drag at
   angle of attack, 309–311
      boundary-layer limitations, 311
      drag-rise factor, 309, 310
      numerical values, 310
Arrow wings, pressure drag, 291, 292
      leading-edge suction factor, 292
      zero angle of attack, 291
Aspect ratio of triangular wing, effect on
   stability derivatives, 395–402
      damping, in pitch, 397–401
      in roll, 396, 397
      dihedral effect, 401, 402
      Magnus forces, 394, 402
      rolling moment due to yaw, 402
      static stability, 395, 396
Average skin friction, flat plate, turbulent flow, 334
Axes, body, general types, 3, 4, 6
   pressure coefficient in, 48
   stability derivatives, 350–353

Base drag, 261–263, 311–323
Base-drag correlation, 312–323
   airfoils, 317–322
      effect, of Mach number, 318
      of Reynolds number, 318, 321
      transitional case, 320–322
      turbulent case, 317, 318, 322
   bodies of revolution, 317–323
      effect, of Mach number, 319
      of Reynolds number, 320, 321
      transitional case, 321
      turbulent case, 319, 320, 322
   cone-cylinder combination, 323
   correlation equation, 316, 317
   theoretical model of flow, 312

Base-drag correlation, variables influencing, angle of attack, 322
   base configuration, 314
   boattail angle, 314, 315
   heating and cooling, 323
   Mach number, 315, 316
   Reynolds number, 315, 316, 318
   tailfins, 322
Bernoulli's equation, 12–14
   compressible flow, steady, 13
      unsteady, 12
   linearized form, 14
Bessel functions, modified, 42, 43, 56, 57
Biot-Savart law, 153
Blunt base, flow behind, 312, 313
   theoretical model of flow, 312
   types of flow, 312, 313
Blunt leading edges, leading-edge suction, 293
Boattail angle, 314, 315
Bodies, of least pressure foredrag, 280–
   287
   of minimum wave drag (see Kármán
      ogive; Newtonian body of least
      pressure foredrag; Sears-Haack
      body)
Body of revolution, base-pressure corre-
   lation, 317–321
   (See also Base-drag correlation)
   linear theory, 34–49
      angle of attack, 37–39
      zero, 34–37
   (See also Slender body of revolution)
Body alone, definition, 113
   in empennage, definition, 403, 404
Body axes (see Axes)
Body vortices, 89–107, 406–409
   body of general cross-section, 94–107
      lift and sideforce due to, 96–101
      motion of vortex pair in presence of,
      94–96
      rolling moment due to, 101–107
   body of revolution, 85–94
      center of pressure, 89, 90
      forces due to, 89, 90
      location of vortex separation, 86
      motion of pair in presence of circu-
         lar cylinder, 91–94
      positions and strengths, 85–89
   coupling effects between, 101
   magnitude compared to wing vortices,
   97, 98
   tailless configuration, 406–409
      damping in roll, 408, 409
      static stability derivatives, 406, 407
Bound vortices, 146
Boundary function, 383
Boundary-layer thickness, 316
Boundary layers, flat plate, 325–330
   static temperature profile, 326
   total pressure profile, 326
   types, 325, 326
   velocity profile, 325, 326
   (See also Skin friction)

Bump on circular cylinder, pressure distribution, 83, 84
Busemann second-order theory, classification, 15
description, 16
hinge-moment coefficients, 247–250
section thickness effects on control effectiveness, 243, 244

Calculative example (*see* Illustrative examples)
Calculus of variations, use in drag minimization, 281, 286
Camber drag, supersonic airfoils, 288, 289
Canard control, definition, 210
Cauchy-Riemann equation, 381
Center of pressure, definition, 20
rectangular wing, 24
slender body of revolution, 69
tangent ogive, 69, 70
triangular wing, 21, 22, 378
Characteristic functions, quasi-cylindrical bodies, 82, 83
Circle, apparent mass, 372
Circular cylinder, motion of vortex pair in presence, 91–94
Circulation, 145
Circulation distribution, effect on wake shape, 168
Classifications of missiles, control surfaces, 2
environment, 2
guidance system, 1, 2
propulsion system, 2
trajectory, 2
trim and control, 2
Combined flow field, 297
Comparison between experiment and theory, damping in roll of triangular wings, 397
lift, and center of pressure of ogive-cylinder combination, 90
of planar wing-body combination, 135, 136
pressure distribution on body of revolution, 85
Complex potential, 27–30
circular cylinder in uniform flow, 29
definition, 27
ellipse, banked with respect to lateral axis, 30
of constant a/b ratio, expanding, 30
in uniform flow, 30
planar midwing and body combination, 30, 115
uniformly expanding circle, 29
Cone-cylinder combinations, base pressure, 323
lift carry-over, 68
lift distribution by slender-body theory, 68
vortex positions, 88

Cone-cylinder combinations, vortex strengths, 87
Cones, equivalent flat plate, 341
laminar skin friction, 340, 341
supersonic pressure foredrag, 276–278
turbulent skin friction, 341
Conformal mapping, 25–30
circle, into ellipse, 27
into planar midwing and body combination, 27
into planar wing, 28
general formulas, 25–30
table, 27, 28
Conical flow theory, classification, 15
description, 17
Conical shock-expansion theory, drag of nonslender noses, 278–280
Control deflection, sign conventions, 211
Control reversal, definition, 212
Control surfaces, bank-to-turn, 3
cruciform, 2
"Control surface" drag method, 264
Convolution theorem, 47
Cooling, effect on transition point, 320
and heating, effect on base pressure, 323
Correlation equations, base pressure, 316, 317
Coupling effects, in all-movable controls, 228–234
angle of attack and angle of sideslip, 125–129, 131
panel loading, for cruciform wing-body combinations, 123, 175, 176
for planar wing-body combinations, 126, 127
angle of attack and thickness, 115, 116, 122, 123
body loading, for cruciform wing-body combinations, 123
for planar wing-body combinations, 115, 116
panel loading, for cruciform wing-body combinations, 122, 123
for planar wing-body combinations, 115, 116
drag of wings alone, 288–290
potential, lift versus no lift for slender body, 53
pressure coefficient, angle of attack and vortices in tail plane, 205–207
slender bodies, angle of attack and thickness, 71, 72, 78
body expansion and angle of attack, 74
lift and sideforce, vortices in presence of body, 101
Covering operations, 358
Cross-coupling derivatives, 213
"Cross-talk" in aerodynamic controls, 213
Cross-wind force, definition, 114
Crossflow drag coefficient, definition, 89
uniformity, 90

Crossflow planes, body of revolution, 40
  definition, 6, 40
Crossflow vortices (*see* Body vortices)
Cruciform empennage, stability deriva-
    tives, illustrative example, 412–416
Cruciform missile, damping in roll, 392–
    394
  lift and sideforce, 391, 392
  mirror symmetry, 359
  rotational symmetry, 359
  stability derivatives, 358–362, 386–
      392, 421–427
  zero, 361
Cruciform wing, circular body, apparent
    mass, 372
  definition, 6
  slender, apparent mass, 386–388
    stability derivatives, gyroscopic,
        390–391
      Magnus, 390, 391
      static, 388, 389
  vortex model, 173–177
    analytical solution, 173, 174
    calculative example, 174–177
      leapfrogging, 176
      vortex positions, 175, 176
    shape of wake, 45° bank, 177
    vortex position, 45° bank, 177
Cruciform wing-body combination, defi-
    nition, 112
  interference, 121–124, 130
    calculative example, 135–137
    coupling between thickness and
        angle of attack, 122, 123
    forces, 124, 130
      effect of bank angle on, 124, 130
    loading, 123, 124
      body, 124
      panels, 123
    moments, 130
    potential function, 122
    velocity components, 123
  vortex model (*see* Cruciform wing)
Cusps, drag curves, 291, 292
Cuts, logarithmic, 62
  source, 99
  vortex, 99
Cylinders, turbulent skin friction, 341

D'Alembert's paradox, 59
Damping, in pitch, cruciform empennage,
    415, 416
  tailless cruciform configuration, 409–
      412
  triangular wing, 399–401
    effect of aspect ratio, 399–401
    slender-body theory, 399–401
    supersonic wing theory, 399–401
    types of pitching motion, 397–399
  in roll, cruciform empennage, 414, 415
  tailless cruciform configuration, 407–
      409

Damping, in roll, triangular wing, 396,
    397
  comparison between theory and
      experiment, 397
  effect of aspect ratio, 397
  slender-body theory, 396, 397
  supersonic wing theory, 396, 397
  in yaw, cruciform empennage, 416
  tailless cruciform configuration, 412
Damping derivatives, definition, 358
Damping parameter of missile, definition,
    252
  effect of altitude, 253
  subcritical, 252
  supercritical, 252
Deadwater region, 86, 263, 312
  slender body, 52, 53
Dihedral effect, definition, 358
  triangular wing, 401, 402
    effect of aspect ratio, 401, 402
    linear theory, 401, 402
    slender-body theory, 401, 402
Dissociated boundary layer, skin friction,
    331
Double-wedge airfoil, supersonic drag,
    289
Double-wedge wing, transition, 337
Doublets, 37, 38
Downwash angle, definition, 144
Downwash lag concept, 411
Drag, slender bodies, due to lift, 52–54
  formulas for, general, 52
    showing Mach-number depend-
        ence, 54, 55
    simplified, 52
    subsonic, 59
    supersonic, 51–55
    Ward's, 51, 52
Drag components for complete missile,
    261–265, 269–341
  base drag, 261, 311–323
  pressure foredrag, 261, 269–311
  skin friction, 261, 323–341
  wave drag and wake drag, 264
Drag curve, analytical properties, 265–
    267
  drag polar, 265, 266
  drag-rise factor, 266
  lift coefficient for minimum drag, 266
  maximum lift-drag ratio, 266, 267
  minimum drag coefficient, 266
  optimum lift coefficient, 266, 267
Drag interference, definition, 294
Drag polar, definition, 265, 266
Drag-rise factor, definition, 266
  effect of leading-edge suction on, 267,
      268
  lifting surfaces of minimum drag, 308–
      310
    arrow, 309, 310
    triangular, 308
  lower bounds on, 303–306
    wing-body vortex drag, 303–306
    wing-body wave drag, 303–306
    wing vortex drag, 303–306

Drag-rise factor, lower bounds on, wing wave drag, 303–306
  physical interpretation, 268, 269

Effective aspect ratio, 22
Ellipse, apparent mass, 372
Elliptical integrals, amplitude, 31
  complete, 31
  modulus, 31
Elliptical potential distribution, 150–151
  horseshoe vortex representation, 151
  wake shape for, 150
Empennage, stability derivatives, 392, 402–416
  (*See also* Tailless configuration)
Entropy gradients, drag of tangent ogive, 278
Environment classification of missiles, AAM, 2
  ASM, 2
  AUM, 2
  SAM, 2
  SSM, 2
  UUM, 2
Equation, of linear aerodynamics, 12
  of motion, missile, impulsive pitch control, 251, 259, 260
Equivalent-body concept, 298, 299

Favorable interference, supersonic "lift catching," 311
Feeding sheet in body-vortex theory, 91
Fineness ratio, 6
Flat plate, boundary layer, 325–330
Flat vortex sheet (*see* Wing-tail interference)
Föppl points, 92, 94
  vortex strength for, 92
Fourier transforms, slender-body theory, 55–58
Free vortices, slender configurations, 96–107
    lift and sideforce, 96–101
    rolling moment, 101–107
    triangular wing, rolling moment, 106, 107
Frequency parameter, 11

Gaps, all-movable controls, 242, 243
  large deflections, 243
  small deflections, 242, 243
Glossary of special terms, 6, 7
Guidance-system types, 1–3
  beam-riding, 2
  command, 1
  homing, 2
    active, 2
    passive, 2
    semiactive, 2
Gyroscopic stability derivatives, slender cruciform wing, 390, 391

Heating and cooling effect on base pressure, 323
Helix angles, 356
Higher-order effects, all-movable controls, 243–247
  angle of attack, 246, 247
  control deflection, 245, 246
  control section thickness, 243, 244
Higher-order stability derivatives, 355
Hinge-moment coefficient, Busemann second-order theory, 247–250
  calculative example, 248–250
  center of pressure shift, 248
  estimation, 247–250
  general approach, 247
Horizontal plane of symmetry, definition, 6
Horizontal reference plane, definition, 210
Horizontal tail, definition, 403, 408
Horseshoe vortex, incompressible, 139, 151
    representation of elliptical potential distribution, 151
  supersonic, 154–156
    versus incompressible horseshoe vortex, 156
    regions of influence, 155
    Trefftz plane flow, 155
    velocities, 154
Hypersonic similarity law, 276
Hypersonic similarity parameter, 276–279

Illustrative examples, angle of attack and of bank, 6
  apparent-mass coefficients of slender triangular wing, 385, 386
  center of pressure of tangent ogive, 69, 70
  downwash field behind planar wing and body combination, 169–171
  drag comparison between Sears-Haack body and Kármán ogive, 285
  drag-curve parameters for double-wedge triangular wings, 293
  forces on tail section of planar wing-body combination, wing-tail interference, 195–197
  hinge-moment coefficient of all-movable triangular control, 248–250
  laminar skin friction, cone, 340–341
    on flat plate with aerodynamic heating, 332–334
  panel forces under combined angles of attack and sideslip, triangular wings, 127–129
  pitching and rolling effectiveness of all-movable controls, planar configuration, 224, 225
  pressure distributions due to bump on circular cylinder, 83, 84
  rolling moment of triangular wing, free vortices, 106, 107

Illustrative examples, section thickness
    effects on control lift effectiveness,
    244
  simple sweep theory, triangular wing
    with supersonic edges, 24, 25
  stability derivative contributions of
    cruciform empennage, 412–416
  tail interference factors for discrete
    vortices in plane of tail panels,
    194
  thickness drag and camber drag,
    double-wedge airfoil, 289
  turbulent skin friction on flat plate
    with aerodynamic heating, 335,
    336
  vortex model for planar wing-body
    combination, 162–165
  vortex paths for cruciform wing at
    45° bank, 174–177
  wing-body interference, planar wing-
    body combination, 136, 137
Impulsive pitch control, damping
    parameter, 252
  effect of altitude on, 253
  response rate, 253
  equation of motion, derivation, 259,
    260
    nondimensional, 252
    solution to, subcritical damping, 252
    supercritical damping, 252
  natural frequency, 252
Included angle, definition, 3, 6
Interdigitation angle, definition, 7
Interference effects, favorable, 311
  nonslender wing-body combination,
    134–137
  static stability derivatives, 403–406
  tailless configuration, damping in roll,
    407–409
  (See also Afterbody effects; Cruciform
    wing-body combination; specific
    types of interference)
Interference factor, tail, 192–194
Interference potential, definition, 113
Isentropic law, 9

Jet control, definition, 210
Jet vane, definition, 210
Jones's criterion, least drag due to lift,
    302–306
  minimum thickness drag, 296, 297
Jones's line pressure source, 17–19, 235,
    236, 337, 338

Kármán ogive, 281, 284–287, 299
  area distribution, 284
  comparison, with Newtonian body,
    286, 287
    with Sears-Haack body, 285
  coordinates, 284
  drag coefficient, 284
  shape compared to Newtonian body,
    286

Kármán ogive, volume, 284
Kutta-Joukowski law, 152

Laminar base flow, definition, 312
  (See also Base-drag correlation)
Laminar skin friction, 330–334
Laplace transforms, slender-body theory,
    41–44
Leading-edge sources, Jones-type, 337
Leading-edge thrust (suction), effect of,
    on drag-rise factor, 267, 268
  effect on, of leading-edge bluntness,
    293
    of trailing-edge sweep, 292
  triangular-wing formula, 292
"Leapfrogging," 172, 176, 177
Lift, general formula for slender bodies,
    48–50
  slender bodies of revolution, 66–68
Lift-cancellation technique, trailing-
    edge controls, 235–237
Lift "carryover," 131
  cone-cylinder body, 68
Lift coefficient for minimum drag,
    definition, 266
Lift ratios, all-movable controls, body
    lift ratio k_B, 217, 218
    panel lift ratio k_w, 217, 218
  wing-body combination, body lift
    ratio K_B, 119, 120, 131–133
    panel lift ratio k_W, 119, 120, 131
    panel-sideslip lift ratio, K_φ, 125–129
Lifting surfaces, cruciform, 2
Line, apparent mass, 371
Line pressure source, subsonic, 17–19
  supersonic, 17–19
  trailing-edge controls, 235, 236
Loading coefficient, definition, 20
Loading distribution, definition, 20
Local skin-friction coefficient, definition,
    329
  laminar, 333, 334
  turbulent, 334, 335
Longitudinal aspect ratio, definition, 305
Lower bound on drag-rise factors, wings
    and wing-body combinations,
    vortex drag, 303–306
    wave drag, 303–306
Lower vertical tail, 403, 405

Magnus stability derivatives, slender
    cruciform wing, 390, 391
  triangular wing, 395, 402
    slender-body theory, 402
    supersonic wing theory, 395, 402
Maple-Synge analysis, cruciform mis-
    sile, 358–362, 421–427
  general method, 358
  triform and other missiles, 362, 363,
    427–431
Maximum lift coefficient, supersonic
    speeds, triangular wings, 246
  wings of other planforms, 246, 247

Maximum lift-drag ratio, definition, 266
Mean-enthalpy method, laminar skin
    friction, 331–334
  first, 331
  second, 332
  turbulent skin friction, 334–336
Method, of apparent masses, 364–374
    (*See also* Apparent mass)
  of characteristics, classification, 15
    description, 16
  of inertia coefficients, 364–374
Midtail empennage, apparent mass, cir-
    cular body, 372
  elliptical body, 373
Minimum drag coefficient, definition, 266
  wings alone, 291
Mirror symmetry, 358
Missile attitude, pitch, 350–353
  roll, 350–353
  yaw, 350–353
Missiles versus airplanes, 1
Modulus of elliptical integral, 31
Moment, slender body of revolution, 69
  tangent ogive, 69, 70
    (*See also* specific configurations)
Multifinned body, damping in roll, 392
  with three or more fins, apparent mass,
    373
Munk's airship theory, 39

N-gonal symmetry, 359
NACA area rule, discovery, 298, 299
  equivalent-body concept, 298, 299
  Kármán ogive, 299
  limitations, 299, 300
  Sears-Haack body, 299
Natural frequency of missile, definition,
    252
  effect of altitude on, 253
Newtonian body of least pressure fore-
    drag, accuracy of drag prediction, 287
  bluntness of nose, 286
  comparison with Kármán ogive, drag,
    287
    shape, 286
  coordinates, 284
  shape, 286
Newtonian impact theory, 15, 285
  drag of nonslender bodies, 278, 280
Nonlinearities, aerodynamic controls,
    242–247
  boundary-layer separation, laminar,
    245
    turbulent, 245
  deflection for incipient separation,
    246
  gap effects, 242, 243
  higher-order effects of angle of
    attack and control deflection, 242,
    245–247
  maximum lifting capabilities, effect
    of planform, 246, 247
  section thickness influence on lifting
    effectiveness, 243, 244

Nonlinearities, wing-tail interference,
    effect of tail height, 197, 198, 202,
    203
  shock-expansion interference, high
    tails, 202, 203
  static stability, planar wing-body
    combinations with high tail, 197,
    198
Nonslender wing-body combination,
    interference effects, 134–137
  calculative example, 135–137
Normal plane, definition, 7, 40
Nose control, definition, 210
Notation, skin friction, 324, 325
  stability derivatives, angles, 351
    axes, 351
    forces, 351
    moments, 351
    velocities, 351

Ogive-cylinder body, vortex positions, 88
  with vortex separation, center of pres-
    sure, 90
    lift, 90
    location, 86
    pressure distribution, 85
  vortex strengths, 87
Optimum lift coefficient, definition, 266
Order-of-magnitude symbol, physical
    meaning, 11

Panel-panel interference, effect on panel
    loading, 126, 127
  cruciform configuration, 126, 127
  planar configuration, 126, 127
  panel center of pressure, 127
  rolling effectiveness of cruciform
    arrangements, 226–228
  triangular wing alone, 127–129
Parabolic arc body, slender, lift distribu-
    tion, 68
Physical model, blunt base flow, 312
Physical plane, definition, 26
Physical properties of air, enthalpy, 332
  Prandtl number, 332
  specific heat, 332
  variation with temperature, 332
  viscosity, 332
Pitch control, coupling between cruci-
    form all-movable controls, 228–234
  definition, 211, 212
  impulsive, equation of motion for, 259,
    260
  maximizing, for cruciform arrange-
    ments, 226
Pitch damping, effect, of $\dot\alpha$, 411, 412
  of q, 410
Pitching effectiveness, definition, 212
Pitching moment, reverse-flow methods,
    wing-tail interference, 201
Pitching-moment formula, slender
    bodies, 48–51
Planar configurations (*see* All-movable
    controls)

Planar midwing, circular body, apparent mass, 372
Planar missile, damping in roll, 392–394
Planar wing-body combination, definition, 112
  downwash field, calculative example, 169–171
    choice of number of vortices per panel, 171
    effect of radius-semispan ratio, 170, 171
  static stability (*see* Tail forces and moments due to wing vortices)
  vortex model (*see* Slender planar wing-body vortex model)
  vortex paths and wake shapes, 166–168
    effect of body on, 167
    effect of circulation distribution on wake shape, 168
      elliptical, 168
      triangular, 168
    Föppl points, 92, 167
    paths, 91–94, 166, 167
Planar wing-body interference, 114–121
  afterbody effects, 118, 123–134, 140
  body lift, 118–120, 130, 139, 140
  center of pressure, body, 134
  complex potential, 115
  coupling between thickness and angle of attack, 115, 116
  lift, carryover to body, 120
    complete configuration, 117, 130
  lift ratios, K_B, 119, 120
    K_W, 119, 120
  loading coefficient, body, 116, 117
    wing, 116, 117
  moments, 130
  panel lateral center of pressure, 119, 121
  panel lift, 118–120, 129, 139, 140
  rule of thumb for lift, 120
  simplified vortex model, 138–140
  span loading, 118
  velocity components, body, 116
    panel, 116
Potential difference, versus span loading, 157, 165
  trailing edge of slender wings, 145, 146
Potential equation, choice of form, 8
  compressible flow, linear unsteady, 10–12
    nonlinear steady, 8–10
  cylindrical coordinates, 35, 41, 42
Potential function, compressible flow, 9
Prandtl-Schlichting equation, 334
Pressure coefficient, definition, 13
  in slender-body theory, body axes, 48
  vortex–angle of attack coupling in plane of tail, 205–207
Pressure drag, components, base drag, 263
  foredrag, 263
  defining integral, 262
  definition, 262
  wings alone, arrow, 291

Pressure drag, wings alone, camber drag, 287–290
  comparison between experiment and theory, 291
  coupling among effects of angle of attack, camber, thickness, 288–290
  cusps in drag curves, 291, 292
  drag charts, 290
  leading-edge suction, 292, 293
  minimum wave drag, definition, 291
  separation, 293
  thickness drag, 287–290
  wings and wing-body combinations of least drag at angle of attack, 302–311
  addition of volume, 311
  arrow lifting surfaces, 309–311
  camber and twist, 305, 306
  design considerations, 306
  drag-rise factor, vortex drag, 303
    wave drag, 304, 305
  Jones's criterion for least wave drag, 302, 303
  triangular lifting surfaces, 306–309
Pressure foredrag, bodies of least drag at zero angle of attack (*see* Kármán ogive; Newtonian body; Sears-Haack body)
  definition, 263
  nonslender bodies, conical noses, 275–280
    correlation by hypersonic similarity parameter, 276–279
    range of applicability of prediction methods, 279, 280
    tangent ogive, 278, 279
    theories for drag, 278, 279
      conical shock-expansion, 278, 279
      Newtonian, 278
      tangent-cone methods, 1 and 2, 278, 279
      Van Dyke second-order, 278, 279
      von Kármán and Moore, 278
  slender bodies of given shape, 269–275
    with circular base, 271, 272
    with cylindrical base, 271
    drag due to lift, circular base, 273–274
    drag formula of Ward, 51, 52, 269
    pointed at both ends, 270, 271
  wing-body combinations at zero angle of attack, 294–296
    drag components, 295
    drag interference, first and second definitions, 294
    quasi-cylindrical body, 295
    rule of thumb, 295
Principal part, 383
Propulsion systems, missiles, 2

Quasi-cylindrical bodies, 80–84
  axes, 80
  boundary conditions, 81
  characteristic functions for circular bodies, 82, 83

Quasi-cylindrical bodies, illustrative
      example, 83, 84
   notation, 80
   potential function, 81, 82
   pressure coefficient, 82
Quasi-cylindrical theory, classification, 15
   description, 17, 80–84
Quasi-cylindrical wing-body drag, super-
      sonic, 295

Reaction jet, definition, 210
Reattachment point, 312
Recovery factor, definition, 327
   flat plate, laminar flow, 331
      turbulent flow, 334
Recovery temperature, definition, 327
Rectangular wings, aerodynamic charac-
      teristics, 22–24
   aspect ratio, classification by, 23
      effective, 22
   center of pressure, 24
   lift-curve slope, 22
   lift distribution, 22, 23
   span loading, 23
Reference area, stability derivatives, 355
Reference axes, stability derivatives, 350
   slender triangular wing, 374, 375
Reference length, stability derivatives,
      355
Regular inscribed polygon, apparent
      mass, 373
Resistance stability derivatives, 355, 356
Response rate of missile pitch control,
      253
   effect of altitude on, 253
Reverse-flow methods, all-movable con-
      trols, rolling effectiveness of planar
      configurations, 221–224
   trailing-edge controls, pitching effec-
      tiveness of planar configurations,
      238–240
   wing-tail interference, 198–201
   (See also Tail forces and moments
      due to wing vortices)
Reverse roll, definition, 224
   panel-panel interference, 226, 227
Ridge-line sinks, Jones type, 337
Roll control, coupling between all-mova-
      ble cruciform controls, 228–234
   definition, 211, 212
Rolling effectiveness, of all-movable con-
      trols, 221–224, 226–228
   cruciform arrangements, 226–228
   planar arrangements, 221–224
   definition, 212, 213
Rolling moment, due to wing-tail inter-
      ference, reverse-flow methods, 201
   due to yaw, triangular wing, 402
      slender-body theory, 402
      supersonic wing theory, 402
   triangular wing, free vortices, 106, 107
Rolling up of vortex sheet, slender wing,
      148–153
   elliptical potential distribution, 150

Rolling up of vortex sheet, "fully-rolled-
      up" condition, 149–151
   horseshoe vortex, 151
   lateral position of vortices, 152
   shape of sheet, 150
   tearing of edges, 148
   vortex strength, 151, 152
Rotary stability derivatives, 355, 357
Rotational symmetry, 358
Rule of thumb, 120, 295
   wing-body combinations, drag, 295
   lift interference, 120

Schulz-Grunow equation, 334
Sears-Haack body, 270, 280–284, 299
   area distribution, 282
   comparison with Kármán ogive, 285
   coordinates, 284
   drag coefficient, 281–283
   volume, 283
Section lift coefficient, 20
Selection rule, Maple-Synge analysis,
      422
Separation, boundary-layer, 245–246
   incipient, 245, 246
   induced by control deflection, 245,
      246
   laminar versus turbulent, 245
   plateau pressure, 245
   drag effects on wings alone, 293
   of vortices from body of revolution,
      85–89
Shock-expansion interference, effect of
      Mach number, 203
   high-tail, planar wing-body combina-
      tion, 202, 203
   physical explanation, 201, 202
   pitching-moment nonlinearity, 203
Shock-expansion theory, classification,
      15
   description, 16
Shock-interference control, definition,
      210
Sideforce formula, slender bodies, 48–50
Sidewash angle, definition, 144
Sign conventions, control deflection, 211
Simple sweep theory, classification, 15
   description, 24, 25
   trailing-edge controls, 240, 241
Singularities, line pressure source, 18
   logarithmic, 62
   slender-body theory, 41
   trailing-edge control, subsonic hinge
      line, 236
Sinks (see Sources and sinks)
Skin friction, 261–263, 323–341
   average value in turbulent flow, 334
   calculative example for flat plate,
      332–336
      laminar flow, 332–334
      turbulent flow, 335, 336
   cones, 340, 341
   cylinders, 341
   laminar, 330–334

Skin friction, local skin-friction coefficient, 329, 333, 334
  definition, 329
  laminar, 333
  turbulent, 334
 nonuniform surface temperature, 341
 notation, 324, 325
 relation to velocity profile, 328
 transition effect on, 337–339
 turbulent, 334–336
 variables influencing, 336–340
 wind tunnel versus free flight, 328
 wing-body combinations, 338
Slender body of revolution, 66–74
 center of pressure, 69
 complex potential, 67
 coupling between pressures due to thickness and angle of attack, 71, 72
 lift, 66–68
 lift-curve slope, 68
 loading, 70–74
 moment, 69
 pressure coefficients, 70–74
Slender-body theory, classification, 15
 range of validity for circular cones, 73
 subsonic, 55–59
  boundary conditions, 56
  d'Alembert's paradox, 59
  drag formula, 59
  evaluation of coefficients in potential function series, 58, 59
  series solution for potential equation, 57
  use of Fourier transforms, 55–58
 supersonic, 40–55
  accuracy of velocity components, 46
  assumptions underlying, 40, 41
  body of revolution at angle of attack, 39, 40
  boundary conditions, 45, 46
  differences between subsonic and supersonic, 40
  evaluation of coefficients in series for potential function, 47, 48
  linearization of boundary conditions, 46
  method of Ward, 40
  order of magnitude of velocity components, 46
  region of validity of series solution, 44
  series form, complex potential, 44
   potential functions, 44
 (See also Slender circular cones; Slender elliptical cones; Slender planar wing-body vortex model; Slender triangular wings; Slender-wing vortex model)
Slender circular cones, 72, 73, 79, 80
 comparison, between drags of circular and elliptical cones, 79, 80
  between slender-body theory and exact theory, 73

Slender circular cones, pressure coefficient due, to angle of attack, 73
  to thickness, 72, 73
Slender elliptical cones, 74–80
 axis conventions, 74
 comparison between drags of circular and elliptical cones, 79, 80
 drag, 79
 lift, 75
 moments, 75
 notation, 74, 79
 pressure coefficient due, to angle of attack, 78
  to thickness, 77
 sideforce, 75
Slender planar wing-body vortex model, 156–166
 center of gravity, 159, 160
 circulation distribution, 157, 158
  table, 158
 illustrative example, 162–164
  effect of bank angle, 165
  initial downwash and sidewash angles, 163, 164
  initial vortex positions, 163
  initial vortex strengths, 163–165
 image vortices, 158–161, 163
 induced velocities, 161, 162
 two-vortices-per-panel model, 160, 161
 vortex path, 162
 vortex strengths, 157, 159
Slender triangular wings, 394–402
 damping, in pitch, 399–401
  in roll, 396, 397
 dihedral effect, 401, 402
 Magnus forces, 402
 rolling moment due to yaw, 402
 static stability derivatives, 395, 396
Slender-wing vortex model, 145–148
 bound vortices, 146
 circulation 145
 effect, of shock waves, 147
  of trailing-edge sweep, 148
 potential difference, 145, 146
 trailing-vortex strength, 145
Slender-wing vortex sheet (see Rolling up of vortex sheet)
Source cut, 99
Sources and sinks, body of revolution, 35–37
 line pressure source, 17, 18
 relation between source strength and body shape, 36
Span-load distribution, definition, 20
 elliptical, 20
Span loading versus potential difference, 157, 165
Speed of sound, 10
Spring constant, missile, 358
Stability derivatives, complete empennage, 402–416
 cruciform empennage, 412–416
 cruciform missiles, 358–362, 421–427
 cruciform triangular wing, 386–391

Stability derivative, cruciform wing-
     body combination, 391, 392
  definitions, 354
  inertial coefficients, Bryson method,
     363–374
  multifinned empennage, 392
  reference axes, 350–353
  slender triangular wing, 374–378, 385,
     386
  triangular wings, effect of aspect ratio,
     394–402
  triform and other missiles, 362–363,
     427–431
  various types, 355–358
Stability derivatives, types, 355–358
  acceleration, 355, 357
  higher-order, 355
  resistance, 355, 356
  rotary, 355, 357
  static (see Static stability deriva-
     tives)
  velocity, 357
Stagnation point, vortices in crossflow
     past circular cylinder, 92, 93
Stagnation temperature formula, 330,
     331
Static stability, of planar wing-body-tail
     combinations (see Tail forces and
     moments due to wing vortices)
  of wing-body-tail combinations, high
     tails, 202, 203
  shock-expansion theory, 203
Static stability derivatives, 357
  cruciform empennage, 413, 414
  tailless configuration, 403–406
  triangular wing, effect of aspect ratio
     on, 395, 396
     slender-body theory, 395, 396
     supersonic wing theory, 395, 396
Static temperature profile, boundary
     layer, 326
Strip theory, classification, 15
  description, 16
Subsonic leading edge, definition, 7
Supersonic area rule, 296, 300–302
  constructural procedure, 300, 301
  drag formula, 300
  limitations, 301, 302
     body shape, 301
  source-strength rule, 300
Supersonic leading edge, definition, 7
Supersonic lifting-line theory, classifica-
     tion, 15
  description, 17
Supersonic lifting-surface theory, 17
Supersonic wing theory, classification, 15
  description, 16, 17
  rolling effectiveness of all-movable
     controls, 227
  trailing-edge control characteristics,
     234–238
  triangular-wing stability derivatives,
     395–402
     damping, in pitch, 399–401
        in roll, 396, 397

Supersonic wing theory, triangular-wing
     stability derivatives, dihedral
     effect, 401, 402
  Magnus forces, 402
  rolling moment due to yaw, 402
  static stability derivatives, 395, 396
Symmetrical wing, definition, 7

Tables, apparent-mass coefficients,
     371–374
  center of pressure of tangent ogive, 70
  classification, of aerodynamic theories
     used in text, 15
     of missiles, 2
  complex potentials for various flows,
     29, 30
  conformal transformations, 27, 28
  coordinates of bodies of least wave
     drag, 284
  direction cosines of body axes, com-
     bined pitch and bank, 4
     small pitch and yaw displacements,
        353
     stability analysis, 352
  nondimensional circulation distribu-
     tion of wing panels, 158
  nondimensional ratios for symmetrical
     deflection of all-movable controls
     mounted on circular body, 218
  slender-body parameters for loading
     due, to bank, 127
     to pitch, 119
  standard conventions and symbols for
     stability derivatives, 351
  values of K_B/K_W, 140
Tail control, definition, 210
Tail effectiveness, definition, 182, 183
Tail fins, effect on base pressure, 322
Tail forces and moments due to wing
     vortices, 194–198
  calculative example, planar wing-body-
     tail combination, 195–198
     effect of tail height, 198
     lift-curve nonlinearity, 197
     moment-curve nonlinearity, 197,
        198
     tail effectiveness, 197
     tail lift, 196, 197
     vortex positions, 195, 196
  calculative method, 194, 195
  reverse-flow methods, 199–201
     basic theorem, 199
     boundary conditions, 200
     lift, on body, 201
        on tail panels, 200
     pitching moment, 201
     rolling moment, 201
Tail interference factor, definition, 192,
     193
  evaluation for discrete vortices in
     plane of tail panels, 194
  methods for calculation, 193
  typical chart, 193

"Tailless" configuration, definition, 403
  stability derivatives, damping in
      pitch, 409–412
    downwash lag concept, 411
    due to $\dot{\alpha}$, 411, 412
    due to q, 410
    wing-tail configuration, 409–412
    damping in roll, 407–409
      body vortex effects, 408, 409
      interference effects, 407, 408
      wing vortex effects, 408, 409
    illustrative example, cruciform em-
      pennage, 412–416
      damping, in pitch, 415, 416
        in roll, 414, 415
        in yaw, 416
      sideforce and yawing moment due
        to yaw, 413, 414
    parts, 403–405
      body, 403, 404
      horizontal tail (fin), 403, 408
      lower vertical tail (fin), 403, 405
      upper vertical tail (fin), 403, 405
      ventral fin, 405
    static, 403–407
      body vortex effects, 406, 407
      interference effects, 403–406
      weathercock stability, 405
      wing vortex effects, 406, 407
Tangent cone methods 1 and 2, drag of
  nonslender noses, 278–280
Tangent ogive, definition, 7
  drag by hypersonic similarity, 278, 279
  effect of rotation term, 278
  slender, center of pressure, 69, 70
Tangent-tail empennage, circular body,
  apparent mass, 373
Theories classification, 14–17
  Ackeret, 15
  Busemann, 15
  conical flow, 15
  method of characteristics, 15
  Newtonian impact, 15
  quasi-cylinder, 15
  shock-expansion, 15
  simple sweep, 15
  slender body, 15
  strip, 15
  supersonic lifting line, 15
  supersonic wing, 15
Theory of residues, use in apparent-mass
  theory, 378, 379, 383
Thickness drag, supersonic airfoil, 288,
  289
Total temperature, definition, 326
Total-temperature profiles in boundary
  layers, 326
Trailing-edge controls, 234–241
  analytical approach to, 234
  definition, 209
  design chart references, 237, 238
  effect of sweep on lift effectiveness,
    241
  lift-cancellation techniques of analysis,
    235–237

Trailing-edge controls, lift effectiveness
    for tip controls, 238–240
    effect of radius-semispan ratio, 239,
      240
  line source-sink analysis, 235, 236
  numerical results, 240
  reverse-flow methods, 238–240
  shock detachment from hinge line, 241
  simple-sweep theory, 240, 241
  subsonic hinge-line pressure singu-
    larity, 236
  tip sector effects, 237
  trailing-edge sector effects, 237
Trailing vortices (*see* Slender-wing vortex
  model)
Trajectory types, ballistic, 2
  glide, 2
  skip, 2
Transformed plane, definition, 26
Transition, double-wedge wing, 337, 338
  wing-body combination, 338
Transitional base flow, 312
  (*See also* Base-drag correlation)
Transitional location, drag due to
  viscous crossflow, 275
  effect on viscous drag, 263, 264
  factors determining, 264
Trefftz plane, definition, 7
  horseshoe vortex, 155, 303
Triangular lifting surface, minimum drag
    at angle of attack, 306–309
    attempts to achieve lower bound,
      309
    drag-rise factor, 308
    lower bound on wave **drag**, 308
    numerical values, 307
    optimum vortex drag, 306, 307
Triangular wing, acceleration derivatives,
    slender wing, 377, 378
  aerodynamic characteristics by linear
    theory, 18–22
    subsonic leading edges, 20, 21
      center of pressure, 21
      lift-curve slope, 20
      lift distribution, 20
      span-load distribution, 20
    supersonic leading edges, 21, 22
      center of pressure, 22
      lift-curve slope, 21
      lift distribution, 21
      span-load distribution, 21
  apparent masses, 385, 386
  double-wedge airfoil section, drag-
    curve parameters, 293
  rolling moment, free vortices, 106, 107
  stability derivatives, 374–378, 385, 386,
    394–402
    (*See also* Triangular wing stability
     derivatives)
  velocity derivatives, slender wing,
    374–377
Triangular wing stability derivatives,
    damping, in pitch, 397–401
    slender-body theory, 399–401
    supersonic wing theory, 399–401

Triangular wing stability derivatives,
    damping, in pitch, types of pitch-
        ing oscillation, 397–399
    in roll, 396, 397
        slender-body theory, 396
        supersonic wing theory, 396, 397
    dihedral effect, 401, 402
        slender-body theory, 401, 402
        supersonic wing theory, 401, 402
    leading-edge suction, 396, 402
    Magnus forces, 402
    roll due to yaw, 402
        slender-body theory, 402
        supersonic wing theory, 402
    static stability, 395, 396
        slender-body theory, 395
        supersonic wing theory, 395
Triform missiles, mirror symmetry, 359
    rotational symmetry, 359
    stability derivatives, 362, 363, 427–431
    zero stability derivatives, 361, 362
Trim and control means, canard, 2
    tail control, 2
    wing control, 2
Turbulent base flow, 312
    (See also Base-drag correlation)
Turbulent boundary layer, virtual origin,
    339
Two-dimensional incompressible vortices,
    153–156
Two-dimensional supersonic airfoil
    theory, drag of wings alone, 288–290
Types of controls, air-jet spoilers, 210
    all-movable, 209
    all-movable tip, 209
    canard, 210
    jet, 210
    jet vane, 210
    nose, 210
    reaction jet, 210
    shock-interference, 210
    tail, 210
    trailing-edge, 209
    wing, 210

Upper vertical tail, definition, 403, 405

Van Dyke's second-order theory, 263,
    278–280, 287
    drag of nonslender noses, 278–280, 287
Velocity profiles, boundary layer on flat
    plate, 325, 326
Velocity stability derivatives, general
    formula based on apparent masses,
    369–371
    slender flat triangular wing, 374–377
Ventral fin, 405
Vertical plane of symmetry, definition,
    210, 211
Virtual origin, turbulent boundary layer,
    339
Viscosity limitation on drag-rise factors,
    305

Viscous crossflow, effect on body drag,
    274, 275
    theory, 15, 85–107
Viscous drag (see Skin friction)
Volume, drag due to, lifting surfaces, 311
von Kármán and Moore method, drag of
    nonslender noses, 278, 280
Vortex cut, 99
Vortex drag, definition, 264
Vortex-induced velocities, calculation,
    153–156
    supersonic horseshoe vortex, 154–156
    two-dimensional incompressible
        vortices, 153, 154, 156
Vortex model, cruciform wing, 173–177
    cruciform wing-body combination, 172,
        173
    planar wing-body combination,
        138–140, 156–166
    slender wing, 145–148
        trailing edge normal to flow, 145–147
        trailing-edge-swept, 148
Vortex pair, mutual indication between,
    149
    in presence, of circular cylinder, 91–94
        asymptotic spacing, 94
        complex potential, 91
        Föppl positions, 92
        paths, 93, 94
        stagnation points in body cross-
            flow plane, 92, 93
    of noncircular body, 94–96
        complex potential, 94, 95
        paths, 96
        transformation, to circular cross-
            section, 95
        velocity, 96
Vortex paths, pair in presence, of circular
    cylinder, 93, 94
    of noncircular body, 96
    planar wing-body combination,
        166–168
    series method of calculation, 177
    variable vortex strength, 96
Vortex sheet, body of revolution, 52, 53
    rolling up (see Rolling up of vortex
        sheet)
Vortex strength, planar wing-body com-
    bination, 139
    trailing vortices, 145
Vortices, horseshoe, 139
    supersonic, 154–156
    two-dimensional incompressible, 153
    (See also Body vortices)

W function, supersonic drag of quasi-
    cylindrical wing-bodies, 295
Wake, definition, 145
Wake drag, definition, 264
Wake shape, cruciform wing at 45° bank,
    177
    effect of circulation distribution, 168
    planar wing-body combination, 168
Ward's drag formula, 52

Wave drag, definition, 264
Wave-drag-rise factor, wings or wing-
    body combinations, 305
Wave-making drag, 265
Weathercock stability, definition, 357
Wing alone, definition, 113
  pressure drag (*see* Pressure drag)
  supersonic drag, lower bounds, 303–306
      vortex drag, 303–306
      wave drag, 303–306
Wing-body combinations, components,
    afterbody, 113
    forebody, 113
    winged section, 113
  drag at supersonic speeds, lower
      bound, on vortex drag, 303–306
    on wake drag, 303–306
  minimum pressure drag, due to angle
    of attack, 302–306
    at zero angle of attack, 269–311
      (*See also* Kármán ogive; NACA
      area rule; Sears-Haack body;
      Supersonic area rule)
  transition, 338
Wing-body interference, 112–143
Wing control, definition, 210
Wing panel, definition, 7
Wing-tail combination in tandem, damp-
    ing in pitch, 409–412
Wing-tail interference, discrete vortices
    in plane of tail, 184–192
  lift on tail section, 189–192
    body, 189, 190
    comparison with flat sheet, 192
    effect on tail height, 192
    span loadings, 190
    tail effectiveness, 190–192
    tail panels, 189, 190
  loading on tail sections, 184–189
    body complex potential, 186, 187
    body loadings, 188

Wing-tail interference, loading on tail
    sections, boundary conditions,
    184
    loading conventions, 185
    panel complex potential, 186, 187
    panel loading, 188
    transformation of tail cross-section
      into unit circle, 185
    types, 188
  flat vortex sheet, 182–184
    intersection with tail panels, 181
    shortcomings of flat-sheet model, 184
    simplified model, 182
    tail effectiveness, 182–184
      definition, 182, 183
      numerical value, 183
      tail span greater than wing span,
        183
      tail span less than wing span, 183
  physical explanation, 181–182
Wing types, cruciform, 6
Wing vortex effects, tailless configura-
    tion, 406–409
    damping in roll, 408, 409
    static stability, 406, 407
Wing vortex strength compared to body
    vortex strength, 97, 98

Yaw control, definition, 212, 213
Yawing effectiveness, definition, 212, 213
Yawing-moment formula, slender-body
    48–51

Zero stability derivatives, cruciform mis-
    sile, 361
    slender, 390
  triform missile, 361, 362
  2-gonal missile with mirror symmetry,
    362
Zero wave drag, 265